Understanding Amateur Radio

BY GEORGE GRAMMER
Technical Director, ARRL

Published by
The American Radio Relay League
Newington, Connecticut

Copyright 1963 by

THE AMERICAN RADIO RELAY LEAGUE, INC.

Copyright secured under the Pan-American Convention

International Copyright secured

This work is Publication No. 22 of The Radio Amateur's Library, published by the League. All rights reserved. No part of this work may be reproduced in any form except by written permission of the publisher. All rights of translation are reserved. Printed in U.S.A.

Quedan reservados todos los derechos

First Edition

Second Printing

Library of Congress Catalog Card Number: 63-10833

$2.00 in U. S. A. proper
$2.25 elsewhere

FOREWORD

The beginner in amateur radio faces such an accumulation of technology that he isn't to be blamed for feeling helpless before it. The ARRL family of publications aims to supply the amateur with up-to-date technical and operating information, but the quantity grows rapidly with the art. There is therefore a definite need for a book that selects those subjects which establish the groundwork for all phases of amateur radio, and which treats them in freer and more leisurely style than is possible in, for example, *The Radio Amateur's Handbook*; the *Handbook* necessarily becomes more concise in treatment and compact in construction as more and more material has to be included. *Understanding Amateur Radio* has been written to meet that need.

"Selection" implies that there are things omitted as well as things included. The things that are omitted in this volume are the more-advanced phases; concepts and techniques that it is best to approach only after the fundamentals are well in hand. Here the accent is on helping you to an understanding of the basic ideas on which circuits and equipment are built.

No book of this nature would be complete—or really useful—if it did not include down-to-earth information on circuit design, construction, testing and adjustment. This aspect has been given ample attention. The equipment described is of a type that has demonstrated wide appeal to beginners of limited technical knowledge and—a feature that the younger generation, especially, will appreciate—with pocketbooks that are not inexhaustible. Most of it has been drawn from the popular Lewis G. McCoy *QST* series for beginners and Novices. A few other items, individually credited, also appeared originally in *QST*. One or two pieces were built especially for this book.

The final chapter, on operating practices, was contributed by Richard L. Baldwin, *QST's* Managing Editor. If you've just received your license and have a station ready to go, read Chapter 16 first! Then look over Chapter 1. After that you're on your own.

Good luck!

JOHN HUNTOON
General Manager, ARRL

Newington, Conn.

CONTENTS

Chapter	1	Setting Up a Station	7
	2	Some Needed Fundamentals	15
	3	How Receivers Work	48
	4	How Transmitters Work	67
	5	What You Should Know About Phone	80
	6	Antennas and Feeders	95
	7	Workshop and Test Bench	119
	8	Building Receivers	136
	9	Accessories for Your Receiver	159
	10	Building Transmitters	169
	11	Transmitting Accessories	208
	12	The Power Supply	225
	13	Modulators and Speech Amplifiers	236
	14	Making Measurements	256
	15	Antennas and Masts	277
	16	Operating Your Amateur Station	296
		Appendix	
		Index	

Chapter 1

Setting Up a Station

Somewhere, sometime, there must have been an amateur who went through the trials of studying for his license, and the tribulations of taking the examination, but who *didn't* get "on the air" at the very earliest possible minute after the mailman rang the door-bell with the "ticket" in his hand. We're still looking for that chap, and some day we may meet him!

The impatience to get going as soon as the bare minimum of equipment is at hand is human. It denotes real interest and enthusiasm—and these you should have. But unless you've had your apparatus well in advance of the license, and have had plenty of time to plan your layout, you'll quickly realize that there's more to a station than just a collection of workable gear. So after the buck fever of the first few QSO's has worn off, sit down and do a little thinking about getting the various and sundry items to work together harmoniously.

Your amateur equipment may be all factory-built, but there is still opportunity for the individual touch in the way you arrange your station. No two operators have exactly the same ideas about it. Nor are any two confronted by exactly the same conditions. One may have a whole room available, another a corner of the bedroom or kitchen, another a spot in the cellar or attic. Each station presents its own problems—what sort of operating table, how to run in antenna leads, where to pick up a.c. power and how to control it, and others like them. In these matters you're pretty largely on your own. The suggestions we have here will help you get started, but if you're like others you'll develop your own preferences. Then, too, part of the pleasure in having a station is the privilege of rearranging it when the mood strikes.

If you do much operating—as no doubt you will in the first phase of your amateur career—the thing to concentrate on is convenience. The controls you use all the time, such as the receiver's tuning knob, want to be within easy reach. Working them for a period of a few hours shouldn't lead to fatigue. Also, you should be able to send and receive with a minimum of lost motion in changing from one to the other. Having to throw a half dozen switches each time doesn't make for snappy contacts!

The Operating Table

In picking out a table or desk on which to set your equipment, don't be stingy with space. You need to have room for writing, both for keeping your log and for jotting down notes during a conversation. Although this can be done with only a foot or so of depth, it's a lot more comfortable if you can have more. The receiver's panel should be about 18 inches from the front edge of the operating table. You also need about that same depth for mounting your key for code work; anything less makes sending a rather tiring chore. As a general rule, the equipment itself won't be more than a foot or so deep, so a table depth of about 30 inches is adequate. A few extra inches, though, make a world of difference in roominess, and are worth having if there's nothing to prevent it.

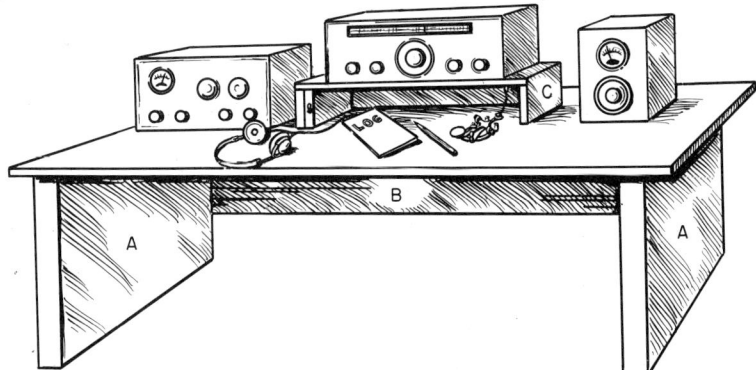

FIG. 1-1—If you build your own operating table, here is a simple design. The top can be ¾-inch plywood cut to a convenient size. The supports, A, are rectangles of 2 × 4, with internal cross-bracing if necessary, covered by ¼-inch plywood. The skirt B, ¼-inch plywood, ties the whole structure together and makes it rigid. Wood screws can be used for fastening, or the sections can be bolted together if there is any thought of having to take it apart later on. The small shelf, C, gives you extra table room. It can be made of ½- or ¾-inch plywood.

Plywood sheet is the material used in the home-built operating table at K3HRF. The construction is simple, and offers plenty of room for both equipment and the operator's elbows.

The table width needed depends, of course, on how much gear has to be accommodated. A 3-foot width will give *you* enough elbow room, but is rather small for the transmitter, receiver, and the accessories that are bound to accumulate. Standard office-desk width—about 60 inches—will take care of the average low- or medium-power station. In fact, such a desk makes a very good operating table. The drawers can be put to good use, obviously.

Homemade Tables

An operating table can be made at home at relatively small cost. A piece of ¾-inch plywood cut to the size you want makes a good top, especially when covered with linoleum or a similar material to give a hard surface for writing. An alternative is an unfinished flush-type door—one measuring roughly 2½ by 7 feet—similarly covered. The supports for the top, in either case, can be made from 2 x 4s. Making table legs that are good and solid isn't the easiest job in the world, and if the table isn't solid it is a poor operating table. You can get around this by using the box-type construction sketched in Fig. 1-1. Form the 2 x 4s into a rectangle of the proper height and the same depth as the table. Covering the rectangle with ¼-inch plywood will give all the stiffness needed.

Another scheme for supporting the top is to use table-high chests of drawers, one at each end. This gives you firm support as well as drawer space, but puts no premium on cabinet-making skill. Look over the unfinished furniture at the local department store or lumber yard for economical items of this sort.

Arranging Equipment

Having ample table-top space doesn't always solve completely the problem of having everything within easy reach. Many amateurs find it useful to spread out vertically as well as horizontally. A low shelf supported above the table will let you put your key, log book, and other accessories of this type *under* the receiver, thus doubling the area within arm's reach. Obviously, in doing this you need to give some thought to the position of the receiver's controls: can they be operated comfortably when the receiver is set up above the table in this way? A shelf height of six inches or so will accommodate many accessories, and also offers a place for mounting small panels for control switches.

One thing *is* essential for code work. The key should be fastened down so it can't move. You can't send decent code if the key keeps sliding away from you every time you bear down on it. And it should be at the right distance from the edge of the table, so your arm can rest comfortably while you send.

Send-Receive Control

A certain amount of coordination is needed in the control of power for the receiver and transmitter, for smooth transition from send to receive and back again. Here a great deal depends on the actual equipment you have and the kind of operating you want to do.

The simplest case is 80- and 40-meter c.w. work with separate antennas for the transmitter and receiver. As pointed out in Chapter 6, the

FIG. 1-2—Send-receive switch.

Equipment Arrangement

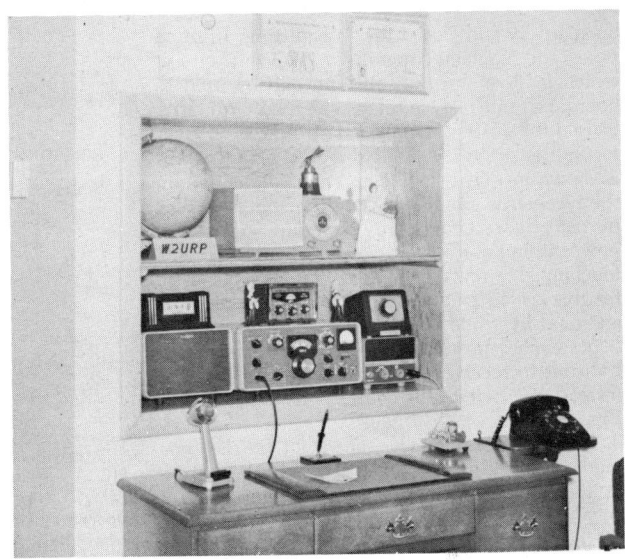

An unusual space-saving arrangement, with the equipment housed in a built-in bookcase. Since the writing desk is part of the normal room furniture this station, W2URP, literally makes no demands on the living room area.

use of separate antennas is not altogether desirable, but it does simplify send-receive changeover. In most low-power transmitters the oscillator is keyed along with the final amplifier. Thus with the key open the transmitter is not generating a signal, and so will not interfere with reception. To send, you merely operate the key. The power is left on both the transmitter and receiver all the time.

The success of this method depends a good deal on your receiver. It is bound to be badly overloaded when you have the key down. This causes **blocking**—a shift in the operating conditions of the tubes in the receiver—and loss of sensitivity, especially near the transmitting frequency. Some receivers don't recover rapidly from blocking, and it may be a couple of seconds before reception is normal again after you've stopped sending. To minimize such effects, use a fairly short length of wire for your receiving antenna, and keep it as far away from the transmitting antenna as you can.

This method won't work for phone if you use a loudspeaker for reception. If you try to transmit with the receiver turned on, sound from the speaker will reach the microphone. This modulates the transmitter and starts a howl—an audio-frequency oscillation arising from what is known as **acoustic feedback**. The speaker has to be silenced when you transmit. The simplest way to do this is to turn off the receiver when your transmitter is on. Communications receivers have provisions for this, in the form of **stand-by** or **muting** terminals. A double-pole double-throw send-receive switch (a toggle switch will do) will serve nicely in this case. Fig. 1-2 shows how to use it. When the transmitter is on the receiver is off, and vice versa. Just how the transmitter may be controlled depends on its circuit; simply substituting the switch contacts for the key may be sufficient. Consult the circuit diagrams of your receiver and transmitter.

Antenna Changeover

Many transmitters have a pair of switch contacts brought out to terminals for controlling external circuits. These are part of the "function" switch, usually, and the external circuit is open

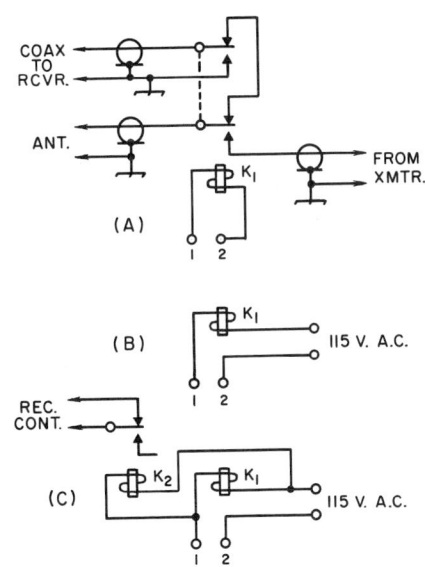

FIG. 1-3—Using an antenna changeover relay with coaxial lines. The circuit at A can be used if the transmitter offers a controlled 115-volt circuit for operating the relay. If the transmitter has only a pair of contacts for relay control, but no power, the circuit at B can be used. The antenna changeover part is the same as in A. C shows how an auxiliary relay can be used with the basic circuit of B, in cases where automatic muting of the receiver is wanted. The relay K1 can be a regular coaxial type, but a less expensive antenna relay (such as the Advance AM/2C/115VA or Potter & Brumfield KT11A, 115 v.a.c.) can be used.

on stand-by and closed for transmitting. In some cases the contacts operate a 115-volt circuit which is "live" on transmit and can be used to operate a send-receive relay. The relay ordinarily is used for shifting the antenna from the receiver to the transmitter. A typical circuit arrangement is shown in Fig. 1-3. The antenna connection on the receiver is grounded when transmitting. This protects the receiver from being damaged by r.f. power from the transmitter, and usually prevents blocking. However, it may not prevent acoustic feedback; you may still have to put the receiver on stand-by while transmitting. This means that *two* switches must be thrown when going from transmit to receive. A second relay can be used instead of a switch; the connections are shown in Fig. 1-3C.

The antenna relay, K_1, need not be the type made especially for coaxial lines, if the length of line between it and the transmitter and receiver is only a few feet. To connect coax line to the antenna terminals on your receiver, tie one of the **doublet** terminals to the ground post and connect the coax braid to this same point. The coax inner conductor goes to the remaining antenna terminal. Of course, if the receiver already has a coax antenna-input fitting, you should use it.

Receiver Muting

Quieting the receiver during transmitting is easy if one of the external receive-standby or muting terminals in the receiver is already connected to the chassis. Then all you need, in addition to the antenna changeover relay, is a d.p.d.t. toggle switch for control. One switch section applies power to the keying relay during transmitting, as shown in Fig. 1-4. The arm of the other section is grounded to both the transmitter and receiver; it simply shorts the key terminals for transmitting, leaving the receiver on standby, and reverses this process for receiving.

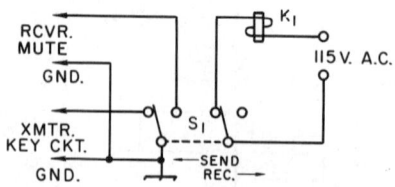

FIG. 1-4—Send-receive switching and antenna changeover with one switch and one relay. The relay is wired as shown in Fig. 1-3.

If your receiver doesn't have such a standby system, the one shown in Fig. 1-5 can easily be put in. R_1 is the manual r.f. gain (or **sensitivity**) control resistor in the receiver. One terminal of this control will be connected to the chassis in practically all receivers. Disconnect this terminal from the chassis and run it to an external terminal as shown in A. When the muting terminals

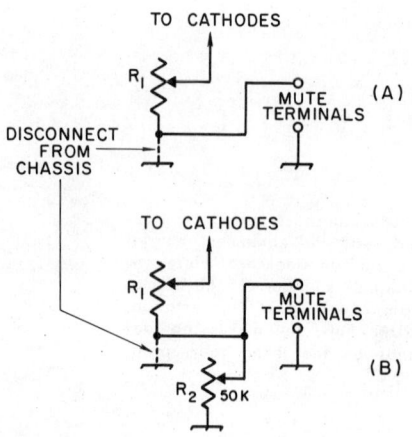

FIG. 1-5—Receiver muting. A cuts off the receiver entirely, desirable for phone work. In B the receiver operates at reduced gain, for c.w. monitoring, while transmitting. In both cases, short-circuiting the muting terminals restores normal reception.

are open the receiver is inoperative; shorting them puts it back in normal condition.

If you're using c.w. you may want to listen to your sending. Although you can leave the receiver completely on, as mentioned earlier, it isn't likely that you'll get a good **sidetone** for monitoring your code, because the receiver will block. However, by cutting back on the r.f. gain it is possible to make most receivers give a good tone. The gain can be reduced by the send-receive switch if an extra control, R_2, is installed as shown in Fig. 1-5B. Adjust it for a good-sounding signal while sending. It is shorted out and the receiver comes back to normal when the switch is thrown to the receiving side.

Other Methods

Control schemes such as those described are well suited to the types of equipment many beginners use when they first get on the air. More elaborate ones have been devised, and you will no doubt run into them in your reading. You'll appreciate better what they're designed to do after you've had some operating experience. We'll mention only one here: the electronic **t.r.** (transmit-receive) switch.

The t.r. switch is a vacuum-tube circuit that replaces the mechanical relay for antenna changeover. Information on how to build one is given in a later chapter. It lets signals through from the antenna to the receiver whenever the transmitter isn't turned on. But whenever any r.f. from the transmitter reaches it, it in effect disconnects the receiver from the antenna. Since there is no mechanical motion the action is practically instantaneous.

C.W. Break-in

The electronic t.r. switch is fine for c.w. **break-**

Station Controls

in operation, since it lets you use the same antenna for both transmitting and receiving. Break-in, strictly speaking, is a send-receive system that lets you hear other signals while you're sending. The advantage of this is that if the operator at the other end has trouble copying you, or wants you to stop sending for any reason, he can simply press his key and you'll hear him.

It doesn't usually work quite as ideally as this in ordinary rag-chewing, unfortunately. When the band is crowded, anyone is likely to open up on the same frequency, so you'll often get an apparent break from some quite different operator who may not even know there's a QSO going on. This can be confusing, both to you and the fellow at the other end. However, real break-in is a useful technique in many circumstances, such as net operating.

What most rag-chewers mean by "break-in" is that they have no switches to throw when they go from receive to send. Separate antennas for receiving and transmitting provide this species of break-in. It isn't real break-in, though, unless the receiving operator can attract the attention of the transmitting operator *while* he's transmitting. For this, the receiver has to be in full operating condition at least between letters in a transmission, if not actually between dots and dashes.

C.W. Monitoring

You can use your receiver for monitoring your sending, as mentioned earlier. However, the receiver has to be tuned to the same frequency as the transmitter, and it may be necessary to reduce the gain. Although you'll often be working stations that are on or very close to your frequency, there will be times when the frequencies are enough different so you can't hear yourself; the receiver tuning will have to be adjusted every time you go from send to receive.

KN1LFH has made an effective desk-top station by putting the principal pieces of equipment in wood compartments. Space below the receiver is used for station controls.

To avoid this, many amateurs use a **keying monitor**—an audio-frequency oscillator keyed along with the transmitter. It is connected into the headphone circuit so you can hear it whenever the key is closed. A simple monitor of this sort is described later in this book.

Such a monitor doesn't tell anything about the quality of the transmitted signal. It is always advisable, therefore, to take an occasional "look" at your actual signal, to make certain that it is up to par. The best way to do this is to disconnect the antenna from the receiver, to prevent overload, and then reduce the r.f. gain to the point where your signal has about the same strength as one coming from a distance. Careful listening then will tell you whether you have key clicks, a

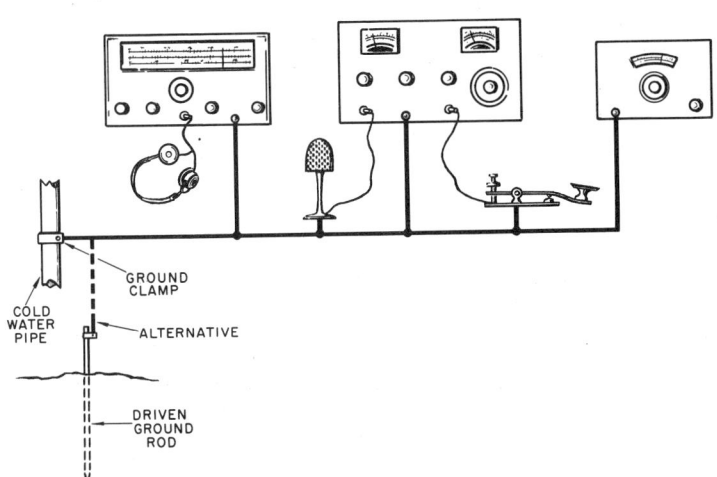

FIG. 1-6—Each piece of equipment should be connected to a good ground external to the equipment itself. The cold water pipe can be used where there is underground water piping. This grounding is an important safety measure.

chirp, or other types of instability that need correcting.

Incidentally, don't depend on reports from other stations for checking your signal quality. The other fellow may not be as knowing about the fine points of keying as you are. Or he may not want to hurt your feelings—or perhaps just wants to avoid the possibility of getting into an argument. Too often these considerations lead to a flattering report that isn't earned. Be your own critic—and be a tough one. Don't be satisfied with anything less than the best-sounding signal you can achieve. You can find enough really good ones on the air to give you a standard to shoot at.

Safety

Your station should offer no hazards, either to you, your family, or your friends, or to property. There are reasons for every accident involving radio apparatus, but never a *good* reason. Take no chances with electricity. Even a low-voltage shock can be serious—sometimes fatal. Practically all manufactured equipment is built so that you can't get your fingers on any "hot" points unless you deliberately take off covers or remove a chassis from a cabinet. But that doesn't mean there cannot possibly be any hazard in using it. There can be, unless you take measures to prevent it.

The worst condition is one where the equipment is seemingly safe but actually isn't, should there be a breakdown inside. Most radio gear is enclosed in metal containers. These offer excellent protection against shock, *provided the cabinet is connected to a good earth ground.* The

An example of a station built for vertical, rather than horizontal, expansion. Equipment that doesn't require frequent adjustment need not be within arm's reach of the operator. The station is W1SUZ.

SWITCH TO SAFETY!

115-volt wiring inside the cabinet is normally isolated from it, but there *could* be a breakdown that would cause the hot side of the line to make contact with the metal. This is dangerous, if there is no external ground connection on the cabinet. *With* such a connection a fuse may blow, but there can't be any voltage lying in wait for you.

The first rule, then, is this: Enclose all equipment in metal so that no "hot" points can be reached, and ground all such cabinets or shields. Do this for *all* your equipment. This goes for microphones, too. Also, one side of the telegraph key—the terminal having the most exposed metal—should be grounded. And don't use a keying circuit that has more than a few volts on it.

Grounds

The general idea is shown in Fig. 1-6. If you live where there is water distribution, make your ground connection to a cold water pipe. These almost invariably go right to ground, and you get a good earth connection through the underground piping. The electrical-system ground is usually made to this same piping. In rural districts where there is no such water system you may have to go to the alternative of a driven ground rod. This is also shown in the figure. Use one at least six feet long (the kind made for television use is satisfactory) or, preferably, two or three spaced several feet apart and connected together. Grounds of this type do not always have as low resistance as might be wished, depending on the kind of soil, and several rods are better than one. Be sure to tie this system to the power company's ground connection at your service entrance, because there can be an appreciable voltage between two such grounds.

Don't depend on the cables that go from one piece of equipment to another for completing your ground connections. Make sure that each piece of equipment is *separately* connected to the ground system. The external ground connection should be the *first* one made when installing a piece of equipment, and the *last* one removed when taking it out of service.

Pull the Plug!

One final point about safety—*never* take a cabinet or dust cover off a piece of equipment, or raise a lid that gives access to the live circuits, without first pulling the power plug. Better a dead circuit than a dead operator.

Fire and Lightning Protection

So far as the equipment itself goes, the same

Safety Precautions

methods that are effective against accidental shock usually will reduce the hazard of fire. If the equipment is in metal cabinets and an internal breakdown causes overheating, the chances are small that the cabinet temperature will get high enough to set fire to anything outside the set. It is a good idea to fuse all 115-volt circuits, since a breakdown generally will cause a fuse to blow and thus—in many cases but not always—will shut off the source of heating power before the faulty component gets dangerously hot.

The lightning hazard from an antenna is much exaggerated; ordinary amateur antennas are no more likely to be hit by a direct stroke than any other object of about the same height in the vicinity. However, an ungrounded antenna system can pick up quite a large electrical charge from a storm in the neighborhood. This can damage your equipment (receiver front ends are particularly susceptible) if you take no precautions against it. The best thing is a grounding switch, as shown in Fig. 1-7. A small knife switch will allow you to ground the feeders when you're not on the air. It will not disturb the normal operation of the feeders (with the switch open, of course) if the lead from the line to the switch contact is no more than a couple of inches long. The grounding switch preferably should be installed where the feeder enters the building. It should be on the outside of the wall, if possible, and should be protected from the weather.

An alligator-jaw clip can be used instead of the switch, if a few inches of flexible lead are allowed for making the ground connection. Whether you use the switch or the clip, don't forget to disconnect the ground when you try to do any transmitting.

Station Accessories

Although the real essentials of the indoor part of an amateur station are the transmitter and receiver (and of course a key or microphone, or

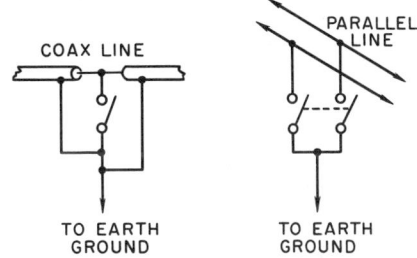

FIG. 1-7—Methods of grounding antenna feeders for protection against surges caused by lightning in the vicinity. A flexible clip lead can be substituted for the switches. The earth ground should be of the type shown in Fig. 1-6.

both), most hams don't stop with these two alone. Other instruments can contribute mightily to efficient operation and to the satisfaction you get out of having a station. We'll just mention a few here; the details of how to build and use most of them will be found in later chapters.

Some auxiliary equipment may turn out to be indispensable—filters for preventing radiation of harmonics, for example, should you run into spurious radiation troubles. Harmonics fall into two general classes; those occurring on frequencies below 30 Mc. or so, and those in the v.h.f. region. The latter are important because they can cause interference with television reception in the immediate vicinity of your station—the **TVI** that you've no doubt heard much about.

The lower-frequency harmonics offer the principal problem for you when you work on 80 or 40 meters. It isn't hard to handle them; a **transmatch** will do the job, and so will a filter designed for the purpose. If you use frequencies from 14 Mc. up, TVI will be your main harmonic difficulty. Combating TVI takes two things: a well-shielded transmitter that won't act like a v.h.f. antenna in itself, and a filter for keeping the harmonics out of the regular antenna system.

"Console" construction as shown by this picture of W5RKS, is both good-looking and thoroughly practical for everyday operation. Sloping panels make for easy reading of meters and dials, and a natural arm position for operating controls. Space underneath is available for accessories and station controls.

A certain amount of measuring gear is useful. Making measurements can be a hobby in itself— and don't make the mistake of thinking that there's nothing more to it than reading a number on the face of a meter. Measurements tell you what's going on, and whether your transmitter and receiver are performing as they should. One thing the FCC regulations require, for example, is a means for determining whether your transmitting frequency is within the proper limits, *and it must be independent of the transmitter itself*. The best way to meet this is with a simple **frequency standard**. A **wavemeter**, which is a "coarse" frequency-measuring instrument, is a worth-while supplement to the frequency standard, but is not a substitute for it.

As an aid in the adjustment of a transmatch for maximum power to the antenna, a device for measuring standing-wave ratio is extremely handy. The **s.w.r. bridge** or **Monimatch** does this.

If you do any building, an instrument for measuring voltage, current and resistance is very helpful. It is also the first thing you reach for when servicing equipment that has developed some fault. Either a **v.o.m. (volt-ohm milliameter)** or **vacuum-tube voltmeter (v.t.v.m.)** is a worth-while investment. Many amateurs find both practically indispensable. Then, too, an oscilloscope is an instrument no advanced station would omit from the list of needed measuring devices. The list could go on and on. You won't need many of those named right away, but as your interests develop along technical lines you'll turn to them sooner or later.

There are lots of other useful auxiliaries— gimmicks for improving reception, monitoring, testing, extending frequency coverage—almost anything you can think of that will help you do better what you're now doing, or help you extend your activities into new fields. Your interests in amateur radio need never be static!

Chapter 2

Some Needed Fundamentals

At this point you'll probably skip to some more interesting part farther back in the book. Who wants to wade through dry "theory," anyway? Transmitters, receivers, antennas, getting on the air—all are far more interesting, and are what attracted you to amateur radio in the first place.

This viewpoint is a reasonable one for the hobbyist, and we have no quarrel with it. What we hope to do in this chapter is to provide you with the answers to questions that, sooner or later, you're going to ask. You'll run into unfamiliar terms, either in reading or in contacts on the air. You'll find that a few concepts thread their way through the techniques of radio communication, and that being familiar with them will save you a lot of head scratching. In other

words, you're going to *need* what we have to offer here—not right now, perhaps, but eventually.

In studying for your license examination you acquired at least a speaking acquaintance with the more elementary electrical and radio principles. They won't get more repetition here than

is actually necessary.* What we want to do is extend them a little farther, so you'll be able to apply them to your own practical problems.

Polarity

You have learned that there are two kinds of electric **charge**, positive and negative. As applied to practical electric circuits, a negative charge means that the charged object has accumulated an excess number of electrons. A positive charge means the charged body has somehow been robbed of electrons that it normally would have.

The negatively-charged body frowns on any attempt to add more electrons to its already swollen total; so, too, the positively-charged body looks with disfavor on any effort to dislodge still more of those electrons it has left. This leads to the rule that objects having charges that are of the same kind repel each other.

On the other hand, a negatively-charged object is quite willing to lose electrons and get back to normal. Similarly, a positively-charged object is eager to acquire some electrons so *it* can get back to normal. Thus, unlike charges attract each other.

This eagerness to get back to normal is described in physics by the word **potential**. In circuits, we're more concerned with the *difference* between the potentials of two charged objects. This difference of potential is expressed in a familiar electrical unit, the **volt**. Since we're dealing with differences, not "absolute" values, there can be a potential difference or voltage between two positively-charged (or two negatively-charged) objects just as readily as between one having a positive charge and one having a negative charge. That is, the **polarity** is *relative*.

A simple example will illustrate. Imagine three charged objects, one (A) having a charge

---

*It is assumed that the reader has a copy of *How to Become a Radio Amateur*, also published by the American Radio Relay League, West Hartford, Conn., and is familiar with the basic material contained in it.

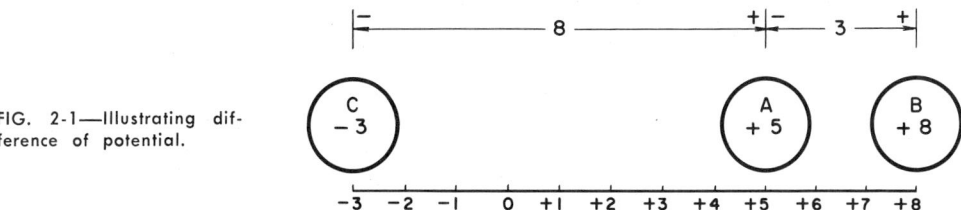

FIG. 2-1—Illustrating difference of potential.

of +5 units, a second *(B)* having a charge of +8, and a third *(C)* having a charge of −3. See Fig. 2-1. The potential difference between *A* and *B* is 3 units, with *A* negative with respect to *B* although *A*'s charge actually is positive. The potential difference between *A* and *C* is 8 units, with *A* positive and *C* negative. You can easily work out for yourself the potential difference between *B* and *C*, and the relative polarity.

This isn't just indulgence in simple arithmetic for its own sake. The idea of relative polarity keeps cropping up all the time in radio circuits. Even though the word "relative" isn't often used in circuit discussions where polarity is mentioned, it is always understood.

Ground Potential

Since we are dealing with differences of potential, or relative potential, it is handy to have something to use as a "reference" potential. This reference should be the same for everybody, to be most useful. Therefore the "something" that has it must be an object whose potential isn't going to be changed by anything we do. The earth itself is ideal—it's so big that anything we do won't make even the tiniest difference in its electric charge. You'll frequently see and hear the term **ground potential**. When we say something is at ground potential we mean that it could be put into electrical contact with the earth without being disturbed electrically. It doesn't actually *have* to be in contact with the earth to be at ground potential—just *could* be.

The metal pan or chassis on which equipment is built is usually at ground potential, although there actually may be no direct connection between the chassis and ground. But the equipment would work just the same if we did make such a connection. It is therefore common to speak of a **chassis ground**, meaning that some part of the circuit is "grounded" to—that is, connected to—the chassis itself. In most circuits a chassis connection is all that is meant when the ground symbol appears. Some confusion can be avoided in this respect if the special chassis-connection symbol is used instead of the earth-connection symbol. The chassis symbol is used in the circuits in this book.

Direct Current

The distinguishing feature of direct current is that the electrons always move in the same direction through the circuit. If the number of electrons passing some point in the circuit is always the same, the current is a steady one and its value is easily specified. You've learned that the unit of current is the **ampere**, and that the relationship between potential difference and current—that is, between volts and amperes—in such a circuit is summed up in **Ohm's Law**:

$$I = \frac{E}{R}$$

where *E* stands for the number of volts, *I* for the number of amperes, and *R* is a property of the circuit itself—its **resistance**. The unit of resistance is the **ohm**. The formula can be transposed to give us two other useful forms

$$E = IR$$

and

$$R = \frac{E}{I}.$$

Going on with a little more review, the power, *P*, in such a circuit is

$$P = EI$$

which, by making a simple substitution, can also be written in two other useful forms:

$$P = \frac{E^2}{R}$$

and

$$P = I^2 R.$$

The unit of power is the **watt**.

These expressions, too, can be transposed: in fact, given any two of the four quantities—voltage, current, resistance, and power—the other two can always be found. It helps to have the formulas handy, since you'll find a need for them more often than you might think. Table 2-I summarizes them.

TABLE 2-I
Ohm's Law Formulas for D.C. Circuits

To find Voltage, *E*	To find Current, *I*	To find Resistance, *R*	To find Power, *P*
$E = IR$	$I = \dfrac{E}{R}$	$R = \dfrac{E}{I}$	$P = EI$
$E = \dfrac{P}{I}$	$I = \dfrac{P}{E}$	$R = \dfrac{E^2}{P}$	$P = \dfrac{E^2}{R}$
$E = \sqrt{PR}$	$I = \sqrt{\dfrac{P}{R}}$	$R = \dfrac{P}{I^2}$	$P = I^2 R$

Power

A word about power: In physics, power is the rate of doing work. (Anything requiring the expenditure of energy is work, in the physical sense; what is done does not have to be useful.) You might guess from the formulas in Table 2-I that in a direct-current circuit the work done is in overcoming the resistance of the circuit and forcing current to flow. You would be right. Resistance is a one-way dead-end street. Energy that goes into it never comes back out in the same form. In a resistor such as you buy in a store the electrical energy put into it comes out

The Fundamentals

as heat energy, raising the temperature of the resistor. The faster the rate at which energy is put in—that is, the greater the power—the more heat generated. "High-power" resistors have to be large in order to handle the heat without reaching a temperature that would burn them up.

An electric heater is a useful device if you want to be warmed by it. In our radio equipment heat is mostly a nuisance—an unavoidable by-product of resistance in the various circuits. Not that resistance isn't necessary in such circuits; the fact is that it has many legitimate uses. However, the heat it produces is useful only in a few cases, such as in heating the cathode of a vacuum tube.

The Nature of Resistance

The formulas in the third column of Table 2-I are really definitions of resistance. The first one ($R = E/I$) must be used with caution, because the name "resistance" is not always applicable. But the other two can be used quite generally. They define resistance as something that has a close relationship with power; that is, as something associated with the expenditure of energy. Energy and power make no distinction between direct and alternating currents. A watt in an a.c. circuit means just the same thing as a watt in a d.c. circuit.

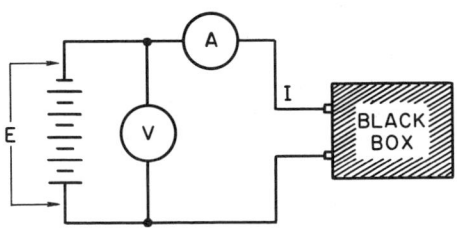

FIG. 2-2—What's in the box?

Let's look at it by the "black-box" method. In Fig. 2-2 there is a battery having an output voltage, E. When it is connected to a pair of terminals on the box a current, I, flows in the connecting wires. We do not know what is inside the box. But on applying Ohm's Law, we find that it seems to contain a resistance, R, equal to E/I. Should we conclude that there is an actual *resistor* in the box? It's possible, of course, that there is. But the most that we can say with certainty is that whatever is in the box is consuming energy from the battery in the same way that such a resistor would.

Now it could be that on opening the box we'd find that it actually contained a radio transmitter, and that the d.c. power from the battery was being converted into radio-frequency power which in turn was being radiated off into space. The transmitter is certainly not a resistor, although before we knew what was in the box it seemed to act like one. Would it be legitimate to call such a device a "resistance" simply because Ohm's Law seemed to say it was one?

The answer is "yes"; a *generalized* definition of resistance is that it is something that uses up power. In using the power it can either convert it to some other form, often one that represents a useful end, or dissipate it internally, usually in the form of heat. It may do both simultaneously. In fact, all electrical devices do dissipate a little of the power themselves. It's just the nature of things for this to happen.

This **apparent** or **equivalent resistance** is a useful concept in dealing with circuits, although it does not physically describe what actually is going on. It is perhaps unfortunate that there isn't some other name given to it. (This isn't the only instance in radio work where you will find one term having several different meanings, or shades of meanings.) But whether resistance is "real," as in a resistor, or "apparent," as in the example above, there is one thing you can bank on: In order to qualify as resistance in circuit operation, it has to obey all the laws that govern ordinary resistance.

Series and Parallel

Throughout your work with circuits you'll be up against **series** and **parallel** connections. These aren't hard to understand. When circuit components are connected in series the current flows through them one after the other. Fig. 2-3A shows this for a number of resistors connected to a source of voltage, E. The same current, I, flows through all of them since it has no other path to follow.

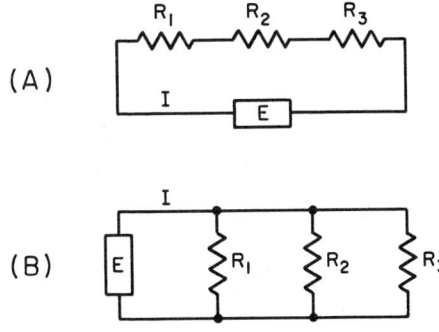

FIG. 2-3—Resistances in series (A) and parallel (B).

If we apply Ohm's Law to such a circuit the voltage measured between the ends of R_1 will be found to be R_1 multiplied by the current; that is, IR_1. This voltage is called the **voltage drop** across R_1. Similarly, the voltage drop across R_2 is IR_2, and the voltage drop across R_3 is IR_3. Obviously, the total of all these voltage drops has to add up to the voltage E coming from the source. By some simple mathematics, this leads to the rule that the total resistance of a number of resistances in series is equal to the sum of the individual resistances. That is,

Total $R = R_1 + R_2 + R_3$ and so on,

for as many resistances as there are in series.

Now look at Fig. 2-3B. Here the resistances have their terminals connected together. The same voltage, E, is applied to each one. The current coming from the source will be divided up among them, some going through R_1, some through R_2, and the remainder through R_3. The currents in each branch have to add up to I, the total current coming from the power source. The current through each resistor will obey Ohm's Law; that is, the current through R_1 is equal to E divided by R_1, the current through R_2 is equal to E divided by R_2, and so on. If we simply divide E by I, we will have a resistance value that will be equivalent to that of all the resistances in parallel. Since the total current is larger than any of the individual resistor currents, this equivalent resistance must be *smaller* than any of the individual resistors. The most important practical case is the one where just two resistances are in parallel. The equivalent resistance can be calculated from

$$R = \frac{R_1 R_2}{R_1 + R_2}.$$

You'll run into many practical applications of resistances in series and parallel. For instance, if you need a certain value of resistance but don't have it all in one resistor, you can make it up by using two or more smaller resistors whose resistance values will add up to what you want. They will be connected in series in this case, of course. Or you may have a couple of resistors of larger value than you want, but which can be connected in parallel to give you a desired smaller value. Sometimes, too, you may want a certain value of resistance that can dissipate more power than can be handled by any single resistor you have. If you need 500 ohms with a safe power dissipation of four watts, for example, you can connect two 1000-ohm, 2-watt resistors in parallel, or two 250-ohm, 2-watt resistors in series.

Alternating Current

The problems of alternating current stem from the fact that the current (or voltage) is changing throughout the **cycle**. In a "steady" alternating current each cycle is like the one before it and also like the one that will follow. But within the cycle there is no such peaceful repose as we find in the behavior of direct current. This continual restlessness leads to all sorts of effects that are absent with d.c. (That is, absent except during those times when the direct current is being started or stopped, or is otherwise subjected to change. These periods can be, and are, ignored in many situations, although not in all.)

First, there is the question of how to assign a value to an alternating current. If we follow the current throughout a cycle, we may find that at successive instants it is increasing until it reaches, say, one ampere. At that instant it starts to decrease, eventually dying away to nothing. Then it reverses itself to do the same thing while flowing in the opposite direction. Next, it starts the whole business over again. At no point does it stay still long enough for us to say, "*that's* the value of the current."

A.C. Amperes and Volts

The clue to settling on a number to use for the current is found in a statement made earlier: Power makes no distinction between a.c. and d.c. A resistor gets just as hot when current flows from top to bottom as when it flows from bottom to top. Thus the power will be the same regardless of the direction of current flow; and since this is so, it doesn't matter how rapidly the current may reverse direction. It follows that we can say we have one ampere of alternating current when that current heats a given resistor exactly as one ampere of direct current would heat it. If the alternating current has the form of a sine wave when plotted on a graph, as in Fig. 2-4, it will have an **effective value** of one ampere when its maximum value during the cycle is equal to 1.41 ampere (the exact figure is $\sqrt{2}$). The same relationship holds for the effective

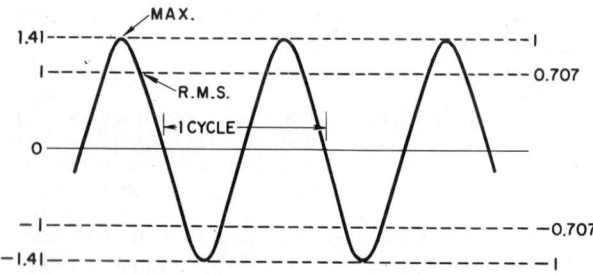

FIG. 2-4—If an alternating current or voltage has the sine form shown here, an effective current of 1 ampere exists when the maximum value is 1.41 amp., as shown by the scale at the left. Figures at the right are in terms of 1 ampere maximum current, in which case the r.m.s. current is 0.707 ampere. The ratio is the same for all sine waves. Other waveforms will have different ratios, in general.

Alternating Current

value of voltage. An alternate term for effective is **r.m.s.** (**root-mean-square**), this name being derived from the method by which such a wave is analyzed mathematically.

The effects associated with alternating current are intimately related to the **frequency** of the current, or the number of cycles per second. The principal one is **reactance,** a term you have met briefly in your studies. Since you'll be running into it all the time, it needs a little more detailed consideration. To appreciate reactance, however, you need first to know a little about energy storage in electric circuits.

Stored Energy

Suppose you carry a stone to an upper floor of your house and place it on a window sill. You probably didn't think of it in these terms, but the fact is that in carrying it up above the ground you have been storing gravitational energy in the stone. The energy stored is equal to the work

you did in carrying it up. If you now push the stone off the sill its stored energy is released, carrying it rapidly back to the ground. This sequence is said to be the result of a gravitational **field,** an invisible something that has been invented to account for an observed effect.

Electrical energy can be stored, too. If you send a direct current through an inductance, a **magnetic field** comes into being around it (Fig. 2-5). This field represents stored energy. If you now open the circuit, all the stored energy comes back. It comes back a lot faster than it went into the field, because it has to get back at the very instant you open the switch. After that it would be too late, since there would be no circuit. If the returning energy is large, it will make itself visible by a fat spark at the switch contacts.

Magnetic Energy

Putting energy into a magnetic field also takes work. One definition of work is that it consists in overcoming an opposing force—gravity, inertia, friction, or what-have-you, in the case of mechanical work. In storing energy in a magnetic field the work done consists in overcoming a force generated in the inductance by the very fact that energy is being stored. This opposition takes the form of an **induced voltage** which bucks the applied voltage. Its value depends not on the actual value of current but on the rate at which the current *changes.* The current changes in value most rapidly at the instant that voltage is applied to an inductance, so at this moment the induced voltage is almost equal to the applied voltage. Then the rate of current change becomes slower and slower, and eventually there is no change that can be measured. At this time the work is complete; the maximum energy is stored in the field, there is no induced voltage, and from then on the resistance of the circuit governs the current flow. Ohm's Law finally prevails.

Electric-Field Storage

You can store energy in a capacitor, too. In this case the storehouse is an **electric field,** not a magnetic field (Fig. 2-6). If you apply a d.c. voltage to a capacitor there will be an instantaneous rush of current into the capacitor to charge it. The only thing that limits the current at the instant of closing the switch is whatever resistance there may be in the circuit. The capacitor itself acts like a short-circuit, at that instant, and all the voltage appears across the resistance. Then as the capacitor "fills up" with electricity—meaning that one set of plates is acquiring an excess of electrons while the other set is being robbed of the same number—the voltage across it rises. Eventually the voltage at the capacitor terminals is equal to the source voltage, and current flow stops. If the source of voltage is then disconnected the capacitor will remain charged to that voltage. The charge will stay there just as long as there is no path by which electrons can travel from one set of plates to the other. A capacitor with very low **leakage** will hold a charge for days on end.

If you connect a resistance to the charged

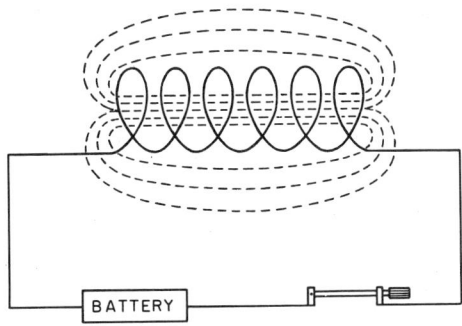

FIG. 2-5—Current sent through an inductance sets up a magnetic field around it. The dashed lines represent the paths along which the field exerts magnetic force.

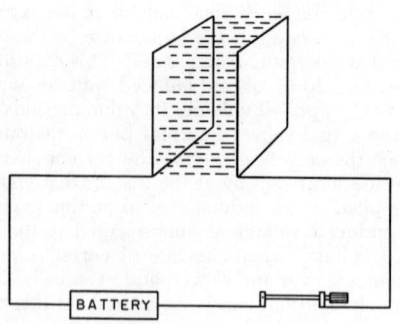

FIG. 2-6—An electric field exists between the plates of a capacitor when a voltage is applied. Dashed lines represent paths of the "lines of force."

capacitor the energy will dissipate itself in heating the resistor. If the capacitance and resistance are both large it may take a long time for voltage to disappear entirely from the capacitor terminals. However, the capacitor can be discharged rapidly into a low resistance or a short circuit. If you touch a wire to the terminals of a capacitor of several microfarads charged to a few hundred volts you'll get quite a spark. (If you touch the terminals yourself you'll get quite a jolt! To avoid danger of this, power supply capacitors have **bleeder resistors** connected across them to drain off the charge.)

Time Constant

The time in seconds that it takes a charged capacitor to lose its charge through a resistance is called the **time constant** of a resistance-capacitance circuit.* You'll meet this term every now and then in practical circuit applications. One example is the automatic gain control used in receivers, taken up in Chapter 3. The larger the capacitance and the larger the resistance, the greater the time constant. An easy rule to remember is that the time constant in seconds is equal to microfarads multiplied by megohms.

There is a time constant associated with inductance and resistance, too, but it works the opposite way. That is, the larger the resistance in series with a given inductance, the shorter the time constant. If we want the current to build up in a hurry in an inductive circuit, we have to use a large resistor in series.

Reactance

Just what does all this have to do with reactance? It goes about like this: From the preceding discussion you've seen that energy is stored in the magnetic field when current through an inductance is increasing, and in the electric field when the voltage across a capacitor is increasing. If the current through the inductance is made to

decrease, energy will come back into the circuit. The induced voltage will tend to keep current flowing in the same direction as the original current. By the same token, if the voltage applied to a capacitor is made to decrease, the capacitor will discharge into the circuit, giving back stored energy.

Now an alternating voltage or current is one which not only reverses its direction periodically, but also is one in which the *value* of the voltage or current is continually changing. Because of this continual change, energy will at times be stored in the magnetic field and shortly thereafter returned to the circuit, if the circuit contains inductance. Similarly with the electric field and capacitance. *All of the energy stored during one part of a cycle is returned by the time the cycle is over.*

Apparent Power

In other words, inductance and capacitance take energy (or power) from the power source only to hand all of it back again. A "pure" inductance or capacitance (i.e., without associated resistance) uses no power. Nevertheless, current does flow in the circuit when voltage is applied. If we multiply the voltage by the current, the same as we do to find power in d.c. circuits, we get a number which seems to represent power. It only *seems* to do so, because no real work is done unless there is resistance. This power is called **apparent power** or **wattless power**. To distinguish it from real power a different unit is used—a **volt-ampere**. One volt-ampere is the same as one watt—except that it doesn't do any work, while a real watt does.

You are undoubtedly curious as to how it is that there can be voltage and current but no power. A detailed examination of what goes on in the circuit is beyond the scope of this book. Briefly, however, it is a matter of timing (for which the technical term is **phase**). The voltage and current don't pull together, as they do in a simple resistance. When one is big the other is likely to be small; or, even, when the polarity of the voltage is positive the current may be negative—that is, flowing in the "wrong" direction. It's something like a tug-of-war in which two

---

*Actually, not all of its charge but 63 percent of it, approximately. There is a mathematical reason for choosing this percentage, but it isn't essential to discuss it here.

Alternating Current

Inductive Reactance

We said earlier that the more rapidly the current changes, the larger the opposing voltage generated in an inductance. A high-frequency alternating current changes more rapidly than a low-frequency one, since there are more cycles per second. Thus the higher the frequency and the larger the inductance, the harder it is for current to flow through the inductance; it meets more opposition. The measure of this opposition is called **inductive reactance**. It is something like the opposition that resistance offers to current flow, and so the unit of reactance is also named the ohm.

Like the wattless watt, though, it is an ohm without resistance. It does act like a real ohm to this extent: Given a fixed frequency, the current through it will be directly proportional to the voltage applied. In other words, we can write for reactance the equivalent of Ohm's Law for resistance:

$$I = \frac{E}{X}$$

where X stands for reactance. But for a given value of inductance, reactance increases with the frequency, so it is not a **constant** like resistance is—unless we specify that the frequency stays constant.

Capacitive Reactance

A capacitor acts in just the opposite way. The more rapidly the applied voltage changes in value, the faster the capacitor stores energy. This means that a high-frequency alternating voltage will put more current into a given capacitor than a low-frequency voltage could. Thus the reactance of a capacitor goes *down* as the frequency increases. Nevertheless, the same formula applies if the frequency stays constant. All we have to remember is that X gets smaller as the capacitance is made larger, and that it also gets smaller as the frequency is made larger.

To distinguish inductive from capacitive reactance the former is usually designated X_L and the latter X_C. Just plain X can mean either one or a combination of both. In the form of equations, the ideas expressed above in words result in

$$X_L = 2\pi f L$$

and

$$X_C = \frac{1}{2\pi f C}.$$

In these formulas, f is the frequency, L the inductance, and C the capacitance. The proper units have to be used.* We won't attempt to explain the factor 2π here because that's a whole topic in itself, and is chiefly of mathematical interest.

Reactances Combined

The "oppositeness" of inductive and capacitive reactance has another important effect. When a coil and capacitor are connected in series in a circuit, one tends to undo what the other is trying to do. This is quite different from placing two resistances in series. The resistances both act the same way, and the total resistance is the sum of the two. But if we put inductive and capacitive reactance in series the total reactance is the *difference*. Conventionally, capacitive reactance is called "negative" and inductive reactance is called "positive." Thus a series circuit might have an inductive reactance of "plus" 15 ohms and a capacitive reactance of "minus" 10 ohms; the total reactance would be only 5 ohms (15 − 10) in that case.

However, reactances of the *same* kind add up just as resistors do. That is, an inductive reactance of 15 ohms placed in series with one of 8 ohms will result in a total of 23 ohms. The same would be true of two capacitive reactances of these same values, except that the sign would be negative.

Also, reactances of the same kind connected in parallel are combined by the same rules that we use for resistances. Not so with reactances of *opposite* kind in parallel! Things begin to get complicated in that case—too much so to be considered in this book, except for one special case, the resonant circuit.

---

*The most convenient units in amateur work are megacycles for frequency, microhenrys for inductance, and micromicrofarads (also called picofarads) for capacitance. With these units the formulas are

$$X_L = 6.28 f L$$

and

$$X_C = \frac{1,000,000}{6.28 f C} = \frac{159,000}{f C}$$

in which 6.28 is the approximate value of 2π. The reactance as given by these formulas is in ohms in both cases.

Resonant Circuits

Since the reactance of an inductance goes up when we increase the frequency, while the reactance of a capacitance goes down, it is reasonable to expect that at *some* frequency the reactances of a given inductance and capacitance will be equal. This is so. The frequency at which it happens is called the **resonant** frequency of the combination.

We're rarely able to ignore resonance in radio-frequency circuits. It's important because at the resonant frequency the inductive reactance is balanced out by the capacitive reactance. This

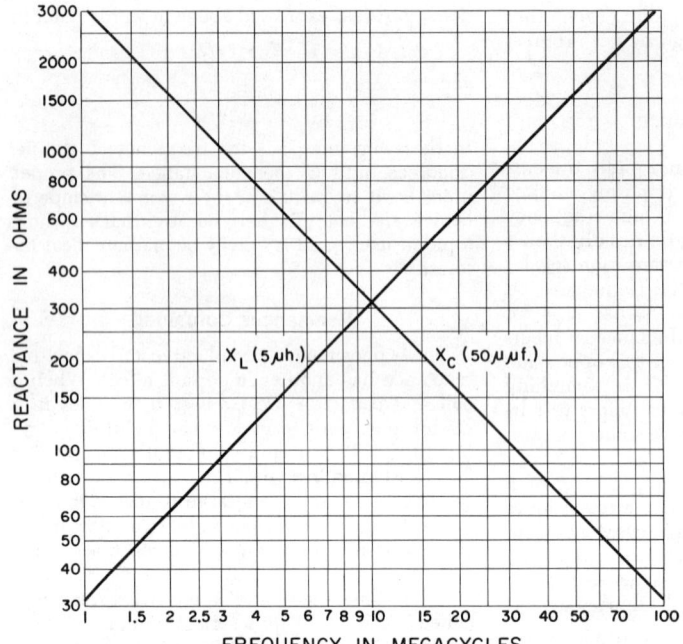

FIG. 2-7—Illustrating how the reactances of a typical coil (5 microhenrys) and capacitor (50 μμf.) vary with frequency.

leaves us with only resistance operating in the circuit. There's more to it than just cancellation of reactive effects, though, as we shall see.

Fig. 2-7 illustrates how reactance changes with frequency. In making up this graph we have chosen 5 μh. for the inductance and 50 μμf. for the capacitance. The scale chosen lets us show a large range of values, of both frequency and reactance, with constant *percentage* accuracy at any point on curves drawn on it. The X_L curve shows that the reactance goes from about 30 ohms at 1 Mc. to over 3000 ohms at 100 Mc. The capacitive reactance does just the opposite —it goes from 3000 ohms at 1 Mc. to a little over 30 ohms at 100 Mc. Other values of inductance and capacitance would have different actual values, but would behave similarly; the ones picked just happen to be convient for study.

The striking thing here is that these two curves cross each other at close to 10 Mc. At this frequency their reactances are equal numerically. In other words, the combination of 5 μh. and 50 μμf. is *resonant* at 10 Mc. Remember that this is only *one* such combination, picked out simply to show graphically how resonance occurs. Theoretically there is an infinite number of combinations that will resonate at any given frequency. Practically, however, we are confined to certain ranges of inductance and capacitance, because of constructional limitations of actual coils and capacitors.

Series Resonance

With respect to the circuit in which they are used, the coil and capacitor may be connected either in series or parallel. This is shown in Fig. 2-8. If we have a source of voltage, E, at the resonant frequency and the two are in series, the current is the same all around the circuit. By the Ohm's Law equations for reactance which we gave earlier, the current will cause a voltage to exist across each reactance. These voltages have been labeled E_L and E_C in the series circuit. Strange as it may seem, they can be many, many times larger than the source voltage, E. In fact, they are usually at least ten times as large, and may be as much as a few hundred times as large.

This can happen because the two reactances cancel each other's effects, since they are equal at resonance. Thus around the circuit there is zero

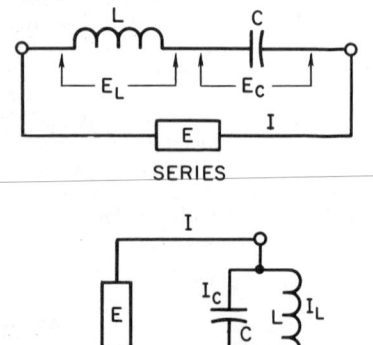

FIG. 2-8—Series and parallel circuits formed by inductive and capacitive reactances, together with a source of voltage, E.

Resonant Circuits

reactance. There is nothing, then, to limit the flow of current except the resistance in the circuit. Although we haven't shown any resistance in Fig. 2-8, there is always some, because no components operate without at least a little power loss. Also, the source of voltage will have an **internal** resistance. But if the total resistance is small, the current will be large, by Ohm's Law. And a large current will result in large, but equal, voltage drops across each reactance. The timing or phase of these voltages is such that the voltage across L has its *positive* maximum at the same instant that the voltage across C has its *negative* maximum. The two voltages always add up just to zero.

We can look at this another way, which may make it seem more reasonable: These unusually large voltages can develop because of energy stored in the reactances. The energy going into the magnetic field of the inductance is energy coming out of the electric field of the capacitor, during one part of a cycle. Then when it has all been stored in the magnetic field, it starts coming back into the circuit and goes into the electric field. This means that a lot of energy can be handed back and forth between L and C without making it necessary for the source to supply any. Of course, the energy "bank account" came from the source. But after an initial surge, the source has only to supply the actual power used up in resistance.

Parallel Resonance

We have a different, but comparable, state of affairs when L and C are connected in parallel. Here there are two current paths, with the same voltage, E, applied to both. The two branch currents, I_C and I_L, each depend only on the same Ohm's Law formula for a reactive circuit. If the reactance is small and the voltage E is large, each branch current will be large. But the same *voltage* is applied to both reactances (in the series circuit it both had the same *current*). So, in this case it is the currents that add up to zero around the circuit. Their phase is such that they cancel each other, in the part of the circuit outside the coil and capacitor. In this case, then, there is no current flowing around the circuit as a whole.

A parallel-resonant circuit "looks like" an open circuit to the source of voltage. Compare this with the series circuit, which "looked like" a short-circuit. The reason for this behavior of the parallel circuit is the same as in the series case—stored energy is tossed back and forth between the inductance and capacitance.

These ideas of "short-circuit" and "open circuit" must be taken with caution. They would be literally true if we could have coils and capacitors without any losses. But these components always do have losses. If the losses are very small the series circuit is *approximately* a short circuit, and the parallel circuit is *approximately* an open circuit. Losses mean that the voltages in the series circuit don't quite balance each other, and the currents in the parallel circuit don't quite cancel each other. Some of the energy is lost each time it is handed back and forth. This lost energy has to be supplied continuously by the source, in order to keep things going at an even rate. The source "sees" a resistance, therefore—a very small one in the case of the series circuit, and a very large one in the case of the parallel circuit.

Impedance

If you've digested what has gone before, you're ready to tackle **impedance**, a word that gets a lot of bandying around in amateur conversations. Its basic definition is simple, but the details are far from being so. In fact, we can't hope to do more than give you a speaking acquaintance with some of them in this book. The inner workings of actual circuits really belong in the field of engineering rather than hobbying. Fortunately, you don't need to know them in order to build and operate amateur equipment.

In broad terms, impedance is a number you get by dividing the voltage applied to a circuit by the current flowing into it. Resorting to the black box again (Fig. 2-9), suppose we measure

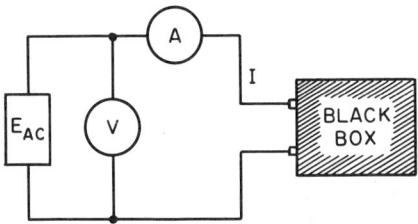

FIG. 2-9—Simple measurements of voltage and current don't give a clue to what the unknown impedance may actually be.

the current I flowing into the box and find it to be ½ ampere when the applied a.c. voltage E is 250 volts. Dividing 250 by ½ gives 500 as the answer. Although this is not a d.c. circuit we say that we have 500 "ohms," since the ohm got established as a unit representing the ratio of voltage to current (that is, E/I) in d.c. work. But we don't know what's in the box, so we can't say that these "ohms" are either resistance or reactance. It would take more than a simple measurement of current or voltage to determine that, because—as we have seen—there is an element of timing or phase that has to be taken into account. The ammeter and voltmeter don't give any information about phase.

The fact is that we could get the same answer whether we had 500 ohms of "pure" resistance or "pure" reactance. But we could also get the same answer if the box contained 500 ohms of something that was a combination of both. Such a combination not only can exist but actually is likely to be more common than either alone. The ohms, then, in this case *could* be either resistance or reactance, but more probably would be impedance, the name for the combination.

You can see why things start to get complicated at this point. In pure resistance the current and voltage are exactly in step. In pure reactance they are out of step by exactly one-quarter of a cycle. These are both very special cases. In a **complex impedance**—one made up of both resistance and reactance—the current and voltage may be out of step in any degree between zero and one-quarter cycle. The number of possible cases is infinite.

Resistive Impedance

It happens that in r.f. work we are concerned mostly with resonant circuits. These, as we have explained, "look like" pure resistances when they are exactly tuned to the frequency. Off the exact resonant frequency the resistance is no longer pure, but in many cases the adulteration of resistance by reactance isn't too great. It is customary, therefore, to use the term impedance rather loosely to mean a **resistive** impedance. A resistive impedance is one in which the resistive effects far override the reactive effects—enough so that the latter can be neglected for practical purposes.

You will see references to impedance in amateur publications, with occasional rules and formulas for one or another special case. Don't fall into the trap of thinking that these rules and formulas are wholly accurate. In most cases they won't be. They are nearly always approximations based on assuming that something can be neglected. This is done to simplify them. Properly used, such rules and formulas can be highly useful in practical work. Occasionally, though, they are misused through ignorance of their hidden limitations. Be cautious, therefore, about applying them to cases other than the one they actually are intended to satisfy.

With this warning, we'll have to move on to some other circuit functions. Just remember that impedance is the ratio of voltage to current in a circuit, and that the timing or phase of the voltage and current are, in general, such that part of the energy is dissipated in resistance and part is stored in reactance.

Q—The Shorthand Number

Another thing you'll hear a lot about and read a lot about is **Q**. This is a number tacked on to coils and capacitors, and circuits formed by them, to give some idea of their characteristics. With its help, and with some simplifying assumptions, a good deal of calculation can be saved in gauging the operation of circuits.

One definition of Q is that it is the ratio of reactance to resistance. Thus in Fig. 2-10 the Q of the combination of X_L and R_L is equal to X_L divided by R_L. R_L may be a separate resistor, as shown. On the other hand, it may be the resistance of the coil itself. This internal resistance can't be separated from the coil, of course, but it *acts* just the same as though it were in series with a lossless coil.

The Q of a capacitor is found in the same way —that is, the Q is equal to the reactance, X_C, divided by the resistance, R_C. Internal resistance here acts the same as in the case of the coil.

Thus the definition of Q applies to components having internal resistance. But it also applies to a circuit formed by the component and an *external* resistance in series with it. In such a case the internal and external resistance have to be added together to find the value of R that is to be used in the formula. Since added external resistance can only raise the total resistance, the Q always goes down when resistance is added in series. There is no way to lower the internal resistance of a coil or capacitor, and thus raise the Q, except by building a better component.

Qs of Components

Capacitors have much better Qs than coils, ordinarily. At least, this is true of the types of capacitors used in radio-frequency circuits where Q is important. These capacitors nearly always are made with either mica or air as the insulator or **dielectric** between the plates. The Qs of such capacitors run well over 1000 as a general rule. This means that their internal losses are very small—in other words, the internal resistance is very small. Coil losses run much higher. A really good coil, one wound with large wire, with spaced turns of rather large diameter (a few inches) and with a minimum of supporting material (such as a few thin strips of insulation) may have a Q as high as 500. The Q of a coil wound of small wire on a form a quarter of an inch or so in diameter, with a powdered-iron **slug** for adjustment of the inductance, may be as low as 25 at high frequencies.

You can see that in Q we have a sort of figure of merit for a coil or capacitor. For a given value of reactance, the higher the Q the lower the internal losses. The Q is important not only because of these losses as such, but because of its effect on the operation of a resonant circuit.

If you've been wearing your thinking cap, you've caught on to one point we haven't men-

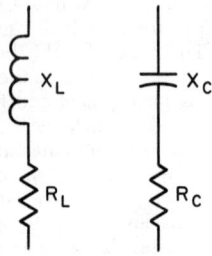

FIG. 2-10—The Q of a component is the ratio of its reactance to its resistance.

Resonant Circuits

tioned. Since reactance varies with frequency, the Q of a coil or capacitor can't be expected to be the same at all frequencies. Unless, that is, its resistance changes in exactly the same way as its reactance. It just happens that over a limited frequency range the resistance of a coil made for radio-frequency use does tend to move along with its reactance. Over such a frequency range —which is usually the useful range for that coil —the Q doesn't change a great deal. But it does change some.

Tuned-Circuit Q

Besides its application to inductance and capacitance singly, Q also is used to rate a resonant circuit. Looking again at the resonant series circuit of Fig. 2-8, the Q can be defined as the ratio of E_L to E; that is, it is the ratio of the **resonant voltage** to the voltage applied to the circuit. Obviously, the Q is also equal to E_C/E, since E_L and E_C are equal (because the reactances are equal) at the resonant frequency.

In the parallel circuit of Fig. 2-8 the Q is equal to I_L/I, or to I_C/I, as you can readily see by analogy to the series circuit.

If we change the frequency slightly, either circuit is no longer exactly resonant. In the series circuit the two reactances no longer balance each other completely; there is some reactance left, and it impedes the flow of current. So the current decreases, and along with it the voltages developed across the coil and capacitor. In the parallel circuit, moving the frequency a little away from resonance—say, a little higher—will cause the current through the capacitor to increase because the capacitor's reactance becomes slightly lower. But the current through the inductance will *decrease*, because the inductive reactance becomes slightly higher. The two currents no longer cancel each other: thus the **line current**, I, becomes larger.

Parallel Resonance

We'll be dealing mostly with parallel-resonant circuits in our amateur equipment. There are two common cases. In the first, shown in Fig. 2-11A, we're interested in the voltage, V, that develops across the terminals of the circuit at and near the resonant frequency. The source of energy is assumed to be a generator, G, for convenience. (Actually, the energy will be introduced into the circuit by some other means in a practical case.) In the second case, Fig. 2-11B, the source of energy, G, is outside the circuit and the thing of interest is the current, I, flowing into the circuit. In both these circuits the Q is determined by the value of the resistance, R, in relation to the inductive reactance of L, just as though these two formed an independent circuit. We can forget the capacitor because its losses are so low, in nearly all cases, that practically all the circuit resistance is in the coil. Thus the Q of the circuit at the resonant frequency is the Q of the coil.

Take Fig. 2-11A first. We will assume a circuit tuned to 3500 kc. The generator, G, has a constant output of 1 volt while its frequency is varied around 3500 kc. If the Q of the circuit is 100, the voltage V at 3500 kc. will be 100 volts. If we change the frequency to, say, 3600 kc., V will be only around 17 volts as shown by the $Q=100$ curve in Fig. 2-12, although there is still 1 volt coming from G. At 3800 kc., V would be only 6 volts, and so on. Most of the resonant rise in voltage takes place quite near the resonant frequency. This is characteristic of all circuits having fairly high values of Q. It means that the circuit "responds" only to frequencies close to resonance. Such a circuit is **sharp**, or **selective**.

(The voltage in this curve is given in relative or percentage terms, since the same ratios will hold for the same circuit no matter what the actual voltage applied through G.)

High- and Low-Q Circuits

Now suppose that we have a circuit tuned to the same frequency, but having a Q of only 10. In terms of *percentage* response, it would look like the one marked $Q=10$ in Fig. 2-12. This response is not sharp like that of the $Q=100$ curve. At 3800 kc. it is still giving 50 percent as much voltage as at resonance, against 6 percent for the circuit having a Q of 100. Such a curve is **broad**, or "non-selective." Both sharp and broad types of response have their uses.

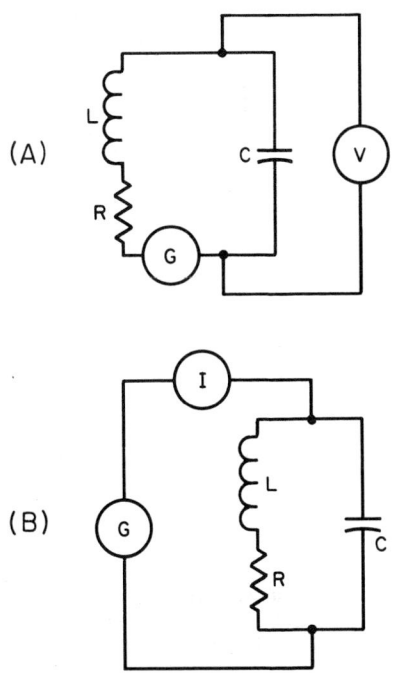

FIG. 2-11—Parallel-resonant circuits as seen from the inside (A) and outside (B). These terms refer to the way in which energy is introduced into the circuit.

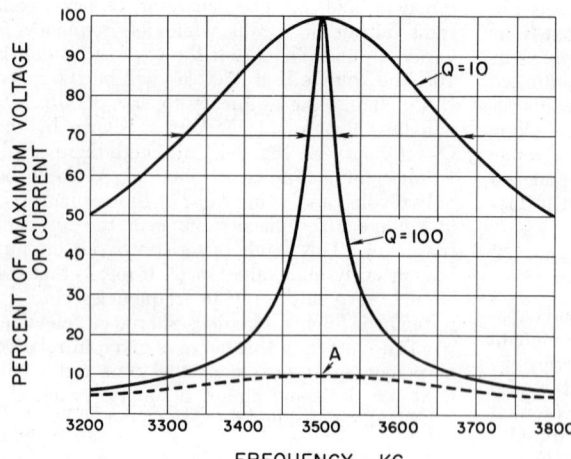

FIG. 2-12—Resonance curves showing the effect of circuit Q on the voltage measured by the circuit of Fig. 2-11A as the frequency of the applied voltage is varied from below to above the circuit's resonant frequency.

Incidentally, if we had 1 volt from G in both circuits, the resonant voltage in the $Q=10$ circuit actually would be only one-tenth as large as that from the $Q=100$ circuit. This is shown by the curve marked A in Fig. 2-12.

Parallel Impedance

The circuit of Fig. 2-11B is supplied energy from the outside rather than the inside, and the generator G is assumed to give a constant output voltage as the frequency is varied. As we saw earlier in the parallel circuit it is the *line current*, I, that changes when the frequency is varied. This current is smallest at the resonant frequency, and rises as the frequency is shifted to either side of resonance.

We could get a rather good idea of how the line current will depend on the circuit Q simply by looking at the curves of Fig. 2-12 upside down. However, it is generally more useful to think of the **parallel impedance** of the circuit, in preference to thinking of current variations. To get the relative impedance variation, we turn the curves right side up again. In other words, the same curves can be used to show either the resonant voltage rise in a series circuit or the impedance rise in a parallel circuit. As given in Fig. 2-12, both are on a percentage basis. The actual values of volts in the series case, or ohms in the parallel case, depend on the actual reactance of L and C at the resonant frequency, and also on the circuit Q*.

The impedance of a parallel circuit operated near resonance is usually considered to be purely resistive. Strictly speaking, this is true only at the exact resonant frequency. As soon as we move the frequency from resonance we get reactive effects. However, in most cases we can ignore them because we operate the circuits so near resonance that the impedance is very close to being a pure resistance.

Bandwidth

Radio-frequency signals in a form useful for communication do not consist of just one single frequency, even though we usually speak of a signal's being "on" such-and-such a frequency. The signal actually occupies a **band** of frequencies. The width, as measured in cycles or kilocycles, depends on the kind of modulation used. An ordinary phone transmission occupies a band at least twice as great as the highest audio frequency in the modulation. This highest frequency might be 4000 cycles, in which case the **bandwidth** of the signal would be twice 4000, or 8000 cycles (8 kc.) If such a phone signal is centered on a **carrier** frequency of 3950 kc., it actually will use a band of frequencies lying between 3946 and 3954 kc.

All of the frequencies within such a band must pass through our selective circuits. Otherwise we wouldn't get the full benefit of the intelligence-bearing part of the signal. It is therefore useful to know just how wide a band a selective circuit will pass. The limits of a circuit's **pass band** are generally taken to be those points where the response in voltage is "down" to approximately 70 per cent of the response at exact resonance. Seventy per cent in voltage represents a 2-to-1 reduction in power, so these points are often called the **half-power** or **−3 db. points**.* They have been marked by small arrows on the curves in Fig. 2-12. By close inspection you can see that the circuit with a Q of 100 has a bandwidth of about 35 kc. while the one with a Q of 10 has a bandwidth of about 350 kc.

---

*In the series circuit, the resonant voltage is Q times the applied voltage. In the parallel circuit, the resonant impedance is Q times the reactance of either L or C (they both have the same value of reactance at resonance).

*Sometimes the bandwidth is based on a 2-to-1 reduction in voltage; that is, a 4-to-1 reduction in power. This is the "6-db. bandwidth."

Resonant Circuits

Circuits with External Loading

To repeat a statement that was discussed at some length earlier in this chapter, resistance is something that uses up energy or power. The object of circuit design is to convey power to a device—the **load**—where some desired use can be made of it. In the tuned circuits so far considered, all the power put into the circuit stayed there. It was used up in the resistance of the circuit itself.

Sometimes the circuit itself actually is the desired load. This is the case with many receiver circuits, where—as in supplying signal voltage to a vacuum-tube amplifier—no power is *required* from the circuit. In a case like this the best use is made of the available power when all of it is used to generate the maximum possible resonant rise in voltage.

However, this is seldom the case in a transmitter. Here we want to get power *out* of the circuit. Tuned circuits are always used, for reasons you will meet later in the section on vacuum tubes. Part of the circuit's job is to see that an amplifier tube is given the kind of resistance load it wants. The actual load—such as an antenna or transmission line—seldom has the value that the tube would like. We'll take just one simple case at this juncture, leaving further discussion for later in this chapter and in Chapter 4.

Load Resistance and Circuit Q

In Fig. 2-13 the generator G represents a vacuum-tube amplifier connected to a resonant circuit, LC. The generator sees a resistance of a value determined by the Q of the circuit and the reactance of L or C. If the Q is low this parallel resistance (between A and B) will be low, as you have seen. Low resistance means that more current will be taken from the generator, assuming that its voltage output is more or less constant. So the lower the Q of the circuit the more heavily the generator is loaded, and vice versa.

Obviously, if we can vary the circuit Q we can adjust the load on the generator to any value we want, within practical limits. An easy way to vary the circuit Q is to connect the actual load across only part of the coil, as shown. If the number of coil turns between A and the tap is small compared with the number between A and B, the current going into the load will be small. As we move the tap up the coil the load takes more current. This has the effect of lowering the circuit Q and thus lowering the resistance between A and B as seen by the generator.

This is only one of many ways in which a load can be introduced into a circuit to vary the Q, and with it the parallel resistance or impedance of the circuit. You'll meet others in the chapters on practical circuit applications. One of the important ones is based on the kind of coupling next considered.

Inductive Coupling

We saw earlier that a changing magnetic field, such as is set up by the r.f. current flowing through a coil, induces a voltage in the coil. This voltage distributes itself on a per-turn basis, if the field around all turns is the same. (It isn't *always* the same, in the kind of coil used in r.f. circuits, for a number of reasons—one of which is the fact that there is no way to keep the field from spreading out in the air.) But here is the interesting thing: The field doesn't care whether the turns in which it is inducing a voltage are all part of the same coil or not. We can have two or more coils in the same field and the voltage in each will be in proportion to the number of turns it has.

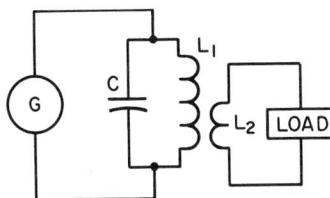

FIG. 2-14—Inductive coupling between the tuned circuit and load. This method also offers a way of changing the circuit Q.

This means that a load can be connected to an entirely separate coil, L_2, as in Fig. 2-14. The second coil could have about the same number of turns as the number across which the load was connected in the tapped circuit, Fig. 2-13. However, in that case it would be necessary for L_2 to be just as close to L_1 as its own turns. This might be done by winding L_2 right over the lower part of L_1, for example.

The advantage of the two-coil arrangement is that there is no direct connection between the load and the power source. This is often convenient in working with vacuum tubes that have to have large d.c. voltages applied to them.

A coil **coupled** to a tuned circuit is often called a **link**. It doesn't have to be wound right over the main coil, actually. If the two are somewhat

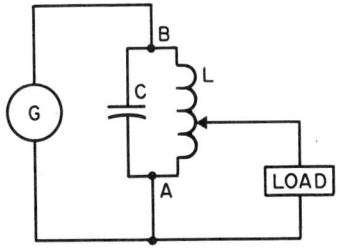

FIG. 2-13—A tuned circuit with load connected. The circuit Q can be changed by moving the tap on the coil.

separated and the link is movable with respect to L_1, the voltage induced in it will be smaller as we move the link farther away. This is called "varying the coupling." It gives smooth adjustment of the loading on the circuit.

The Transformer

Two coils used in this way form what is called a **transformer**. The transformer principle—that a coil in a changing magnetic field has a voltage induced in it whether or not there is a direct connection between it and the source of energy—is valid at any frequency. It is widely used at the power-line frequency, 60 cycles, as you no doubt already know.

Low-frequency transformers are made differently from those used in the r.f. circuits we have been considering up until now. Coils for r.f. circuits are wound on insulating forms. Coils for 60-cycle transformers, and for frequencies corresponding to the air-vibration rate of sound, up to around 15,000 cycles, are wound on iron cores (Fig. 2-15). They also have

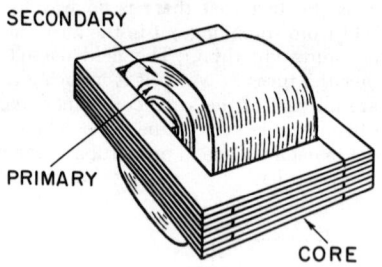

FIG. 2-15—The iron-core transformer, used for frequencies up to 15,000 or 20,000 cycles. The coils are wound on a "leg" of the core midway between the outsides. Sometimes the primary is wound over the secondary instead of the reverse order shown. The core consists of thin sheets called laminations.

many more turns than we find in r.f. coils. The iron core and the large number of turns are necessary at such low frequencies. This is because the magnetic field changes slowly at, say, 60 cycles per second as compared with 7,000,000 cycles (7 Mc.) per second. (In fact, the rate at the latter frequency is over 100,000 times faster!) So to get much induced voltage we have to use many turns and also use iron. The iron increases the strength of the magnetic field many, many times as compared with its strength in air.

Turns Ratio

The iron has another advantage. The magnetic field finds it much easier to stay in it than to leak out into the air. So all we have to do is wind the coils around the iron core and they are all in essentially the same magnetic field. This means that the voltage per turn will be the same in all of them. Let's see how it works out.

Suppose we have a 60-cycle transformer having two coils, one of 200 turns and the other 400 turns. If we connect the 200-turn coil to a 115-volt source, the voltage induced in it will be 115 volts. But since the 400-turn coil is in the same field and has twice as many turns, the voltage induced in it will be 230 volts. Simply by winding on the right number of turns we can get any voltage we want, within constructional limitations. This is the way power transformers—the transformers that heat the cathodes of our tubes and supply the power for the plates—are made.

Primary and Secondary Power

But there is another aspect. The transformer can't manufacture power itself. In fact, it will turn some of the power supplied to it into heat —heat in the windings and heat in the core. This is lost power, but it can't be helped; the wire and iron aren't perfect from an electrical standpoint. If they *were* perfect, all the power put into the **primary** winding—the one connected to the power source—could be taken from the **secondary** or output winding. However, for rough estimates we usually assume that we *can* take as much power from the secondary as is put into the primary. This is approximately true, because a well-designed transformer will turn only a small percentage of the total power into heat.

Now the consequence of the fact that the secondary power is never more than the primary power is this: If the secondary voltage is larger than the primary voltage, the secondary current has to be proportionately smaller than the primary current, and vice versa. In other words, the current multiplied by the voltage has to be the same in both windings (neglecting those heat losses, of course, which show up as additional power going into the primary). So if the secondary voltage of the transformer in the example above is twice the primary voltage, the primary current will have to be twice the secondary current.

Impedance Transformation

One thing about transformers of this type that you need to keep in mind is that the currents depend on what is connected to the secondary winding. If a 2300-ohm resistor is connected to the 230-volt secondary winding the current will be 0.1 ampere. The current in the 115-volt primary winding will be 0.2 ampere, by the rule just mentioned. The source of power is "seeing" a load of 0.2 ampere at 115 volts. This load looks like 115/0.2, or 575 ohms, although the actual secondary load is 2300 ohms. If we connect a 230-ohm resistor to the secondary, the secondary current will be 1 ampere and the primary current will rise to 2 amperes. In that case the primary sees a load of 57.5 ohms.

The load "looking into" the primary is not something that is a fixed property of the transformer. It is determined by the *secondary* load

Transformers

resistance and the ratio of turns on the primary and secondary windings.

By choosing the proper **turns ratio** for the windings, we can make any given load resistance look like some value that is more suitable for our power source. This is summed up in a rule: the impedance ratio of the transformer is proportional to the *square* of the turns ratio. In the transformer used in the example above, the turns ratio was 2 to 1, so the impedance ratio was 4 to 1. (We talked in terms of resistance rather than impedance because it's simpler, and in most cases we do deal with simple resistance rather than impedance generally.)

Impedance Matching

Fig. 2-16 shows a couple of examples of impedance transformation and also gives the general rule. Transformers having the required turns ratio are used for changing **impedance levels** in audio-frequency circuits when the actual load does not **match** the resistance that would be optimum for the generator. A familiar example is the output transformer used between an audio power tube and a loudspeaker. The speaker will have some low value of impedance or resistance, such as 4 ohms, while the power tube wants a load resistance of possibly 5000 ohms for its best operation. The transformer turns ratio would be chosen accordingly, thus bringing about a match between the speaker and tube.

Incidentally, the resistance that is matched is the *optimum* load resistance for the source of power, not the resistance that may be in that source (and all sources of power *do* have internal resistance). We *could* match the source's internal resistance, and such matching does in fact result in the highest possible power output. However, it also results in poor efficiency and, in the case of devices such as vacuum tubes and transistors, a great deal more distortion than is wanted. The *optimum* load resistance considers these factors, and is chosen to give us the best possible all-around compromise.

Shielding

It is worthwhile at this point to get off the main track of coupling for a moment and take up another aspect of it that you will use continually. This is the matter of **shielding**.

If you put a coil inside a metal enclosure, the enclosure will act like the secondary of a transformer. The coil itself is the primary. The voltage induced in this "secondary" will cause a current to flow in it. The current in turn sets up a field that opposes the primary magnetic field that produced it. If the resistance of the enclosure is low, the two fields will just about cancel each other *outside* the enclosure. Thus the enclosure is a **shield** which prevents the coil from being magnetically coupled to other circuits outside it. Preventing such stray coupling is highly necessary in many circuits.

This method works fine at radio frequencies, but to get the same result at audio frequencies

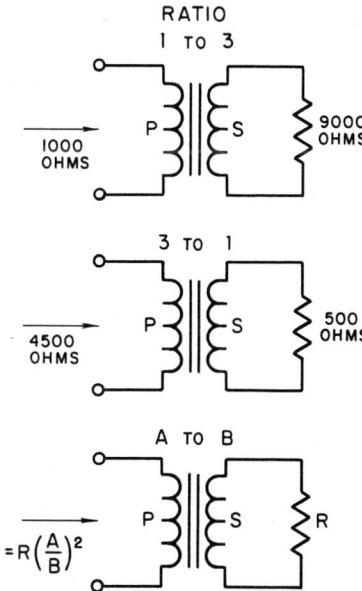

FIG. 2-16—Examples of impedance transformation. The name "primary" (P) always is applied to the winding to which the source of power is connected. The secondary, S, is the winding from which power is taken.

the shield would have to be very thick. In this frequency range we depend on the iron core to confine the field by offering it the most favorable path. An iron enclosure, not necessarily one of low electrical resistance, also helps for the same reason.

Electric, as contrasted to magnetic, fields are easily confined inside metal enclosures, even thin ones. This is true at all frequencies if the resistance of the metal is low. The metal simply short-circuits the electric field so it can't get through.

Coupled Tuned-R.F. Circuits

In Fig. 2-14 one r.f. circuit, that formed by L_1 and C, was tuned. The second, L_2 and the load, wasn't. The secondary circuit *can* be a tuned one, however. When both circuits are tuned to the same frequency more current will flow in the secondary, because the reactance of the capacitor cancels the reactance of the coil.

One result of this greater current flow is that the two coils do not have to be so near each other for transferring a given amount of r.f. power from the primary to the secondary. That is, **loose coupling** can be used. Another is that two tuned coupled circuits are more selective than one. Both of them have resonant Qs, and thus both will respond most strongly to just one frequency. This gives us a way of increasing selectivity in receivers. Extra selectivity is often useful in transmitters, too, because transmitters are prone to generate frequencies we don't want along with the one we do want. These spurious

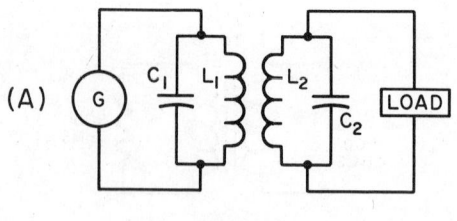

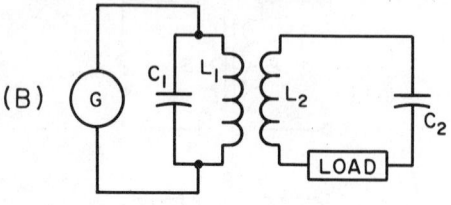

FIG. 2-17—Inductively-coupled tuned circuits. Circuit A is used for coupling to high values of load resistance—of the order of thousands of ohms. B is used for low load resistances—100 ohms or less, usually.

frequencies can't be allowed to go out with the intended signal.

Two common types of inductively-coupled resonant circuits are shown in Fig. 2-17. The arrangement at the top is almost universally used in receivers, where the load often is a very high —almost immeasurably high—resistance. Here we are interested in getting the largest possible voltage from the secondary circuit. The lower circuit is used when the load is a low resistance. It is often found in transmitting circuits.

Coupling and Q

The way these circuits operate depends principally on their individual Qs, including the effect of loading on the Qs. If both circuits have high Qs—50 or 100 or more—the coupling between them can be very loose indeed, even when the maximum power is being transferred from the primary to the secondary. The resulting selectivity will be quite high. On the other hand, if the Qs are low—say in the neighborhood of 10 each—the coupling between the two coils must be much tighter for optimum power transfer, and the selectivity will be lower.

The high-Q **tuned transformer** is the kind we want for our receivers. The low-Q one is more useful in transmitters, where large amounts of power must be handled and we can't afford to lose much of it in the circuits themselves. A circuit loaded by a useful resistance such as an antenna has to work at relatively low Q so that most of the power will go into the load instead of being burned up in heating the coil.

Coefficient of Coupling

The degree of coupling between two coils is expressed by a number called the **coefficient of coupling**. At this stage it isn't essential for you to know its technical definition. It is sufficient to note that a very small coefficient of coupling will suffice for maximum power transfer if the two coupled circuits have high Qs. That is, the coils can be relatively well separated. If the circuit Qs are low, the coupling coefficient must be larger, meaning that the two coils will have to be rather close together.

Selectivity of Coupled Circuits

What happens if we vary the coupling between two tuned circuits? If the coupling is very loose, varying the frequency applied to the primary circuit will cause the secondary response to go through the values shown by curve A in Fig. 2-18. The curve is sharp—good selectivity—but we haven't transferred all the possible r.f. energy from the primary to the secondary.

If we now increase the coupling to the point where the secondary response is as shown by curve B, we are getting the maximum possible energy transfer. This is called the point of **critical coupling**. The curve has the same general shape as A, but is less selective. If the coupling is increased still farther the circuits are said to be **overcoupled**. An overcoupled response, shown by curves C and D, always shows two "humps," or points of maximum response. These are about equally spaced from the true resonant frequency. The dip in the center of the curve is small if the circuits are just beyond critical coupling, as in C. The more the circuits are overcoupled, the deeper the dip and the farther apart the humps become.

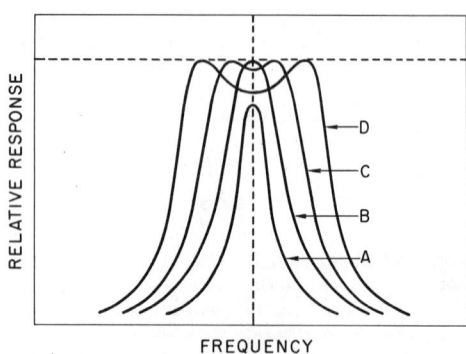

FIG. 2-18—Typical response curves obtained at several degrees of coupling when the frequency applied to two coupled circuits is varied. Both circuits are tuned to the same frequency, indicated by the vertical dashed line. The actual shapes of curves like these depend on the circuit Qs.

Overcoupling gives a **bandpass** effect that is often useful. The response curve is approximately flat-topped if the circuits are not too badly overcoupled. This is fine for passing signals that have appreciable bandwidth.

Vacuum Tubes

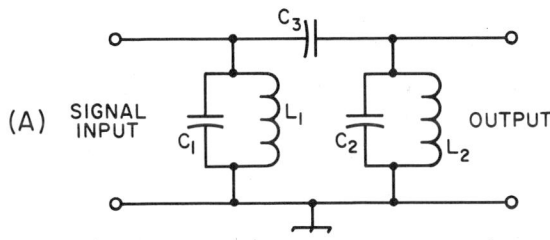

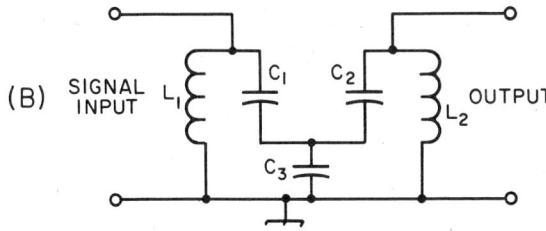

FIG. 2-19—Capacitive coupling between tuned circuits. A and B are frequently called "top" and "bottom" coupling, respectively.

Other Types of Coupling

In some receivers you may find circuits similar to those shown in Fig. 2-19. There is no inductive coupling between the tuned circuits L_1C_1 and L_2C_2. In fact, they may be shielded from each other. The coupling is through a capacitor, C_3, connected between the two circuits.

In A, C_3 connects the "hot" sides of the two circuits. The coupling coefficient depends on the value of this capacitance. If the circuit Qs are high, C_3 will be quite small when the coupling has its critical value. At even very low radio frequencies just a few $\mu\mu f$. will suffice.

In B, the circuits are coupled at the low-potential side. C_3 is common to both circuits, and the voltage developed across it by current flowing through it and C_1 introduces energy into the circuit completed by C_2 and L_2. Only a very small voltage is needed for critical coupling, so the capacitance of C_3 in this circuit is very large compared with the capacitances of C_1 and C_2.

Fig. 2-19B is probably the one more frequently used. By switching in various values of C_3, the coupling—and thus the bandwidth—of the tuned transformer can easily be changed to suit the bandwidths of various types of signals.

Vacuum Tubes

You're already familiar with what it is that makes a vacuum tube "tick." Here we want to pick up some additional details that will be helpful in your reading and in discussions with other amateurs.

The **diode** or two-element tube is a simple structure, having just a cathode and plate. Its only use is **rectification**. Rectification, the ability to conduct current in one direction but not in the other, has two principal roles in amateur equipment. One is the **detection** of received signals. The second is converting a.c. into d.c. in power supplies. Circuits for these purposes are discussed in other chapters. The difference between the two cases is emphasized by the differences in size and construction of the tubes employed in them. These differences in turn result from the large differences in the power levels that the tubes have to handle.

Power-Handling Ability

A diode used as a detector in a receiver is called on to handle only microscopic amounts of power. It can therefore be quite small. The type 6AL5 tube is a good example—it has *two* diodes in the smallest bulb used for regular receiving tubes. In fact, a couple of diodes often will be found in tubes which are primarily triode or pentode receiving-type amplifiers. The plates are just tiny cylinders surrounding an extension of an indirectly-heated cathode.

On the other hand, a diode (or double diode, which is more usual) for a power supply will invariably have a rather large bulb and much bigger plates. We may need 30 or 40 watts of d.c. power for an ordinary receiver, and often much more for a small transmitter. The rectifier tube has to handle this.

The diode rectifier, whether it is simply detecting a signal or serving to transform a.c. into d.c. in a power supply, doesn't do it 100 per cent efficiently. Some of the power going into the plate or plates doesn't go on through to the load. A voltage must exist between the plate and cathode if current is to flow through the tube. If this voltage (called the **tube voltage drop**) is multiplied by the current, we have the power that is lost in the tube itself. This power heats the plate. If the current is large the power heating the plate will be large, too, especially if there has to be a rather large voltage between the plate and cathode to make the large current flow. To keep the plates at a reasonable temperature it is necessary to make them big. And to get the heat out into the air quickly the tube has to have a bulb with plenty of surface area.

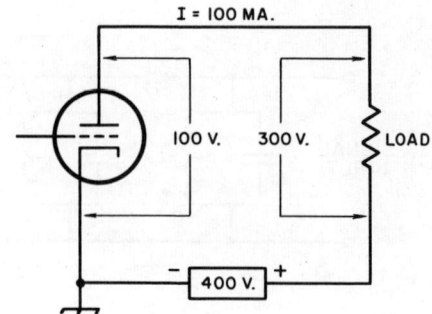

FIG. 2-20—Power from the plate supply divides between the tube and the load.

Plate Dissipation

You can see what a difference this makes by comparing the 6AL5 with, say, a 5U4G. (The latter is the type of rectifier commonly used in television receivers where a considerable amount of d.c. power is needed.) If the tube is to have long life, it must have the ability to get rid of a lot of heat. This ability is summed up in the **rated safe plate dissipation** of the tube.

The safe plate dissipation—meaning the number of watts that can be allowed to heat the plate without damaging the tube—is an important factor in *any* type of tube, not just the diode rectifier. No matter what the tube type, the plate heating comes from the same cause—the tube voltage drop and the plate current. One of the objectives in designing power tubes is to make large plate currents flow with relatively little

"MUST ... GET RID OF A LOT OF HEAT"

voltage between the plate and cathode, thus reducing the power lost in the plate.

Fig. 2-20 may help make this clearer. Although we haven't shown anything connected to the grid, this might be an amplifier circuit. The plate-supply voltage is 400 and the current is 100 ma. The voltage drop in the load is 300 (which means that the load resistance is 3000 ohms, as you can easily figure from Ohm's Law). This leaves 100 volts between the plate and cathode of the tube. The power in the load is 300 volts $\times$ 0.1 ampere, or 30 watts. The power lost in the tube plate is 100 volts $\times$ 0.1 ampere, or 10 watts. You can easily see why we want to make most of the voltage to be "used up" in the load rather than in the tube. The load power is *useful* power. The tube power just heats the plate.

Fig. 2-20 is pretty highly simplified, in terms of ordinary tube operation. The actual current would be varying—radio or audio frequency—when the tube is amplifying. Also, the load probably won't be a resistor, if any power is being handled. Think of the load as something that *acts like* a resistance in that it consumes energy. For example, it might be a loudspeaker coupled to the tube through a transformer.

Plate Efficiency

Whenever an attempt is made to get much power from a tube, the relationship between power delivered to the load and power lost in the plate is very important. The measure of it is called the **plate efficiency** of the amplifier circuit. Plate efficiency is the ratio of the output power to the d.c. power input to the plate.

In the above example, the d.c. input is the sum of the plate dissipation and the load power; that is, 10 + 30 = 40 watts. Thus the plate efficiency is 30 divided by 40, or 0.75 (75 per cent.)

Tube Characteristics

A **characteristic** is something that tells you what kind of operation you can expect from a tube. Probably the most important of these is one having the rather technical-sounding name of **transconductance**. It simply means the change in plate current for a given change in grid voltage. A tube can amplify because a change in the voltage applied to the grid can cause the plate current to change. Thus the bigger the plate-current change for a given grid-voltage change, the better the tube is as an amplifier. (We really should say "in its intended application," because other things than transconductance can affect the amplification—for example, the frequency at which the tube works.)

Characteristics related to the transconductance are the **amplification factor,** and the **plate resistance.** The amplification factor, often called the **mu** (μ) tells us something about the relationship

Vacuum Tubes

between plate voltage and grid voltage. Suppose we change the grid voltage by one volt and observe how much the plate current changes. Then we change the plate voltage by whatever amount is necessary to bring about the same plate-current change. Perhaps the plate voltage will have to be changed by 40 volts to do it. Then the amplification factor is 40 volts divided by 1 volt, or 40. Triodes can be built with amplification factors running from less than 10 to over 100.

If the plate-cathode circuit of the tube were like an ordinary resistor, the plate resistance would be equal to the plate-cathode voltage divided by plate current it caused. However, that leads only to a "static" value, not of much use. We're more interested in the effect of a *change* in plate voltage on the plate current, since such changes are typical of what goes on in the tube when it is amplifying. So the plate resistance is found by taking current readings at two different plate voltages and dividing the *difference* in plate voltage by the *difference* in plate current. This figure corresponds to the internal resistance of a generator. The plate circuit of a tube acts just like such a generator when amplifying.

Making Tubes Work

Reduced to bare essentials, a tube amplifier circuit is shown in Fig. 2-21. The source of a.c. signal to be amplified is represented by the generator G. The amplified signal appears in the load. The d.c. voltage which is applied to the plate to make it attract electrons from the cathode region is of course essential. Without it the tube cannot function. The tube itself does not supply power to the load, although it is sometimes handy to think of it as a generator. It simply converts d.c. power, taken from the plate power supply, into a useful kind of output power in the load. It can do this because the grid is able to control the plate current.

Note that a **grid bias** source is shown in the grid circuit. This bias is a d.c. voltage applied to

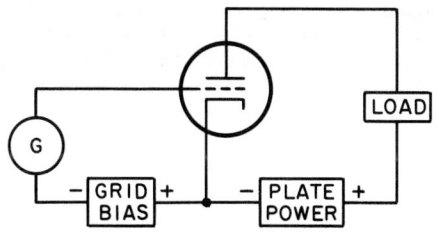

FIG. 2-21—Essentials of an amplifier circuit.

the grid. It is needed in the great majority of cases in order to make the tube operate in some desired fashion. It isn't always essential; some tubes are designed so they can operate without it.

When grid bias is required, the d.c. voltage applied to the grid is always negative. This has the effect of decreasing the plate current, as you already know. The signal voltage to be amplified is an a.c. voltage. When it rises in the positive direction it cancels part of the steady bias voltage on the grid. When it swings in the negative direction it adds to the negative bias voltage. This is shown in Fig. 2-22. We say that the signal is "superimposed" on the grid bias.

Fig. 2-22 shows a signal having an instantaneous peak of 10 volts superimposed on a negative d.c. voltage of 20 volts. At the positive signal peak the *actual* voltage at the grid is 10 volts negative with respect to the cathode, while on the negative signal peak it is 30 volts negative with respect to the cathode. At some time during each cycle the instantaneous grid voltage passes through every possible value between -10 and -30 volts. The plate current of the tube follows these changes, being largest when the grid is at -10 volts and smallest when the grid is at -30 volts.

Negative-Grid Operation

In this example the grid voltage never reaches

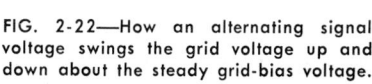

FIG. 2-22—How an alternating signal voltage swings the grid voltage up and down about the steady grid-bias voltage.

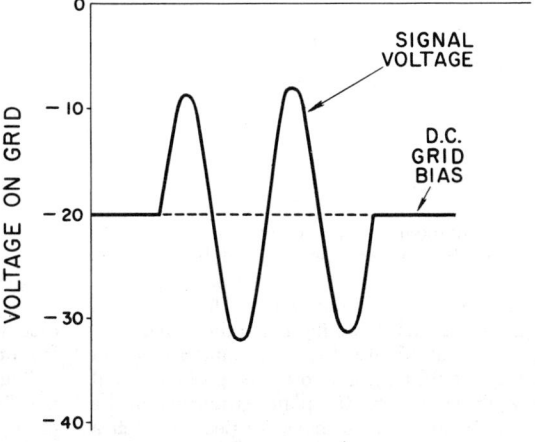

zero—that is, it does not reach the point where there is no voltage between the grid and cathode—even instantaneously. As you know, a negatively-charged electrode in a tube will not attract electrons. So if we always keep the grid negative, no current will flow in the circuit formed by the cathode, grid, signal source, and bias source. This is important, because a tube operated in this way does not take any power from the signal source. (Both current and voltage are needed to make power.) This is fine for the purpose of amplifying weak signals, because really weak signals can't supply appreciable power.

The disadvantage of keeping the grid entirely on the negative side is that it limits the range over which the plate current can be made to change. We can illustrate this by a simple graph, Fig. 2-23. This shows the plate current that will

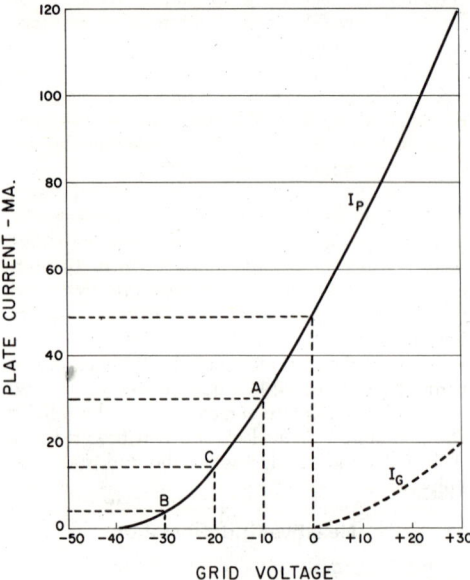

FIG. 2-23—A curve such as this shows how the plate current of a tube will change when the grid voltage (that is, the voltage between the grid and cathode) is changed. The plate voltage is assumed to be constant in making up such a curve.

flow for each voltage we put on the grid. (The plate voltage is assumed to have some fixed value suitable for the tube.) If this is the tube which is biased as shown in Fig. 2-22, you can see that the instantaneous plate current at the positive peak of the 10-volt signal would be about 30 ma., while on the negative signal peak the plate current would be about 5 ma. This is a total change of 20 ma. If we increased the signal voltage to a peak of 20 volts, the instantaneous grid voltage would be just zero on the positive signal peak, at which time the plate current would be almost 50 ma. On the negative peak of such a signal the grid voltage would reach −40 and the plate current would be zero. This almost-50-ma. change is the most we can get if the grid is to stay negative all the time. But as the curve shows, the plate current keeps going up when the grid is made *positive* with respect to the cathode.

Grid Power

If the signal source could drive the grid positive, we could get a bigger swing in plate current and thereby increase the tube's output. But the instant the grid goes positive, current starts to flow in the grid circuit, as shown by the curve marked "I_G." This means that the signal source has to supply real power whenever the grid is positive. If the source can supply the power, well and good. Some sources can and some can't.

The Cutoff Point

Fig. 2-23 shows that the plate current gets smaller and smaller as we make the grid bias more negative. At −40 volts bias the current is practically zero—that is, it is **cut off**. Making the grid voltage still more negative can't do more than that to the plate current. But there are actually advantages to setting the bias "beyond cutoff" for certain types of operation, particularly in r.f. power amplifiers.

In still other types of amplifiers there are advantages to setting the bias at, or not quite to, the cutoff point. In either case the benefit lies in increasing the plate efficiency of the amplifier, as compared with the type of operation just discussed in which the plate current flows throughout the signal cycle. Just how this comes about is a rather technical subject which you can find treated in more advanced books. For the present, it is enough to know that operating a tube with bias near or beyond cutoff does help increase the plate efficiency of the amplifier.

Distortion

The increased efficiency has its price. The output signal can't be exactly like the signal applied to the grid if the plate current is cut off during part of the cycle. In other words, there is **distortion**.

Sometimes there is a *lot* of distortion. But in many kinds of amplifiers distortion doesn't matter. It matters in an audio amplifier. It does not matter much in many types of r.f. power amplifiers. You'll find more on this subject in subsequent chapters.

Voltage and Power Amplification

This brings up the point that there are two kinds of amplification in practical use. In one case we try to get the largest possible *voltage* from the amplifier. This is worthwhile when that voltage is to be amplified, in turn, by a following amplifier stage (**stage** is the name given to one of a number of amplifiers used consecutively;

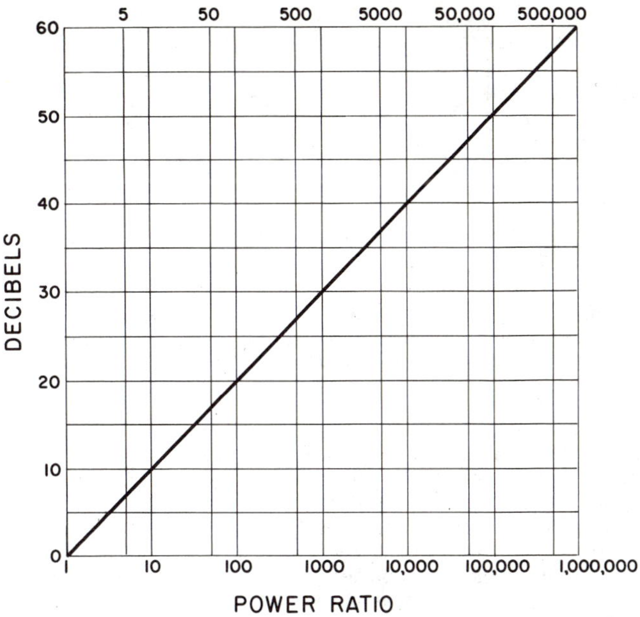

Vacuum Tubes

FIG. 2-24—This chart shows the number of decibels corresponding to power ratios from 1 to 1 (0 db.) to 1 million to 1 (60 db.).

consecutive amplifiers are said to be in **cascade** *if* the grid of the following tube is never allowed to "go positive."

The second kind is called **power amplification**. Here we're interested in getting as much *power* as possible from the tube. This often requires driving the grid of the tube positive. Thus power is used up in the grid circuit in order to produce the most output power in the plate circuit.

Voltage amplification is the ratio of the output voltage to the signal (grid) voltage that produces it. Power amplification, similarly, is the ratio of the power output to the signal (grid) power that produces it.

The Decibel

Note that power amplification is a *ratio* of two powers. In communication work you'll find that power ratios often mean much more than the actual value of the power. A signal having a power, or **power level**, of one watt will sound, let's say, twice as loud when its power is increased to two watts. To make it sound twice as loud again we would have to increase the power to four watts. To make it twice as loud again we would have to increase it to eight watts. Each step in the loudness scale requires *multiplying* the power by some constant factor—the factor 2 in this example.

There is a mathematical system for changing these ratios into simple numbers that can be added. If we use it we don't have to go through a lot of multiplication to arrive at an estimate of how much we increase signal strength by increasing power. It is handy to make the unit of such a system one that stands for a just-noticeable increase in signal strength. The unit in use is called the **decibel**, abbreviated **db**. It stands for multiplying the power by 1.26, approximately. If the power is increased by 5 decibels, for example, it means that there have been 5 successive just-noticeable increases in signal strength.

An easy relationship to remember is that a power increase of 3 decibels is almost exactly the same as multiplying the original power by 2. Fig. 2-24 is a chart showing the number of decibels corresponding to various power ratios.

The decibel works both ways—that is, it gives the ratio of a *decrease* in power level just as readily as it gives the ratio of an increase. All you do is subtract instead of adding. If you cut your transmitter power down to one-half—that is, divide it by 2—you have decreased it by 3 db. A number of decibels with a minus sign in front of it means a *loss* of power of that ratio. If the sign is positive, or no sign is shown, a *gain* in power is meant.

Amplifier Gain

The gain of an amplifier, whether it has just one stage or many, can be expressed either as a simple ratio of voltage or power, or given in decibels. Usually you will have no trouble in seeing what is meant, because it is customary to speak of a **voltage gain** of 1000, or a **power gain** of 250, or a **gain** of 30 db.—using the appropriate actual numbers, of course.

Watch out for gains given in decibels, though, when an amplifier contains voltage-amplifier as well as power-amplifier stages. The decibel measure is sometimes used loosely—and often incorrectly—to express a voltage gain. Always remember that the decibel stands for a *power* ratio.

Amplifier Classifications

We've mentioned that there are several different types of amplification, and it's appropriate

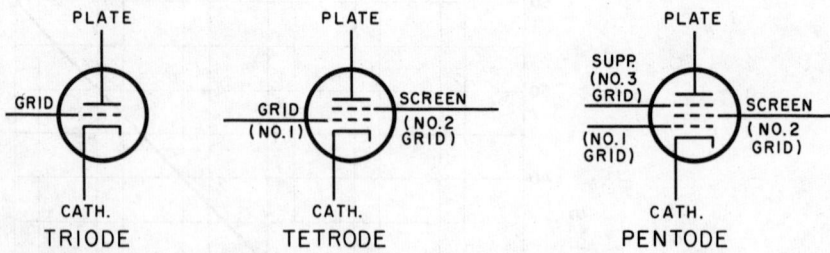

FIG. 2-25—Grids in multigrid tubes are numbered in order, starting with the one next to the cathode. In some pentodes the suppressor grid (No. 3) is internally connected to the cathode and does not have a separate base pin. It may be omitted in drawing the symbol in that case, since no wiring is required for it.

at this point to sort them out. First, there is the kind discussed in connection with Fig. 2-22. Its distinguishing feature is that plate current flows throughout the signal cycle—the grid is never, even instantaneously, driven negative beyond the cutoff point. An amplifier operated in this way is called a **Class A** amplifier. Since its plate current is driven *down* just as much as it is driven *up* by the signal, the *average* plate current stays just about the same whether or not a signal is on the grid.

Now suppose that the bias is set at cutoff, or slightly to the positive side of cutoff so that just a little plate current flows when there is no signal. During the negative half of the signal cycle the plate current stays cut off. During the positive half, the plate current will be approximately in proportion to the amplitude of the signal. That is, the average plate current *changes* when the signal is applied, and the bigger the signal voltage the greater the plate current. This is called **Class B** amplification.* You'll see applications of it in later chapters.

There is a third general type—the **Class C** amplifier. Here the bias is set well beyond the cutoff point—up to twice the bias required for cutoff, usually. A large grid driving voltage is used—so large that making it larger wouldn't cause any further increase in plate current. This is called driving the tube to **saturation**.

Class C operation gives the highest plate efficiency that can be obtained. It is useful in r.f. power amplifiers that are used for code work, or are to be modulated for phone. It is used everywhere in transmitters except in one case—where a *modulated* signal must be amplified. Class C operation destroys the amplitude modulation on a signal applied to its grid. A Class A or Class B amplifier must be used for amplifying a modulated signal.

Screen-Grid Tubes

It would be nice if we could get very large swings in plate current without having to drive the grid positive. This would mean large power output from the tube with no **driving power**. Modern tube design has gone a long way toward this goal. It isn't done with triodes, however. It hinges on having another electrode, the **screen grid**, in the tube.

The screen grid—the name is generally shortened to **screen**—is a grid-like structure placed between the regular grid and the plate. The element we have known so far simply as the "grid" is now called the **control grid** to distinguish it from the screen. A positive voltage is applied to the screen, and this element's job is one of attracting as many electrons as possible from the region around the cathode—but to do it in such a way that the electrons will continue on through to the plate and not land on the screen itself. Not all of them do get by the screen, so there is a current in the screen circuit. However, it is small compared with the plate current.

Other than helping along the flow of electrons to the plate, the screen takes no part in amplification in ordinary circuits. This sort of tube structure makes it possible for large plate currents to flow when the control grid voltage is zero. The presence of the screen, with its positive charge, makes it unnecessary for the control grid to "go positive" to make the plate current large. Thus the control grid can be operated in the negative region and no power will be taken from the signal source. (You can get still more plate current by making the control grid positive, just as in a triode, but the point is that you don't have to, in order to get a large output.)

Tetrodes and Pentodes

A tube having a cathode, control grid, screen grid and plate is called a **tetrode**. (If the cathode is indirectly heated, there will be a heater element, too, but it is counted as part of the cathode.) When a screen grid is placed in a tube the electrons that travel through the screen get up to quite high speed before they reach the plate. In fact, they hit the plate so hard that they often knock electrons loose from the plate itself. These splashed-out electrons are called **secondary electrons**, and the process by which they are released from the plate is called **secondary emis-

---

*An intermediate value of bias also can be used, so that the operation is partly Class A and partly Class B. This is called Class AB operation.

Vacuum Tubes

sion. Once free, they often are attracted to the screen because of its positive charge.

The secondary-emission current flowing to the screen doesn't help in maintaining good tube characteristics. To get rid of it a third grid, the **suppressor grid,** is often introduced. This grid is near the plate and is usually operated at the same potential as the cathode. It shields the plate from the screen, so the secondary electrons don't "feel" the positive charge on the screen. As a result, they fall back on the plate for the most part. Tubes of this sort are called **pentodes.**

Modern tetrodes of the **beam** type (where the control-grid and screen-grid wires are lined up so that electrons can pass freely through both) are designed to be relatively free from secondary-emission effects. They usually have **beam-forming plates,** connected internally to the cathode. These plates at least partially shield the plate from the screen, and help guide the electron flow *to* the plate in desired paths inside the tube.

Remote-Cutoff Tubes

In any tube, making the grid bias more negative reduces the transconductance, so the amplification also decreases with larger bias. In an ordinary tube the transconductance decreases almost uniformly with the negative bias, until the plate-current cutoff point is reached. Varying the grid bias offers us a way of controlling the amount of amplification.

However, this method also can cause distortion and associated effects. To get around it, many of the small pentodes made for use in receiver amplifier circuits are given a rather peculiar characteristic. The transconductance "tails off" more and more as the bias is made more negative, so it takes a rather large negative bias to cut off the plate current. For example, a **sharp-cutoff** tube of regular design might reach cutoff with −10 volts on its grid. An otherwise identical tube with **remote cutoff** (or **variable-μ**) might require −40 or −50 volts to cut off the plate current. This characteristic is useful when the tube is required to handle large signals and the bias is varied to control the gain.

Cathode Bias

In most cases, it isn't necessary to use a separate battery or power supply just to provide negative bias for the control grid. The bias can be stolen from the plate supply by a simple circuit trick. This is shown in Fig. 2-26. Bias obtained in this way is called **cathode bias.** There will be a voltage drop in resistor R_1 because the plate-to-cathode current has to flow through it. The polarity is such that the cathode is "raised above ground" (the negative terminal of the plate supply is "ground" in this circuit) by a positive voltage equal to the voltage drop in R_1.

Note that in the upper drawing the grid is connected back to ground through the signal source, G. But since the cathode is more positive than ground, the voltage from the grid to cathode is negative. Thus the grid is given a negative bias with respect to the cathode.

The Cathode Resistor

The bias can be set at a desired value by a simple application of Ohm's law. If you know the bias you want and also know the total current that will flow to the cathode at that bias, the value of R_1 is found by dividing the bias voltage by the cathode current. If you need a bias of 5 volts and the cathode current will be 10 milliamperes at that bias, R_1 will be 5 volts divided by 0.01 ampere (10 milliamperes), which is 500 ohms. Note that the plate current must be converted from milliamperes to amperes if you want the resistance to come out in ohms.

Cathode bias isn't entirely free. The voltage between the plate and cathode, which is the *effective* plate voltage, is reduced by the amount of the bias. In most cases the loss of plate voltage is of no consequence; it is only a few volts out of some hundreds.

Grid Return Circuit

The lower drawing in Fig. 2-26 works just the same way as the upper one. It is used when the signal source, G, won't pass direct current, or when for some reason it is not desirable to have

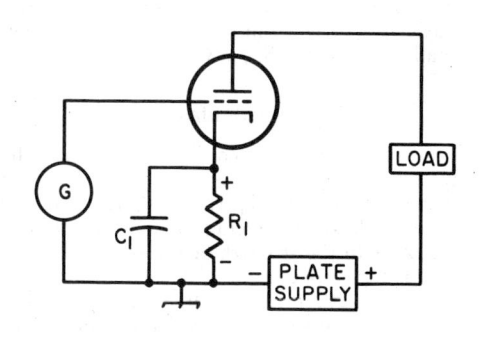

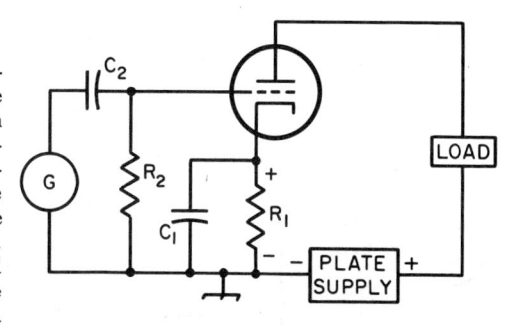

FIG. 2-26—Negative grid bias obtained from a resistor between cathode and ground. Two types of "return" circuits for the grid are shown. The lower one isolates the grid (for d.c.) from the signal source.

it do so. Resistor R_2 substitutes for it, so the grid is still at ground potential for d.c. The grid is insulated for d.c. by the **blocking capacitor,** C_2. To avoid putting an undue load on the signal source, the resistance of R_2 must be large compared with the load resistance that G can handle. In many circuits R_2 will be as much as a megohm or so. The capacitance of C_2 must be large enough so that for the a.c. signal its reactance is very small compared with the resistance of R_2. In r.f. circuits a few hundred $\mu\mu$f. is generally enough, but at low audio frequencies it sometimes takes 0.1 microfarad or more.

Cathode Bypass

C_1 is the **cathode bypass capacitor.** Since R_1 is in the plate-cathode circuit, some of the amplified signal voltage will be developed in it, just as a d.c. voltage was. It happens that this amplified signal voltage in R_1 *opposes* the signal applied to the grid. This has the effect of reducing the amplification, and is called **negative feedback.** It has its uses, but if it is not wanted we have to prevent any signal voltage from being developed across R_1. This is done by bypassing R_1 for alternating currents. To be effective, C_1 must have low reactance compared with the resistance of R_1. Since R_1's resistance usually is fairly small, the capacitance must be large. Values of 0.01 μf. are customary if the frequency being amplified is a radio frequency in the 3–30 Mc. range. Larger values are used at audio frequencies (several microfarads at least) but smaller ones will suffice at v.h.f.

Fixed Bias

Cathode bias is especially useful when the average plate current of the tube does not change much, if at all, when a signal is applied to its grid. If the average plate current does change with the strength of the signal, the bias does not stay put at one value but rises as the signal strength increases. In cases like this it may be necessary to use a battery or special power supply for bias, so the bias won't change with the signal level. Bias of this sort is called **fixed bias.** But for many applications, such as the amplifiers used in receivers and many of those used in low-power r.f. and audio stages in transmitters, cathode bias is quite satisfactory.

Grid-Leak Bias

There is another way of making a tube furnish its own grid bias, shown in Fig. 2-27. If you look at this circuit without knowing what goes on, you may wonder how there could be any grid bias at all, since no source of bias is shown. The answer is that this method, called **grid-leak bias,** only works when the signal source drives the grid positive with respect to the cathode.

Remember that a current flows in the grid-cathode circuit whenever the grid is positive with respect to the cathode. This current charges the capacitor C_1, called the **grid capacitor** or **grid blocking capacitor.** When the signal voltage is making the grid positive it attracts electrons. These collect on the capacitor plate connected to the grid, giving that side of the capacitor a negative charge. If R_1 is large, this charge "leaks" off slowly through it, so the net voltage at the grid is negative. Thus the tube gets an average negative d.c. bias through its own grid-current flow. The actual bias can be found by multiplying the resistance of R_1 by the direct current flowing through it.

Grid-leak bias depends entirely on grid current; there is no bias if the grid isn't getting any signal. This means that the source of signal has to supply power for it. Also, the signal has to be large, in order to drive the grid into the positive region. However, an advantage is that the bias adjusts itself to the signal level. The bigger the signal the greater the bias. This is just what is wanted in such devices as oscillators, since the bias automatically controls the tube operation to keep it in an optimum region. Grid-leak bias also is found frequently in r.f. power amplifier circuits, where the automatic feature also is useful.

The resistance of the **grid leak,** R_1, may be as low as a few thousand ohms in tube circuits handling a lot of power. Values as high as 100,000 ohms may be used in oscillator circuits of very low power. The value of C_1 depends on the frequency being amplified. C_1 has to have enough capacitance so that it won't lose much charge during each cycle of the signal frequency. A few hundred $\mu\mu$f. usually will suffice at radio frequencies, where grid-leak bias is mostly used.

Voltage Amplifier Circuits

We have been saying right along that the load for a vacuum-tube amplifier doesn't have to be an actual resistor. In fact, it is usually something else—a tuned circuit (which may in turn be loaded, as you have seen earlier in this chapter), a speaker with a transformer, or a variety of other devices. Nearly always, though, it "looks like" a resistance to the amplifier tube. This is necessary for getting power from the tube.

There is one type of amplifier that does use a resistor in its plate circuit. It often has the form

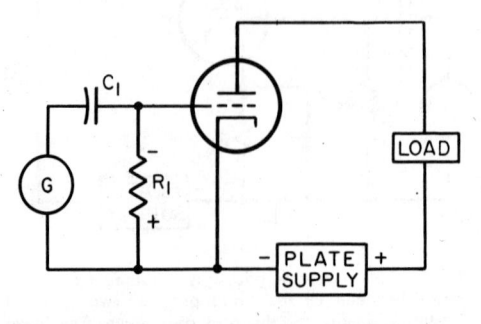

FIG. 2-27—Grid-leak bias circuit.

Amplifiers

FIG. 2-28—Typical resistance coupled voltage-amplifier circuit for audio frequencies. R_3 is the plate load resistor, C_3 is the plate bypass capacitor, and C_4 is the output coupling capacitor.

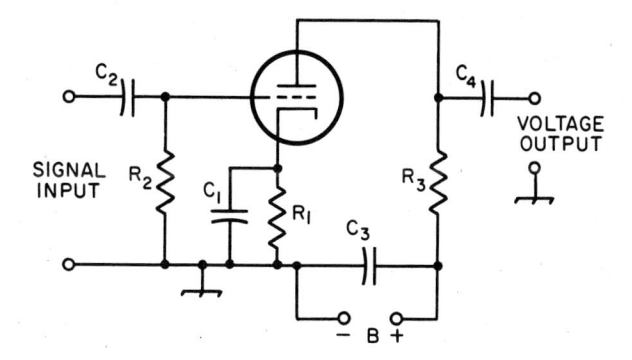

shown in Fig. 2-28, and is called a **resistance-coupled amplifier**. This type of circuit is used almost exclusively for audio-frequency amplification; for reasons explained later it is not useful at radio frequencies. Resistance-coupled amplifiers are always voltage amplifiers, because the power that may be developed in the load resistor can't be taken out.

Transformer Coupling

Transformers can be used in audio voltage amplifiers, too, as in Fig. 2-29. The transformer itself often steps up the output voltage by having a secondary winding with more turns than the primary. (This step-up is not considered to be amplification. There is actually less *power* in the secondary than in the primary, since there are losses in the transformer itself.) A transformer step-up is used only when the secondary is connected to a following amplifier tube whose grid is never driven positive.

The amplifiers you find in receivers between the antenna and the detector are also voltage amplifiers. We'll have more to say about them in Chapter 3.

Bypassing and Decoupling

You're already familiar with the uses of the components in the circuits of Fig. 2-28 and 2-29 except for C_3. C_3 is a **plate bypass capacitor**. It is used to give the alternating signal current in the plate circuit a good low-impedance path to the cathode. We don't want this current to flow through the B supply if we can avoid it, especially when several amplifier stages are connected to the same supply.

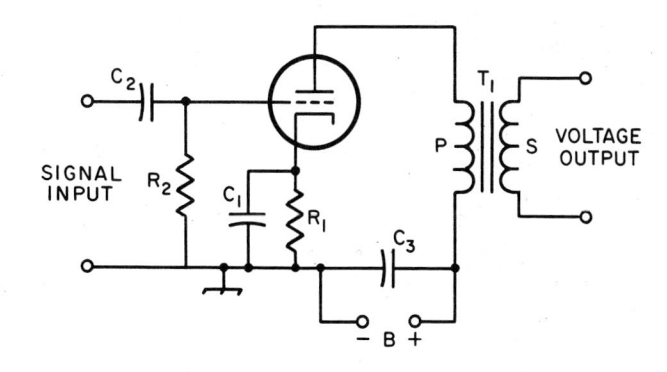

FIG. 2-29 — Transformer-coupled amplifier. An audio transformer is substituted for R_3 and C_4 of Fig. 2-28.

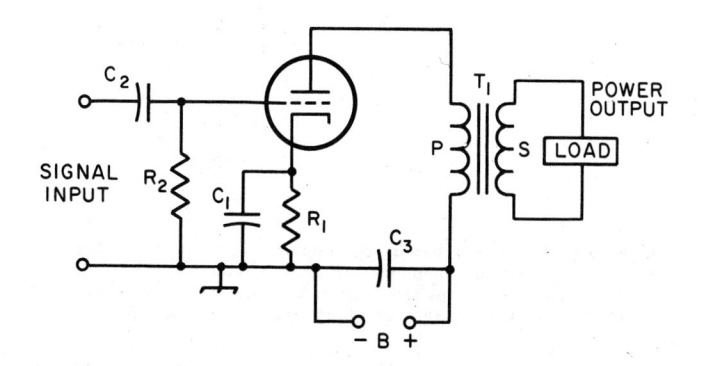

FIG. 2-30—Why plate circuits must be by-passed.

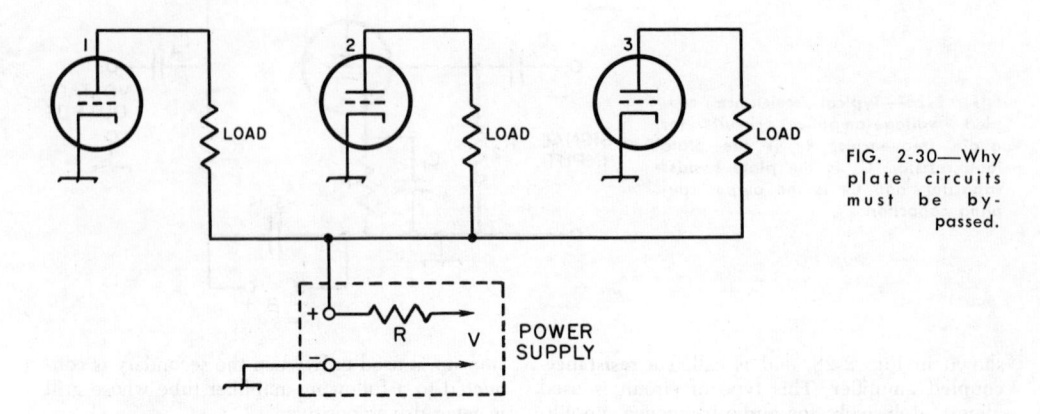

Fig. 2-30 illustrates why. Just enough detail is given in this figure to show that if the power supply has an internal resistance, R (and all power supplies do have some internal resistance), this resistance will be part of all three amplifier plate circuits. For example, the highly-amplified current from amplifier 3 will flow through R and cause a signal voltage to develop across it. This voltage, in turn, is introduced into the plate circuits of stages 1 and 2. Such **common coupling** can lead to lots of trouble.

Fig. 2-31 gives the remedy. If the bypass capacitor, C, has a reactance of not more than a few ohms at the signal frequency, the signal current in each stage will be confined to that stage and not wander into other stages in the amplifier. Assurance is made doubly sure by connecting a **decoupling resistor** on the power-supply side of the bypass capacitor as shown. Usually this resistance is several hundred times the reactance of the bypass capacitor—values in the neighborhood of 1000 to 5000 ohms are common—so any tendency for the signal currents to get off their proper paths is rather thoroughly discouraged.

Power Amplifiers

When power is wanted, as for running a speaker, we don't care particularly how much the grid signal *voltage* is amplified. Here we have to provide a load that represents the optimum resistance for the tube to see as a *power* amplifier. It is done by matching the speaker (or other) load resistance to the optimum value through a transformer, as described earlier.

A typical circuit is shown in the lower drawing of Fig. 2-29. It looks almost like the voltage-amplifier circuit above it, but a different type of transformer is used. Also, the tube's operating conditions are chosen to give the most power output rather than the most voltage output.

Push-Pull

A favorite form of circuit where large power outputs at audio frequency are needed is the

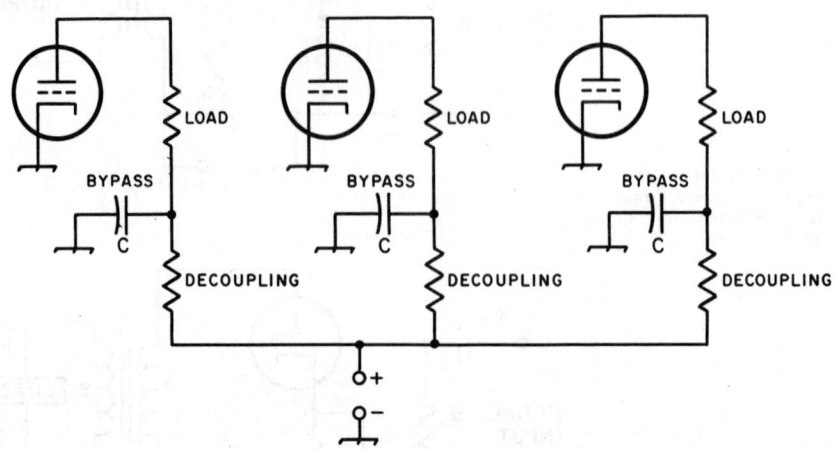

FIG. 2-31—A separate bypass capacitor on each stage, backed up by decoupling resistors, will effectively prevent common coupling through the plate supply. The decoupling resistors are not always used, in which case the "brute-force" effect of the bypass capacitors is depended upon to keep the signal currents from wandering into the wrong stages in the amplifier.

Amplifiers

FIG. 2-32—Typical push-pull amplifier circuit. This circuit can be varied in a great many ways, depending on whether the tubes operate as Class A, AB, or B amplifiers. If the tubes aren't driven into grid current, resistance coupling can be used to the signal source. In voltage amplifiers the output side may also be resistance coupled, but the transformer T_2 is always required if power is to be taken from the circuit.

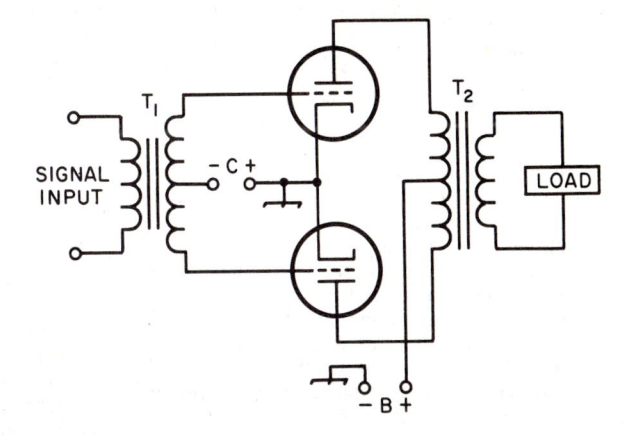

push-pull arrangement shown in Fig. 2-32. It is so called because what one tube does is always the opposite of what the other tube is doing. The secondary winding of the input transformer, T_1, is tapped at the center and the outer ends of the winding connect to the two grids. If the upper end of the winding is positive at some instant, the lower end is negative by exactly the same amount of voltage at the same instant. Thus when one tube's plate current is increasing because the signal voltage on its grid is going more positive, the other tube's plate current is decreasing because the voltage on *its* grid is going more negative.

This see-saw action opens the way to getting a great deal of audio power out of a pair of tubes, in relation to the size of the tubes. The reason is that this type of circuit reduces even-harmonic distortion, by a large margin, as compared with a single-tube circuit. Both tubes may still distort, but the distortions are balanced out, by the push-pull connection, in the output transformer, T_2. In fact, the tubes can be biased down practically to the plate-current cutoff point, so that each takes plate current only when the grid is driven by the positive half of the signal cycle, being idle during the negative half cycle. This is an extreme case of distortion, corresponding to half-wave rectification in each tube—yet the two half cycles, one from each tube, combine to give a composite output with low distortion.

A push-pull amplifier in which the tubes are biased near cutoff is a form of Class B amplifier. It operates at very good plate efficiency. A pair of tubes each rated to give 5 watts audio output as single-tube Class A amplifiers usually can develop 35 to 40 watts of total output when working together in push-pull.

Interelectrode Capacitances

The circuits just discussed are useful at low frequencies—those through the audio range. At radio frequencies they don't work at all well. To understand why, we have to take a closer look at tube structure.

A triode tube has three elements, all metallic and quite close together. Each pair of electrodes actually has the characteristics of a capacitor. The capacitances are only a few $\mu\mu f.$, but you will remember that the reactance of a capacitor keeps going down as the frequency goes up. The reactances of these **interelectrode capacitances** are so high at audio frequencies that they usually can be ignored. Not so at radio frequencies. There the reactances are considerably smaller than the load resistance we would need for appreciable voltage amplification. The tube capacitances actually bypass the alternating current around the load resistance. This not only ruins the amplification but introduces other very undesirable effects which are beyond our present scope.

FIG. 2-33—How the input and output capacitances of a tube can be absorbed by the input and output tuning capacitances.

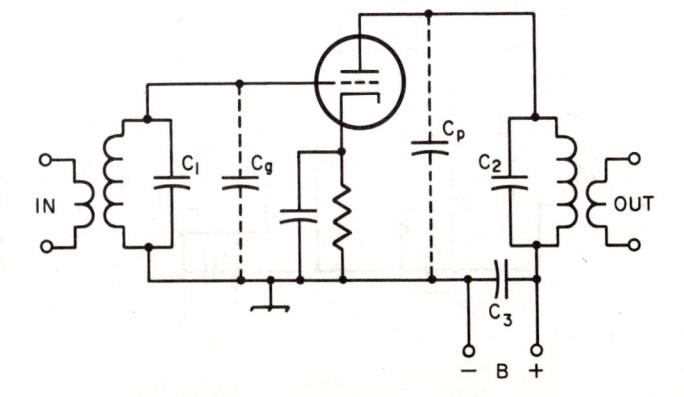

Why Tuned Circuits Are Used

For the moment, consider just two of the interelectrode capacitances—the one between the grid and cathode and the one between the plate and cathode. At radio frequencies the former would tend to short-circuit the input circuit, if it were resistance-coupled, and the latter the output circuit. We can't eliminate the capacitances from the tube, because they are part of its physical structure. So we take the opposite tack, and make them go to work for us. This is done by using *tuned* input and output circuits.

In Fig. 2-33, C_g represents the grid-cathode capacitance and C_p the plate-cathode capacitance of the tube. You can see that C_g is in parallel with C_1, the tuning capacitor of the input circuit. Likewise, C_p is in parallel with C_2, the tuning capacitor of the output circuit. (C_3, the plate bypass capacitor, is so large compared with these last two that it is the same as a short-circuit for the radio frequency.) Thus the interelectrode capacitances become *part of* the tuning capacitances, and they do no particular harm aside from setting a lower limit on the total tuning capacitance that can be used. This is not so important at the lower amateur frequencies, but it does require consideration at v.h.f.

Using tuned circuits allows us to get normal amplification, besides providing the selectivity that is needed at radio frequencies. We'll have more to say on the latter point in the chapter on receivers.

Grid-Plate Capacitance

We got around the **input** and **output** capacitances of the tube fairly easily by using tuned circuits. But Fig. 2-33 ignores one very important capacitance—the one between the grid and plate. Radio-frequency current can and does flow through it. In fact, if the grid and plate tuned circuits are adjusted to about the same frequency —as they have to be to give amplification—the grid-plate capacitance offers an ideal path for the amplified r.f. energy in the plate circuit to feed back into the grid circuit. When this happens a sort of merry-go-round action takes place.

The fed-back energy adds to that already present in the grid circuit. This is amplified to give still more r.f. energy in the plate circuit, and so on. The over-all result is that the tube *generates* radio-frequency power instead of just amplifying it. It becomes an **oscillator**. Fig. 2-33 is in fact an oscillator circuit, not an amplifier circuit. It is known as a **tuned-grid, tuned-plate** oscillator.

Screen-Grid Amplifiers

Fortunately, the grid-plate capacitance can be reduced to negligible size. This is done by placing a shield between the grid and plate, forming a screen-grid tube. Screen-grid tubes such as are used in the r.f. amplifiers in receivers usually also have a suppressor grid; that is, they are pentodes. A screen-grid amplifier circuit is shown in Fig. 2-34. The suppressor nearly always is connected to the cathode as shown, although sometimes it is connected to ground.

The screen grid has to have a positive voltage on it if the tube is to work properly. Generally, however, less voltage is needed by the screen than by the plate. The screen voltage is obtained from the plate supply through R_1, the **screen dropping resistor**. Screen current flowing through R_1 causes a voltage drop across R_1 that subtracts from the plate-supply voltage, leaving just the proper voltage on the screen. The right resistance value can easily be found if you know the current taken by the screen at its operating voltage. For example, say the plate-supply voltage is 250, the screen requires 100 volts, and the screen current is 10 ma. under normal operating conditions. (Figures of this sort can be obtained from published tube data.) Then R_1 must "drop" 250 — 100 = 150 volts. Its resistance is therefore 150 divided by 0.01 ampere, or 15,000 ohms.

If the screen is to shield the plate from the control grid and thus eliminate the grid-plate capacitance, it must be at the same r.f. potential as the cathode. The **screen bypass capacitor**, C_4, ensures this by short-circuiting to the cathode any r.f. that might show up on the screen. Like other bypass capacitors, it must have low reactance at the frequency being amplified. A capaci-

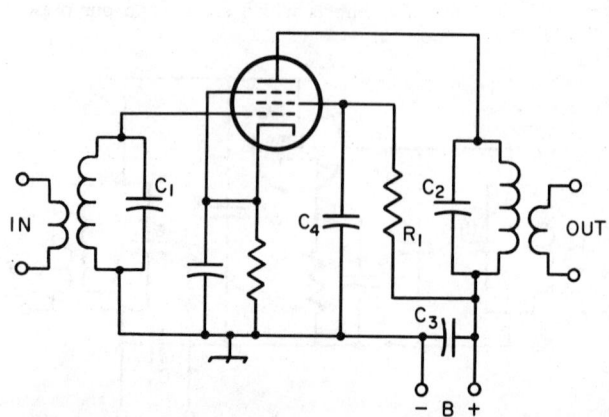

FIG. 2-34—Screen-grid amplifier circuit for r.f. amplification. C_1 and C_2 are the grid and plate tuning capacitances and C_3 is the plate bypass capacitor. Cathode bias is shown. R_1 is the screen dropping resistor and C_4 is the screen bypass capacitor.

Amplifiers

FIG. 2-35 — Grounded-grid amplifier, using the grid of a triode to shield the output circuit from the input circuit.

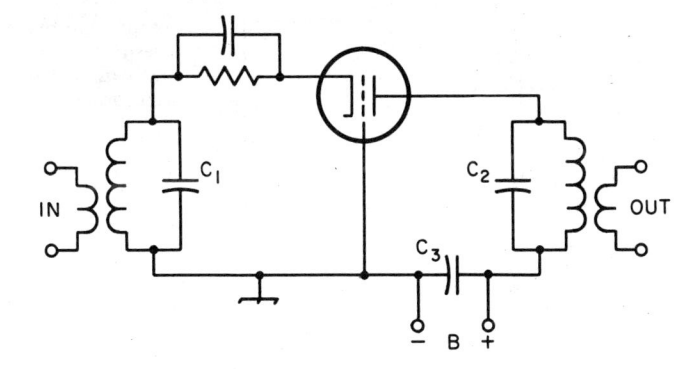

tance of 0.01 µf. is typical for r.f. amplifier circuits.

Screen-grid amplifiers are used both in receivers and transmitters. The circuits are basically the same whether the tube is a miniature receiving type or one that can handle hundreds of watts. There may be differences in the way the grid bias is obtained, or in the method used to supply screen voltage, but these are details that will vary even in receiver amplifiers.

Grounded-Grid Amplifiers

The idea of using a grid to shield two elements from each other is used in another kind of r.f. amplifier circuit. In a triode the grid will shield the plate from the cathode *provided* the grid has no r.f. voltage on it. It must, like a screen grid, be grounded for r.f. (It can still be at some different d.c. voltage than ground, though, just like the screen grid.)

Of course, if the grid is grounded the cathode can't be. So the signal voltage must be applied between the cathode and ground, as shown in Fig. 2-35. This circuit is like Fig. 2-33 except that the cathode and grid have changed places. The input signal voltage is still *between* the grid and cathode. But now the grid shields the plate (output) from the cathode (input) and very little r.f. energy can travel from the output side to the input side. Thus there is much less tendency for the tube to go into self-oscillation. The **grounded-grid amplifier** is used in both receivers and transmitters. In receivers it is used principally at v.h.f.

One of the peculiarities of this circuit, which is getting to be rather popular in amateur transmitters, is that the plate current has to flow through the input circuit to reach the cathode. Thus the input and output circuits are actually connected together, and a closer study would show that they operate as though they were connected in series. The upshot of this is that the source of input signal supplies part of the output power. This is quite different from the way an ordinary amplifier operates. The signal source must be prepared to supply quite a bit more power than would be required of it when the amplifier's cathode is grounded.

Only a few types of triodes have been designed with grounded-grid operation in mind. In those that have not been, the grid is not always as good a shield as we might want. Generally, tubes with large amplification factors will be the best in this respect, because the openings in the grid are smaller in such tubes. A number of screen-grid types are useful if the control grid and screen are both grounded for r.f.; in these, the shielding is likely to be very good.

Grounded-Plate Amplifiers

The plate of a tube can be grounded at the signal frequency, too. The circuit is as shown in Fig. 2-36, in a frequently-used form. An amplifier operated in this way is usually called a **cathode follower**, because the "hot" side of the output circuit is the tube's cathode. The plate bypass capacitor, C_3, grounds the plate for the signal frequency; the normal d.c. plate voltage is used.

The output is taken from between the cathode and plate. R_1 is thus the plate load resistor as well as the cathode bias resistor. C_1 and R_2 are in the circuit for the same purpose as the equivalent components in Fig. 2-26. C_2 is the output coupling capacitor.

Since the input signal voltage is applied between the grid and ground, it has to be larger than the output voltage. If it weren't, there would be no signal between the grid and cathode. The cathode follower doesn't give any voltage ampli-

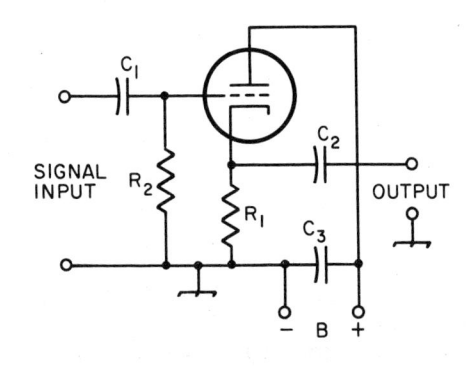

FIG. 2-36—The cathode follower, or grounded-plate amplifier.

up through several megacycles. Another is that it doesn't require a critical load resistance. The circuit can be looked upon as the equivalent of a transformer for stepping down from a very high impedance to a very low impedance. Suitably-designed cathode-follower circuits can work into loads as small as 100 ohms, or even less.

Oscillators

You already know that a tube can generate power without having its grid circuit driven by an external source. It can do this because more power can be obtained from its plate circuit than is needed by its grid circuit to produce the plate power. To start a circuit oscillating we need to feed power from the plate back to the grid in such a way as to reinforce the signal on the grid. This is called **positive feedback**, as distinguished from the *negative* feedback that *opposes* the grid signal. With enough positive feedback, any small irregularity in the plate current will supply a signal that will start oscillations going at the frequency to which the tube's circuit is tuned.

Positive feedback can be developed in so many ways that we can't begin to cover all of them here. We've already mentioned the tuned-grid tuned-plate oscillator (Fig. 2-33). Two other general types of circuits are so commonly used that you should be able to recognize them on sight. One, the **Hartley** circuit of Fig. 2-37A, belongs to the **magnetic** or **inductive feedback** class of circuit. Alternating plate current flowing through the lower part of L_1 induces a voltage in the upper part, which is connected to the grid. The second general type, the **Colpitts** circuit of Fig. 2-37B, uses **capacitive feedback**. Here the plate-circuit energy is fed back by introducing it across C_2, which is part of the tuning capacitance of the circuit. This coupling sets up an r.f. voltage across the whole circuit. The voltage that, as a result, develops across C_1 is applied to the grid.

Another circuit you will meet now and then is shown in Fig. 2-37C. It isn't basically different from the Hartley, but instead of tapping the coil it uses inductive coupling through a **tickler coil** tightly coupled to the tuned-circuit coil.

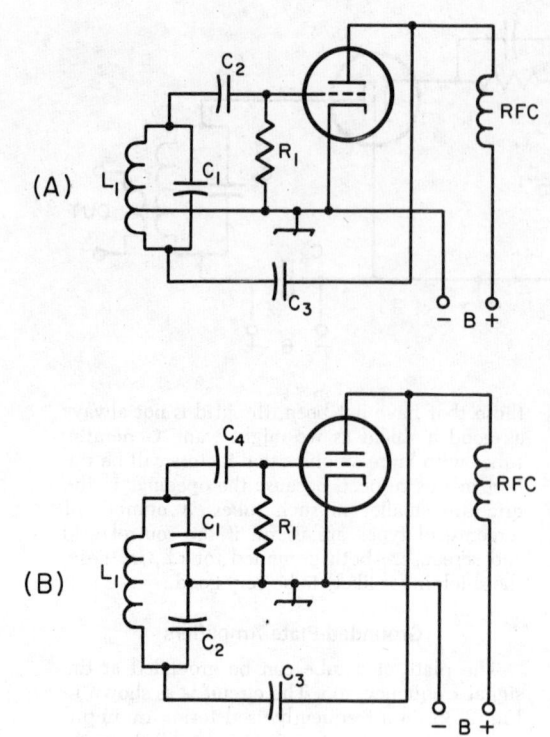

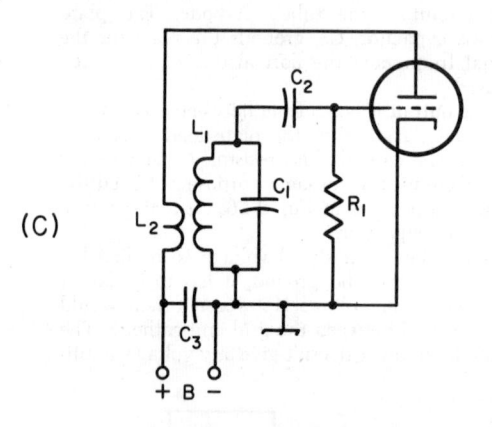

FIG. 2-37—Some common oscillator circuits. A—Hartley; B—Colpitts; C—tickler.

The Tank Circuit

The tuned circuit is often called the **tank circuit** of an oscillator or r.f. power amplifier. The reason is that it stores several times as much energy as it is called on to give up each radio-frequency cycle. This is necessary if the oscillator is to operate with good stability (the subject of stability is treated in later chapters; it is an extremely important oscillator characteristic). This high ratio of stored-to-used energy is summed up quite simply by saying that it corresponds to a circuit of high Q.

All three of the circuits shown are grid-leak biased. R_1 is the grid leak or grid resistor in each case. C_2 is the grid capacitor in A and C; in B it is C_4.

fication; in fact, there is actually a voltage loss. However, there is *power* amplification.

Earlier, in discussing cathode biasing, we mentioned that the cathode resistor is bypassed in order to avoid unwanted negative feedback. In the cathode follower we have *all* of the output voltage fed back into the input circuit to oppose the signal voltage. This has its good features, although at the price of a loss in voltage amplification. One benefit is that the amplifier will work over an extremely wide frequency range without tuned circuits—from very low audio frequencies

Semiconductors

Series and Parallel Feed

The circuits at A and B in Fig. 2-37 are known as **parallel-fed** circuits. The d.c. power for the tube is fed to the plate through a radio-frequency choke. In this type of circuit the plate is "hot" with r.f. voltage, and its output would be short-circuited by the power supply if the choke were not used. The choke offers a very high impedance to r.f. current, so little or none of the current gets back into the power supply. Instead, it flows through C_3, which connects the plate to the oscillating circuit. C_3 is called a **blocking capacitor** because it blocks off the d.c. voltage so it won't be present on the tank circuit.

Parallel feed is very often used in transmitting circuits. It is an excellent safety measure; there is no high d.c. voltage on any of the r.f. circuit elements that might be touched in regular operating.

Fig. 2-37C is a **series-fed** circuit. Here the d.c. plate current flows through the tickler-coil part of the r.f. circuit. C_3 is a plate bypass capacitor in this case. No r.f. choke is shown in the plate-supply lead because in many cases it isn't strictly needed. However, a choke is sometimes used as extra insurance against r.f. leakage into the plate supply.

Other Oscillator Circuits

In building equipment requiring oscillators, you have your choice of these and several other circuits. With equal care in design and construction all of them are capable of giving equally-good results, so you may wonder why designers don't settle on one and forget about the others.

The answer is that one circuit may be better adapted than another to a particular application, often for reasons of mechanical construction which you will appreciate as you become more familiar with actual apparatus. Also, there is the rather undefinable matter of the designer's tastes in circuits—after all, blondes, brunettes and red-heads all have their rooters!

As a Novice, you must use a crystal-controlled oscillator in your transmitter. We'll defer talking about crystal oscillators until Chapter 4.

Semiconductors

One of the pioneer devices in radio communication was the crystal rectifier. This was a mineral crystal—usually silicon or galena—capable of detecting radio signals. Shelved for years by the more sensitive and reliable vacuum tube, which could amplify as well as detect, the crystal was revived in quite recent times and studied more thoroughly. The result has been a series of devices, classified broadly as **diodes** and **transistors**, based on the peculiarities of electron movement inside a class of solid materials called "semiconductors."

Semiconductors are crystalline substances—principally silicon and germanium—which during manufacture have been "doped" so that they either have free electrons in excess of those needed by the atoms in the crystal, or are short of enough electrons to fill all the atoms. Materials with an excess of electrons are called **N** type; those with not enough are called **P** type. An atom lacking an electron is said to have a **hole**. When an electric field is applied, an electron can jump from a "full" atom to a hole in another, leaving a hole in the atom which the electron left. We can think of this as the movement of a hole rather than the movement of an electron. In N-type material the electrons move and in P-type material the holes move.

How the Diode Works

The rectifying action of a semiconductor is based on this dissimilar action in the two types of material. Fig. 2-38A shows a **P-N junction** formed by placing N and P materials in contact. If a battery voltage is applied as in B, the electrons in the N material are attracted to the positive battery terminal and the holes in the P material move toward the negative terminal. Since there is no movement *through* the junction,

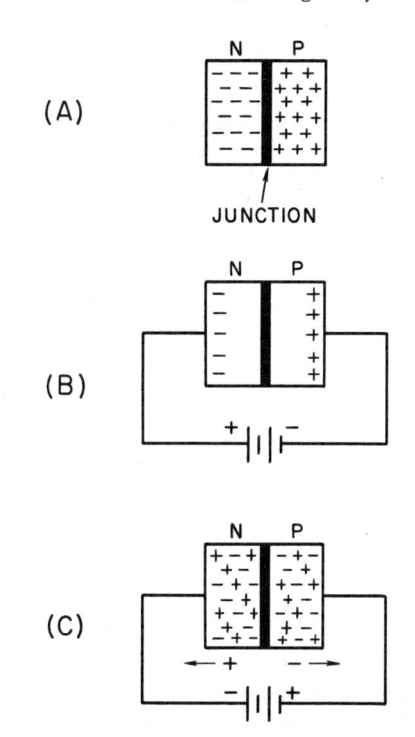

FIG. 2-38—The junction diode, a semiconductor rectifier. B shows the nonconducting condition (reverse bias); C shows the conducting condition (forward bias).

no current can flow in the circuit. However, if the battery polarity is reversed as in C, holes move across the junction toward the negative battery terminal and electrons move across it toward the positive terminal. With this battery polarity, therefore, a current flows through the junction and the circuit.

The semiconductor rectifier is not a perfect insulator when the voltage is applied in the reverse polarity, as in Fig. 2-38B. A small reverse current can flow, when the diode is **back biased**. In the forward direction, Fig. 2-38C, it takes only a couple of volts to cause very large currents to flow—much less than it does in a vacuum-tube diode.

Ratings

Aside from the reverse current, which is of the order of a thousand times less than the forward current under normal conditions, the principal limitation of the semiconductor diode is the fact that it cannot stand very high reverse voltages. Diodes suitable for power-supply use can be obtained with reverse-voltage or **peak-inverse voltage** ratings up to several hundred volts, but in vacuum tubes such ratings may be in the thousands of volts. Also, the semiconductor diode can't stand any overload in this respect; it will usually break down and ruin itself if the reverse voltage is just a little over the rating. Its advantages are small size, ability to conduct large currents with low voltage drop, and (as a result of this) very little lost power in the rectifier.

Point-Contact Rectifiers

The junction in the diodes we have just been discussing is a region of appreciable area. It can pass large currents. Another type of rectifier uses a **point contact** or **catwhisker**, Fig. 2-39. In the usual type the semiconductor is N-type material and during manufacture of the diode a very small amount of P-type is formed under the point contact.

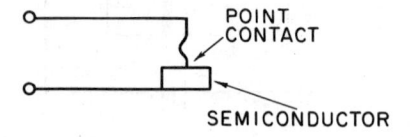

FIG. 2-39—The point-contact semiconductor diode is like the old-time crystal detector with a catwhisker.

The area between the two types of material is so small that the point-contact diode has very low capacitance—generally only about one micromicrofarad. This makes it especially suitable as a rectifier for radio frequencies, since very little current is shunted around the rectifier by the capacitance. This type of rectifier is not used in power-supply circuits except where only small currents—less than 50 ma. or so—are needed.

Reverse voltage ratings run from 30 to about 100 volts.

The Diode Symbol

The interpretation of the circuit symbol used for the semiconductor diode is sometimes confusing. As shown in Fig. 2-40, the arrowhead in the symbol corresponds to the plate in a vacuum-tube diode and the line corresponds to the cathode.

If you apply a d.c. voltage to the diode with the polarity indicated in the drawing, a large current will flow. If the polarity is reversed, the current will be very small—provided, of course, that the diode is in good condition! Thus, looking through the diode from the plus end of the battery, the direction in which the arrowhead

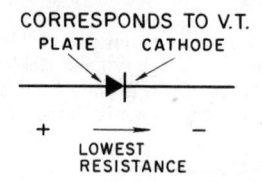

FIG. 2-40—The semiconductor diode or contact-rectifier circuit symbol and its meaning.

points is the direction in which you will measure the lowest resistance (highest current).

Keep the analogy to the v.t. diode in mind in interpreting the diode connections you see in circuit diagrams and you'll have no trouble.

The Transistor

A **transistor** is formed by putting two diodes back to back in a three-layer structure as shown in Fig. 2-41. A sandwich is formed with one type of material as the "meat" and with the other type as the "bread." The in-between material, a thin layer, is called the **base**. The outer ones are called the **emitter** and **collector**. As in Fig. 2-41 shows, the material arrangement can be either P-N-P or N-P-N. The circuit symbols for the two types are also shown, along with the battery polarities that will cause current to flow in the circuit formed by the base and emitter, and that formed by the collector and emitter.

If you compare these arrangements with the simple diode of Fig. 2-38 you can see that if the battery voltage E_2 is larger than E_1, the collector-to-base diode is reverse biased and no current can flow between the collector and base. However, current *will* flow through the emitter-base diode with the polarity shown for E_1. The holes move from the emitter to the base through the diode junction and the electrons move in the opposite direction. Some of the moving holes get over into the collector-base region, giving the collector more than it needs, and the extra ones are attracted by the negative voltage on the collector. Thus there is a current flowing directly

Semiconductors

FIG. 2-41—The two fundamental types of transistors. Unlike tubes, transistors can be made to work with either polarity of applied voltage, as shown by the symbolic drawings at the bottom. The battery polarities shown are those used for normal biasing in amplifier circuits.

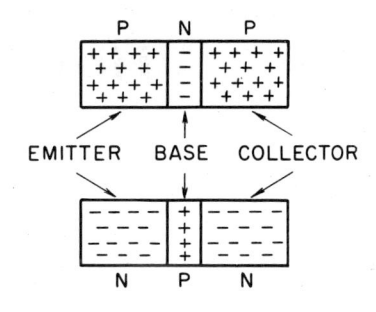

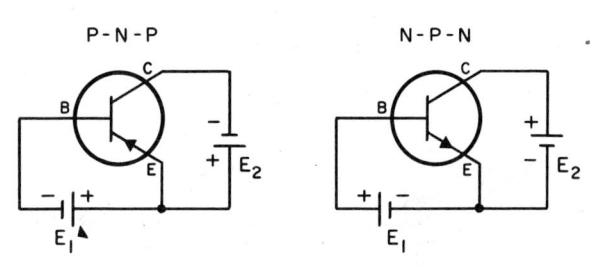

through the transistor from the emitter to the collector. It doesn't show up in the external base circuit at all, and it actually is many times larger than the emitter-base current. Furthermore, the *amount* of collector current is directly related to the base current. If the base current is zero, so is the collector current. The larger the base current the larger the collector current.

Amplification by Transistors

All this adds up to the fact that a transistor is an amplifier, just as the three-element vacuum tube is an amplifier. There are quite a few similarities between tubes and transistors, but also quite a few differences. One difference is that a current has to flow in the signal input circuit (the base-emitter diode) if there is to be any collector current. This means that the transistor can't amplify without taking power from the source of the signal to be amplified. A vacuum tube doesn't *require* such power although, as you saw earlier, some types of operation do call for it. Another point is that the transistor diodes, like ordinary semiconductor diodes, are not perfect; there are leakage currents with reverse voltage. The vacuum in a tube is a well-nigh perfect insulator when there is no electronic conduction.

A third point is that the transistor is essentially a high-current device that works at very low voltages, while the tube is just the opposite.

All these things mean that there is quite a difference in the way circuits have to be handled in using transistors, as compared with tubes. But the circuits for amplifiers, oscillators and the like aren't too different in principle. The transistor has to have a proper load for its output (collector) circuit in order to give optimum amplification, for example. Amplification with transistors ordinarily is expressed in terms of power gain, since the signal-input circuit always consumes power.

Transistors and Tubes Compared

It helps in considering the operation of transistor circuits to compare the three elements—emitter, base and collector—of a transistor with the three elements of a triode tube. The emitter corresponds to the cathode of the tube, the base to the grid, and the collector to the plate. If you draw a triode tube circuit and then substitute a transistor using this correspondence, you'll have a workable transistor circuit. Usually it won't be the best possible circuit, because there are details that should be different for optimum performance. One of these is the way the emitter-base bias is obtained, but this and similar subjects are beyond our scope here.

The transistor is a highly efficient device because, first, it has no hot cathode and thus no heater to consume power, and second, because it has very little internal voltage drop when carrying large currents. For a given power output, the power lost in internal heating is therefore very low. This is fine for battery-operated equipment where the power drain is quite important.

Chapter 3

How Receivers Work

By the time you've received your license you will have done a good deal of listening on the amateur bands. You should be pretty familiar with your receiver's controls and have a fair idea of how to operate them. It is a sad fact, though, that many an operator never gets the results his receiver is capable of giving. Having no clear idea of what the receiver can do, he fails to handle it to best advantage.

To get a little perspective on this receiver business, let's get down to the bare bones of receiving necessities and then go on from there. You can't hear, see, feel, smell or taste a radio signal. It has to be converted into something that human senses can recognize. The sense of hearing is used in most amateur communication, so the signal must somehow be converted into an appropriate sound.

The Simplest Receiver

If you are young enough—or old enough— you've probably played with crystal receivers for standard (amplitude-modulated or "a.m.") broadcasting. These usually have no more than a half-dozen parts, but even here there are only three really essential ones—an **antenna** or length of wire to pick up some of the r.f. energy in a broadcast signal, a mineral crystal or **detector** for converting the r.f. energy into an electrical current that varies at audio frequency, and a head-

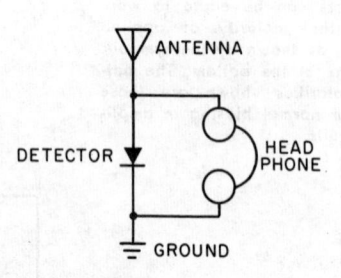

FIG. 3-1—The simplest receiver—just an antenna, detector, and headphones.

phone. The latter simply converts the audio-frequency current into sound waves.

The detector is really a rectifier; it takes the radio-frequency current and changes it into a direct current. The *intelligence* in the signal is carried by the *variations* in its signal strength or **amplitude**, and the direct current that comes out of the rectifying detector follows these variations. In a phone signal (or a broadcast of voice and music) these variations occur at an audio-frequency rate, as shown in Fig. 3-2. Once the variations are converted into sound by the headphone, the signal becomes intelligible.

In a sense, everything else in an actual receiver is a superstructure leading ultimately to this basic process of detection. But that doesn't

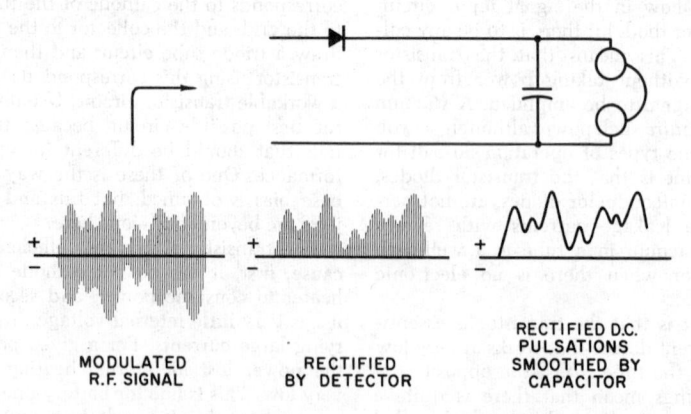

MODULATED R.F. SIGNAL   RECTIFIED BY DETECTOR   RECTIFIED D.C. PULSATIONS SMOOTHED BY CAPACITOR

FIG. 3-2—The operation of the diode rectifier in detecting a modulated r.f. signal. The r.f. signal itself makes no impression on the headphones, but rectifying it gives us a direct current that varies just in the same way that the modulation varied the strength of the original signal. The high-frequency pulsations in current that result from rectification are filtered out by the capacitor, leaving only the audio-frequency variations in the direct current to actuate the phones. Another way of looking at it is that the high-frequency pulsations are bypassed by the capacitor, but since the more-slowly-varying direct current cannot pass through the capacitor, the current flows through the headphones to produce a sound.

Receiver Principles

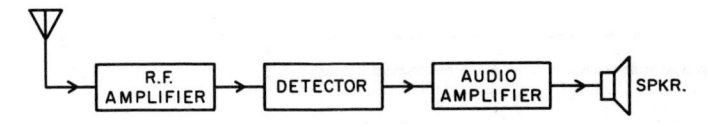

FIG. 3-3—A simple receiving setup in block-diagram form. The r.f. amplifier builds up the signal strength before detection, while the audio amplifier increases the strength of the audio signal resulting from detection. A receiver of this type is not very useful, practically, but illustrates an important point—amplification can and does take place in a receiver both before and after detection. Splitting up the amplification is necessary because there is a limit to the amount that can be used, without instability, on any one frequency or band of frequencies.

mean that the superstructure is unimportant. Far from it! Without the superstructure actual communication by radio would be practically impossible. For it is the superstructure that lets you hear a very weak signal even though the air is filled with thousands of others at the same time, many of them far stronger. In other words, the rest of the receiver is there in order to give you high sensitivity to weak signals, and high selectivity to let you hear the weak ones while eliminating all the ones you don't want.

Even the simple crystal set usually takes a timid step in that direction. It often has at least a crude system of tuning, or building up the response to the radio frequency of a desired signal while reducing the response to other frequencies. But it can go only a very limited distance in that direction, and more elaborate measures are necessary.

Amplification

The energy that produces the sound coming out of the headphone of a crystal set is all furnished by the radio signal itself. In other words, it is energy that travels from the transmitting station to the receiving antenna. If we had to depend solely on transmitted energy for producing the sound waves that ultimately come from the receiver, the distances over which we could communicate would be very limited indeed.

The answer to this is *amplification*. In broad terms, an amplifier is a device that uses locally-supplied energy to magnify the signal into a bigger reproduction of itself. The signal can be amplified either before or after detection; in the first case it is amplified at a radio frequency, in the latter case at audio frequency. Generally a combination of both is used, because each method has its advantages.

The Limit of Amplification

Since the energy extracted by the receiving antenna from passing radio signals is extremely small, the amount of amplification used in receivers is very great—a factor of some millions, usually. You might be tempted to think that this amplification process could be carried on indefinitely, so that even the weakest imaginable signal could be brought up to usable strength. Unfortunately, it cannot. The thing that limits the usable amplification is noise. This noise is from randomly-varying electrical currents generated in nature (**static**) and in electrical circuits and devices, including the amplifiers, used in the receiver itself. It occurs at all frequencies, both radio and audio, and is inescapable. This doesn't mean that there are no means available to make it less bothersome, though. There *are* such means, and receivers make use of them. One of the big objectives in designing good receivers is to improve the **signal-to-noise ratio**. Every effort is made to amplify the signal as much as possible while amplifying the noise as little as possible.

So much for a little deep-down background. Some of the technical points that bear on these matters have been discussed in Chapter 2. Now let's take a look at an actual receiving arrangement.

FIG. 3-4—The amount of amplification that can be used is eventually limited by noise. The strengths of the signals at A and D in this drawing are below the strength of the noise and would be heard poorly, if at all. The signal at C is well above the noise and would be easily "copied."

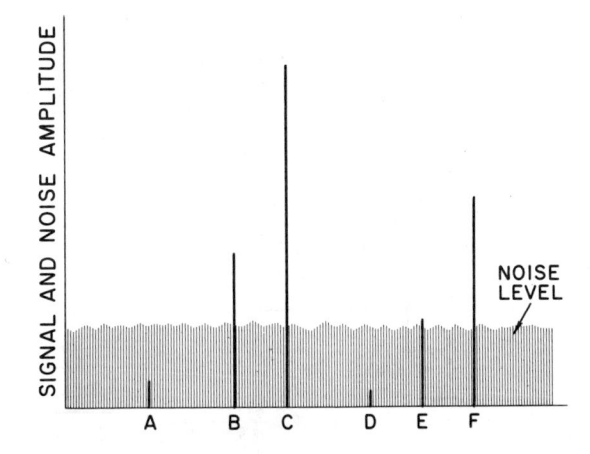

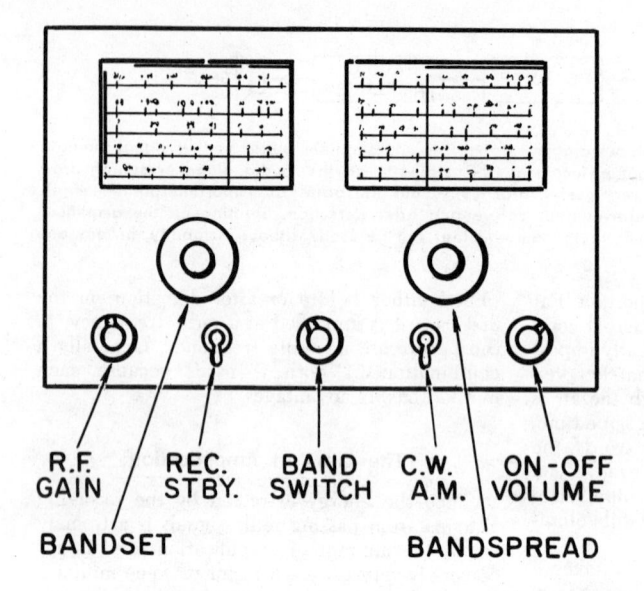

FIG. 3-5—Your receiver may not look exactly like this, but probably has the same controls. (If it is an amateur-bands only receiver it won't have the bandset tuning.) Even very unpretentious communications receivers should have the controls shown in the bottom row, although sometimes the names will differ.

Receiver Controls

Fig. 3-5 is not an attempt to represent the panel of any particular brand of receiver. It is intended to show the controls you have at your disposal in a minimum-design communication receiver. (We say "minimum-design" because a receiver with fewer functional parts than are shown in the block diagram of Fig. 3-6 hardly warrants the "communication" label.) There will be variations, of course. Actual receivers for amateur use may cover only the amateur bands instead of offering continuous tuning over the high-frequency radio spectrum. Such sets will have only one tuning dial and knob. And many receivers will offer features not shown on our sample panel.

A "two-dial" receiver as shown in Fig. 3-5 usually lets you listen anywhere between about 550 kc. and 30–40 Mc.—sometimes even up to 54 Mc. This continuous coverage is advantageous if you like to listen to what's going on between the amateur bands. If you have such a receiver, you know that you have to set the **bandset** or **general-coverage** dial rather carefully when you want to make a ham band come out right on the bandspread dial. There is a point to be learned from this. Tune in a phone station that is reasonably steady—that is, not fading particularly. Tune through it with the bandspread dial. Note the dial readings at which the signal just appears and just disappears as you tune through. Also, note how much of an arc you turn through with the bandspread knob as the station comes in and goes out. Now do the same thing with the general-coverage dial and knob. If you were well up in the spectrum, say at 14 Mc., tuning with the bandset dial was probably a hair-trigger operation. It may not have been critical at all with the bandspread knob.

Bandspread and Tuning Rate

At times, critical tuning is confused with selectivity. The two are not the same thing. You may call the tuning sharp on your general-coverage dial and broad on the bandspread dial, because that's the way it *looks* to you. But to the *receiver*, it is merely a question of how fast you tune through the signal. One dial is faster-tuning than the other, that's all. The ability of the receiver to separate stations, so that one can be brought in while another is excluded, is not affected by these tuning tricks.

Prove this to yourself by trying the same tuning test on a station that is under some interference. If anything, the bandspread dial—which

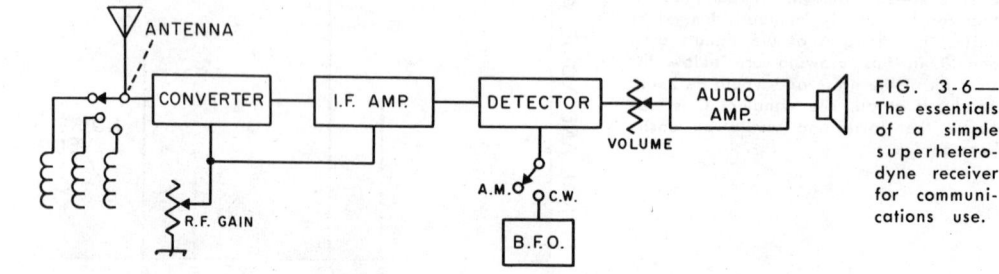

FIG. 3-6—The essentials of a simple superheterodyne receiver for communications use.

Receiver Principles

makes the signals seem broad because you tune through them slowly—actually will help you get rid of interference that you might not be able to tune out with the "sharp" bandset dial. If this is so, it's simply because the bandspread knob is easier for you to adjust. It lets you reach the proper tuning with a minimum of effort and strain.

So don't be one of those who confuse bandspread, or tuning rate, with selectivity. Actual selectivity is quite different from these surface effects. The quest for selectivity is the thing that led to the type of receiver in universal use—the **superheterodyne** or "superhet." The superhet, alone among receivers, offers the same kind of selectivity in any part of the spectrum. You'll find that you can eliminate an interfering signal the same number of *kilocycles* away from the desired signal, no matter whether the signal is on 1000 kc. in the broadcast band, or on 30,000 kc. on the other end of the h.f. spectrum. Watch out for that tuning rate, though—it probably won't be the same as at these two extremes. This may mean quite a difference in how the receiver handles. But ease of handling—or its opposite—is not a measure of selectivity.

Selectivity

If you have read the section on tuned circuits in Chapter 2 carefully you'll recall that the bandwidth of a tuned circuit gets wider as the resonant frequency is raised, other things being equal. We might have a circuit with a bandwidth of 10 kc. at 500 kilocycles, for instance. A comparably good circuit at ten times the frequency, 5000 kc. or 5 Mc., would have the same *percentage* bandwidth, but ten times the actual number of kilocycles. In this example, the bandwidth would be 10 × 10 = 100 kc.

Now it happens that the bandwidths of the signals we want to receive do *not* increase with the operating frequency. A phone signal requires no more actual bandwidth at 144 Mc. than it does at 4 Mc., or at any lower frequency still. Its bandwidth is determined by the audio frequencies in the human voice, not by the radio frequency at which the signal goes out. Therefore, we can use exactly the same selectivity, in kilocycles, at 144 Mc. as at 4 Mc. There will be times when it will be just as much needed on one frequency as on the other.

Adjacent-Channel Selectivity

We can't get that uniformity of selectivity in all amateur bands if we try to do it at the radio receiving frequency, because of the way tuned circuits behave. The logical answer, then, is to pick some one frequency where we can get the kind of selectivity we want, and then convert the incoming signal to that selected frequency. This is just what the superhet does. It takes a signal at, say, 21 Mc. where the inherent selectivity of the circuits would be poor, and changes it into a signal at perhaps 455 kc. The converted signal is identical in all respects, except frequency, to the original one. Or the signal might be in the 3.5-Mc. band, or the 28-Mc. band; in every case it would be changed into one at 455 kc. Its principal amplification would take place at that frequency. The selectivity of a 455-kc. circuit—or series of such circuits, as actually would be used in a practical receiver—is very much greater than that of any circuits in the receiver's front end. (The **front end** is the part that handles the incoming signal at its "real" frequency.) So for all practical purposes the selectivity of the receiver is exactly the same at 28 Mc. as at 3.5 Mc.—or at 455 kc.

The selectivity here considered is what is known as **adjacent-channel** selectivity. It is measured by the receiver's ability to single out one of a group of signals very close to each other in frequency. If the edge of the band occupied by one signal just touches the edge of the band occupied by another, the two signals are on adjacent channels. It is obviously very greatly to our advantage to have the ability to "copy" one such signal without interference from the other.

Visualizing Selectivity

The workings of selectivity are not always understood. You may have heard a fellow ham remark that "so and so's carrier is broad as a barn door"—usually about some nearby station with a powerful signal. The fact is that an unmodulated carrier has no width at all. It is just one single frequency. How, then, does the idea get abroad that an unmodulated signal can be "wide"?

Fig. 3-7 shows two selectivity curves that are reasonably typical of a modern superhet receiver. At 6 db. down from the maximum response one has a width of 500 cycles (0.5 kc.) and the other has a width of 5000 cycles (5 kc.). The numbers along the sides of the curves show the bandwidth in kilocycles at the point indicated; for example, the bandwidth of the right-hand curve is 6 kilocycles at 20 db. down.

Fig. 3-8 shows the transmitted bandwidths of "ideal" signals of various types. Actual signals would not be confined entirely within the bands shown, but if the transmitter is well designed

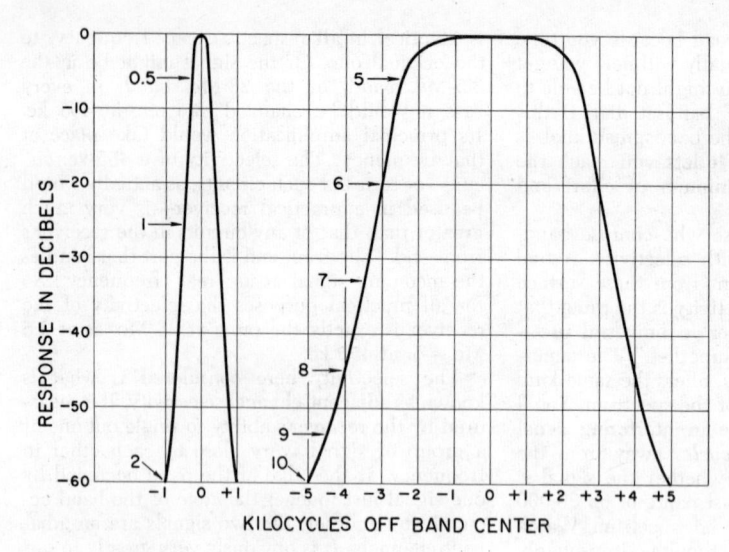

FIG. 3-7—Two selectivity curves. The narrow one at the left is very good for code reception but does not pass a wide-enough band of frequencies for good reproduction of voice signals. The curve at the right is a band-pass type intended for phone-signal reception.

and properly operated nearly all of its emission should be inside the limits. Perhaps the single-sideband signal at C is too ideal in this respect. However, a practical s.s.b. transmitter should not have any components less than 30 db. down in the "other" sideband. Such a signal is represented at D by the dashed line cutting off at −30. There should be nothing above this line in the transmitted signal.

Cardboard "Tuning"

These two figures will give you the basis for an instructive and possibly amusing half hour. Trace Fig. 3-7 on a piece of paper and paste the tracing on a card. Then cut out the parts *inside* the selectivity curves. Place your tracing on Fig. 3-8 so that its −60 axis is on top of the −60 axis of Fig. 3-8. Now slide the tracing sidewise, keeping the two axes together. As the openings pass over the "signals" of Fig. 3-7 you can see the response build up, as the leading edge of the curve slides over the "signal"; it reaches a maximum at the center of the curve, and then decreases as the trailing edge of the curve passes over it. Finally it disappears altogether from the cut-out, and that signal is "tuned out."

This is a representation of what actually happens as you tune through a signal with your receiver. Tuning is simply the process of moving the pass-band of the receiver across a signal (or series of signals, when more than one is present). If you repeat this experiment a number of times with each selectivity curve on each type of signal, and observe carefully what is happening to the response, you will soon get a good idea of what selectivity really means. You will see, for instance, that the unmodulated carrier at A "seems" to have width; it appears to be 10 kc. wide with the broad curve but only 2 kc. wide with the sharp curve. In other words, what you

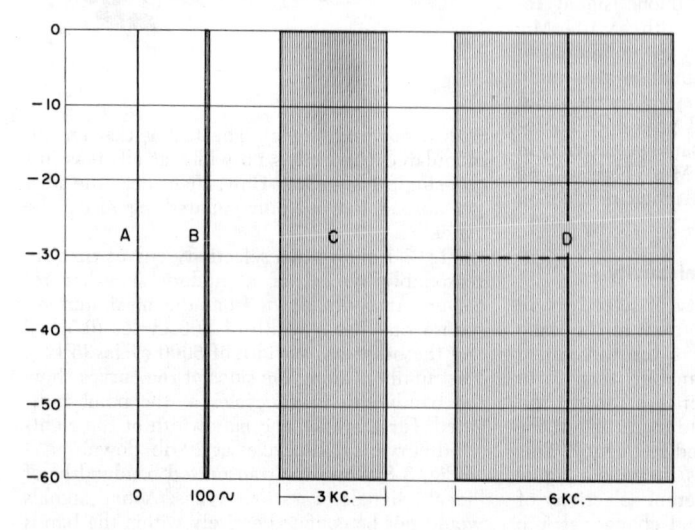

FIG. 3-8—Bandwidths of ideal signals of various types. A—unmodulated carrier; B—keyed c.w. signal; C—single-sideband signal; D—amplitude-modulated phone signal. This drawing and Fig. 3-7 are to the same scale, and you can use them as described in the text.

Receiver Principles

FIG. 3-9—In converting to an intermediate frequency we have two choices for the local oscillator frequency, as shown. But either one will bring about response to an undesired frequency or image.

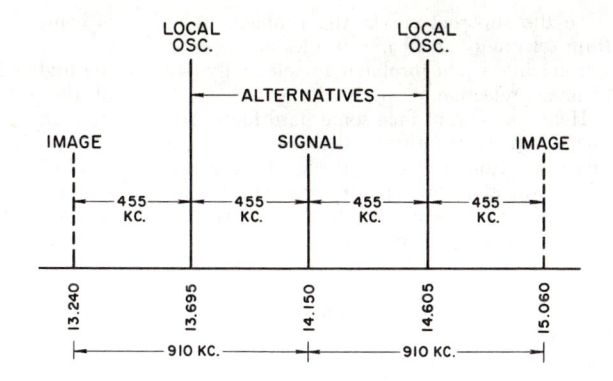

observe in tuning across an unmodulated signal is *not* the width of the signal itself. Its width is zero. What you actually observe is the response curve of your receiver's selective circuits.

Bandwidth and Skirt Selectivity

In tuning across a phone signal you can see that the sharp curve will not bring in the entire signal by any means. You need greater bandwidth for receiving phone. Note, however, that although the a.m. phone signal is actually 6 kc. wide, it will *seem* wider as you tune through it because you can still hear (that is, see) part of it even when you're far from having it centered in the wider curve.

One other thing this little experiment will do for you is give you some appreciation of the importance of the *shape* of the selectivity curve. Some curves drop off relatively slowly on the sides or **skirts**. You'll hear a signal over a wider arc of the tuning knob with such a curve. If the skirts drop down almost vertically the signal comes in quickly and goes out quickly as you tune through it. Try making up a few curves of various skirt shapes, each with the same bandwidth at 6 db. down, and see for yourself. You'll soon come to realize the value of **skirt selectivity**.

Frequency Conversion

Now back to the superhet and frequency conversion. There is another advantage to the frequency-conversion process besides the constant selectivity. An amplifier having a given number of stages operating at a relatively low radio frequency—below 500 kc., say—can be designed for considerably more gain, with stability, than a similar amplifier at high frequencies. High selectivity and high gain go hand in hand at low frequencies. Thus the superhet solves a major receiving problem in high-frequency reception.

However, frequency conversion also brings in some new problems that weren't there before. They aren't insurmountable, but they do demand care in design and construction.

There is only one known method of frequency conversion that is at all useful for the purpose. This is the process known as **heterodyning** or **beating**. To get a desired new frequency—known universally as the **intermediate frequency** in a superhet receiver—it is necessary to "mix" a third frequency with the signal frequency. The **mixing** process (which differs considerably from ordinary amplification) leads to an interesting result: two new frequencies appear, one equal to the difference between the original two, and one equal to the sum of the original two. These are in addition to the two original ones, which also appear in the output of a mixer. In the superhet receiver the one we almost always want is the *difference* frequency.

Here is an example: Let us say that our intermediate-frequency amplifier operates at 455 kc. There is a signal on 14,150 kc. that we want to convert to 455 kc. Thus 455 kc. will be the difference between 14,150 kc. and a new frequency that we have to introduce in order to effect a conversion.

You can readily see that there are *two* frequencies—13,695 kc. and 14,605 kc.—that will meet the specification. Either one is 455 kc. from 14,150 kc.; the former is 455 kc. lower and the latter 455 kc. higher in frequency. We can use whichever we choose of these two **local-oscillator** frequencies. (They are called local-oscillator frequencies because they are generated by an oscillator in the receiver itself and do not come from outside as the signal does.)

Image Frequencies

The fact that such a choice exists is one source of trouble in a superhet. For if either of two frequencies will produce a desired output or i.f. (intermediate-frequency) signal, it is also true that each local-oscillator frequency can generate the same difference from two frequencies, one above and one below it. For example, if we should choose 14,605 kc. for the local-oscillator frequency in the case mentioned above, a signal on 15,060 kc. also would be 455 kc. from the local oscillator. Thus we have a new source of interference—so-called **image** signals, or signals on the other side of the local-oscillator frequency. As you can see from the example, these image signals are always spaced from the desired signal by exactly twice the intermediate frequency (15,060 − 14,150 = 910 kc., which is twice 455 kc.).

So the superhet solves the problem of uniform selectivity at all frequencies at the cost of introducing a *new* problem in selectivity—that of image rejection.

Here we have to face some hard facts. These image signals are always the same number of kilocycles from the desired signals, regardless of the signal frequency. In percentage, therefore, the image gets closer to the desired signal the higher we go in the spectrum. Unfortunately, there is no way to reject the image signal except by using enough selectivity at the desired-signal frequency. It might seem that we're right back where we started from when we went to the superhet arrangement. And indeed we are, except for one thing: the image interference is always distant, in frequency, from the desired signal by twice the intermediate frequency. It is signals on that frequency *only* that can interfere. And we can have considerable control over their rejection. Obviously, the farther the image is from the desired signal the less response a given tuned circuit will have to it, since the response drops off as you move away from the resonant frequency. So we can put the image frequency as far away as possible, by using an intermediate frequency that is fairly high.

Front-End Image Response

Fig. 3-10 shows relative response of a typical circuit to signals 900 kc. and 3000 kc. off the resonant frequency of 14,150 kc. The two i.f.s,
of course, would be 450 kc. and 1500 kc. There is a 10-db. improvement in image rejection at the higher intermediate frequency, just because of the way a tuned circuit responds to off-frequency signals.

Actually, at least two such circuits in cascade would be needed to give adequate image rejection at high frequencies, even with the 1500-kc. intermediate frequency. It is for this reason, among others, that receivers of any pretensions at all always have an r.f. amplifier stage preceding the mixer. There will be one such tuned circuit associated with each of these stages.

Double Conversion

But again we've arrived at a happy answer only at the price of new difficulties. Intermediate-frequency amplifiers at 1500 kc. and higher can't be built to have the same selectivity and gain per stage as at frequencies below 500 kc. —at least, not when conventional tuned circuits are used. So we generally find that still *another* frequency conversion is needed—from a first intermediate frequency that is used because it results in good image suppression, to a second intermediate frequency that gives us the close-in or adjacent-channel selectivity we want. This is known as the **double-super** or **double-conversion** arrangement.

Of course, still another local oscillator is necessary, operating on the right frequency to give the appropriate difference. Taking again the signal on 14,150 kc. and assuming a first i.f. of 1500 kc., we need a first local-oscillator frequency of $14,150 + 1500 = 15,650$ kc. Out of the mixer comes 1500 kc., and suppose we want to convert this to 455 kc. To do so we need a local oscillator at either $1500 + 455 = 1955$ kc. or at $1500 - 455 = 1045$ kc. Usually the higher frequency would be used, because of the next crop of troubles that the superhet brings with it—**spurious responses**.

Birdies

The image signal we talked about above is a prime example of a spurious response—so called because it is not wanted, even though it is quite legitimately generated. Unfortunately there are still other types. Each of the local oscillators generates harmonics along with the frequency we want from it. If there are two such oscillators there are endless possibilities for harmonic combinations that will mix to give a difference frequency equal to the i.f., or perhaps to give a spurious signal at the actual frequency to which the receiver is tuned. These **birdies** can show up in any tuning range, and usually do. They can be kept to a minimum in the ham bands by a judicious choice of the first intermediate frequency. But the more oscillators there are going simultaneously in the receiver the more birdies there are—and it is not just a simple proportion!

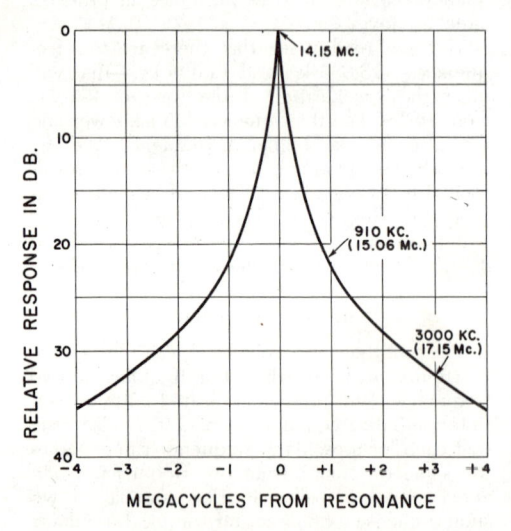

FIG. 3-10—This is typical of the selectivity of a single circuit using a good coil at a signal frequency of 14.15 Mc. An image signal at 15.06 Mc. (455-kc. i.f.) will get through, because the response of the circuit at that frequency is only a little over 20 db. below the response to a desired signal on 14.15 Mc. Changing to a 1500-kc. i.f. will add 10 db. to the suppression of the undesired image. In practice, a receiver using either of these i.f.s would have at least two such tuned circuits, theoretically doubling the suppression (in decibels) in either case.

Receiver Circuits 55

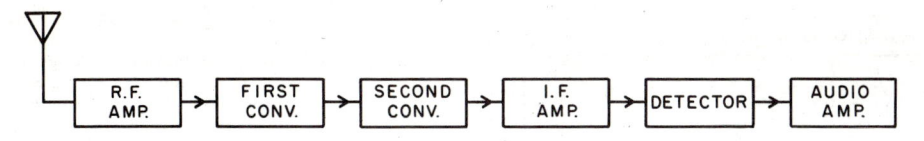

FIG. 3-11—Block diagram of the basic double superheterodyne. Although the r.f. amplifier stage may sometimes be omitted in simple receivers, its use distinctly improves the receiver performance both in sensitivity and image suppression.

Altogether, the superhet receiver would seem to be a kind of Pandora's box, with all sorts of undesirable things popping out of it. Nevertheless, the advantages in selectivity and gain, over a very wide tuning range, do outweigh these side effects—provided care is used in design and construction to minimize the spurious responses of various sorts. A good receiver will reduce them to the point where they are rarely noticeable. We have emphasized them here to make you appreciate the fact that they do exist, that no receiver is *wholly* free from them. A little knowledge on this point may save you the embarrassment of having to eat your words after complaining vociferously about some station being off frequency or invading a ham band—when the signal actually was a "phantom" conjured up by the receiver itself!

Identification

There are ways of recognizing birdies and images for what they are, although sometimes it takes a good deal of experience. Spurious signals generated by the receiver itself will be unmodulated (or nearly so—sometimes there may be a little hum on a birdie) and will be there all the time. They will not usually be affected by the antenna, so if you take off the antenna and the signal is still there as big as ever, you can blame it on the receiver.

Images, of course, are signals coming from outside, and will disappear along with the "legitimate" signals when you disconnect the antenna. However, you can often spot them. If the image signal happens to be close enough in frequency to an actual signal to make an audible beat tone or heterodyne, there is an easy test. Vary the receiver tuning back and forth slightly. If the beat tone changes, one of the signals is an image. If the tone does *not* change, both signals are "legitimate." Of course, the character of the signal can be a clue, too; we don't expect to find commercial or broadcasting stations working in bands that are exclusively amateur world-wide. But in other parts of the world some sections of the amateur bands are shared with commercial and government users, so unless you know your frequency assignments pretty well, and can measure frequency pretty accurately, you can't always depend on the nature of the signal itself to tell you whether it is real or spurious.

A Closer Look at Circuits

By now you should have a fairly good idea of the over-all operation of a superhet receiver. There are lots of details in receiver circuit design that make for better reception or more convenient operation. It isn't within the scope of this book to try to cover them all, so we'll confine ourselves to touching on features that you will find in practically any receiver designed for amateur use.

If you will look back at the block diagram in Fig. 3-6, you'll see that there are four divisions in the main line—converter, i.f. amplifier, detector, and audio amplifier. We've already discussed the common form of detector, the diode rectifier, to some extent. A practical detector circuit probably would use a tube diode (or a diode section in a multipurpose tube) and would look about as shown in Fig. 3-12. The rectified current develops an audio-frequency voltage across R_2 by the process described in connection with Fig. 3-2. This audio signal is handed on to the audio amplifier through the coupling capacitor C_3. The capacitor is used to block off the d.c. part of the rectified r.f. signal so it can't put a steady bias on the grid of the following audio voltage amplifier.

The intermediate-frequency signal is taken from the secondary of the i.f. transformer (we'll have more to say about these transformers below), and the rectified current flows through R_1 and R_2 in series. If we were concerned only with

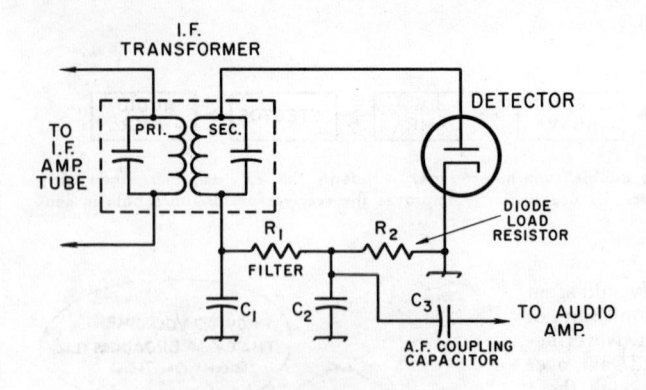

FIG. 3-12—A typical diode detector circuit. C_1, R_1 and C_2 form a resistance-capacitance filter to smooth out the intermediate-frequency pulsations in the rectified output of the detector. Representative values would be R_1, 47,000 ohms; R_2, 0.22 to 0.47 megohm; C_1 and C_2, 100 to 250 µµf. each; C_3, 0.01 to 0.1 µf. The rectified current develops an audio-frequency voltage in R_2 which is coupled to an audio amplifier through C_3.

audio, C_2 would be adequate for smoothing out the r.f. pulses as shown in Fig. 3-2. However, this smoothing is not generally adequate for another purpose—eliminating any intermediate-frequency pulses that might reach the grid of the audio amplifier and upset its operation. R_1 and C_1 are added to improve this i.f. filtering. They aren't always used, but it is good practice to install them.

I.F. Amplifiers

Intermediate-frequency amplifier circuits are pretty well standardized. A representative circuit is shown in Fig. 3-13. The tube is always a screen-grid pentode, generally of the remote-cutoff type. The screen-grid minimizes feedback from the plate circuit to the grid circuit and thus keeps the i.f. amplifier from breaking into self-oscillation. (This is discussed in more detail in Chapter 2). A remote-cutoff tube is used because its characteristics lend themselves well to a convenient form of gain control—varying the transconductance of the tube by varying the bias on its grid.

The numbered capacitors are there to keep the i.f. current where it should go—in other words, they are bypasses. The resistors all back up the capacitors by offering a poor path for the current. In fact, that is the sole reason for using R_4, which is purely a decoupling resistor. R_2 and R_3 do some decoupling, too, but also have another job—R_2 is a cathode biasing resistor, and R_3 drops the B-supply voltage to the proper voltage for the screen of the tube. The decoupling function of R_1 is important, but this resistor also is part of the automatic gain control circuit which will be taken up later.

I.F. Transformers

Now about those i.f. transformers. For intermediate frequencies below 500 kc. they are almost always double-tuned, meaning that there are two independent tuned circuits, coupled together somehow so that energy can be transferred from the primary side to the secondary side. The circuits are, of course, all tuned to whatever intermediate frequency is being used. These i.f.s. may range from 50 kc. to 500 kc.,

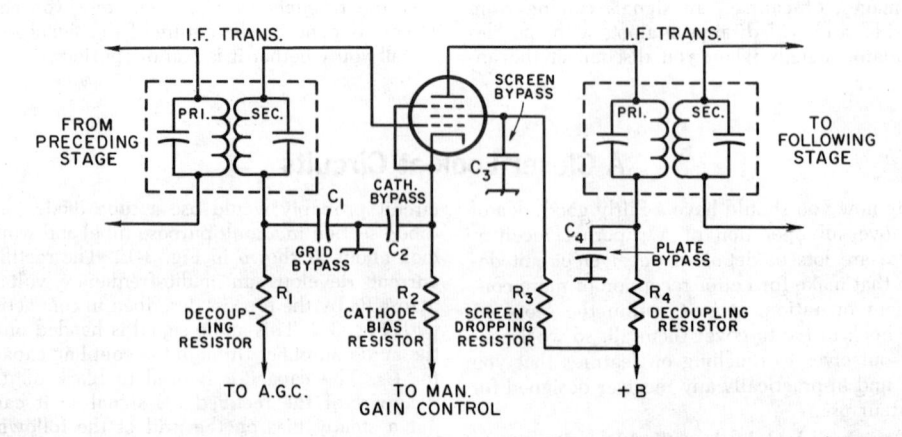

FIG. 3-13—An i.f. amplifier. The tube is a remote-cutoff type screen-grid pentode. The i.f. transformers would be designed for the particular intermediate frequency chosen. C_1, C_2, C_3 and C_4 will generally have the same value, ranging from 0.01 to 0.1 µf. The higher values are used at the lower intermediate frequencies. The value of R_1 is a matter of designer's choice. R_2 is usually a few hundred ohms, depending on the type of tube, while R_3 will be of the order of 50,000 to 100,000 ohms, again depending on the tube. R_4 is a few thousand ohms in most cases.

Receiver Circuits

depending on the designer's objectives. The lower the frequency the greater the selectivity, as a general rule. Some of the frequencies you will find in use are 50, 85, 100, 262, 455, and 500 kc.

In double superhets the first i.f. usually will run between 1500 kc. and 3000 kc., and will be followed by a low-frequency second i.f. in the range mentioned above. A high-frequency (i.e., 1500-3000 kc.) i.f. may use single-tuned transformers, in which the secondary circuit is tuned but the primary is simply a coil closely-coupled to the secondary coil. However, double-tuned transformers are common in this range, too.

I.f. transformers are mounted in shield cans and the tuning is screwdriver-adjusted. The tuning may be done by varying either the capacitance or the inductance in each circuit. Inductance tuning by means of an iron slug is probably the most common in current practice. The characteristics of the transformers affect both the gain and selectivity of the i.f. amplifier. On both counts the transformer should have high-Q coils. Selectivity is highest with loose coupling, but the gain is highest with critical coupling (see Chapter 2). In some special cases transformers will be overcoupled in order to get a selectivity curve which is flat-topped—that is, has fairly uniform response throughout a desired band—but which also has reasonably steep sides. See Fig. 3-15.

Adjustable Selectivity

The more advanced receivers will offer you several degrees of selectivity, these being obtained in many cases by changing the coupling between primary and secondary circuits in the i.f. amplifier. Ordinary transformers, though, do not have any such built-in provision; the coupling usually is fixed by the manufacturer and is not readily adjustable. The transformers shown in Figs. 3-12 and 3-13 have inductive coupling between the coils, but other methods of coupling are sometimes used, especially in amplifiers having various degrees of selectivity. This is because capacitive or resistive coupling offers a very convenient way of varying the coupling and thus the selectivity.

Tunable I.F. Amplifier

Double superhets are sometimes built with a first i.f. amplifier that is tunable over a range about equal, in the total number of kilocycles covered, to the width of an amateur band. This is in contrast to the conventional arrangement where the receiver tuning is done in the front end by varying the tuning of the high-frequency local oscillator along with the tuning of the signal-frequency circuits. The advantage of the tunable i.f. is that it allows us to use a fixed-frequency oscillator, usually crystal-controlled, for each amateur band. This results in a high degree of stability, and about the same stability on all bands. It also gives uniform bandspread on all bands, since all the tuning is done with one set of circuits.

Examples of this method of building receivers are given in Chapter 8. It is an especially useful method in v.h.f. receivers, where it is hard to get highly-stable oscillator operation except with crystal control.

Linear and Nonlinear Circuits

Moving on toward the front end of the receiver brings us to the first block in Fig. 3-2. This is where the incoming signal is changed

FIG. 3-14—A typical i.f. transformer and shield can. Most present-day units are "permeability tuned" by moving a powdered-iron core in and out of universal-wound coils. The capacitors are silver mica (for good stability with changes in temperature) and are fixed in capacitance.

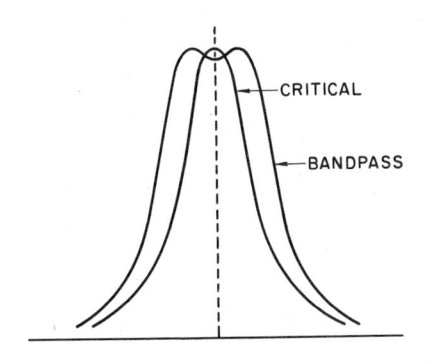

FIG. 3-15—The shape of the selectivity curve depends on the coupling between tuned circuits in the intermediate-frequency amplifier. The curve marked "critical coupling" has a rounded top and marks the degree of coupling which just transfers the maximum energy from primary to secondary. The "overcoupled" curve gives a flattened top or band-pass effect which is desirable for uniform reception of a band of frequencies such as the band occupied by a voice transmission.

over to the intermediate frequency. We've already dealt with the process in general terms. There are three requisites: a tuning arrangement for bringing in the signal, a source of r.f. (an oscillator) of the proper frequency to generate an i.f. signal when mixed with the incoming signal, and a device for bringing about the actual mixing with as good efficiency as possible.

Good mixing calls for a **nonlinear** device—one in which the current is not in simple proportion to the applied voltage. Or, in other words, one which does not follow Ohm's Law. Ordinary conductors and resistors are **linear**. When we apply two different voltages to linear circuits there is no mixing in the sense we mean the word here. Each voltage causes its own current to flow according to Ohm's Law; the two currents simply exist simultaneously in the circuit. But in a nonlinear circuit this is not so; one current affects the other. We might say, with accuracy, that one **modulates** the other, so that as one rises it causes the other to rise along with it, and vice versa. The result of this, as we have seen earlier in this chapter, is that when two currents of different frequency exist in a nonlinear circuit two new *additional* frequencies are generated.

Vacuum-tube amplifiers are, or can be, reasonably linear. That is, the output obtained from the plate circuit will be a rather good reproduction of the actual signal applied to the grid. Also, the amplitude of the output signal will stay in fixed proportion to the amplitude of the input signal over the usable operating range of the amplifier. A weak signal is amplified just as many times as a strong one, and vice versa.

Mixing

Now consider the diode rectifier. Since rectification makes use only of one-half of an a.c. cycle—the half during which the voltage at the anode is positive with respect to the cathode—one side of the signal is missing entirely in the output. This is nonlinearity with a vengeance. A diode, therefore, will mix two signals applied to it simultaneously. It can be used as a frequency converter. However, it hardly ever is so used—at least not in the frequency range where tubes having grids will work. (The diode *is* used at u.h.f. and s.h.f.) It isn't used mainly because it gives no gain, and we can get amplification along with mixing if we use triode or multigrid tubes. The only trick is to choose operating conditions that make the multielement tube act like a diode rectifier while it is also amplifying. It isn't hard to do.

Mixer Circuits

A typical circuit is shown in Fig. 3-16. This is only one of several that could be used; the principle is essentially the same in every case. The mixer tube is shown here as a pentode, but a triode could be used. The value for R_1, the cathode bias resistor, is selected to generate enough bias to bring the operating point near plate-current cutoff. Then the positive swings of the signal at the grid cause the plate current to increase more than the negative swings cause it to decrease. Another way of looking at this is that the amplification will not be the same for signals of all strengths, but will change with the amplitude of the signal.

The oscillator in Fig. 3-16—again, this is only one of many circuits that could be used—supplies the local r.f. voltage with which the signal is to be mixed. This voltage is coupled to the grid of the mixer through a small capacitor, C_8. The r.f. voltage from the oscillator varies the

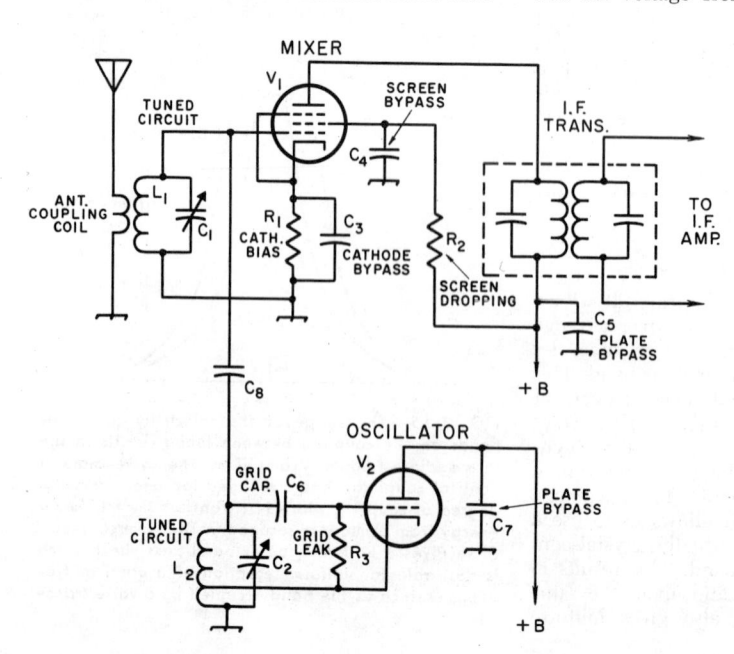

FIG. 3-16—Oscillator and pentode mixer circuit for converting the signal frequency to the intermediate frequency. L_1C_1 is a circuit tuned to the signal frequency and coupled inductively to the antenna. L_2C_2 sets the frequency of the local oscillator and is tuned so that the difference between it and the signal frequency is equal to the intermediate frequency. R_1 will usually be a few thousand ohms, or perhaps less. R_2 is several thousand ohms in the average case. R_3 may range from 20,000 to 50,000 ohms or so. C_3, C_4, C_5 and C_7 usually will be 0.01 µf. C_6 is generally between 100 and 250 µµf. C_8 is of the order of a few µµf. V_1 is almost always a sharp cut-off pentode and V_2 is usually a triode, as shown.

Receiver Circuits

amplification in the mixer at an r.f. rate, since the oscillator voltage is in effect swinging the operating bias up and down. An incoming signal applied to the mixer grid will experience this rapid variation in amplification in the plate circuit, resulting in a beat at the intermediate frequency.

A mixer differs from an ordinary amplifier in another respect. Its plate circuit is arranged so that only the i.f. beat is amplified and all the other frequencies are ignored. The transformer tuned to the intermediate frequency takes care of this. The other frequencies include the signal frequency, the oscillator frequency, and the unused beat frequency (remember, there are two beats). Generally the difference beat is used, so the sum beat disappears in the output circuit.

Mixer Tubes—Pulling

Any tube can be used as a mixer, but there are special types that have been developed just for this purpose. Although the details of their operation are different, the principles are much the same. A feature is that there are separate grids for the signal and oscillator **injection** voltages. The oscillator voltage on the **injection grid** varies the amplification in the tube—essentially the same thing that is done in the circuit of Fig. 3-16. The advantage of the special mixer tube is that the two separate grids (which usually also are shielded from each other) help to isolate the oscillator from the signal circuit, and vice versa.

Isolation of the signal and oscillator or injection circuits helps minimize some undesirable effects that accompany the simple circuit of Fig. 3-16. One of these is oscillator **pulling**. As you can see in Fig. 3-16, the signal circuit, L_1C_1, is coupled to the oscillator tank circuit, L_2C_2, through capacitor C_8. This means that tuning L_1C_1 will also change the tuning of L_2C_2; in other words, the frequency of the oscillator will be changed if L_1C_1 is adjusted. This is undesirable, because we want the oscillator frequency to be as independent of everything except its own tuning as it can possibly be. If it isn't, the receiver will be unstable, because the oscillator frequency is the most critical frequency in the system. If the coupling capacitance can be eliminated, tuning L_1C_1 will not affect the oscillator frequency.

The special mixer tubes permit this separation. The capacitive coupling between the oscillator and signal circuit is practically negligible in these tubes. There is still a certain amount of electronic coupling—that is, interaction because both the signal and injection grids are in the same electron stream inside the tube. Over-all, though, there is a considerable improvement in isolation as compared with the circuit of Fig. 3-16.

The Oscillator

We've said above that the oscillator frequency is the most critical one in the receiver. Therefore one of the most important features of the local oscillator is its stability. If the frequency "wanders" much from heating (**drift**), or is affected by voltage changes or other factors, it will be hard to keep a signal tuned in properly, especially if the selectivity is high.

Here's an illustration. Suppose you have 500-cycle selectivity for c.w. reception. If you have the signal tuned "on the nose" and there is a sudden change in line voltage that shifts the oscillator frequency 300 cycles, the signal will be moved almost out of the pass band of the receiver. A 300-cycle change from such a cause (and there are many other causes) isn't excessive, especially when the oscillator is operating at 30 Mc. or so. Just varying the gain controls on some receivers can do the same thing.

One of the distinguishing features of a really good receiver is the oscillator stability it offers. Good stability is easier to achieve in a small

FIG. 3-17—A converter circuit, using a single tube designed for the purpose. Circuit designations correspond to those of Fig. 3-16. Combining the oscillator and mixer in one tube in this way is economical and you will find it in many of the less-expensive receivers. However, it is generally considered that the separate mixer and oscillator give better performance. Frequently a converter tube is substituted for the pentode mixer in Fig. 3-16, in which case the oscillator voltage is applied to the No. 1 grid (the one next to the cathode) while the signal goes to the No. 3 grid as shown above.

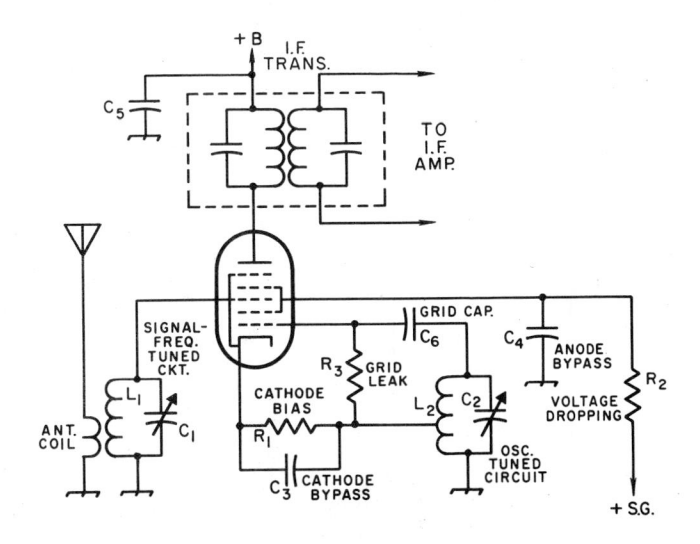

tuning range than in a large one, which is one of the advantages of amateur-band-only receivers. Stability is primarily a matter of good design; we'll have more to say on this subject in Chapter 8. The circuit used is relatively unimportant; it's how the circuit chosen is handled that counts. The crystal-controlled oscillator is best of all, but it requires a tunable intermediate-frequency amplifier for covering the amateur bands. You'll also find circuits of crystal-controlled oscillators in Chapter 8.

Converter Circuits

Tubes designed for mixer operation usually also are designed to take on the oscillator's job. This lets one tube handle the complete frequency conversion process. Customarily, the tube is called a **mixer** when a separate oscillator is used. It then becomes a **converter** when it combines oscillation with mixing.

A typical converter circuit is given in Fig. 3-17. A Hartley-type oscillator is shown, but other oscillator circuits can be used. The tube operates with a "hot" cathode—that is, the cathode is not at ground potential for r.f.—in most converter circuits. This is done so that the screen grid, which is the oscillator anode or plate, can be grounded for r.f., thus maintaining the shielding between the control grid and the plate of the tube. Aside from the hot cathode, the operation of the mixer part of the circuit is quite similar to that of Fig. 3-16.

R.F. Amplifiers

Inexpensive receivers may go directly from the antenna to the converter or mixer, as indicated in Figs. 3-16 and 3-17. However, a converter generates more internal noise, and gives less gain, than a **straight-through** amplifier. We can make a better receiver by putting an r.f. amplifier ahead of the mixer or converter. One such stage is usually enough to build up the signal to a level that will overcome the converter noise.

An r.f. stage also adds to signal-frequency selectivity. This additional selectivity results from the fact that there is another tuned circuit associated with the amplifier. It helps a great deal in suppressing image response. (In Fig. 3-18 the output circuit, $L_3 L_4 C_6$, would substitute for the antenna coil and $L_1 C_1$ in either of the mixer-converter circuits just described.)

Even so, if there is only one intermediate frequency and it is of the order of 455 kc., images still will tend to be bothersome at 14 Mc. and higher frequencies. Two r.f. stages have been used on this account in some of the more expensive amateur receivers of the past. The tendency today, especially in amateur-band receivers, is to go to the double superhet. With a high first-intermediate frequency one r.f. stage usually is enough for image suppression.

R.F. Transformers

The r.f. transformers in r.f. amplifier stages usually have only one circuit—the secondary circuit—tuned. Theoretically, the primary circuit could be tuned, just as in an i.f. stage. The advantage would be additional selectivity. However, these signal-frequency circuits cannot be tuned once and then left alone, as the i.f. circuits can. Their tuning has to be varied to cover a fairly wide range. Also, they are usually **ganged** —that is, the tuning capacitors are mounted on a common shaft and all tuned together. The practical difficulty of keeping ganged double-tuned circuits properly aligned is so great that it is rarely attempted, and single-tuned circuits are used instead.

The circuit shown in Fig. 3-18 is typical of what is used in receivers operating up through 30 Mc., and even 60 Mc. However, there are better amplifier circuits for v.h.f., so we often find considerably different arrangements in receivers for 50 Mc. and above. A popular r.f. amplifier circuit in the v.h.f. range is the **cascode**, which you will see in a practical form in Chapter 8. Grounded-grid amplifiers also are used. In either of these circuits the tube or tubes will be triodes rather than screen-grid tubes, because triodes generate less internal noise than screen-grid tubes. The difference between the two tube types isn't so important below 20 Mc. or so, because the tube noise is usually masked by noise reaching the antenna from outside.

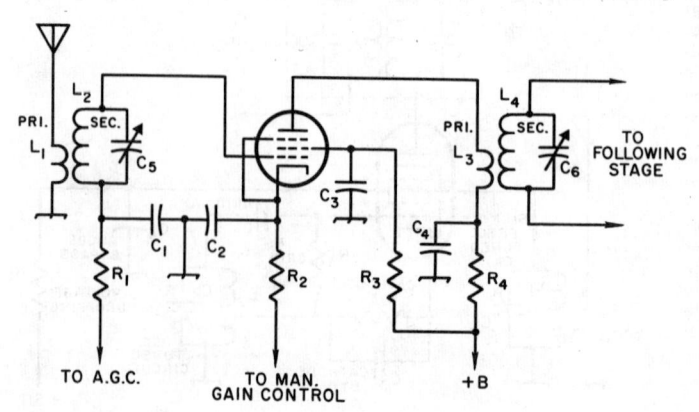

FIG. 3-18—R.f. amplifier. Compare this with Fig. 3-13; except for the tuned circuits, it is practically the same. Since it operates at frequencies above 1500 kc., the bypass capacitors can be smaller; 0.01 µf. is a common value. The resistor values are the same as in Fig. 3-13. Note that in the r.f. transformers only one circuit is tuned; the primary in each case is simply a coil closely coupled to the tuned secondary.

Receiver Circuits

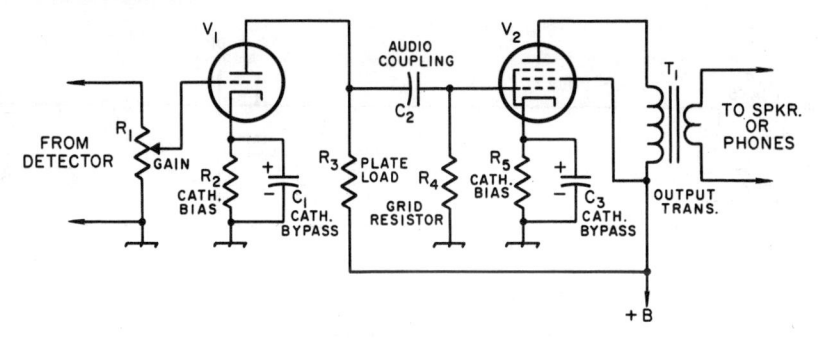

FIG. 3-19—The audio amplifier. A two-stage amplifier having a pentode power-output tube is standard practice in communications receivers. V_1, usually a high-μ triode, may be one section of a double triode or the triode section of a diode-triode combination tube. Typical circuit values are: R_1, ½ to 1 megohm; R_2, 1500-2000 ohms; R_3, 50,000 to 250,000 ohms; R_4, ½ megohm; R_5, a few hundred ohms, depending on tube type; C_1 and C_3, 10 μf. or more; C_2, 0.01 to 0.1 μf. C_1 and C_3 are always electrolytic capacitors.

Audio Amplifiers

Audio amplifiers in communications receivers are usually pretty straightforward. The voice-frequency range is about all they're called on to handle—that is, about 200 to 3000 cycles—so there are no stringent requirements to meet. The output tube is practically always an audio pentode such as a 6AQ5. It is ordinarily preceded by a resistance-coupled voltage amplifier using a high-μ tube. This combination has ample gain for following the detector in a superhet. It will turn out an audio power of a couple of watts.

Fig. 3-19 is a representative circuit. The audio gain control is in the grid circuit of the first tube in this circuit. In some receivers you will find that the gain control is used as the load resistor in the diode detector circuit (replacing R_2 in Fig. 3-12). In such case a fixed resistor of a half megohm or more will replace R_1 in Fig. 3-19. The control is equally effective either way.

Automatic Gain Control

If you will go back to the early part of this chapter and look again at Fig. 3-2, you will recall that after detection we have a direct current varying at a rate that transforms the modulation into an audio frequency. Only the *varying* part of the detector's output is used for reproducing the intelligence on the signal. The direct current, which is indicated in Fig. 3-2 by the fact that all of the output is on one—in this case, the positive—side of the axis, contributes nothing to the audio output.

However, this direct current isn't useless. The amount or amplitude of the rectified d.c. is proportional to the intensity of the signal reaching the detector. A big signal results in a large current, relatively speaking, and a small signal produces only a small current. If we take this varying d.c. and smooth off the audio-frequency variations, we will have a current that is practically in direct proportion to the *average* strength of the signal—that is, proportional to its carrier strength only.

A.G.C. Voltage

A circuit such as is shown in Fig. 3-20 will do this. It is similar to Fig. 3-2 in all but the last part. Fig. 3-20 also gives a bit more actual detail. The signal applied to the detector comes from an i.f. amplifier through a tuned transformer. Note also that the detector connections in this circuit are reversed, as compared with Fig. 3-2. If we just want to detect or **demodulate** the signal it does not matter which way the detector is connected, since the audio-frequency variations will be the same either way. However, when we want to use the d.c. output for **automatic gain control** (a.g.c.) we must have a *negative* voltage with respect to chassis. (This is because the grids in the amplifier stages whose gain is going to be controlled are always themselves negative with respect to chassis, and the more negative they are the less the gain.) This negative a.g.c. voltage is then used to add bias to the amplifiers and reduce their gain.

The stronger the signal the greater the negative a.g.c. voltage and the more it reduces the gain. A weak signal will develop very little a.g.c. voltage and the gain will stay high. This is the sort of action we want; it tends to keep the audio output amplitude constant over a wide range of signal strengths. It helps prevent overloading of the receiver—which causes distortion—as well as saving our ears from the sudden blasts that would occur if we tuned unsuspectingly into a very strong signal.

The a.g.c. voltage is applied to the tubes whose gain is to be controlled as shown in Fig. 3-21. It is good practice to use an *RC* filter—R_1C_1, R_2C_2, and R_3C_3 in Fig. 3-21—at each tube These filters help keep r.f. from leaking from the grid circuits back through the a.g.c. line. They also give the set designer some additional freedom in choosing different time constants for different controlled stages. This is sometimes advantageous, to take care of certain types of signal fading.

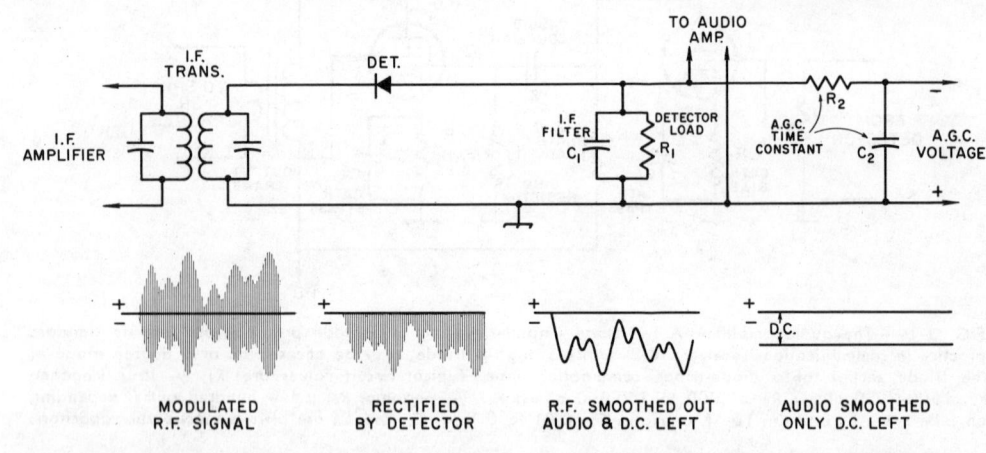

FIG. 3-20—Automatic gain control circuit. The final d.c. output of the rectifier, after the audio-frequency variations have been smoothed off by R_2C_2, is used to control the gain of r.f. and i.f. stages. R_1 and C_1 would have values similar to those of Fig. 3-12 (the extra filter circuits probably would be used, also). R_2 and C_2 depend on the desired a.g.c. time constant. For fast a.g.c. action R multiplied by C will be small; for a slow time constant the RC product will be large. These are relative terms, of course; actual values are of the order of 1 megohm for R_2 and 0.1 µf. for C_2.

A.G.C. in C.W. and S.S.B. Reception

There are times, as in c.w. or s.s.b. reception with many receivers, when it is necessary to shut off the a.g.c. This is done by short-circuiting the output of the a.g.c. rectifier. The simple s.p.s.t. switch in Fig. 3-21 does this.

The reason why the a.g.c. in many receivers cannot be used for c.w. and s.s.b. reception is that the beat-frequency oscillator must be used for detecting these two types of signals. The b.f.o. voltage is introduced into the detector circuit, as discussed below, and is generally quite high in amplitude. Since it will be rectified by the detector, it produces a strong a.g.c. voltage that would drastically reduce the receiver gain.

If a.g.c. is to be used on c.w. or s.s.b., the a.g.c. rectifier must be completely isolated from the b.f.o. voltage. This can be done by using a separate i.f. amplifier stage between the a.g.c. rectifier and the regular i.f. amplifier. The general idea is shown in block form in Fig. 3-22.

Manual Gain Control

In the familiar broadcast receiver the r.f. and i.f. gain is always controlled automatically by an a.g.c. circuit. If an amplitude-modulated phone signal were the only kind we ever handled in a communications receiver, a.g.c. would suffice for our amateur receivers, too. But for the reasons just given, the a.g.c. cannot always be used. Nevertheless, we have to be able to reduce the gain when a strong signal is being received.

An audio gain control alone doesn't suffice. The over-all gain, particularly in the r.f. and i.f. stages, has to be very high in order to bring in weak signals. If the receiver always operated with the gain "wide open," a strong signal would

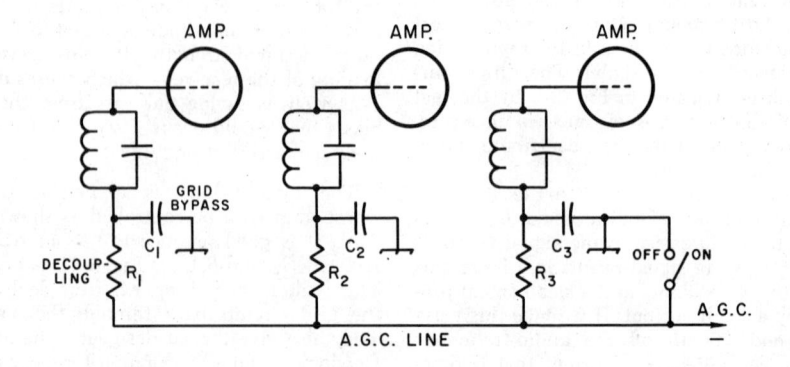

FIG. 3-21—A.g.c. voltage is applied to r.f. and i.f. stages through RC filters. R_1, R_2 and R_3 all have the same functions, as do C_1, C_2 and C_3. Typical values are 0.01 µf. for the capacitors in r.f. stages, 0.1 µf. in low-frequency i.f. stages. Resistors will be of the order of 50,000 to 100,000 ohms.

Receiver Circuits

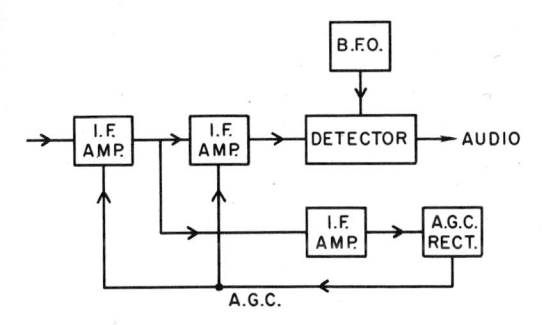

FIG. 3-22—A separate i.f. amplifier for the a.g.c. rectifier isolates the rectifier from the b.f.o. voltage at the detector circuit. This allows use of a.g.c. for c.w. and s.s.b. reception. The circuits, shown in block form here, are basically similar to those previously discussed.

build up to such proportions that some stage ahead of the detector would be overloaded. Overload is always accompanied by undesirable effects which, in a nutshell, make the receiver seem to generate its own interference—in addition to the distortion that accompanies overloading.

Thus we find **manual** r.f. and i.f. gain control in receivers, meaning that the total gain up to the detector can be changed by turning a knob. The gain is varied by changing the grid bias on a number of amplifier stages, just as in the case of a.g.c. The simplest method, and one most commonly used, is shown in Fig. 3-23. The grids actually are at the same d.c. potential as the chassis, but the cathodes are "raised" above the chassis by increasing positive voltage in order to reduce the gain. This is the same as applying a negative voltage to the grids, since the bias that counts is the bias voltage between the cathode and grid, not between the chassis and grid.

The values of resistance in the circuit are usually chosen so that anywhere from 20 to 50 volts of cathode bias is available. The minimum bias, with the arm of R_1 at the chassis end of the resistor, is the value determined by the fixed cathode resistors. The gain is highest at this point.

A good manual gain control system will reduce the gain to such an extent that even the strongest signals will not be heard. Usually, you can cut the gain back enough so that you can monitor the signal from your own transmitter—which is about the strongest signal your receiver is likely to get.

Beat-Frequency Oscillators

The signal from an ordinary phone transmitter is like the ones we get on the standard broadcast band. The detector gives us an audio-frequency output that is the same as the original voice or music. If the speaker is silent for a moment, or there is a pause in the music, we hear nothing except possibly a little hum or background noise.

A conventional code or c.w. transmission is like that more-or-less silent background. The signal has no audio-frequency modulation; a c.w. signal isn't permitted to have any on amateur frequencies below 50.1 Mc. The power simply is turned on and off to form the characters of the International Morse Code. If you listen to a code signal of this type with a diode detector, you'll hear nothing except a change in the background noise when the signal is keyed on and off.

To make such a signal audible we again make use of beat frequencies. In this case, though, the beat is in the audio frequency range rather than coming out of the mixer at some radio frequency in one of the i.f. ranges. Just as in the case of the superhet converter, we need a local oscillator and a mixer. The local oscillator is almost, but not quite, on the same frequency as the intermediate-frequency signal. The difference between the two frequencies is usually somewhere between 200 and 1500 cycles. The beat frequency is selected to suit the tastes of the operator, who picks an audio tone that he happens to like.

B.F.O. Circuits

Circuits similar to those shown in Figs. 3-16 and 3-17 could be used, with minor differences. L_1C_1 would be the secondary of the last i.f. transformer, and L_2C_2 would be a circuit just about the same as L_1C_1. Instead of i.f. output as shown in these circuits we would have audio output, so a resistance-coupled output circuit would be

FIG. 3-23—Manual gain control circuit of the type commonly used. Other schemes are possible; for example, a negative bias supply can be built into the receiver and an actual negative gain-control voltage applied to the grids through a potentiometer.

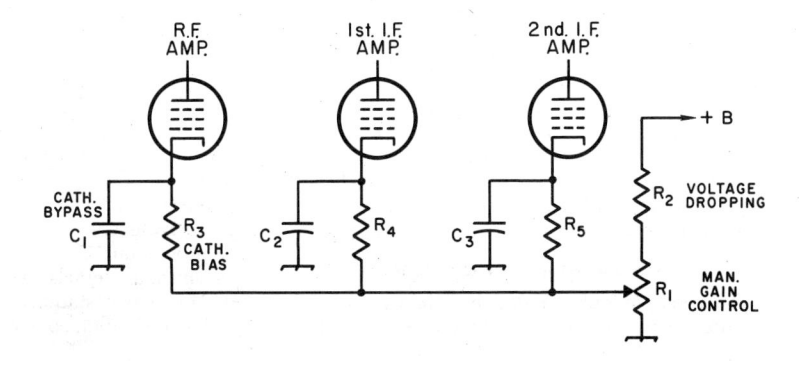

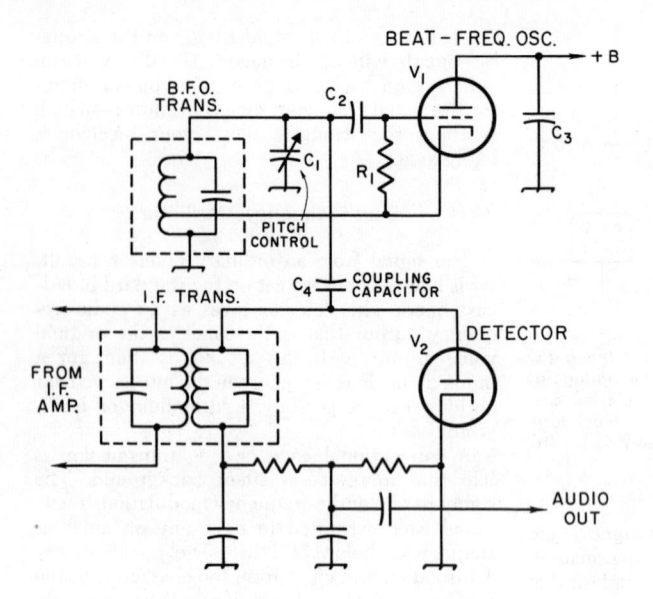

FIG. 3-24— B.f.o. and diode detector. The various components have the same functions and values as in the circuits previously discussed. C_1, the panel pitch control, is usually a small (25 to 50 $\mu\mu f.$) capacitor for tuning the b.f.o. over a small frequency range. C_4 is generally quite small—only a few micromicrofarads.

used. It would feed directly into the first audio amplifier stage.

However, it is much more common practice to use a diode detector and simply inject some voltage from the **beat-frequency oscillator** (**b.f.o.**) into its circuit. The diode gives us a.m. phone detection (with a.g.c. if we want it) with the b.f.o. turned off, and gives us c.w. reception with the b.f.o. turned on. In the latter case, as we have seen, the a.g.c. should be turned off. The arrangement over-all is a highly convenient one. A typical circuit is given in Fig. 3-24.

Since the difference between the frequencies of the i.f. signal and the b.f.o. is quite small, the b.f.o. circuit is usually put up in an i.f.-type shield can and has the same general construction. The assembly is called a **b.f.o. transformer**. However, it's rarely an actual transformer, but is simply a tuned circuit.

For best c.w. (and s.s.b.) reception the voltage injected into the detector circuit by the b.f.o. always should be considerably greater than the signal voltage which beats with it. This results in cleaner code reception and helps improve selectivity. Unfortunately, in many receivers the b.f.o. voltage is rather small. In such a case you have to keep the r.f. and i.f. gain down to the point where the signal amplitude is considerably lower than the b.f.o. amplitude. This means that you have to depend on high gain in the audio amplifier to give you volume. For c.w. and s.s.b. reception with such receivers the optimum method is to turn the audio gain to maximum and adjust the signal level by the manual r.f. gain or "sensitivity" control.

Some Hints on Operating Your Receiver

Knowing the operation of the various receiver sections we've described should help you to arrive at an explanation of some effects that may puzzle you. We gave you a few hints earlier on how to recognize birdies and images. Another thing to watch out for is **cross-modulation**. You may be listening to a distant phone station when a nearby phone in another part of the same band opens up. The voice of the home-town operator suddenly comes in along with the distant one. Don't jump to the conclusion that your local friend has a broad signal. The trouble may be entirely in your receiver.

Cross-Modulation and Overloading

This cross-modulation of one signal by another occurs when some stage in the receiver, usually near the front end, is overloaded by a strong signal. **Overloading** simply means that a tube is being driven by a signal that is too big for it to handle. Its operation becomes nonlinear, and mixing occurs. The only remedy is to cut down the strength of the undesired signal so it can't overload anything. A good a.g.c. system will help take care of this, but in severe cases even it may fall down. When this happens there's nothing much to do but use an antenna that brings in less signal, at least while the interference is present.

In receiving c.w. signals you don't have the a.g.c. protection. Cross-modulation here shows up in spurious signals as you tune across the band. When two stations happen to have their keys down simultaneously, you hear bits and

Receiver Operation

pieces of signal that don't make recognizable code. A related effect, when you have a strong transmitter nearby, is a change in receiver gain with the keying of the interfering signal. The background noise and received signals seem to "bounce." Both things are caused by overloading. Reducing the r.f. gain by means of the manual control is the remedy.

C.W. Reception

The manual r.f. gain control is an important control in good c.w. reception. As we said earlier, the incoming signal *must* be weaker, at the detector, than the signal from the b.f.o. If it isn't, strong signals will sound mushy, and will tend to wash out weaker signals that you might otherwise be able to copy.

Set your audio gain all the way up. Then adjust the r.f. gain so that strong signals give a good clean-sounding beat tone. If your receiver is properly designed, you'll have all the volume you can use. Keep the r.f. gain down to the point where there is no mushiness, but don't cut it so far that you lose the weak ones in the noise. This, unfortunately, does happen if the receiver designer hasn't paid attention to the fine points of c.w. reception. With a receiver that is really well designed, you'll never lose signals in the noise, no matter how far down you go on r.f. gain.

Setting The B.F.O.

The b.f.o. **pitch control** shouldn't be used for tuning. Tuning is the function of the regular tuning control. The b.f.o. control should be used to set the pitch to one you like—or to change the tone if you get tired of listening to the same one over a period of operating—*after* the signal is properly tuned in.

To set the b.f.o. properly first turn it off. Also turn off the a.g.c. Then turn your selectivity control, if there is one, to maximum and tune in a steady signal. Adjust the tuning so the background hiss you hear when the transmitter's key is down has the lowest tone. You'll recognize this when you get the swishing sound as you tune through the signal. Then let the tuning alone, turn on the b.f.o., and adjust the pitch control for the beat tone you like.

The more selectivity your receiver has for c.w. reception the more important this method of setting the pitch control becomes. If you use the pitch control at some random setting and then tune a signal to the tone you want, the signal may actually be well down on the side of the selectivity curve. Result: the desired signal is weak, and a signal that happens to be right on the peak of the selectivity curve will cause far more interference than it should. *Tune on the nose first, then set the b.f.o. control!*

S.S.B. Reception

Receiving methods for c.w. and s.s.b. are practically alike. In both cases the audio output is a beat frequency generated by the signal beating against the b.f.o. The difference is that with c.w. there is only one tone, but with s.s.b. the tones are voice audio frequencies. Also, in c.w. reception you can choose any tone you happen to like, but in s.s.b. reception the tones have to be just right. If they aren't, the result is the well-known "Donald Duck" sound. S.s.b. tuning is really critical, because if you are more than a few cycles off tune you lose intelligibility.

An s.s.b. signal will occupy a band 2500 or 3000 cycles wide, so you should use the receiver's 3-kc. bandwidth setting, if it has a selection of bandwidths to offer. To start, turn off the b.f.o. and tune in an s.s.b. transmission with the audio gain control set well up. Adjust the manual r.f. gain control for moderate audio output. Tune in the signal so it sounds loudest. You won't be able to understand it at this point, but you can tell when you're getting the most output. Then let the tuning control alone, turn on the b.f.o., and adjust the pitch control carefully until you have intelligible speech from the signal. After this, don't touch the pitch control; do your band scanning with the regular tuning control, always tuning for clear speech.

Upper and Lower Sideband

With 2- or 3-kc. selectivity this setting of the b.f.o. will suffice for the sideband—upper or lower—that was in use by the transmitting station. If you happened to set the pitch control on a station using the lower sideband, for example, it will be in the right place for any station using lower sideband. You won't be able to clear up a station using the upper sideband; this will require a new adjustment of the b.f.o. pitch control. The same method is used for either sideband.

Generally, you will find that s.s.b. stations working near the high-frequency end of a phone sub-band will be using lower sideband while those at the low-frequency end will be using upper sideband. It's a convenience to find the proper b.f.o. settings for both. Mark or otherwise identify the settings so you can return to them at will.

S.s.b. tuning is a matter of a little practice. If you can visualize what's going on as you vary the

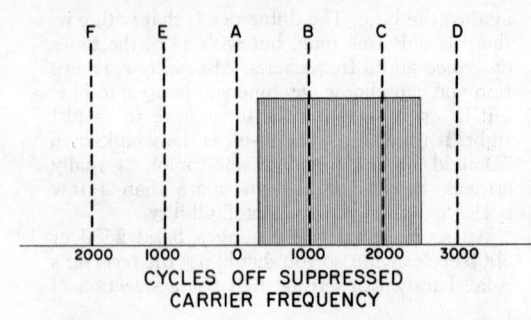

FIG. 3-25—The shaded block represents the band occupied by a single-sideband signal using the high (upper) sideband. The suppressed carrier frequency is represented by A. As you tune your receiver into such a signal from the high-frequency side—that is, as the receiver tuning moves from D toward A—the beat tones seem to descend in pitch. Then when your receiver tuning is between B and C the voice frequencies are on both sides of your b.f.o. and the beat tones are lowest. As you approach A the pitch seems to rise, and right at A the voice becomes intelligible.

Coming from the other side (approaching the signal from the "other" sideband area) all beat tones are high pitched, becoming lower as you near A.

In detecting the lower sideband these effects are reversed.

tuning control it should help you get the knack. Fig. 3-25 shows what happens.

The Product Detector

Many receivers incorporate two detectors, one —a diode—for a.m. reception and a second for c.w. and s.s.b. reception. The alternative detector is called a **product detector,** which is just another name for a converter-type circuit. As we said earlier, a mixer or converter system can be used for c.w. reception, although until recently the diode and b.f.o. combination has been rather universal. The product detector has its b.f.o. too, of course.

From practical experience, the advantages of the product detector over the diode seem to be somewhat nebulous. It is still necessary to have a large ratio of b.f.o. voltage to signal voltage for good c.w. or s.s.b. reception. A properly-operating product detector will give little if any response to an a.m. signal if the b.f.o. is turned off, which is why receivers using it also incorporate a diode detector. But a diode with a strong b.f.o. will give equally good reception on c.w. and s.s.b.

F.M. Reception

Occasionally on the lower frequencies, but more often on the v.h.f. bands, you will hear **frequency-modulated** phone signals. It isn't hard to identify them. The receiver should have its controls set for a.m. reception—a.g.c. on, audio gain at the level you like, manual gain control all the way up. If you have an S meter and tune for maximum reading, you'll find that at this exact tuning setting the modulation will practically disappear. To get maximum volume you have to tune off a bit to one side.

This is called **slope detection.** It means that the incoming signal's carrier frequency is tuned down on the side of the receiver's selectivity curve. Thus you might set the carrier frequency at point A on the curve, Fig. 3-26, by detuning a bit. Then as the station's frequency swings between B and C with modulation, the frequency swings are converted to a change in the amplitude of the signal reaching the detector. In other words, this type of tuning converts the f.m. to a.m. A.m. is the only type of modulation the or-

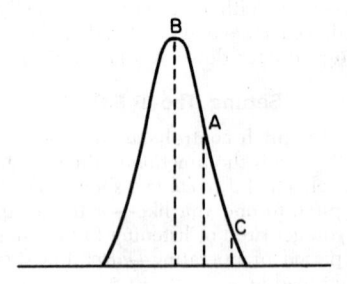

FIG. 3-26—Slope detection of an f.m. signal with an a.m. receiver. The signal, represented by the band between B and C with the carrier at A, has to be tuned off to one side of the setting used for a.m. reception.

dinary diode detector—which is what you find in most receivers—can detect.

Slope detection obviously isn't an ideal system, since you have to tune the receiver into a region where it is highly susceptible to noise and interference. There are better f.m. detectors. These are sometimes available in the form of adapters that can be used with a regular receiver.

Chapter 4

How Transmitters Work

If you're like most beginners you'll have built, or otherwise acquired, a rather simple crystal-controlled transmitter. Very possibly it will have been a commercial kit which you assembled yourself. Most such kits, as well as designs you may see in magazines and in books such as this one, are basically similar. The cabinets and panels may not look alike, and there may be quite a few differences in detail. But essentially they consist of a **crystal oscillator**, a **buffer amplifier** which is also a **frequency multiplier**, and a **final amplifier**. Sometimes the oscillator and buffer may be combined in one tube or stage. Often, too, there is provision for phone operation, but that will not concern us in this chapter.

A possible panel layout of such a set is shown in Fig. 4-1. You'll no doubt find the same controls on the transmitter you own, although the labels may differ slightly. There will be a buffer tuning control—often labeled **drive**—and two final-amplifier controls. One of these is the **loading** and the other is the **tuning** control. These three controls are the ones used in adjusting the transmitter for maximum output on a given frequency.

The other controls on the panel are not used in the tuning procedure. The two most important ones are the **function switch**, for turning the transmitter on and off, selecting c.w. or phone, and so on, and the **band switch** for selecting the proper tuned circuits for a given band.

Tuning the Transmitter

Let's start out on 40-meter c.w. The band switch is set to "40" or "7 Mc."—whatever is indicated on the panel. Plug in a crystal of the frequency called for in the transmitter's instructions for this band; usually you have the option of using either 80-meter or 40-meter crystals when the output is on 7 Mc. (Chapter 10 has a table of crystal frequencies that you can use, depending on the class of license you have.) The antenna is already connected, probably—or maybe you have a **dummy** antenna. An ordinary 115-volt lamp of the 40-watt size is a practical dummy (see Chapter 11), and it's a good idea to use one when you're rehearsing your transmitter tuning. The function switch has been on **standby** so the tubes have been warming up to operating temperature. And of course the key has been plugged into the key jack.

Your function switch may have a **tune position**. If it doesn't, there should be an equivalent setting which will keep the input to the final amplifier within safe limits when you make preliminary adjustments. See the instructions for your transmitter on this point. Whatever the label, set the switch there and set the meter switch to read the final-amplifier grid current.

Now press the key and the meter pointer should move up the scale. Adjust the drive control for the highest possible reading and then, if necessary, turn the control one way or the other to bring the current down to the value recommended for the tube used in your particular transmitter.

Driving Power

In adjusting the drive control you have been tuning the buffer stage of the transmitter. This control is indicated in the block diagram of Fig. 4-2, which shows the principal divisions of the transmitter.

The **drive** is the radio-frequency power that is going to **excite** the final amplifier. The final stage amplifies this power to a considerably larger value. The grid current read by the meter is a measure of the power fed to the amplifier grid.

FIG. 4-1—The front panel of your transmitter could look like this, although it probably doesn't duplicate the layout exactly. Most low-power transmitters have the controls shown.

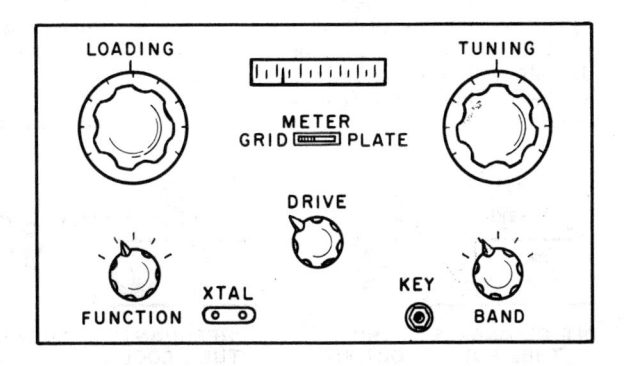

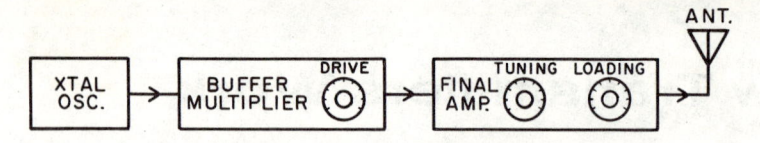

FIG. 4-2—In block-diagram form, the r.f. section of many low-power transmitters will have these three stages.

The reason why a particular value of grid current is specified is that this value will result in the most effective use of the amplifier tube's capabilities. You can put too much power into the amplifier's grid as well as too little. Too much driving power will just heat the grid, without giving you any more output than you would get with exactly the right amount. In fact, you may even get *less* output if you over-drive an amplifier.

Tank-Circuit Resonance

Now let up on the key and throw the meter switch to "plate." Set your function switch to c.w. In the next few adjustments you want to hold the key down only long enough to see what the meter does. First, set the loading control at minimum loading. If the amplifier uses a pi-network tank circuit, as most of them do, this will be the maximum-capacitance setting of the loading capacitor.

Now press the key momentarily. The meter will very likely go up to a high reading—possibly off scale. Don't hold down the key more than a second or so at a time. In these short key-down intervals, swing the plate tuning control rapidly through its range. At one point the current will dip down to a fairly low value, but it will rise rapidly if you move the knob just a little to either side. The minimum point is the one where you have the amplifier tank circuit in resonance with the frequency applied to the amplifier's grid.

You should *always* keep an amplifier tank tuned to resonance. An off-resonance tank circuit is just about like a short circuit on a battery—the current is large, but all the power stays in the battery. So with the amplifier; the input to the plate is high, but the power isn't going anywhere but into the plate of the tube. If this short circuit exists for more than a few seconds the tube will get excessively hot. You can ruin a tube in a hurry on this account if you aren't careful.

Loading the Amplifier

Now you have the plate current at a minimum point, but the input isn't very great. The tube is operating safely, but there is very little power going into the antenna. You need to bring the input power—the plate current at resonance is an indicator of this power—up to normal for the tube and transmitter. The loading control gives you the means for doing it.

With the key open, advance the loading control a little way toward maximum loading. Then close the key momentarily, and again adjust the tuning control for minimum plate current. Changing the loading control detunes the amplifier tank circuit a bit, so the circuit has to be brought back to resonance by the tuning control. If you don't do this retuning you'll at least partially short-circuit the amplifier and the efficiency will be low. And low efficiency means that the tube will get unnecessarily hot.

After retuning, you'll observe that the plate current is now somewhat higher than it was when the loading control was at minimum. But it is still below what you're shooting for, probably. Set the loading control a little farther toward maximum and go through the same process again. Keep it up until you reach the recommended value of current—never forgetting to readjust the tuning control for minimum current after every change in loading. If you have a lamp dummy antenna its behavior as you go through this adjustment process should be quite illuminating (in more ways than one!).

Overloads

You may even be able to continue the loading/tuning process until you reach a value of plate current higher than is recommended for the tube. It isn't a very good idea to do it, though. True enough, you may have more input and the tank circuit may be in resonance. But if you can actually observe the r.f. output you may find that it has started to *decrease* after you passed

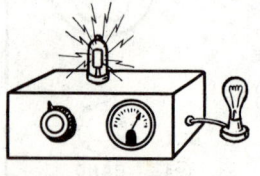

OFF RESONANCE
TUBE HOT
NO OUTPUT

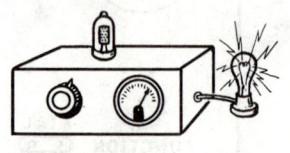

RESONANT
TUBE COOL
OUTPUT

FIG. 4-3—The plate milliammeter is showing the same current in both cases, but there's quite a difference in the way the tube is operating. The plate meter reading means only one thing—an overheated tube—if the plate tank circuit isn't tuned to resonance.

Transmitter Operation

through the point of optimum loading. This, too, causes the tube to get hotter than it should.

There is never any point in running more input than you actually need for getting maximum output. If you don't have a meter for indicating the r.f. output, by all means try out your transmitter on the lamp dummy. Since the lamp will be brightest when the output is greatest, you can readily find the value of plate current that gives you the most output. Then when you hook on an actual antenna you should adjust your loading for that value of plate current. It may not be the value as given in the book. Nevertheless, it will be the *best* value, provided you don't exceed the recommended figure. Although it may be possible to get more output by running the input higher than the rating, it won't do the tube any good. Tubes, like humans, can be overloaded and will temporarily bear up under the strain. But in the long run overloading usually leads to breakdown.

Retouching

When you have the transmitter tuned up for maximum output, switch back to read grid current. Probably it will have dropped while you were tuning the amplifier. This is quite normal; the higher the plate current, the lower the grid current goes, usually. If it is below the recommended value, readjust the drive control to bring it back. Then read the plate current again. Giving the amplifier its normal drive probably will have made the plate current go down a little, showing that the amplifier is operating more efficiently. You can now touch up the loading and tuning adjustments to bring the input up to par again. If you're using the lamp dummy, very likely you'll find that the lamp is brighter now even though the input is the same. In fact, you don't need to look at the grid current if you have the lamp. Simply adjust the drive control, while jockeying back and forth with the amplifier loading and tuning controls, until the lamp gets brightest. (Don't exceed the recommended plate current, of course.) *Then* check the grid current. This is the value to use when you go on the regular antenna.

Although we chose 40 meters for starting, exactly the same procedure should be used on other bands. Most transmitters operate under optimum conditions around the 7-Mc. range, and the current values you'll get should be nearest "normal" on this band. As you go higher in frequency the plate current, at resonance, may not dip so much at minimum loading. This simply means that the circuits become less efficient at the higher frequencies. You can't expect that the output will be quite as high at, say, 21 Mc. as it will at 7 and 3.5 Mc., when you use the same plate input in all cases.

The instructions with your transmitter should have outlined the basic tuning procedure as we have just described it. It is doubtful that they included all of the detail, and almost certain that there was no mention of comparing input and output. The instruction books for commercially-built gear, especially, attempt to make the whole process as simple as ABC. The reasons are obvious. But right now you're interested in learning a more-advanced "how," and also in picking up some of the "why." *Why* does the amplifier plate current dip at resonance? *Why* does the dip become smaller as the loading is increased? *Why* does the loading change when the controls are varied?

The reasons are to be found in the behavior of oscillator and power amplifier circuits, so we'll take a closer look at what goes on in those circuits. As oscillators, frequency multipliers and whatnot are just special applications of amplifiers, we'll begin with power amplification.

Power Amplification

In Chapter 2 you've seen how it is that a tube can amplify. In a radio-frequency power amplifier we can use a circuit arrangement that is practically identical with the circuits used in receiving-type amplifiers. The thing that makes the power amplifier different isn't the circuit but the way the tube is *operated*. The more power we want, the larger the voltages and currents, and the larger the tube and the components.

We have to introduce a much-misused word here—**efficiency**. Its actual meaning is simple—it is the ratio of two powers. The first is the useful power output of the device; the second is the power input that produces the output. If you put 80 watts into the plate circuit of an amplifier and get 50 watts of r.f. output, the efficiency is 50 divided by 80, or ⅝. In percentage, it is 62.5 per cent.

Hams often describe a piece of equipment as "efficient" when what they really mean is that it

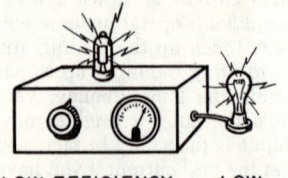

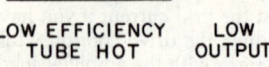

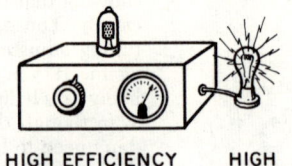

FIG. 4-4—But keeping the plate tank properly tuned isn't always enough. If the plate efficiency isn't high, the power burned up in the plate of the tube may be over the tube's rated ability to dissipate it.

works well. An r.f. stage in a receiver may work to the owner's complete satisfaction, but it is certainly not efficient in the technical sense. While the power input to it can easily be measured with ordinary instruments, the power output cannot—it's too microscopic. The efficiency is almost zero.

Plate Efficiency

What we are concerned with in power amplifiers is plate efficiency. As explained in Chapter 2, plate efficiency is important because the difference between the output and input stays right in the plate of the tube. In the example above, the difference, 30 watts, between 80 watts input and 50 watts output is "lost" in the plate. If the tube is built to stand having 30 watts lost or dissipated in its plate, well and good. But what if the tube is one that can dissipate only 15 watts safely? The extra 15 watts will cause the plate to heat excessively, and if the tube is forced to operate under such conditions for a length of time it may be permanently damaged.

If the amplifier efficiency can't be improved so the plate loss will be smaller, there is only one thing to do—decrease the input to the plate until the loss is no more than the plate-dissipation rating of the tube. At the same efficiency, 62.5 per cent, we should have to reduce the input to 40 watts with the 15-watt tube. This would give an output of 25 watts and the power lost in the plate would be just 15 watts.

Clearly, we need to know what sort of operating conditions will result in good plate efficiency. In a receiver's r.f. amplifier, efficiency doesn't matter. The tube has to handle only small signals and so only needs a small d.c. plate input. It is designed to operate safely with this input even when there is no output at all. This isn't true of tubes used in transmitting power amplifiers. In such amplifiers, the tube either delivers or it overheats—unless we're satisfied to keep the input down to no more than the rated plate dissipation. This is not very desirable, because the output would be small, too.

The Plate Load

To begin with, for reasonably efficient operation we have to have the right load resistance for the tube's output (plate) circuit. (Chapter 2 has discussed the necessity for a load.) A *tuned* plate circuit is needed in order to avoid short-circuiting the output at radio frequencies (also discussed in Chapter 2), so we might as well make the circuit do as many jobs as we can. One of these jobs is to take whatever resistance the actual load may have and transform it to the value the tube should be getting. Fig. 4-5 shows a few of the many circuits that could be used. In all three, the radio-frequency choke and the blocking capacitor C_1 are simply practical details having nothing to do with the resistance transformation.

Fig. 4-5A is the easiest to understand. The load is tapped across part of the tuned-circuit coil, L_1. The coil acts a good deal like an ordinary transformer. The r.f. voltage that the tube develops across the tuned circuit is larger than the voltage appearing between the taps. As explained in Chapter 2, a resistance connected to the low-voltage side of a transformer is changed into a larger value of apparent resistance on the high-voltage side. The resistance represented by an antenna or feed line is usually rather low, while a power tube generally wants a load somewhere between 1000 and 3000 ohms for optimum operation. If the tap is placed at exactly the right point, the tube will "see" just the right load.

The right load can't be defined exactly. It's usually a matter of compromising between several conflicting factors. And it isn't necessary to think of it as a certain number of ohms. The amplifier's plate current is a measure of the load on the tube, so we adjust the circuit for some predetermined value of plate current (always remembering that the circuit must be tuned to resonance when checking the plate current). The current may be a value set up by the tube manufacturer, representing his idea of optimum operating conditions. Or it may be chosen on some other basis. You're usually given the trans-

Power Amplification

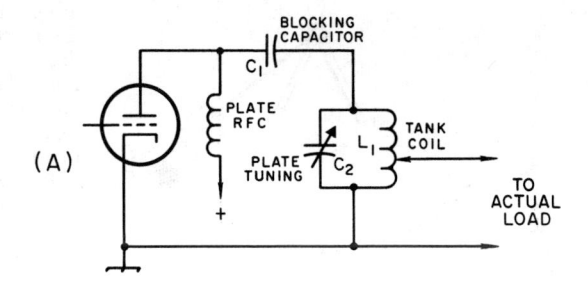

FIG. 4-5—Amplifier tank circuits with three methods of adjusting the load on the tube. The pi network (C) is the one most often used in present-day transmitters. The tank-circuit constants, C_2, C_3, L_1, depend on the frequency in use, the plate voltage and plate current of the tube, and the value of operating Q selected. C_1, the plate blocking capacitor, usually is 0.001 μf. or more; its value can vary within fairly wide limits. The plate r.f. choke is ordinarily 1 to 2.5 mh. for frequencies up to 30 Mc. Smaller values are commonly found in v.h.f. amplifiers.

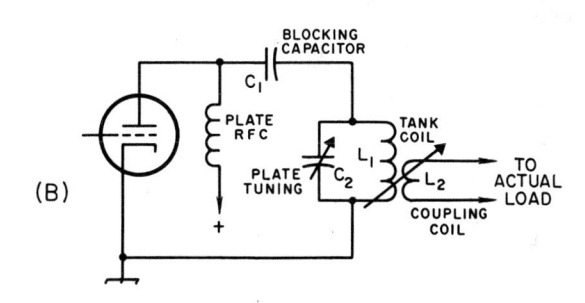

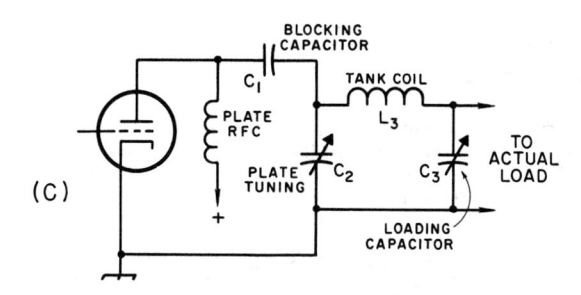

mitter designer's figure in the operating instructions that go with the transmitter.

The circuit of Fig. 4-5B does exactly the same job as Fig. 4-5A. Instead of a tap on the tank coil L_1, the loading is varied by changing the coupling between L_1 and L_2.

The Pi Network

Fig. 4-5C shows the pi network, currently the most popular type of final-amplifier tank circuit. It, too, transforms the actual load resistance into a value that is more suitable for the tube. It's really a resonant circuit just like the other two. C_2 is the plate tuning capacitor and C_3 is the loading capacitor. In ordinary transmitters C_3 is much larger than C_2; it has to be, if the circuit is to transform a low actual load resistance into a high one for the tube.

A rough way of looking at the operation of this circuit is that C_3 bypasses some of the r.f. current that otherwise might flow through the load. If the capacitance of C_3 is large, the load on the tube is "light"—low plate current—because a large capacitance bypasses most of the

current and little goes to the load. As the capacitance of C_3 is made smaller, more of the current goes through the load and less through C_3, so the amplifier is loaded more heavily—larger plate current.

It should be apparent that C_3 will have an effect on the tuning of the circuit when its capacitance is varied. Each time C_3 is changed, C_2 has to be readjusted to restore resonance. This is what we did in going through the amplifier tuning procedure.

Off-Resonance Input

We've been emphasizing right along the importance of keeping the plate tank circuit in resonance. Harking back to Chapter 2, in the section on tuned circuits you've seen how the parallel impedance of the circuit varies with frequency. It is quite high at resonance, but drops rapidly when the tuning is varied either side of resonance. The parallel impedance of the plate tank circuit is what the amplifier tube sees as a load. When you have the tank circuit in resonance the tube sees a simple resistance. A resistance uses

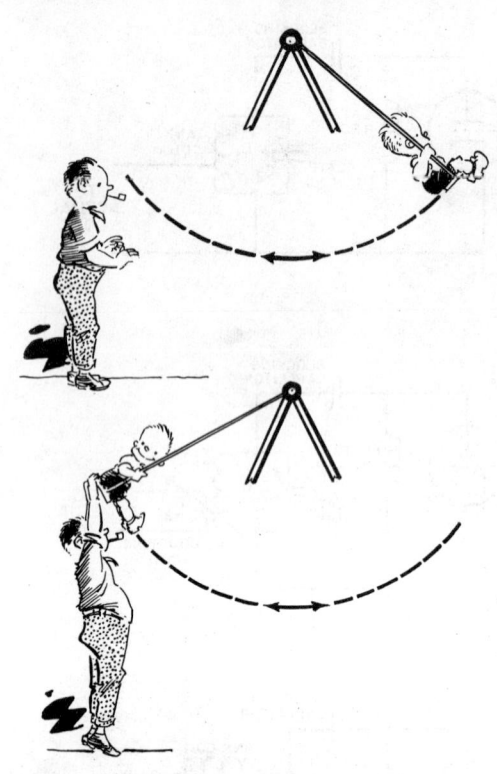

FIG. 4-6—The child's swing gets a push near the peak of travel once each oscillation. There is a close parallel to this in the way a high-efficiency (Class C) amplifier works. A pulse of plate current "kicks" the tuned tank circuit once each cycle. For most of the time the r.f. energy in the tank is oscillating on its own. If you've ever pushed a swing, you'll appreciate that it takes much less energy to build up a large amplitude (travel) by the method shown than it would if the swing were pushed and pulled throughout the entire period.

Note that although the push is quite short, the oscillations are smooth and regular. The amplifier tank circuit also smooths out the pulses that it gets from the tube, keeping the oscillations going at a uniform rate.

up the output power, and the more it takes from the tube the less of the total input is left to heat the plate. So we try to let the tube "see" a resistance that will take everything the tube is capable of giving.

When the circuit is detuned from resonance, the tube no longer sees a simple resistance. The load has the characteristics of a resistance *smaller* than the resonant value, shunted by an inductance or capacitance. (Whether it is inductance or capacitance depends on whether the detuning is above or below the resonant frequency.) Some of the r.f. current flows through this equivalent inductance and capacitance, and some through the resistance. But the part that flows through the inductance or capacitance—for convenience, we can substitute the general term reactance for either—does no useful work. It simply flows around the circuit and through the tube. Worse, the reactance quickly becomes quite small as we detune, so the tube tries to supply a larger current. Thus the input—that is, the plate current—goes up. More input and less output add up to a lot of tube heating, and the result can easily lead to a trip to the radio store to replace the tube.

Beware of assuming that a tube is properly loaded when the plate meter reads the recommended value of current! The plate current has a real meaning *only* when the tank circuit is kept tuned to resonance.

Class C Operation

Proper loading is very important, but it isn't the whole story. To operate at good efficiency a tube has to be driven properly, too. A complete explanation would get us into technicalities that we want to avoid in this book. However, it goes about like this:

To get large output, the tube has to be driven so that the plate current hits very high instantaneous values—three or four times what the plate meter shows. But these large values of plate current must exist only during a small part of an r.f. cycle. Over at least half, and preferably somewhat more than half, of each cycle there should be no plate current at all. Making the plate current flow in this way will give us a large amount of output, compared to the d.c. input, along with only a relatively small amount of plate heating. A tube so operated is called a Class-C amplifier.

Grid Current

The plate current can be cut off during half or more of the cycle by using a large-enough value of negative grid bias. The bias should be at least the value that will cut off the plate current when there is no r.f. driving voltage on the grid. Then in order to get large peak values of plate current the r.f. voltage has to swing the grid voltage rather far in the positive direction—usually, it will overcome the bias completely and make the peak grid voltage actually positive with respect to the cathode. When this happens, a rectified current flows in the grid circuit. This is the current measured by the meter when it is switched to the grid-current position. If the bias has some known value, the amount of grid current tells us whether or not we are driving the grid far enough positive.

Driving Power

The grid current represents power used up in the grid circuit—some in the grid itself and some in the bias arrangement. Usually, this power is put to work. The grid current is made to flow through a resistance—the grid leak—to develop a negative voltage that is used for biasing the tube. This saves having to have a special bias-voltage source. (A separate bias power supply is sometimes used for various reasons that

Power Amplification

FIG. 4-7—If the grid currents, I, are the same, and if the value of R is properly chosen, the bias voltage, E, will be the same in both circuits. The power used in the grid leak, R, is equal to the voltage multiplied by the current, from Ohm's Law. With the battery bias the polarity is the same, E is the same, and I is the same. Therefore the power is the same as when the grid leak is used. The battery *absorbs* power instead of supplying it. As a result, it will experience exactly the same heating effect as the resistor, although its temperature will not rise as much because it is usually much larger physically.

The two methods are equivalent when the voltages and currents are the same, but with the grid-leak resistance R, the bias voltage depends on the grid current. The battery-bias voltage remains the same regardless of grid current (neglecting the effect of the internal resistance of the battery). Both systems have their advantages.

need not concern us at the moment.) Thus we can get "free" bias, if we want to, whenever there is grid current. It is simply a matter of using a grid resistance that, by Ohm's Law, will have a voltage drop equal to the required grid bias when the recommended value of grid current flows through it.

Of course, the power in the grid resistor isn't actually free. It was originally r.f. power that came from the tube that drives the amplifier. But whether the amplifier bias is obtained from a grid leak or from a separate voltage supply, the driving tube has to supply the same power to the amplifier's grid circuit. The bias source makes no difference. Whenever there is grid current there is a demand for power from the driver.

Some amplifier tubes take more driving power than others, for the same plate input and output. Modern tetrode and pentode power tubes are very obliging in this respect. Many of them will deliver outputs of 50 to 100 watts when only a fraction of a watt is used up in the grid circuit. This is one reason why tubes of this type are so popular in amateur transmitters of all sizes.

Tank Circuit Q

In the type of tube operation just described the plate current flows in **pulses** that are rather short compared with the length of one r.f. cycle. The plate current has no resemblance to a sine wave. This means that it has in it a lot of frequencies, all of them harmonics of the driving frequency. A sine wave is the only kind that has all its energy on just one frequency, and a *single* frequency is what we want to get from the r.f. amplifier. The harmonics have to be removed from the output somehow. In some cases this removal process may take several steps. The first step always is to make use of the selectivity of the tuned plate tank circuit.

The selectivity of a tuned circuit is expressed by its Q, as discussed in Chapter 2. Since the circuit is handling power, the Q is determined more by the loading than by the resistance of the coil. In other words, the **operating Q** is the Q that counts, not the Q of the coil alone. In effect, the tuned circuit (which we may consider to have negligible losses by itself) is shunted by the load resistance that the amplifier tube sees. This makes the operating Q fairly low, as compared with the Q of the coil.

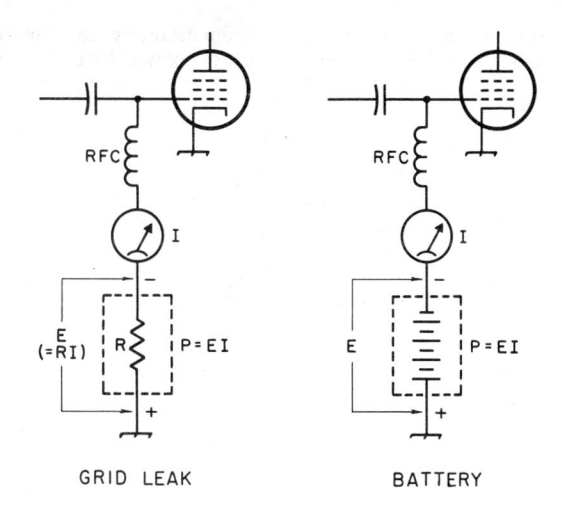

In practice, the operating Q of a tank circuit is determined by two things. The first is the ratio of plate voltage to plate current; this is very closely related to the optimum value of load resistance for the tube. The second is the capacitance of the plate tuning capacitor, the one that connects directly from the plate to the cathode of the tube. For a given plate voltage/plate current ratio, the bigger the capacitance the higher the Q.* Long experience has shown that the tank Q should be of the order of 10 or 12 in a power amplifier of the type we have been considering. With such a Q value, the r.f. voltage at the terminals of the tank circuit will be a rather good sine wave. It won't be a *perfect* one, but will be close enough to it so that the power in harmonics will be very small compared with the power in the fundamental frequency.

Suppressing harmonics in this way is necessary for efficient amplifier operation; the power lost in the plate is lower when the r.f. output voltage is a sine wave. Besides, power in harmonics doesn't help a bit in strengthening the signal on the fundamental frequency, which is the one we actually use for communication. And

---

*A formula for the tank capacitance required is

$$C = 300\frac{QI_B}{fE_B}$$

where C is the capacitance in $\mu\mu f$, f is the frequency in megacycles, I_B is the full-load plate current in milliamperes, and E_B is the plate voltage. For example, if Q is to be 12, f is 7 Mc., I_B is 100 ma., and E_B is 600, the required capacitance is 86 $\mu\mu f$. This formula is an approximation, but a highly-accurate calculation isn't necessary for the purpose, since the Q value isn't critical.

a final consideration, at least as important as the others, is that harmonic-frequency power that gets into the antenna and is radiated—even if it's only a very small fraction of the fundamental power—sometimes can cause interference with other services. The tank circuit alone probably won't be able to take care of this last problem; additional selectivity usually has to be inserted between the transmitter and the antenna. An antenna- or line-matching network—a **transmatch**—is an excellent device for getting such selectivity. It has other useful features as well, as described in later chapters.

Flywheel Effect

The tank circuit of an amplifier has often been compared to the flywheel on an engine. Both the tank circuit and the flywheel are driven by short, regularly-timed bursts of energy from the primary source of power. The flywheel converts the pulsating mechanical power into a smooth rotating motion. It does this by storing up energy (in inertia) during the power bursts, and releasing it uniformly during the in-between periods. The heavier it is the greater the energy storage and the smoother the resulting motion. A heavy flywheel has high Q.

So, too, with the tank circuit. It stores electrical energy during the bursts of plate current and releases it uniformly during the periods when the plate current is cut off. (The stored energy is represented by r.f. current circulating around the tank formed by the coil and capacitor.) The result is a smoothed output that can be expressed as a single frequency, just as the power from the flywheel can be expressed in terms of so many smooth rotations per minute.

Stabilizing the Amplifier

Chapter 2 explained that the screen grid in a tetrode or pentode shields the plate from the grid so that very little of the amplified energy can feed back to the grid circuit from the plate. Unfortunately, the shielding isn't quite perfect, in practicable transmitting tubes. There is always a remnant of capacitance between the plate and grid, offering a feedback path that cannot be avoided. The shielding usually is good enough so that the feedback is not serious at frequencies from 7 Mc. on down, but at higher frequencies it may be great enough to set the tube into self-oscillation. An amplifier that tends to go off into oscillations of its own is said to be **unstable**. We don't want instability in our amplifiers, because a self-generated oscillation may be outside the amateur band. Also, self-oscillation may result in inefficient tube operation.

You will find, therefore, that many practical amplifier circuits include a provision for preventing self-oscillation. Such amplifiers are said to be **neutralized**. Neutralizing circuits take a little bit of r.f. from the grid circuit and feed it

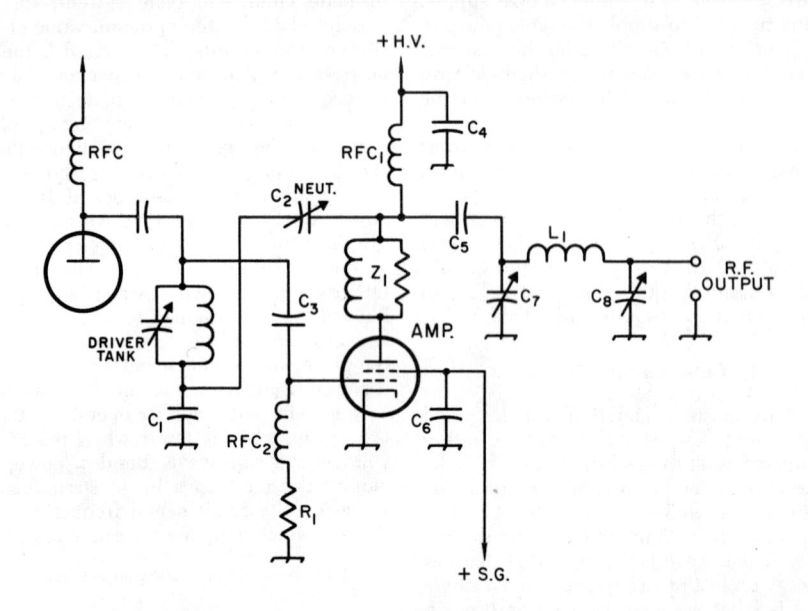

FIG. 4-8—A typical amplifier circuit, with a neutralizing system and v.h.f. parasitic suppressor. The pi-network tank circuit is similar to Fig. 4-5C. C_4, the plate-choke bypass, and C_6, the screen bypass, are not at all critical, values from 0.001 to 0.01 μf. being used. RFC_2 is usually 1 to 2.5 mh. R_1 depends on the tube, and is determined by the required negative grid bias and grid current. C_3 is generally 100 μμf. or more. C_1 is of the order of 500 μμf. for most amplifier tubes. C_2 is very small—a few μμf. at most. Oftentimes it suffices just to use a couple of inches of wire mounted close to the tube; the capacitance between the wire and the tube's plate is enough to stabilize the amplifier.

Power Amplification

to the plate circuit to balance out the energy that is fed through the tube from the grid to the plate.

Neutralizing

Fig. 4-8 shows such a circuit. The driver tank plays an important part in the neutralizing. C_1, at the bottom of the driver tank, is not a complete bypass for r.f., as you usually would expect it to be in a plate circuit. Instead, its capacitance is such that there is a small r.f. voltage across it. The "hot" side of C_1 is connected to the plate of the amplifier through a very small capacitance, C_2, the **neutralizing capacitor**. When C_2 is adjusted to the right value the r.f. voltage from the top of C_1 to the amplifier tube's plate has just the same amplitude as the r.f. voltage from the grid of the amplifier to its plate. If you trace out the circuit you'll see that the first voltage is from the bottom of the driver tank while the second comes from the top of the tank. That is, the two voltages are of opposite polarity or phase, and so they balance each other at the amplifier plate.

A neutralizing circuit of this sort is known technically as a **bridge**. The bridge has the property of being balanced in both directions, so it is equally as well balanced for energy fed from the plate to the grid as from the grid to the plate. Thus there is no feedback and no **feedthrough**. The latter term applies to energy fed from the grid to the plate. The description above was based on neutralizing the feedthrough, because it is a little easier to visualize the circuit operation that way. However, once you understand it you can readily appreciate how it works for energy going in the opposite direction.

Parasitic Oscillations

There is another type of instability that nearly always needs attention in power amplifiers. The neutralizing circuit will work as just described so long as the connecting leads, including those inside the tube itself, have inductances that are small enough to be considered negligible at the operating frequency. With small power tubes this is true enough at frequencies up to perhaps the 50-Mc. band. Larger tubes have to have longer leads associated with them simply because the tubes are bigger. With such tubes the upper limit at which lead lengths can be considered to be negligible is in the neighborhood of 30 Mc.

The lead inductances, along with stray capacitances, make the neutralizing circuit (which is quite simple in principle) very complex in practice. The upshot is that the sought-for balance can't be achieved. A neutralizing circuit may do a very adequate job up to 30 Mc. or so, but may fall down badly in preventing self-oscillation at frequencies over perhaps 60 Mc. Even though the amplifier may have only been *intended* for use up to 30 Mc., the possibility of self-oscillation at still higher frequencies can't be ignored. Most modern tubes will "take off" in a **parasitic**

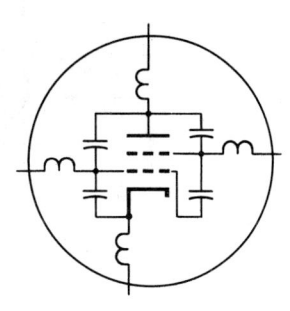

FIG. 4-9—At v.h.f., the tube looks something like this. The internal leads have enough inductance to become an important part of the total circuit, and the tube capacitances are not "get-at-able." Ordinary neutralizing circuits fall down when confronted with a situation like this.

oscillation at v.h.f. unless something is done to prevent it. The better the tube the more likely it is to do so.

These v.h.f. oscillations can be prevented by Z_1 in Fig. 4-8. Z_1, consisting of a small coil in parallel with a resistor, is called a **parasitic suppressor**. The coil serves to couple the v.h.f. oscillator circuit to the resistor, and the value of the latter is chosen so that it loads the v.h.f. circuit so heavily that the tube is unable to get oscillations started. The coil size and resistance usually have to be found by experiment for each layout.

Buffer Amplifiers

As its name implies, a **buffer amplifier** is one interposed between two stages to keep them apart—usually, to separate a power amplifier from an oscillator. When an oscillator is coupled to an amplifier the oscillator's frequency is likely to be affected by any changes in the amplifier's operating conditions. Examples of such changes would be keying the amplifier, tuning its plate circuit, or adjusting the loading on it. All of these react on the oscillator, changing its frequency because they change the load on the oscillator.

A buffer amplifier is supposed to absorb these reactions and keep the load on the oscillator constant. Of course, the buffer also can serve to amplify the oscillator's output when that output isn't sufficient to drive the final amplifier.

Buffer circuits are similar to those used with any amplifier, with emphasis on good isolation between the buffer's input (grid) and output (plate) circuits. In small transmitters, additional amplification usually isn't really needed, because the final amplifier tube always is a tetrode that requires very little driving power. In such sets the buffer stage is included in the circuit primarily because it is needed for frequency multiplication. When operating on the same frequency as the oscillator its plate circuit often is untuned, or only approximately tuned.

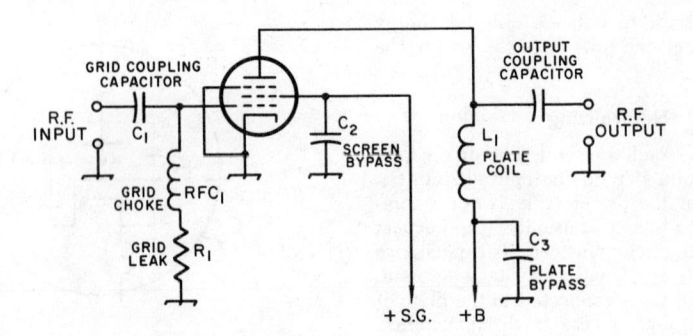

FIG. 4-10—An untuned buffer stage. Values in a circuit such as this are similar to those used in tuned amplifiers, except for the plate coil. This is often an r.f. choke, especially if only a small amount of power is needed from the buffer. If the inductance of L_1 is made to resonate with the tube and stray capacitances, the Q is low and the circuit is broad enough to work over a band without readjustment. This broad-banding is sometimes helped out by shunting L_1 with a fixed resistor. Because the power gain is low and the circuit is used almost exclusively at low frequencies (3.5 or 7 Mc.) neutralization is rarely needed.

Untuned Buffers

A typical untuned buffer circuit is shown in Fig. 4-10. Capacitive coupling to the preceding and following stages is almost invariably used in a buffer-multiplier stage because it is the simplest coupling method. Also, it happens that the grid circuit of an amplifier tube "looks like" a resistance of a few thousand ohms, and so is about the right value to be a good load for a smaller amplifier or an oscillator. Therefore it isn't necessary to use one of the output circuit arrangements shown in Fig. 4-5, especially since the power input to a small buffer-multiplier is low enough so that we needn't worry much about plate efficiency. This is a spot where convenience can be allowed to over-ride optimum design.

Tubes used in such stages are small pentodes or tetrodes, like the 6CL6, 12BY7, and similar types. They give, relatively, a lot of output with only a small amount of driving power. The plate coil, L_1, often is simply an r.f. choke. The inductance value isn't critical—anything from 1 to 2.5 millihenrys is generally used. Since the plate circuit isn't tuned, the amplifier is not at all efficient in developing power—but we can get by because not much power is needed. In some transmitters L_1 will be a simple coil having an inductance that will be resonant, in conjunction with the tube capacitances, about the middle of the band on which the oscillator is operating. To broaden the tuning of such a circuit you may find L_1 shunted by a resistor of some thousands of ohms.

Frequency Multiplication

In describing the operating conditions for efficient amplifier operation earlier in this chapter, we mentioned that the pulse-like waveform of the plate current meant that not only the fundamental frequency but harmonics of the fundamental were present. In the **straight** amplifier —that is, one whose output is on the same frequency as the grid drive—we try to get rid of these harmonics. However, there are times when we *want* an output frequency that is some simple multiple of the fundamental. In this case the amplifier becomes a **frequency multiplier**. All this term means is that we tune the amplifier's plate circuit to some desired multiple or harmonic, instead of to the fundamental frequency. That way we can get power output at the harmonic frequency. The circuit, Fig. 4-11, doesn't *look* different. The difference is principally in the tube's operating conditions.

In the discussion of amplifier operation we

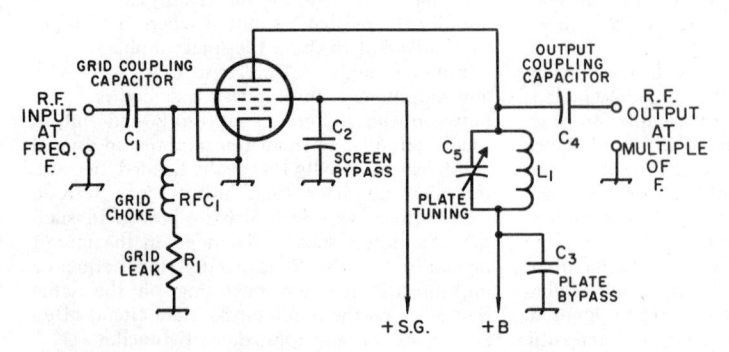

FIG. 4-11—A frequency multiplier circuit is like the ordinary amplifier circuit. However, the plate tank is tuned to twice or three times the driving frequency so that the output is on a harmonic of the fundamental frequency. Even greater multiples can be used, but with considerably-reduced output.

Oscillators

saw that the tube should be biased to cut-off or beyond, so that the plate current flows in pulses lasting only a half cycle or less. For equally-efficient frequency multiplication this idea has to be carried to extremes. When doubling frequency, for example, the grid bias should be large enough so that the plate-current pulses will not last more than a *quarter* of a fundamental-frequency cycle. When tripling frequency, plate current should not flow more than about a *sixth* of a fundamental cycle. Neither type of operation is ordinarily practicable, because both require an inordinately large value of negative grid bias and extremely large r.f. driving voltages at the grid. Good practical design lets the frequency multiplier operate at relatively low efficiency and then, if necessary, uses an extra straight-through amplifier to build up the harmonic power. This results in better over-all economy than attempting to get the greatest possible output directly from the frequency multiplier.

Multiplier Efficiency

In practice, the efficiency of a frequency multiplier may be 40 per cent or so when doubling and 20 to 30 per cent when tripling. The output drops off quickly at higher harmonics. Now and then you'll find a quadrupler, when the following tube takes very little driving power, or when there are reasons for being contented with below-par operation of the amplifier. One such reason, for example, is to provide *some* output on a band that otherwise couldn't be covered by the design.

The efficiency and output of a frequency multiplier will improve if the negative grid bias is made larger. This bias is usually made as large as possible, consistent with the driving power available for the multiplier. The practical application of this is that the grid-leak resistance is greater than it would be for straight amplification. Multiplier grid resistors often will have two or three times the resistance of the grid leak used with the same tube as a straight amplifier.

A frequency multiplier does not need to be neutralized. Since its grid and plate circuits do not operate on the same frequency, it would take very exceptional circumstances to make the stage go into self-oscillation at its operating frequency. V.h.f. parasitic oscillations can occur, though. The treatment for them is the same as in the case of the straight amplifier.

Oscillators

The oscillator is the first stage of a transmitter, but we've left it until last because it is really an amplifier of a special type. The tube amplifies the r.f. voltage on its grid just like any other amplifier. The difference is that the driving power is fed back to the grid from the same tube's plate output circuit. There is still plenty of power left over because there is more available from the plate than is needed by the grid. The power the grid doesn't need is used for driving the following stage in the transmitter.

The several types of oscillator circuits were discussed briefly in Chapter 2. In this chapter we'll concentrate on crystal-controlled oscillators, since this type of oscillator is required in transmitters used by Novices. The self-excited variety is discussed at some length in Chapter 10.

The differences between the various crystal-oscillator circuits are in the methods used for getting the feedback that establishes and maintains self-oscillation.

Crystal Oscillator Circuits

A **quartz crystal** is a mechanical vibrator that behaves like a resonant LC circuit. In fact, it can be substituted for an actual coil and capacitor in almost any circuit where the crystal's resonant frequency can be used. There are two limitations, though: First, we can't have more than two terminals to the circuit, since there are only two crystal electrodes—in other words, a crystal can't be tapped in the way we would tap an ordinary coil for the Hartley oscillator. Second, the crystal is an open circuit for direct current, so series feed can't be used with it. Crystal-oscillator circuits have to observe these restrictions.

Two arrangements are commonly used for getting feedback. One is equivalent to the tuned plate-tuned grid circuit. It uses the crystal as the tuned grid circuit and has a separate LC tuned circuit connected to the plate. This is shown in Fig. 4-12A. The second, known as the **grid-plate** oscillator, is equivalent to the Colpitts circuit. For feedback it uses a pair of capacitors in series, with the crystal connected to their outside terminals as shown in Fig. 4-12B.

There are several variations of the grid-plate circuit. Sometimes the capacitor **divider**, $C_4 C_5$, is not obvious, because one or both of the capacitances may be the interelectrode capacitances of the oscillator tube. If C_4 and C_5 are both omitted, the circuit is called a **Pierce** oscillator. Another version of the Pierce circuit, using a tetrode or pentode tube, is discussed below.

Oscillator Tubes

The circuits in Fig. 4-12 have been simplified as much as possible to illustrate the principles. In actual practice, tetrodes and pentodes are often used instead of triodes. These multielement tubes will give considerably greater output without driving the crystal any harder. It is desirable

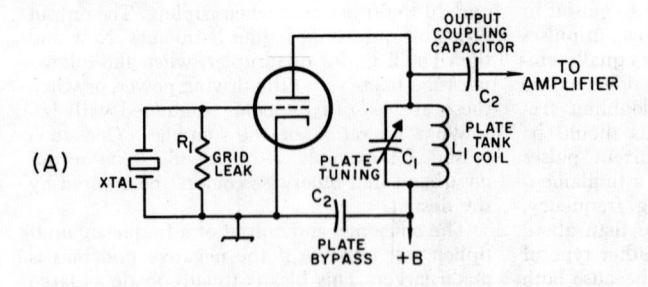

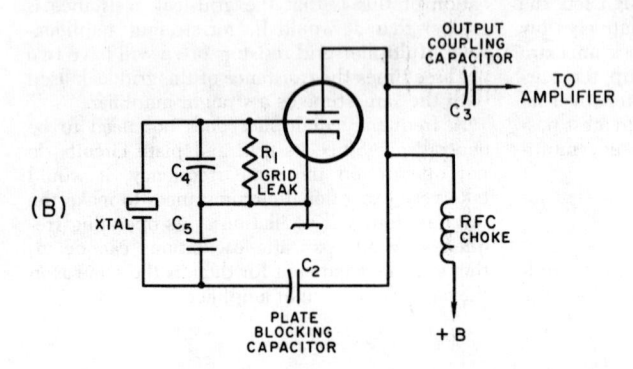

FIG. 4-12—The tuned-plate tuned-grid crystal oscillator circuit at A is used principally in very low-power applications, such as in crystal-controlled converters for receivers. The Colpitts-type circuit at B is used, in various forms, in most transmitters. In both circuits C_2 is of the order of 0.001 µf., not critical, and C_3 is 100 µµf. or so. R_1 can be anywhere from 10,000 to 50,000 ohms; it, too, is not especially critical. C_4 and C_5 depend on the type of tube. With most tubes currently used as transmitting crystal oscillators, C_5 will be 100 to 500 µµf. and C_4 may range from zero to 15 or 20 µµf. C_4 is sometimes made variable for feedback adjustment.

to keep the crystal drive as low as possible because the harder the crystal vibrates the warmer it gets. Like all circuit components, it responds to a change in temperature with a change in its constants—in this case, a change in its resonant frequency. This in turn causes the oscillator frequency to drift. Further, if the crystal is driven too hard it will be in danger of shattering itself. Once cracked or broken it is useless.

Electron-Coupled Circuits

The circuit in Fig. 4-13 is a popular one in current transmitters. It takes advantage of another feature offered by the tetrode or pentode tube—**electron coupling** to the output circuit. In this circuit the screen grid of the tube acts like the anode or plate of the triode circuit shown in Fig. 4-12B. However, the screen is bypassed directly to chassis, putting it at ground potential for r.f. (although not for d.c.). Thus it acts as a shield between the oscillator portion—consisting of the control grid and cathode along with the screen—and the actual plate or output circuit of the tube. Thanks to this shielding, there is very little capacitive coupling between the oscillator portion and the output portion. Output power is developed in the plate circuit because the electron current to the plate is modulated by the oscillator portion.

In a good tetrode or pentode ("good" meaning, here, one designed for r.f. applications) whatever may be going on in the plate circuit has little effect on the control grid, so this electron coupling helps to isolate the output circuit from the actual oscillator—a definite aid to stability. The isolation isn't perfect, but it is much better than it would be with capacitive coupling from the oscillator circuit itself to a following amplifier.

Tetrode Pierce

Another circuit you may see frequently is the tetrode or pentode Pierce oscillator, shown in Fig. 4-14. The simplicity of this circuit is its outstanding feature. At first glance there would seem to be no provision for feedback, but actually the oscillating circuit is the same as Fig. 4-12B, with the screen of a tetrode or pentode replacing the plate of the triode shown in the latter figure. C_4 and C_5 are inside the tube; C_4 is the capacitance between the control grid and cathode, and C_5 is the capacitance between the screen grid and cathode. If these capacitances have about the right ratio—they do, with most tubes—the circuit will oscillate without the external capacitors C_4 and C_5. However, the feedback cannot be controlled without adding at least one external capacitor.

In the tetrode or pentode Pierce circuit the coupling to the plate is not just electronic; there is also coupling through the screen-to-plate capacitance of the tube because the screen is "hot" for r.f. Thus this circuit does not offer the isolation that Fig. 4-13 does. Using untuned output coupling, as shown by the plate r.f. choke in Fig. 4-14, helps relieve this somewhat. A tuned circuit can be used, just as in Fig. 4-13, but is

Oscillators

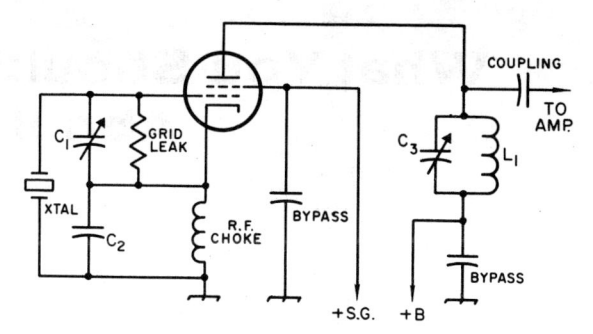

FIG. 4-13—An electron-coupled crystal-oscillator circuit. C_3L_1 can be tuned to the crystal frequency or to a harmonic of it, if harmonic output is wanted. If the tube has appreciable grid-plate capacitance, tuning C_3L_1 exactly to the crystal frequency will "kill" the oscillator, so the tank usually is tuned a little on the high-frequency side of exact resonance. A tube with very good isolation between the grid and plate circuits will continue oscillating no matter how C_3L_1 is tuned. Tuning the tank to a harmonic usually has little or no effect on the oscillations, regardless of the type of tube used. A tetrode is shown, but if the tube is a pentode having the suppressor-grid connection brought out to a base pin, the suppressor should be grounded. The r.f. choke inductance is not critical; 1 to 2.5 millihenrys is a representative value.

useful only when tuned to a harmonic of the oscillation frequency.

The tubes in Figs. 4-13 and 4-14 are tetrodes, but pentodes can be used just as well. The best types are those having the suppressor grid brought out separately to a base pin. This allows the suppressor to be grounded, giving the screen an assist in shielding the oscillator from the output portion of the circuit. Types with the suppressor connected internally to the cathode also can be used, but the isolation is not as good.

Harmonic Output

The circuits of Figs. 4-13 and 4-14 are not restricted to giving output at just the crystal frequency. With a tuned plate circuit it is possible to get usable output at harmonics of the crystal frequency as well. The harmonics are actually generated by the oscillator portion—practically all oscillators generate harmonics along with the fundamental frequency—and are simply "picked off" in the plate circuit by appropriate tuning. The higher the harmonic the lower the output, compared with the output at the fundamental frequency. However, the second harmonic (twice the fundamental) and the third (three times the fundamental frequency) are quite usable in most crystal oscillators.

Thus the tetrode or pentode oscillator can be a frequency multiplier as well as a fundamental-frequency generator. This makes it possible to design quite simple circuit arrangements for low-power transmitters. Several examples are shown in Chapter 10.

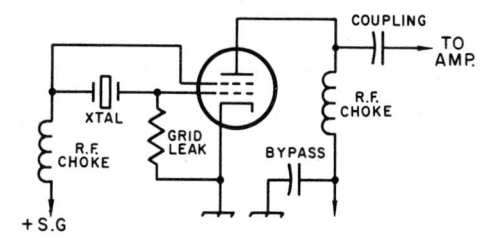

FIG. 4-14—The electron-coupled Pierce circuit. This circuit takes fewer parts than others, but lacks the input-output isolation that is a feature of the circuit in Fig. 4-13. It can be used for harmonic output by substituting a tank circuit, tuned to the desired harmonic, for the r.f. choke in the plate lead.

Other Details

In this chapter we've concerned ourselves with how the radio-frequency circuits of the transmitter work. There are more things in a transmitter than this, but they exist primarily in order to make the r.f. circuits operate, and to operate in a *useful* way. Power is needed, of course, and it is the function of the power supply to furnish the right voltages and currents. Devices for making measurements of one type or another are necessary for checking the operation. Also, there has to be some way of making the r.f. output of the transmitter convey a message to the receiving point; the keying and modulation circuits do this.

These features of the transmitter are taken up in other sections of this book, under appropriate chapter headings.

Chapter 5

What You Should Know About Phone

Ninety-nine out of a hundred amateurs will think of phone transmission whenever the word "modulation" crops up. The subject of modulation is a lot broader than that. However, the principles are the same whether the radio transmission is modulated by the human voice, by pulses (as in radar), by any of the varied forms of coded signals that are used in many communication systems, or by a telegraph key waggled by a Novice operator. The process of **modulation** adds a *message* to what would otherwise be a plain, single-frequency radio signal that, in itself, tells you nothing except that it exists.

The Voice Band

It is possibly easier to grasp the essence of modulation when it is put in terms of phone. You are aware of the fact that sound is a vibration in the air, a vibration usually made up of many frequencies lying within the range we call **audio frequencies**. The lower and upper limits of this range are not the same for all listeners; what you can hear depends on the condition of your hearing apparatus. The extremes are usually taken to be about 15 or 20 cycles at the low end and 15,000 or 16,000 cycles at the high end.

But long experience, backed by many tests, shows that a frequency range of about 200 to 3000 cycles will contain all the frequencies needed for making your voice understandable to a receiving operator. The actual range generated by a person talking may be much greater, but if those frequencies below about 200 and above about 3000 cycles are eliminated—filtered out— there is little, if any, loss in the speaker's ability to make himself understood. That is, voice **intelligibility** is high when the frequencies are kept within that range.

Microphones

In phone transmission the first step is to translate the air vibrations into electrical vibrations having the same form. The device that does this is the familiar **microphone**. Mostly, two general types are used by amateurs. One is the **carbon microphone** shown in Fig. 5-1A. When a sound strikes the **diaphragm**, which is usually a thin metal sheet of circular shape, the diaphragm vibrates right along with the air, following the air vibrations pretty faithfully. The diaphragm is held rigidly around its rim, so the movement is greatest at the center, as shown by the small double arrow. The alternate back-and-forth movement varies the pressure on loosely-packed carbon grains in the **button**. This changes the resistance in the circuit formed by the button, battery *BT* (a few volts) and the primary of transformer *T*. In turn, the varying resistance causes the current in the circuit to vary. Theoretically, the current will change in proportion to the vibration of the diaphragm, and in practical microphones of this type it makes a reasonably good stab at doing so.

Since the resistance of a collection of carbon grains is rather low, the impedance of the transformer primary has to be low, too, in order to take maximum advantage of the change in current. Values are usually around 100 to 200 ohms. Thus the primary will have only a few turns, relatively speaking. The secondary can have, and usually does have, a very large number of turns. This steps up the microphone's output voltage by a large factor. It isn't difficult to get 25 volts or so (peak value) from the secondary of the microphone transformer.

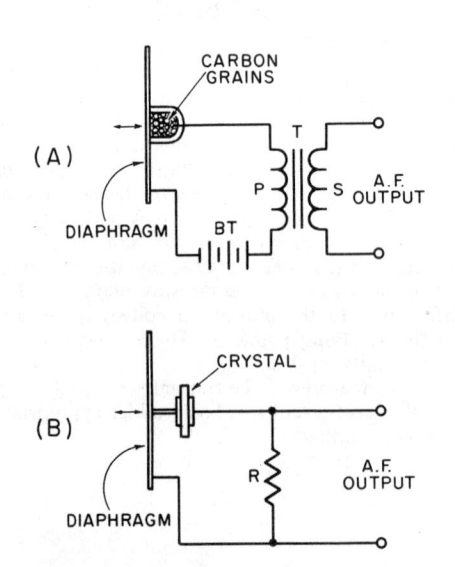

FIG. 5-1—Carbon and crystal microphones and their connections.

Modulation

The carbon microphone is widely used in mobile and portable work because of its high output and the fact that batteries usually are available to run it. Its **quality**, or ability to reproduce a voice faithfully and with little distortion, isn't especially good. The second type of microphone, Fig. 5-1B, is better in this respect. It makes use of a property of crystals that we found useful for selectivity in receivers and for frequency control in transmitters. Here, however, the process is reversed—that is, the crystal is set into vibration by sound waves, and the vibration generates a voltage. (In the uses of crystals discussed in Chapters 3 and 4 the application of an a.c. voltage caused the crystal to vibrate.) Crystals for transmitters and receivers are cut from quartz, but those used in microphones are made from Rochelle salts or certain ceramic materials. The former is generally called a **crystal microphone** while the latter is known as a **ceramic microphone**.

The voltage generated by the crystal or ceramic microphone is very small—only a few hundredths of a volt—and only a microscopic current can be produced. Because of this the microphone can't work into a transformer but instead has to work into a very high-impedance circuit. Usually it is connected directly to the grid circuit of a high-gain amplifier tube. The resistor R in Fig. 5-1B is the grid resistor for such a tube. Its value ordinarily will be a megohm or more.

Changing Audio to Radio Frequency

Now that we have the sound frequencies, or waves, translated into corresponding electrical waveshapes, what next? We can't put these low frequencies on the air directly. Even the highest audio frequency—about 15,000 cycles—would have a radio wavelength of around 20,000 meters. This is certainly nowhere near any radio frequency the amateur regulations let us use.

The answer is that the band of voice frequencies has to be moved up in the radio-frequency spectrum to some amateur band where we can make use of it. This is where modulation comes in. Suppose we want to operate in the 75-meter phone band, the one that lies between 3800 and 4000 kc. Further, suppose that we could get into that band simply by adding some frequency in it to the voice range of frequencies, 200 to 3000 cycles. It's easier to do the arithmetic if we change the cycles to kilocycles; in kilocycles, the voice range is 0.2 to 3 kc. If we select 3900 kc. as the frequency to add, we should have a signal occupying the band from 3900 plus 0.2 to 3900 plus 3, or 3900.2 to 3903 kc.

Simple, but there's a catch; there isn't any way to do it directly. Nature insists on doing things symmetrically. We can add the voice band to 3900 kc., all right, but *at the same time* we find that we can't avoid subtracting the frequencies, too. So we wind up not only with the band 3900.2-3903 kc. but also with *another* band from 3900 minus 0.2 to 3900 minus 3, or 3899.8 to 3897 kc., as in Fig. 5-2. Instead of a band a little less than 3 kc. wide we find ourselves with one having a total width of 6 kc.—3897 to 3903 kc., in fact.

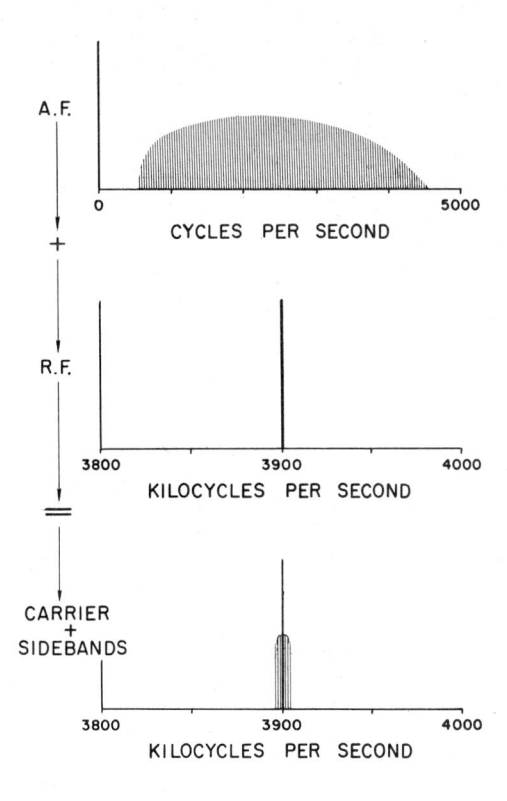

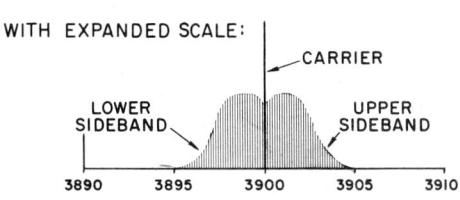

FIG. 5-2—Modulation with a band of audio frequencies sets up two sidebands, centered on the carrier frequency.

Modulation Sidebands

Every modulation system gives us these "sum and difference" frequencies. So the actual frequency band occupied by a modulated signal is always twice what is actually needed for transmitting the message. This is rather unfortunate, because it means that each signal so modulated takes up twice as much room on the air as it really needs.

The message content of the signal is referred to as the modulation, while the frequency we added in order to get the signal into a ham band is called the **carrier**. In the example, the carrier is the 3900-kc. frequency and the bandwidth of the signal is 3897–3903 kc. Since the carrier is between two equal bands, the signal is said to consist of a carrier and two **sidebands**. The sideband from 3900 to 3903 kc. is called the **upper sideband**, being the higher of the two in frequency, while the sideband from 3900 to 3897 kc. is called the **lower sideband**.

Now hold on tight while we go around a sharp curve. Our center frequency, the carrier, actually doesn't carry anything. All of the message is

"..DOESN'T CARRY ANYTHING"

contained in *either* of the sidebands. We could throw away the carrier, or the carrier and one of the sidebands, and still get the complete message. It takes special receiving methods to do it, but it can be done. In fact, it's being done right along. Why, then, speak of a "carrier"? The answer is that the name arose a long time ago because of a particular system of modulation in which it *seemed* as though something was done to the center frequency to make it carry the modulation. The actual fact is that the carrier isn't affected at all by the modulation, if the modulating is done properly.

Amplitude Modulation

The modulation system that gave rise to the carrier idea is called **amplitude modulation (a.m.)** It is a system for doing just what we have described above. But when we come to the actual mechanics by which the signal is modulated it is easy to *think* of it from a different viewpoint. That viewpoint is this: that the amplitude of the carrier (that is, the value of the carrier's current or voltage) is made to follow faithfully the instantaneous changes in the audio-frequency voice voltage. The general idea is shown in Fig. 5-3. In the amateur bands, the carrier frequency is thousands of times greater than the audio modulating frequency, so the carrier will go through a great many cycles during one cycle of even the highest audio frequency we want to transmit. This is shown by the shading in the figure; we couldn't begin to draw in actual r.f. cycles because there would be far too many to be printed.

Now it does seem here as though something actually is being done to the carrier. But this is only because it isn't possible to draw a picture of more than one aspect of modulation at a time. The picture you see in Fig. 5-3 is really a composite one showing the *result* of the action of three separate frequencies in a circuit that will pass all of them. The three are the carrier frequency, the upper side frequency and the lower side frequency. They all add together in such a way as to give the appearance of a single frequency (the carrier) whose amplitude is changing in just the same way that the signal doing the modulating is changing. But appearances are sometimes deceptive, as you know. It would be more accurate to say that the actual modulation process is one of mixing, which you have met in your reading about receivers in Chapter 3. Nevertheless, it's easier to grasp some things about amplitude modulation with the help of a picture such as Fig. 5-3, and we'll take advantage of it. Never forget, though, that actually the carrier does *not* vary in amplitude, and that the modulation is all in the two sidebands.

Modulation Percentage

One of the things that it's easier to visualize by the carrier idea is **percentage of modulation**. In Fig. 5-3 the modulation has just the right value to make the carrier amplitude go to zero

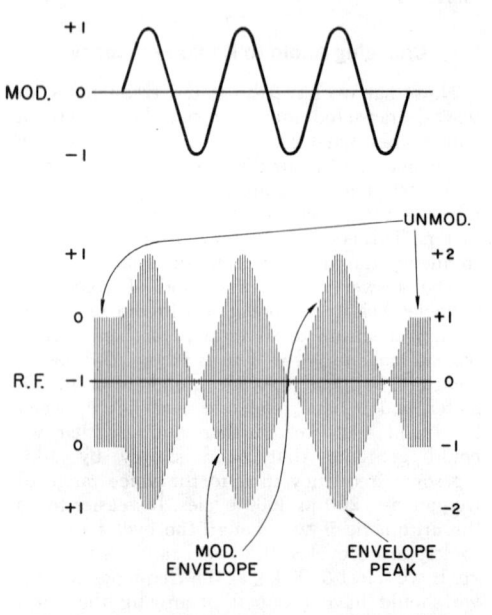

FIG. 5-3—The audio frequency waveform in the upper drawing is superimposed on the r.f. signal, as shown by the outline, or modulation envelope. Note that the modulation envelope is duplicated on the lower side of the r.f. axis. This isn't really a separate envelope; it just looks that way because when the r.f. amplitude is changed, it changes symmetrically both above and below the zero axis. The positive and negative half cycles of r.f. have to be equal.

Amplitude Modulation

on the downswing and to twice its unmodulated value at the peak of the upswing. This is "100 per cent" modulation.

Now look at Fig. 5-4. Here we have the same carrier, but the modulating signal is only half as big as in Fig. 5-3. The r.f. amplitude goes down to one-half the unmodulated value on the downswing, and up to 1.5 times the unmodulated value on the upswing. This is 50 per cent modulation. The per cent modulation is the ratio of the maximum swing of the modulation (either up or down) to the amplitude of the unmodulated carrier. With voice the modulation percentage is continually varying, because some sounds are loud and others are not, all within a word; also, everyone raises and lowers his voice volume within a sentence.

Overmodulation

Can there be more than 100 per cent modulation? Obviously the r.f. output can't be reduced below zero. All we can do is make it *stay* at zero for a more-or-less long period of time. Fig. 5-5 illustrates this. There is a gap during which there is no output. This is bad. The outline of the r.f. signal—the **modulation envelope**—now differs considerably from the waveshape of the modulation. (In Figs. 5-3 and 5-4 the modulation envelopes were identical to the modulating signal.) One result of this is that the signal heard by the receiving operator isn't the same as what was intended to be transmitted. In other words, the modulation is distorted.

Worse, the bandwidth of the transmission has been made greater by this **overmodulation** in the downward direction. This doesn't show in the drawing, of course, because the drawing doesn't show the *frequency* aspects. A simple way of looking at it, however, is that because there is distortion there must be new modulation frequencies present—harmonics. You've been introduced to these in earlier chapters. Since harmonics are always *multiples* of the frequency from which they were generated, they are the same as *higher* frequencies added to the modulation. Higher frequencies mean greater bandwidth. Downward overmodulation increases the bandwidth beyond what is necessary—sometimes far beyond—and thus can lead to unwarranted interference with other stations.

You can modulate upward as far as your transmitter can go. Theoretically there is no limit. Practically, though, most transmitters can't go very far upward, and when they can't the peaks of modulation become **flattened**. The over-all result, in its effect on the bandwidth, is just about the same as equivalent overmodulation downward.

Power in Amplitude Modulation

The reason why most transmitters can't be modulated very far upward will become clear on examining what happens to the power in a modulated signal. Suppose that the transmitter output without modulation is 100 watts. What is it when the signal is modulated 100 per cent as in Fig. 5-3? By inspection, you can see that the power will be zero at the bottom of the downswing. And since the amplitude at the peak of the upswing is twice the unmodulated amplitude, the power at this instant is *four times the unmodulated amplitude,* or 400 watts. (Remember that amplitude refers to current or voltage; when the voltage doubles the current also doubles, in an ordinary circuit.) When the power is averaged over one or more cycles of the modulation frequency, a mathematical study will show that the power in a 100-per-cent modulated sig-

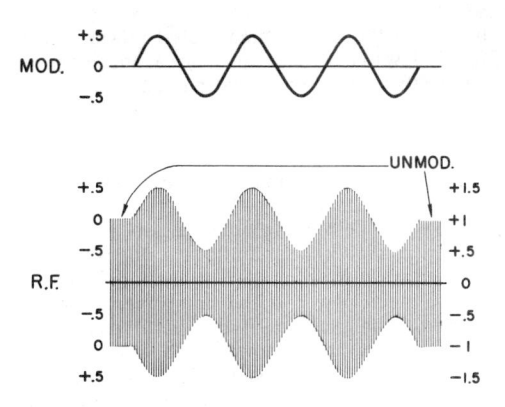

FIG. 5-4—In this case the carrier is not fully modulated, as shown by the fact that the envelope peak is not twice the unmodulated-carrier amplitude while the "valley" in the envelope does not reach zero.

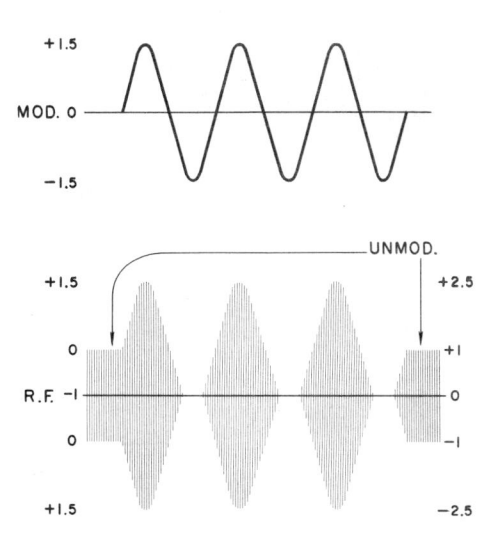

FIG. 5-5—Attempting to modulate more than 100 per cent gives a modulation envelope that is only a distorted reproduction of the original. It also leads to greater signal bandwidth and "splattering" into adjacent channels.

nal is 1.5 times the unmodulated power. Thus the 100 watts without modulation increases to 150 watts with 100 per cent modulation.

Voice Power

The figure of 1.5 times is actually true only of single-tone modulation—that is, modulation by a sine wave. Voice waveforms are not sine waves. For the same *peak* amplitude as a sine wave, an average voice waveform will contain only about half as much power as the sine wave. It would be nearer the truth to say that with voice modulation the average power in a 100-per-cent modulated signal is about 1.25 times the unmodulated power. *But,* the power at the top of the envelope (called the **peak-envelope power**) of a 100-per-cent modulated signal is always four times the unmodulated power. Any two waveforms that have the same peak amplitude have the same peak power, even though the powers averaged over one complete cycle may differ widely. It is these peaks that we must be prepared to furnish during modulation.

The extra power in the modulated signal doesn't come out of nowhere. It has to be supplied somehow in the modulation process. The extra power required is proportional to the square of the modulation percentage. Thus the 50-per-cent modulated signal of Fig. 5-4 would have

"...EXTRA POWER CAPABILITY"

only one-fourth as much extra power as the 100-per-cent modulated one of Fig. 5-3. The latter, for the 100-watt unmodulated power in the example, required 50 extra watts when modulated; modulated as in Fig. 5-4, only 12.5 extra watts, would be needed. On the other hand, the overmodulated signal of Fig. 5-5 is hitting an up-peak that is 2.5 times the amplitude of the unmodulated signal. The peak power needed here is 2.5 squared, or 6.25, times the unmodulated power. Few transmitters are designed to have that much extra power capability!

All the message content of a signal is in the modulation. The higher the modulation percentage the more **talk power** we transmit to the receiving end. The goal is 100 per cent—but without overshooting.

Modulation Circuits

Since modulation is really a mixing process, any of the mixer or converter circuits described in Chapter 3 can be used for generating a phone signal. Instead of introducing two *radio* frequencies into a tube circuit we simply introduce one radio frequency (the carrier frequency) and the voice band of audio frequencies.

Mixer circuits used in receivers are designed to handle a small signal and a large local-oscillator voltage. This means that the percentage of modulation is low. In a transmitter we want to get as close as possible to 100 per cent modulation, and we also want more power output. For these reasons modulator circuits differ in detail from receiving mixers, although much the same in principle.

Screen Modulation

One method that is particularly popular in low-power kit transmitters is **screen modulation**. The circuit shown in Fig. 5-6 is reduced to the bare essentials, on purpose. You can easily get lost in the details of the various screen modulation circuits in use, and thus not be able to see the forest for the trees. The r.f. circuit is that of an ordinary r.f. power amplifier—any type so long as it uses a screen-grid tube. The difference is in the way the tube is *operated.*

If we had a perfect screen-grid tube the output amplitude (voltage or current to the load, not the power) would be perfectly proportional to the voltage on the screen. If we should double the screen voltage, for example, the output amplitude also would double; if we reduced the screen voltage to zero the output would go to zero. This sort of operation is shown graphically in Fig. 5-7. The relative amplitude of the r.f. output voltage or current is shown along the left-hand scale. Up to the point marked A the plot of amplitude versus screen voltage is a

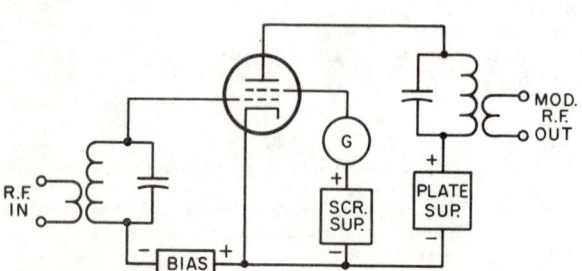

FIG. 5-6—Stripped-down screen modulator circuit.

Screen Modulation

straight line, a fact that leads to calling such a characteristic linear. The corresponding power output, which varies with the square of the amplitude, is shown by the second curve.

Flattening

In practice, the amplitude characteristic won't be a perfectly straight line. However, if the tube's operating conditions are properly chosen the characteristic will be reasonably linear. But if the screen voltage is increased too far the line will start to bend over horizontally. In this region, to the right of A in the figure, the modulation will begin to be distorted. Therefore we have to operate in such a way that the screen voltage doesn't get into this flattening region. In the graph shown, this would mean that the screen voltage shouldn't go over 400 volts.

Now let's relate this to what we learned earlier about power in a modulated signal. If we are to modulate 100 per cent, *point A represents the modulation up-peak*. It is *not* the unmodulated level. Without modulation, the amplitude can only be one-half the peak-envelope value, as shown in Fig. 5-3. So without modulation we have to reduce the d.c. screen voltage to point *B* on the graph, that is, to 200 volts. The power is now down to one-fourth of its peak value, which is where it should be.

Now if a source of audio-frequency power is connected in series with the d.c. screen supply as shown in Fig. 5-6 the audio voltage will add to the 200 volts d.c. on the screen. If the audio voltage has a peak value of 200 volts, it will swing the screen from *B* to *A*, back to *B*, then down to zero, and back to *B* during each a.f. cycle. This modulates the signal 100 per cent, just like Fig. 5-3.

Carrier vs. Talk Power

But suppose we try to set the unmodulated amplitude somewhere between *B* and *A*—at point *C*, say—in order to fool ourselves into thinking we have a bigger signal. If we do, we can no longer modulate 100 per cent, because we can't swing the screen voltage beyond *A* without flattening. At point *C* the d.c. voltage on the screen is 300, so we can swing up only 100 volts without going beyond the flattening point, *A*. This is only half the swing we had from point *B*, meaning that we have only half as much amplitude variation and only one-fourth the talk power. The *unmodulated* power has increased in the ratio of $(300)^2$ to $(200)^2$, or 9 to 4—over twice as much. There is a bigger carrier, to be sure, but to get it we've had to sacrifice the thing we want—talk power.

The Plate Meter

Unfortunately, talk power doesn't show on the transmitter's meters. Usually, the meter reads the modulated amplifier's plate current, which only gives an indication of d.c. input to the plate. Don't be misled by the meter; it has an innocent face but a lying tongue when it comes to modulation.

Actually, the d.c. input to a normally operated amplifier *doesn't change* when it is modulated. It stays just the same at any modulation percentage up to 100 if the **modulation characteristic** is linear. This is because the plate current swings down just as much as it swings up, during each audio cycle, so the *average* current doesn't change.

One thing you *can* tell by the plate meter: if its reading changes when you talk, you're not modulating linearly. You've got distortion and possibly overmodulation.*

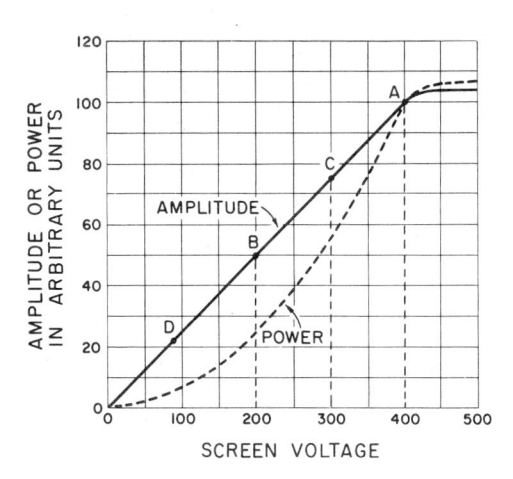

FIG. 5-7—Idealized screen-modulation characteristic. This shows the relationship between screen voltage (which varies instantaneously with the voice signal), output r.f. voltage and current amplitude, and output power.

A Practical Circuit

The simplest and probably the best of the practical circuits for screen modulation is shown in Fig. 5-8. You don't need to be concerned about the part above the dashed line. This is merely an ordinary r.f. amplifier circuit, and it doesn't matter what circuit you use so long as you can adjust the plate-circuit loading and have some way to vary the amplifier tube's grid drive. The modulation part is enclosed in the dashed box.

With small power tubes such as the 807 and 6146 (or a pair of them) as r.f. amplifiers, the modulator tube, V_2, can be any audio power tetrode or pentode of the kind found in receivers. The 6AQ5 is typical of the type of tube that can be used. It operates here just the same way that it would in a receiver; it's simply an audio amplifier. The coupling transformer T_1 should be one designed to go with the modulator tube selected. A driver-type transformer is suitable. The secondary-to-primary turns ratio should be 1 to 1

---

*There is an exception to this (controlled carrier) that we'll take up a little later on.

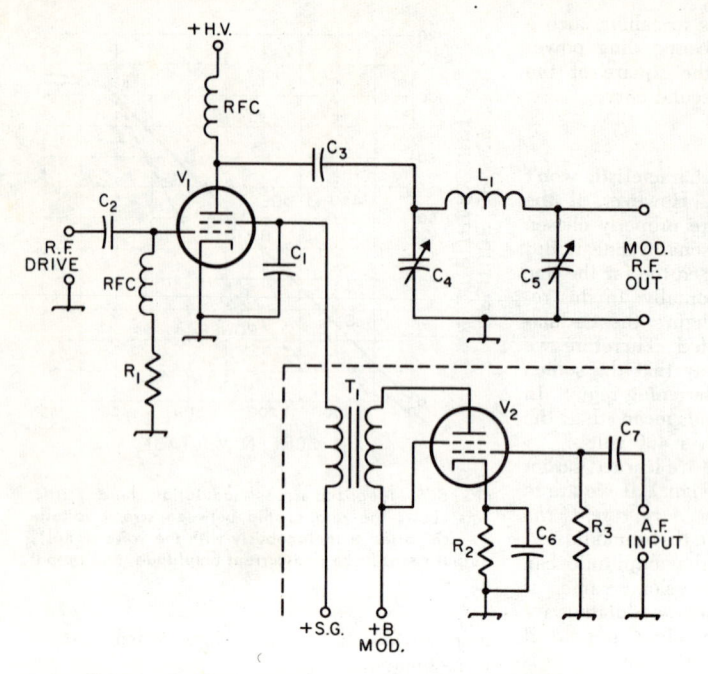

FIG. 5-8 — A practical screen-modulation circuit. Circuit values and adjustment are considered in the text.

or perhaps as high as 1.5 to 1; it shouldn't be necessary to go higher.

Operating Conditions

Now we have an ordinary r.f. amplifier and an ordinary a.f. power amplifier. What makes a modulation system out of them? The answer is the operating conditions of the r.f. amplifier.

First, the grid leak R_1 should have the right resistance for operating the tube as a c.w. amplifier. You can get the recommended grid bias, grid current, and screen voltage for c.w. operation from the published tube data. Suppose the tube is a 6146, to be operated with 600 volts on the plate. Typical operating conditions are given as 180 volts on the screen, a grid-leak resistance of 24,000 ohms, and a grid current of approximately 2.8 milliamperes. Use these values of grid resistance and screen voltage, and tune up the plate tank circuit to get a plate current of 125 ma. with 2.5 to 3 ma. grid current. You should have a dummy antenna (a 40- or 60-watt lamp will do) that will give you some idea of how your adjustments affect the power output.

Adjusting for 100 Per Cent Modulation

Now comes the important part. Decrease your grid drive a little, readjust the plate tuning and loading for the same 125 ma. plate current, and watch the dummy antenna for any change in power output. There probably won't be an observable change if you made a *small* change in grid current, although the output may actually *increase* if you started out with around 3 ma. Cut back on grid current a little more and readjust the plate tank again, always keeping the plate current the same. Continue this until you see a small, but definite, *decrease* in output. This is the point you want. Keep the grid current there.

Now reduce your screen voltage to one-half its c.w. value—to 90 volts in this case. Don't touch anything else. The plate current should drop approximately to half, too; that is, to between 60 and 65 ma.* The lamp dummy will be dimmer. Now you're ready to talk into the microphone. The lamp will flash a bit brighter when you talk. Watch the r.f. amplifier's plate meter while you increase the speech gain, or while you increase your voice level. When you go high enough there will be a flicker in plate current on voice peaks. That's far enough. Any more will put you into the flat region of the curve in Fig. 5-7.

Power

Why did we pick 125 ma. as the plate-current value to use? Choosing the right plate current is a vital part of the process. Look again at Fig. 5-7. Point A on this curve represents ordinary c.w. operation. (The screen voltage used in making up the graph is different, of course; the actual screen-voltage scale will depend on the tube type.) Point B, half the c.w. screen voltage, is where we set the operating point for 100-per cent modulation. Now at this point the output amplitude has dropped to one-half its c.w. value. Although the graph doesn't show plate current, the plate current also will drop to one-half its

---

*The plate current may not drop to half when a lamp is used as a dummy load, because the resistance of the lamp changes when the current in it is changed. This step really should be taken with an actual antenna, or with a dummy antenna that does not change resistance appreciably with current.

Linear Amplifiers

audio power for plate modulation has bedeviled designers ever since there was modulation. There have been innumerable attempts to get around it. Most of them have been like shooting stars—a brief moment of glory and then forgotten. The trouble with most such attempts has been either that they were unworkable, or that they would work but were so complex to get into operation that they were impractical, in terms of amateur communication.

The system you will find in several kit-type low-power transmitters, **controlled carrier screen modulation,** is not really an attempt to equal the r.f. efficiency of plate modulation. It is an effort to squeeze a little more out of screen modulation by taking advantage of the **duty cycle** of speech modulation. The theory is that when you are not talking the d.c. input to the modulated tube can be reduced, because the maximum possible power input is not needed at that time. Then when you do talk the d.c. power swings up, giving you a bigger signal than you would get with the "straight" system described earlier. When not modulating, the input is low and the tube runs cooler.

This system requires some way of automatically controlling the d.c. voltage on the screen. Instead of setting the unmodulated carrier at *B* in Fig. 5-7 it might be set instead at *D*, where the input to the plate will be safely below the rated plate dissipation of the tube. Then, when you modulate, the d.c. screen voltage must swing up to *B* so you get the benefit of 100 per cent modulation over the entire usable length of the curve; that is, from zero to *A*.

Peak-Envelope Power

To show what this might mean in terms of output, let's take the 6146 again. It has a rated maximum input of 90 watts in c.w. operation. With "straight" screen modulation an reach 90 watts peak-envelope input only if the unmodulated carrier level is set at 45 watts. But since the plate efficiency is only about 33 per cent, 30 watts of the 45 would go to heat the plate—more than the rating. However, if the input is reduced to, say, 20 watts when we're not modulating, the tube will run cool enough to stay within ratings.

The difference between 90 watts and 75 watts peak-envelope input isn't great. It makes a difference of only 2 or 3 watts in effective carrier output—that is, 15 watts against 12.5. But it is about all that can be accomplished with controlled-carrier screen modulation unless the screen voltage is swung farther up than the c.w. rating, or unless considerably more than 600 volts is used on the plate. The initial adjustment of the carrier control and modulation percentage is a hopeless task without special test equipment such as an oscilloscope. Cut-and-try adjustment only leads to swinging the d.c. input too high on modulation peaks. This looks wonderful on the plate meter (or even on an r.f. output meter) but results in a signal with a low modulation percentage and, often, excessive distortion and bandwidth.

Linear Amplifiers

Since the audio-frequency power needed for plate modulation is expensive, how about amplifying a phone signal *after* it has been modulated? Then the modulation could be done at a very low power level, where audio power is cheap.

It is perfectly possible to amplify a modulated signal. It's done all the time in receivers, for example, before the signal reaches the detector. The amplifier has to be one that will not distort the modulation envelope, which must be preserved in just the form that it had in the modulated stage.

Therein lies the difficulty in using amplification after modulation. An amplifier that won't distort the modulation envelope—called a **linear amplifier**—has rather poor plate efficiency when amplifying an amplitude-modulated signal. Low distortion and high plate efficiency just don't go hand in hand in this case.

In fact, the linear amplifier has to operate in just about the same way as the screen-modulated amplifier. With the linear amplifier, the modulation envelope of the signal applied to the control grid has to swing up and down about the carrier level, to produce an amplified modulated signal in the plate circuit. In the screen modulator the audio voltage on the screen has to swing up and down about the carrier level to produce a modulated signal in the plate circuit. In both cases the plate efficiency at the unmodulated-carrier level is about 33 per cent.

Whether you screen-modulate a tube or use it as a linear amplifier, you can run just the same input and get the same r.f. output. For example, the 6146 as a linear amplifier will give you 12 or 13 watts of carrier output—the same output that we saw was possible with screen modulation. There is no power advantage one way or the other. Screen modulation is the preferable system because it is easier to adjust.

If you're tempted by the thought of a linear amplifier to follow your screen-modulated 6146 or 807, consider for a moment: The d.c. input you can run on the linear will be about 1½ times the rated plate dissipation of the tube or tubes you pick. The carrier output will be about ⅓ of the d.c. input. If you want to get a carrier output of 100 watts, for instance, you will have to run a d.c. input of 300 watts to the linear amplifier. The plates of the tube or tubes you use in it will have to dissipate 200 watts. This is a pretty large order, and it isn't especially cheap. You'd probably do better to get your 100-watt carrier by plate-modulating an amplifier that would be operating at c.w. (Class C) efficiency and running only about 150 watts input.

It's the r.f. *output* that counts, not the d.c. *input*. Before buying a transmitter, or deciding on a design of your own, take a critical look at this business of modulation methods and their related inputs and outputs.

Getting Rid of the Carrier

In either the grid-modulated amplifier or the linear amplifier, it's the unmodulated carrier that establishes that 33 per cent plate efficiency. A linear amplifier can take a lot more power if it doesn't have to handle a carrier. Earlier we saw that the carrier doesn't actually carry anything. It would be nearer the truth to say that it just goes along for the ride—and it's an expensive passenger.

"...JUST GOES ALONG FOR THE RIDE"

What we set out to do, in generating a radiotelephone signal, was to move an audio-frequency voice band of perhaps 200 to 3000 cycles up into the radio spectrum where it would be useful to us. We saw (Fig. 5-2) that in the practical process of doing so we wound up with a carrier and a second complete voice signal. The carrier, 3900 kc. in the example we used, had one purpose, initially: added to each frequency in the 200–3000 cycle voice band, it got us into the 75-meter phone band. After having done so it is no longer essential.

The Balanced Modulator

The carrier can easily be eliminated from the transmitted signal. It takes two tubes in a rather special form of modulator circuit, one of the many varieties of which is shown in Fig. 5-10. This particular one was picked because you're already familiar with the principles of screen modulation.

Compare this circuit with Fig. 5-8 and you will see a number of similarities. The plate circuit, consisting of blocking capacitor C_3 and the pi-network tank circuit $C_4 C_5 L_1$, is just the same as in Fig. 5-8. (The plates of V_1 and V_2 are simply connected in parallel.) C_1 is the screen bypass (one on each tube, in Fig. 5-10) and R_1 is the grid-leak resistor. The difference between the two circuits is that V_1 and V_2 have their control grids driven with push-pull r.f., and the screens are modulated by push-pull audio. This requires balanced circuits in both cases. The operating conditions—grid drive, plate loading, d.c. screen voltage—are chosen with the object of getting linear modulation, just as in the ordinary circuit.

How the Balanced Modulator Works

What is the consequence of putting push-pull r.f. on the grids and connecting the plates in parallel? If there is no audio signal on the screens, the two tubes each amplify equally. But when V_1's grid is driven positive by the r.f. input, V_2's grid is driven negative, simultaneously. The tubes are driven out of phase. Their outputs are likewise out of phase, and since the plates are connected directly together the output voltages buck each other. In fact, there is no r.f. output at all.

But now suppose that a push-pull audio signal is applied to the screens. When V_1's screen goes positive, V_2's screen goes negative. The two tubes are being modulated oppositely. You can visualize this by referring to the screen-modulation characteristic of Fig. 5-7. If the voltage on V_1's screen goes from B to C while the voltage on V_2's screen goes from B to D, you can see that V_1's output has increased while V_2's has decreased. There is now some output—equal to the *difference* between the outputs represented by C and D on the graph. The maximum difference, and thus the maximum output, is of course obtained when the audio voltage modulates the tubes over the entire linear part of the characteristic—that is, up to A and down to zero.

The Balanced Modulator

FIG. 5-10—One form of balanced-modulator circuit. The carrier bucks itself out in the output circuit because the grids are driven in push-pull while the plates are in parallel. Only the modulation variations —the sidebands— are present in the r.f. output circuit.

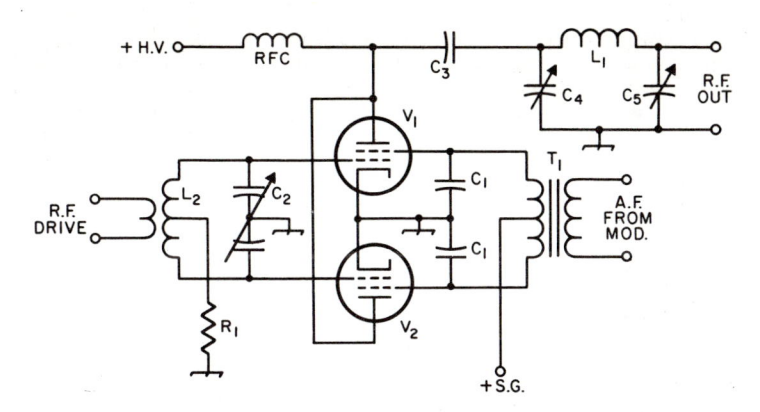

In ordinary modulation the r.f. driving signal applied to the grid is amplified to become the carrier. But here the carrier is made to buck itself out. Only the sidebands—both of them— are left in the output of this **balanced modulator**.

Effect on Plate Heating

Now for the important bearing this has on linear-amplifier operation. Note that there is *no* output when you don't modulate. A following linear amplifier doesn't have to work at all during these silent periods. Thus there is none of the plate heating that occurs when an unmodulated carrier has to be amplified at low plate efficiency. The heating that does occur depends entirely on the intensity of the sidebands at any instant, and this in turn depends on voice intensity. At maximum output from the balanced modulator, corresponding to 100 per cent modulation in Fig. 5-7, a linear amplifier will work at its c.w. efficiency. The plate efficiency does go down at lower voice levels—but the input goes down right along with it.

This, plus the cooling-off periods when you don't happen to be modulating, means that the linear amplifier can handle a peak-envelope power about twice as great as the c.w. input, given the proper operating conditions. In round figures, the peak-envelope power output can be as much as ten times the straight screen-modulated carrier output—and with less tube heating.

We have described the operation of a balanced modulator in terms of putting two regular modulators together. However, eliminating the carrier from the modulator output means that we don't really have to consider it when we set the d.c. screen voltage. The d.c. operating point can be anywhere between *B* and zero, Fig. 5-7, on both tubes. (There would be no point in setting it between *B* and *A*.) Moving the operating setting down toward zero makes higher output possible, and also increases the plate efficiency of the modulator. The latter point isn't hard to appreciate: If there is no output when you're not modulating, all the d.c. input goes to heating the plates of the tubes. But if the d.c. plate input is made very small by using zero, or close to zero, screen voltage, the tubes won't get hot.

Sideband Detection

Here we need to digress a moment from the main subject of this chapter, modulation, because you've undoubtedly been wondering why it is that anybody transmits a carrier when it eats up so much of a transmitter's power capacity. (You may also have been wondering how come "sideband" transmitters are so costly if it's as simple as all you've seen so far.)

The answer is to be found in the receiver, not the transmitter. Here you could easily become lost in a maze of technical details. They're interesting, but not within the scope of this book. We'll take a mountain-top view of the scene.

Recall again that the original problem of phone transmission was to move the voice band of audio frequencies up into the r.f. spectrum. What we wanted to get was simply an r.f. band having a width of 2800 cycles, the difference between 3000 and 200 cycles. In getting such a band, 3900.2 to 3903 kc. in the example, we had to take some other things too.

Imagine for the moment that we do transmit just the band 3900.2 to 3903 kc.—the upper sideband. When this band gets into an ordinary detector, such as the diode used in most receivers, what comes out doesn't sound like human speech. To get the audio back into its original form we have to reverse the process we used for generating the r.f. signal in the first place. That is, we have to *subtract* the 3900 kc. we added. Then we come out with the original audio.

Reinserting the Carrier

In reception, this subtracting is done by combining the 3900.2–3903 kc. signal with a 3900-kc. signal (supplied by the receiver) in the detector circuit. The detector mixes the two signals and lets us take the difference frequency, which is the audio output. (There is also a sum frequency, but it simply disappears in the bypassing; we have no use for it.) To get natural-sounding speech output from the detector the re-introduced frequency, in this case 3900 kc., has to be within a few tens of cycles of the exact frequency used in the transmitter—the **suppressed carrier** frequency.

This requires pretty high frequency stability in both the transmitter and receiver, and also puts a premium on skill in operating a receiver. Now you can see the advantage in sending the carrier along with the sideband. When the transmitter furnishes the local-oscillator signal for correct demodulation, *anybody*—even a small child—can tune the receiver. In amateur radio, though, sending the carrier to save the receiving operator a mite of effort is paid for many times over in wasted transmitter power.

Double Sideband

What if *both* sidebands are transmitted, instead of just one? In such a case we have to get *exactly* on the suppressed-carrier frequency. Both sidebands will give audio output, and the two audios won't be the same unless the reintroduced carrier frequency is exactly right.

Furthermore, the two audio outputs have to be exactly in step with each other. We wouldn't want them to cancel each other instead of adding. It could easily happen. In fact, the requirements are almost impossible to meet without rather elaborate receiving methods. If only one sideband is transmitted, any receiver that has a beat-frequency oscillator can demodulate the signal correctly.*

Sideband Techniques

We'll have to confine ourselves to a brief mention of the various aspects of single-sideband transmission and reception in this book. They're more advanced subjects that you'll come to later in your amateur career.

The balanced-modulator circuit we discussed (Fig. 5-10) can be used to generate a double-sideband suppressed-carrier signal just as it stands. A linear amplifier following such a modulator can work at high input with low tube heating. The signal cannot be received satisfactorily with a receiver that passes both sidebands, for the reasons just given. But if the receiver can pass one sideband while rejecting the other, the signal can be detected with no more difficulty than a single-sideband signal. In effect, the receiver converts the signal from double-sideband to single-sideband. However, the *transmitted* signal is not so converted. It thus takes up twice the spectrum space that the receiver can use.

Eliminating one sideband at the transmitter calls for circuits that are considerably more complex than those used in ordinary a.m. phone transmission. There are at least two basic ways of doing it, known as the **filter** and the **phasing** methods. The filter method makes use of selective circuits—much the same idea as the ways used for getting selectivity in a receiver. It is shown in block form in Fig. 5-11. The phasing method makes use of the properties of the sidebands themselves. By special circuits it is possible to balance out one sideband and let the other through, in what amounts to a kind of neutralizing system. With both methods it is customary to start out with a balanced modulator so the carrier is eliminated before the "undesired" sideband is suppressed.

Reduced Carrier

The carrier doesn't *have* to be suppressed in the s.s.b. system. Nor does it have to be completely suppressed in the double sideband system using a balanced modulator. In both cases it can be transmitted with reduced strength to serve as a **pilot carrier** for the receiving operator. But besides the power advantage we've talked about, there is another reason for eliminating it: If carriers are not transmitted there will be no heterodyne "whistles" set up by them. **Heterodynes,** caused by two or more carriers that happen to be close enough in frequency to give an audio-frequency beat note in the detector—much like beat-note reception of c.w. signals—are by far the most potent cause of interference in phone work. So eliminating the carrier not only gives you a more effective transmitter but helps you in reception, too. The price you pay for this is small, and one in which you can take some pride—developing skill in operating your receiver.

Frequency Modulation

You're no doubt aware of **f.m.—frequency modulation**—because it is the kind of modulation used for the sound channel that accompanies TV, and also for v.h.f. sound broadcasting in the 100-Mc. region.

The amplitude-modulation systems considered so far in this chapter operate by **making the amplitude of the r.f. signal vary in the same way**

---

*In the example given here, the locally-introduced frequency was assumed to be identical with the suppressed-carrier frequency, to keep the details simple. In actual reception with a superhet receiver the incoming signal would be converted to an intermediate frequency, and an equivalent suppressed-carrier frequency would come from the regular b.f.o.

Frequency Modulation

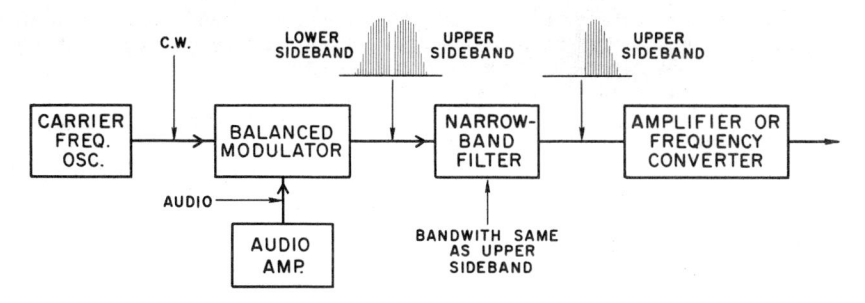

FIG. 5-11—Generating a single sideband signal by the filter method. The filter frequently uses quartz crystals or mechanical resonators to allow a 3000-cycle (approximately) band of radio frequencies to pass through, while suppressing all other frequencies to a high degree. This drawing shows the upper sideband passing through while the lower sideband is suppressed. The sideband selection could be reversed by moving the carrier frequency about 3 kc. so the lower sideband would fall within the filter's passband.

that the amplitude of the audio-frequency voice currents varies. In frequency modulation the varying amplitude of the a.f. signal causes the *frequency* of the r.f. signal to vary correspondingly. The amplitude of the r.f. stays just the same whether or not there is modulation.

The job of an f.m. modulator, therefore, is to convert a.f. amplitude variations into r.f. output frequency variations. The simplest circuit, known as the **reactance modulator**, has rather interesting behavior. The plate-cathode circuit of an amplifier tube is connected across the transmitting oscillator's tank circuit, as in Fig. 5-12. Its grid is also connected to the same tank circuit, through capacitances C_1 and C_2. When C_1 and C_2 have the right values, the tube's r.f. plate current is out of step with the r.f. voltage that the tank puts in its plate. This, as we have seen in Chapter 2, is also true of the current and voltage in a coil or capacitor. In Fig. 5-12 the tube simulates the action of a coil in parallel with the oscillator tank coil, L_1. The gain of the amplifier is varied at an audio-frequency rate by an audio voltage also applied to the control grid. This swings the tuning of the oscillator tank at the same a.f. rate, and with it the oscillator frequency.

Bandwidth

The **swing** in frequency with modulation can be almost any amount we please, within reason. **Wide-band f.m.** usually means a frequency **deviation** of about five times the highest audio frequency. For example, if the highest audio frequency is 3000 cycles, wide-band f.m. would mean that the frequency would shift 5 times 3000, or 15,000 cycles (15 kc.), up and down about the unmodulated carrier frequency. Thus a signal on 53,000 kc. would swing between 52,985 and 53,015 kc. at maximum modulation, and would occupy a band 30 kc. wide. This is too much bandwidth to be permitted on the lower frequencies, because of interference. On such frequencies amateurs must use **narrow-band f.m.**, which is defined as an f.m. signal which occupies no greater bandwidth than a normal a.m. signal. An f.m. signal meets the

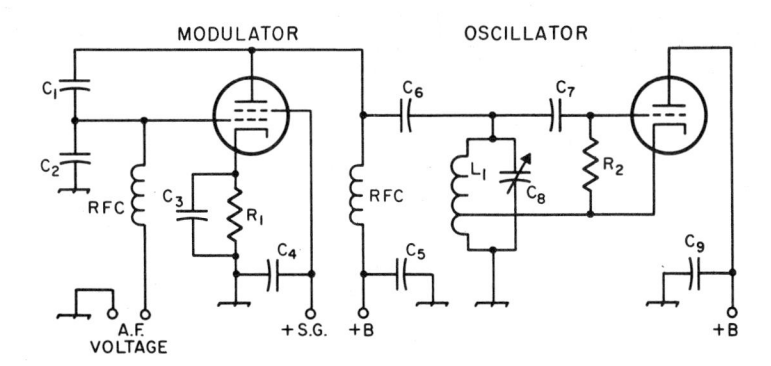

FIG. 5-12—Reactance modulator circuit. The r.f. chokes in the modulator circuit prevent short-circuiting the r.f. on the grid and plate of the modulator tube. R_1 and C_3 are the normal cathode resistor and by-pass capacitor for biasing the tube as an amplifier. C_4 and C_5 are screen and plate bypasses, respectively. C_6 is a blocking capacitor between the modulator plate and the oscillator tank circuit C_8L_1. Other components in the oscillator circuit have their usual functions; any type of tuned oscillator may be used.

narrow-band requirement if its deviation is held down to about 0.6 times the highest audio frequency—that is, to about 1800 cycles deviation.

Frequency modulation requires special methods of reception if its possibilities are to be fully realized. Chapter 3 goes into this aspect briefly. Amateurs use f.m. principally at v.h.f., because it helps to reduce interference in nearby broadcast receivers. Being designed for a.m. reception, these receivers are insensitive to f.m. signals.

Phase Modulation

Phase modulation (p.m.) is closely related to frequency modulation. It is hard to tell one type of signal from the other, when you hear them on the air, unless you are well acquainted with the peculiarities of each system. The difference is principally in the way the two types of modulators respond to various audio frequencies. Phase modulation can be converted to frequency modulation, and vice versa, in relatively simple audio circuits.

F.M. and P.M. Amplifiers

Both f.m. and p.m. can be put through frequency multipliers without affecting the linearity of the modulation. The frequency deviation is multiplied in the same ratio as the carrier frequency is multiplied. Also, the amplifiers can work at high efficiency—actually, they run Class C, just like a c.w. amplifier. However, *narrow-band* f.m. doesn't put any more power in the sidebands than you can get with screen modulation. Add the fact that reception is inefficient with receivers not designed for f.m., and you have the principal reasons why f.m. is used but little on the lower frequencies, in spite of its apparent advantages.

Chapter 6

Antennas and Feeders

Some years ago a widely-used textbook on radio engineering began a chapter on antennas with this statement: "An understanding of the mechanism by which energy is radiated . . . involves conceptions which are unfamiliar to the ordinary engineer." Obviously radiation must be a stiff subject. So in this book we simply ask you to accept the well-known fact that energy *is* radiated in the form of electromagnetic waves. We won't attempt to explain why.

In studying for the Novice license you were introduced to wavelength and frequency. The formula is:

$$\text{Wavelength in meters} = \frac{300}{\text{Frequency in Mc.}}$$

In Fig. 6-1 the transmitter is generating a radio-frequency voltage, indicated by the sine wave in the upper drawing. When the voltage is applied to an antenna, energy is radiated into space and travels away with the speed of light. As shown by the lower drawing, it covers a certain distance—one **wavelength**—in the time the voltage takes to go through one cycle.

Current in a Wire

This relationship between wavelength and frequency has a very practical use. Suppose we connect an r.f. ammeter in the center of a wire having a length L, as in Fig. 6-2. Further, suppose that by some means we introduce r.f. energy of adjustable frequency into the wire. If the frequency is gradually raised, it will be found that the current indicated by the ammeter will also rise, at first. But after reaching a maximum at some frequency, f, the current will start to go down again if we continue to raise the frequency. This is the sort of thing we found to happen in an LC circuit, as discussed in Chapter 2. The wire, in fact, acts like a resonant circuit. It is tuned to the frequency, f, for which its length is equal to one-half wavelength. If the wire is 40 meters long, for example, it would be resonant at the frequency for which 40 meters is one-half wavelength. From the formula above, this would correspond to a resonant frequency of 300/80, or 3.75 Mc.

A wire such as this is called a **dipole**, when its length is of the order of a half wavelength, or less. One exactly a half-wavelength long is called a **half-wave dipole**. Very often, the simple term "dipole" is used when a half-wave dipole actually is meant.

Two Practical Points

Before going farther, it is well to translate this into a more familiar unit of length, the foot. Converting units and changing to a *half* wavelength gives us

$$\tfrac{1}{2}\text{ wavelength in feet} = \frac{492}{\text{Frequency in Mc.}}$$

or

$$\text{Resonant frequency in Mc.} = \frac{492}{\text{Length in feet}}.$$

The second point is this: These formulas are not quite accurate for an actual wire. They apply only to a wave traveling in space. In a practical half-wave antenna the difference amounts to

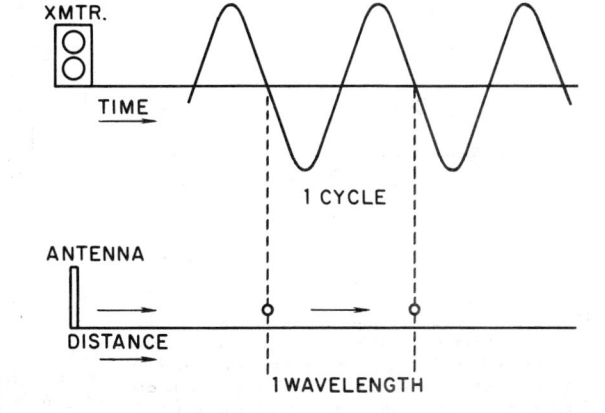

FIG. 6-1—One wavelength is the distance that radiated energy will cover, traveling at the speed of light, during one cycle of the radiated frequency.

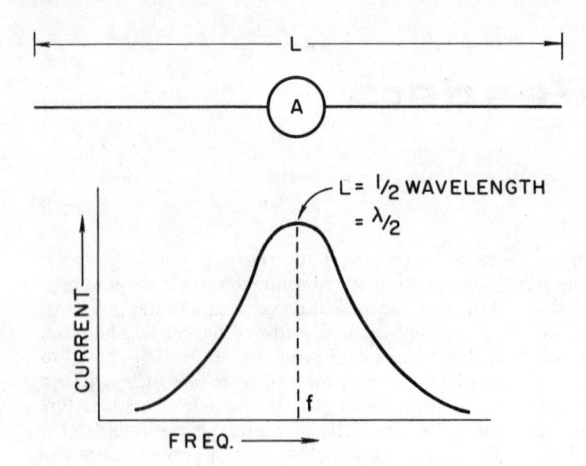

FIG. 6-2—The r.f. current at the center of a wire is highest when the wire length is equal to one-half wavelength.

about 5 per cent, on the average. Thus an *average* formula for resonant length would be

$$\tfrac{1}{2} \text{ wavelength in feet} = \frac{468}{\text{Frequency in Mc.}}$$

Remember that this is only an average. In a particular case actual resonant length might differ by a few per cent from the length given by this formula. The difference is usually small enough to have little practical effect.

Electrical Length

Electrically, the length given by the last formula is a half wavelength because it is a *resonant* length, even though it is physically short of being a half wavelength in space. We can account for the difference in length by the fact that energy does not travel quite as fast along a wire as it does in **free space**.

In some cases, as you will see later when we get to transmission lines, there can be quite a marked difference between electrical wavelength and free-space wavelength. When you see the length of an antenna or line expressed in terms of wavelength you can safely assume that an electrical measure is being used, unless it is made plain that free-space measure is meant.

Enter Time

In Chapter 2 we dealt with circuits that offered a complete path around which electrical energy could move. A wire such as is shown in Fig. 6-2 doesn't offer any such path. How is it that current can flow in it?

In the "closed" circuits of Chapter 2 it was assumed, without our having said it in so many words, that electrical energy traveled around the circuit so rapidly that its action could be taken to be instantaneous. In the circuits we use in transmitters and receivers for frequencies up to 30 Mc., at least, this is a satisfactory assumption. As long as the circuit is small compared with the wavelength (the wavelength corresponding to the frequency we happen to be using) the action is instantaneous, for all practical purposes.

But an antenna such as the wire in Fig. 6-2 is *not* small compared with the wavelength. If the length L is one-half wavelength, a length of time equal to one-half cycle of the applied frequency is needed for energy to go from one end of the wire to the other. Imagine a voltage applied to the left-hand end of the wire at an instant when the voltage is at the positive peak of the cycle. A voltage impulse will go along the wire to the right, reaching the end one-half cycle later. But at this instant the *applied* voltage has moved on to its *negative* peak.

Standing Waves

In other words, when the left-hand end of the wire is negative the right-hand end is positive, and vice versa. Also, when the voltage reaches the end of the wire it comes to the end of the track. There is no place for it to go except back over the same path. The energy is **reflected** from the end. In going back, it combines with energy— from a later part of the cycle—that is going out.

Fig. 6-3 shows what happens in a wire one-half wavelength long. All the components of

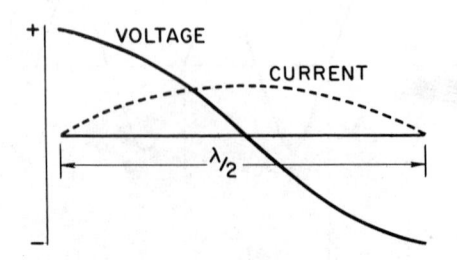

FIG. 6-3—The current and voltage along a half-wave wire have different values at all points along the wire. When plotted as shown above, the graphs are wave-like in shape, and since their positions are fixed with respect to the wire they are called standing waves.

Antenna Fundamentals

voltage, those traveling out and those reflected back, add up to make a **standing wave** of voltage. If we could go along the wire with a meter for measuring r.f. voltage we should find that the voltage is highest at the ends of the wire and is practically zero at the center. Between the ends and the center it gradually decreases. When plotted against length, as in Fig. 6-3, it is like part of a sine wave.

Polarity

The plus and minus signs on the scale at the left can be somewhat misleading. One end of the antenna isn't always positive, and the other end isn't always negative. In fact, both ends alternate between positive and negative each r.f. cycle. What the picture tries to show is that *when* the left-hand half of the wire is positive the right-hand half is negative. The reverse is also true. The voltages in the two halves of the antenna always have opposite polarity.

On the other hand, if we went along the wire with a meter for measuring the r.f. current, we should find that the current is zero at the ends. This you might expect, since current can't flow off the wire into space. The current gets larger as we move toward the center, and is largest right in the middle of the wire. In the drawing, the current is shown entirely on the plus side of the scale. Again this shouldn't be taken literally; it actually goes from positive to negative and back again each cycle. The picture means that the current is always flowing in the *same* direction, at any given instant, throughout the entire length of a half-wave wire.

Nodes and Antinodes

The point where the amplitude of a standing wave passes through zero is called a **node**. Thus in Fig. 6-3 the standing wave of voltage has a node at the center of the antenna. The standing wave of current has two nodes in this figure, one at each end of the wire.

A point of maximum amplitude is called an **antinode** or, sometimes, a **loop**. (Properly, the term loop refers to the entire segment of the standing wave between two nodes.) The standing wave of voltage has antinodes at the ends of the wire in Fig. 6-3, while the current antinode is at the center.

Note that where there is a current antinode there is a voltage node, and where there is a current node there is a voltage antinode. Also, an antinode of current is one-quarter wavelength away from a current node; similarly with the voltage. These two statements are true, in general, of all standing waves along wires.

Longer Wires

The tuned circuits you met in Chapter 2 were resonant at just one frequency, that for which the inductive and capacitive reactances were equal. An antenna isn't quite so simple. If the things shown in Fig. 6-3 happen when a wire is a half wavelength long because of the *time* it takes energy to surge back and forth, it seems reasonable to expect that another half wavelength of wire added to the first will see a repetition of these same events. And so it is. There will be a repetition each time a half wavelength is added.

Fig. 6-4 shows the **current** and **voltage distribution** when the wire is two half-waves (or one wavelength) long and three half-waves (1½ wavelengths) long. At the ends of each half-wave section the voltage is high and the current is zero. In the middle of each such section the current is high and the voltage is zero. But there is a difference between two adjacent half-wave sections. You can see that when the voltage at the left end of the first section is positive, as shown, the voltage at the left end of the second section is negative. It has to be the same as the voltage at the righthand end of the first section, of course, since the two sections are connected together. Also, when the current in the first section is positive, as shown, the current in the next section will be negative. That is, the currents in adjacent half-wave sections flow in opposite directions. This is called a **phase reversal**.

Phase

In the third section, shown in the bottom drawing of Fig. 6-4, there is again a phase reversal. This brings the phase relationships in this section back to exactly what they are in the first

FIG. 6-4—Harmonic resonance. The upper drawing shows the standing waves on a wire one wavelength long; the lower shows them on a wire 1½ wavelengths long.

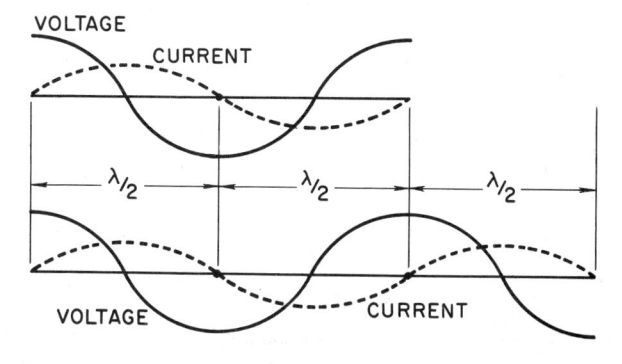

section. In other words, *alternate* half-wave sections have identical standing waves of current and voltage on them. They are said to be **in phase**. *Adjacent* sections are **out of phase**. This goes on no matter how many half-wave sections are added to the wire.

Harmonic Resonance

Each of these sections is just as much resonant to the applied frequency as another. In effect, we have two resonant antennas end-to-end in the upper drawing of Fig. 6-4, and three in the lower drawing. These are called **harmonic resonances,** since they occur at the same frequencies as the harmonics of a fundamental frequency. That is, they are integral (whole-number) multiples of the fundamental.

In the case of an antenna, the fundamental frequency is the one for which the entire wire length is equal to one-half wavelength. For example, an antenna that is a half wavelength long at 7150 kc. will be two half-waves long at 14,300 kc. (second harmonic), three half-waves long at 21,450 kc. (third harmonic), and so on up the scale. The actual multiples are approximate, not exact, integers. The resonant frequencies will differ slightly from exact harmonics. The reasons are the same as given earlier, in the discussion of the length of a practical antenna.

Grounded Antennas

A half wavelength is the shortest length of wire that will be resonant to a given frequency, if the wire is simply considered by itself. However, if we connect one end of the wire to earth, the grounded end is no longer "free." We can't raise the potential of the earth itself, so the voltage at the grounded end has to be zero. On the other hand, we *can* make current flow into the earth. Thus the earth can be made to act as a substitute for one half of the half-wave antenna.

Fig. 6-5 shows this. The current is large at the earth connection, and decreases to zero at the open end of the antenna. The voltage is zero at the bottom and has its greatest value at the top. But the length *L* for this antenna is only a *quarter* wavelength, at resonance. So a grounded antenna need be only half as long as a dipole antenna to be resonant at the same frequency.

Antenna Impedance

Impedance, as it was defined in Chapter 2, is equal to voltage divided by current. When the current and voltage both change as we move along the antenna, as they do in Fig. 6-3, the impedance also is different everywhere along the antenna. Therefore, if we want to talk about antenna impedance we have to specify the point at which it is measured.

The customary place to measure the impedance of a simple antenna is at the center of the wire. In Fig. 6-6 an r.f. generator, *G*, is inserted in series with the antenna at its center. The voltage from the generator will cause a current, *I*, to flow; this current has the same value on both

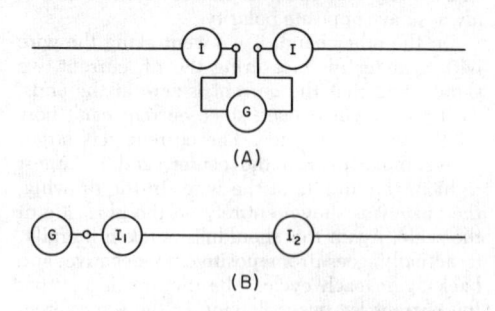

FIG. 6-6—A half-wave wire driven at the center behaves like a series-resonant circuit. One driven at the end acts like a parallel-resonant circuit.

sides of the terminals. The antenna behaves like a circuit having resistance, inductance and capacitance in series. At the resonant frequency of such a circuit the inductive and capacitive reactances cancel each other, as you saw in Chapter 2, leaving only the resistance. This is also true of the antenna. Thus at its resonant frequency the antenna "looks like" a simple resistance, and it is at this frequency that the current is largest. A half-wave antenna has a resistive impedance, measured at this point, in the neighborhood of 70 ohms. It is rarely exactly 70 ohms in any practical case, because the actual resistance depends on the same factors that affect the resonant frequency.

If the frequency is moved off resonance the impedance rises, just as it does in a series *LC* circuit. It also becomes complex—there is reactance, now, along with the resistance.

Now suppose the r.f. generator to be connected to one end of the antenna, as in Fig. 6-6B, with one ammeter at the end and the other at the center. As the frequency is varied the current I_2 will reach its highest value at resonance, where the antenna is a half wavelength long. But the current I_1 at the terminal where the generator is

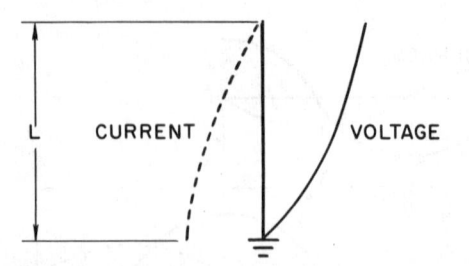

FIG. 6-5—Grounding one end of the antenna chops off one-half of the standing wave—that is, the length *L* need be only a quarter wavelength for the antenna to be resonant.

Antenna Impedance

connected will be *smallest* at this frequency. As seen by the generator, the antenna is just like a parallel LC circuit. That is, at resonance its impedance is maximum, and is a simple resistance. As the frequency is moved away from resonance the current I_1 increases; the impedance at this point becomes smaller and is again complex, containing both reactance and resistance.

Impedance Values

Antennas are usually fed with r.f. power either at the center or the end. Thus the two cases illustrated by Fig. 6-6 have some practical importance. The resonant impedance at the end is much more dependent on the thickness of the antenna conductor and other such factors than is the impedance at the center. Values can range from a few hundred to several thousand ohms. The thicker the conductor the lower the resistance as viewed from the end. At the center, the effect of conductor thickness on the resistance, at resonance, is relatively minor.

The impedance of a grounded antenna usually is measured between the earth and the bottom of the antenna. Like the center-fed antenna with free ends, the grounded antenna acts like a circuit having *L, C* and *R* in series. As the antenna is only half as long, for the same resonant frequency, the resistance is only half as great. That is, it is in the neighborhood of 35 ohms, for an antenna a quarter wavelength long.

This assumes a "perfect" ground—one that has extremely low losses at the operating frequency. Ordinary ground is far from perfect, and the earth connection usually adds quite a considerable amount of resistance to the system—often as much as 25 ohms. The ground resistance can be reduced by burying a large number of wires, having a length of about a half wavelength, going out from the base of the antenna like the spokes of a wheel. To be effective, though, a really large number of them—several dozen—has to be used.

The Nature of Antenna Resistance

In Chapter 2 it was emphasized that resistance, as defined in broad terms, is something in which power is used up—usefully or otherwise. The resistance of an antenna divides into two parts, one useful and one not. The useful part is called **radiation resistance**. The power used up in this resistance is the power actually radiated into space from the antenna. The non-useful part of the resistance is represented by losses, partly in the conductor (because of its ordinary resistance at the operating frequency), partly in insulation associated with the wire, and partly in conductors and dielectrics close enough to the antenna to be in a strong electromagnetic field. These are lumped together and often called the **ohmic** resistance. Power dissipated in ohmic resistance is turned into heat.

Since only the power used up in the radiation resistance is useful, we want the radiation resistance to be much larger than the ohmic resistance. It is the *ratio* of the former to the latter, rather than the actual values in ohms, that is of interest. We may measure different values of total resistance at different points along a given antenna, but the ratio of the two components of the resistance does not change. In other words, it does not matter where power is introduced into the antenna; the same proportion will be radiated, and the same fraction lost, in every case.

Why Impedance is Important

Since it is only the *ratio* of radiation resistance to ohmic resistance that counts, you would be justified in concluding that the actual value of resistance is unimportant. This is so in the antenna itself. But another factor must be taken into account. Somehow, r.f. power must be put into the antenna before there can be any radiation. In feeding power to the antenna the actual antenna resistance—or impedance—*is* important.

R.f. circuits using practical components work at best efficiency when the impedance level is between perhaps 25 and 2000 ohms. These are not exact limits by any means, but do indicate the general range. If the impedance is only an ohm or two, or is many thousands of ohms, the losses in the circuits themselves may be far greater than the power that can be delivered through them to a load. And between the plate of the transmitter's final-amplifier tube and the antenna itself there must be circuits—often several of them. Each exacts its toll of power.

The resistance of a half-wave antenna is about 70 ohms, as we have mentioned. This value is well within the optimum range for minimizing the losses in any circuits we may use to match the antenna to the final amplifier. Furthermore, it is nearly all radiation resistance. Ohmic resistance amounts to only a few per cent of the total if the antenna is mounted in a clear spot.

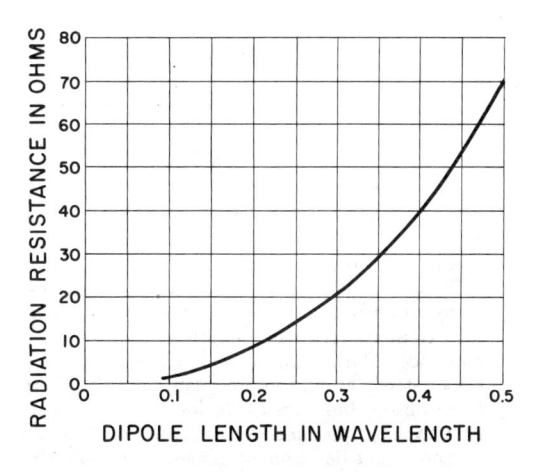

FIG. 6-7—Radiation resistance measured at the center of an antenna as the length of the wire is varied. Lengths here are in terms of free-space wavelength.

However, the radiation resistance decreases if the antenna is shortened. For example, if a dipole is a quarter wavelength long its radiation resistance as measured at the center is only about 14 ohms, as shown in Fig. 6-7. If the length is shortened to one-eighth wavelength the resistance drops to around 4 ohms.

Coupling Losses

If the same power can be put into all these values of resistance, all of the power will be radiated. However, the "if" is a big one. The half-wave antenna is resonant, and so needs no tuning. The shorter antennas are not resonant; their impedances have large amounts of reactance along with resistance. In order to put power into a short antenna the reactance has to be "tuned out," by adding the same value of reactance, but of the opposite kind, at the antenna terminals. A short antenna has capacitive reactance, so inductive reactance has to be added to cancel it, as in Fig. 6-8. But coils inherently have resistance, and a coil of the size needed for tuning a 1/8-wave antenna, for instance, will have more resistance than the radiation resistance of the antenna itself. As a result, more power is used up in heating the coil than is radiated by the antenna.

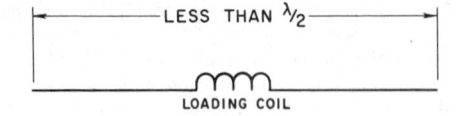

FIG. 6-8—Inductive loading of a short antenna to make it resonant. The shorter the antenna the greater the inductance required. The term loading, as used in this connection, has nothing to do with the type of loading (for power transfer) discussed in Chapter 2. It dates from early radio times, and refers to tuning a circuit—usually by adding inductance—to a lower frequency than the one to which it is naturally resonant. The natural resonance in this case would be that of the wire without the coil.

Aside from considerations such as these, there is nothing sacred about the resonant length. The antenna will radiate just as well whether or not it is resonant. However, it will not *get* all the power output of the transmitter if it is so far off resonance that the tuning apparatus uses up an appreciable portion of the power.

Beginners often take antenna resonance far more seriously than it warrants. A small departure from the resonant length is of little consequence. The resistance and reactance change rather slowly around the resonant point, so there is no observable increase in loss if the antenna isn't exactly resonant. As a matter of fact, an antenna can't be resonant at more than one single frequency. Yet it isn't by any means necessary to use different antennas for each frequency in an amateur band.

Directivity

Offhand you might think that the strength of the signal radiated from an antenna would be the same in all directions—up, down, and to all sides. It isn't. The radiation is stronger in some directions than in others. This comes about because the ends of the antenna always have opposite polarity, and because the antenna is not just a point but has a length that isn't small compared with the wavelength.

You can think of it as a case of timing, or phase. The electromagnetic field from one part of the antenna doesn't reach a distant point at the same time as the field from another part. In an extreme case, the fields reaching such a distant point may even get there with the same amplitude but *opposite* polarity. Then they add up to zero; there is no radiation in that direction. Or, in another direction, the fields may reach the distant point with the same amplitude and the *same* polarity. Being "in phase," they add together to give the strongest field the antenna is capable of producing. In still other directions, neither of these conditions is met completely, so the strength of the signal has an intermediate value.

Directive Patterns

This rather complex operation is summed up in what is called the **directive pattern** of the antenna. The pattern is a graph showing the relative strength of the radiation in all directions. We can't show a pattern completely on a sheet of paper, since the paper has only two dimensions, while the antenna actually radiates into all the space surrounding it. Antenna patterns usually are a "slice" or cross section of the full pattern.

Fig. 6-9A shows typical cross-sectional patterns for a half-wave dipole. The arrows marked 1, 2 and 3 show, by their length and direction, the relative strength of the radiated field. Don't forget that this drawing is a slice; in order to visualize the complete pattern you would have to imagine that the pattern rotates around the antenna wire, in and out of the paper, to form a doughnut with a point, not a hole, in the middle. Then when you turn the antenna on end, as in B, a slice at right angles would give you just a circle, as shown.

Taking these two patterns together, you can see that a *horizontal* half-wave antenna will radiate best directly upward and downward (if you are looking at the antenna from the side) and won't radiate at all directly off the ends. If you imagine yourself *over* the antenna in A, it radiates best at right angles to the direction in which the wire runs. On the other hand, if you are looking directly down on a *vertical* antenna, as in B, the antenna is radiating equally well in all directions. These last directions, of course, are along the ground, going around the compass.

If the antenna is shorter than a half wavelength the patterns will still have much the same

Antenna Patterns

(A)

ANTENNA WIRE

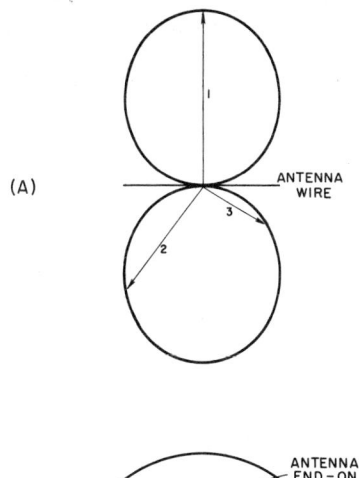

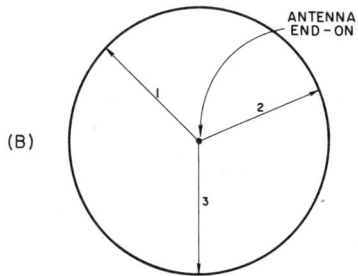

(B)

ANTENNA END-ON

FIG. 6-9—Cross-sections of directional pattern of a half-wave antenna. A—in the plane in which the wire lies; B—in a plane cutting through the center of the wire at right angles to it.

shape. However, if the length is two or more half wavelengths there are rather drastic changes. Figs. 6-10A and 6-10B show, respectively, the patterns for the "full-wave" and "three half-wave" antennas whose current and voltage distribution are shown in Fig. 6-4. The maximum radiation is no longer broadside to the wire but goes off at an angle, as you can see by comparing these drawings with Fig. 6-9A. These, too, are cross-sections of a solid pattern that you can visualize by imagining the cross-section drawing to be rotating around the antenna.

The Earth's Part

Since the antenna radiates in all directions, some of the energy must go toward the ground. The earth acts more-or-less like a huge reflector for radio waves. The rays hitting it bounce off much like light rays from a mirror. These reflected rays combine with the direct rays from the antenna at a distance. The result is that the directive pattern of the antenna is modified by the presence of the earth "mirror." Just what the mirror does depends on the height of the antenna above it, and whether the antenna is horizontal or vertical.

Fig. 6-11 shows a couple of typical cases for a half-wave antenna. The patterns at the left show the relative radiation when you view the antenna from the side; those at the right show the radiation pattern you would "see" when you look at the end of the antenna. Changing the height from one-fourth to one-half wavelength makes quite a difference in the upward radiation —that is, the radiation at high angles. The **radiation angle** is measured from the ground up.

Fig. 6-12 shows what happens to the pattern of a vertical half-wave antenna sitting on the ground. Here the maximum radiation is along the ground.

Lest you take these pictures too seriously, we have to warn you that the ground isn't like the mirror on your wall. It's pretty foggy, as a matter of fact. In other words, it isn't by any means the perfect reflector that these pictures assume it to be. The fogginess is principally the result of energy losses; a fairly husky proportion of the wave energy striking the ground is used up in the ground resistance. The principal effect of this is that you don't get the radiation at very low angles that Fig. 6-12 would lead you to expect. Practically, there isn't a great deal of difference between horizontal and vertical antennas in this respect, if the horizontal is a half wavelength or more above the earth.

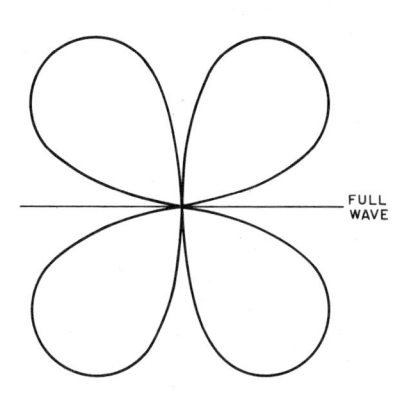

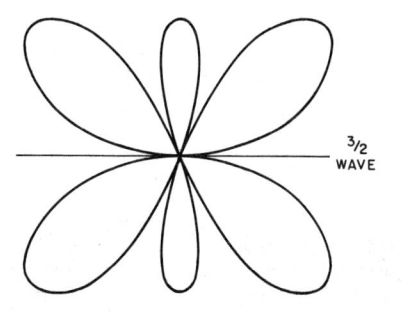

FULL WAVE

3/2 WAVE

FIG. 6-10—Cross-sections of directional patterns of (A) a full-wave antenna and (B) one having a length of 1½ wavelengths. The cross-sections correspond to the one in Fig. 6-9A, in relationship to the antenna wire.

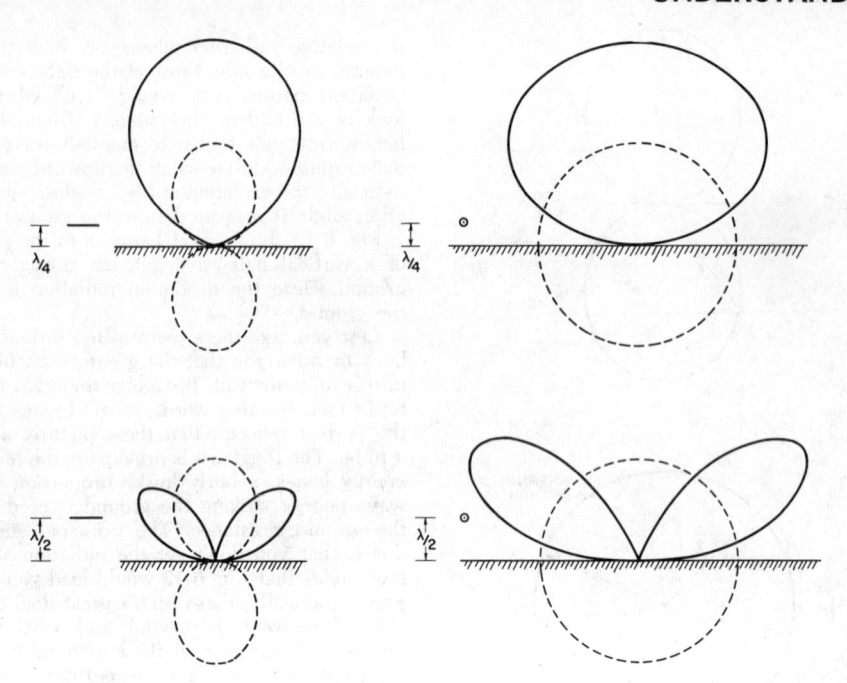

FIG. 6-11—Effect of the ground on the radiation from a horizontal half-wave antenna, for heights of one-fourth and one-half wavelength. Dashed lines show what the pattern would be if there were no reflection from the ground.

Wave Paths

From what we've just said you may have concluded—and rightly—that most of the radiation from the antenna goes up toward the sky. This being so, how does a signal get back to earth at a distance?

Fortunately, there is a reflector in the sky. At least, there is one that operates pretty regularly at frequencies below about 30 Mc. It doesn't operate at v.h.f. except now and then on the 50-Mc. band. This is the reason why the v.h.f. bands differ from those at the lower frequencies in the distance ranges that normally can be covered. On v.h.f. it takes highly-developed equipment for working very far beyond the optical horizon as determined by the antenna heights. (It can be done, though.) But on those frequencies for which the sky reflector works, communication is possible over very long distances even with low power.

The Ionosphere

The sky reflector is really not a reflector, technically speaking. It is a region in the upper atmosphere where the paths of radio waves are bent so the signals come back to earth. This region is known as the **ionosphere**. In the ionosphere, energy from the sun breaks up or **ionizes** the thin atmosphere into electrically charged particles which collect in several separate layers. The two principal ones are at heights of about 60 miles and 150 miles, respectively. These layers have the same effect as mirrors, but mirrors of a special kind. They are better mirrors for longer wavelengths than for short ones, and are also better for waves striking them at a glancing angle than for waves hitting them head on.

Even an elementary recital of the various effects that are associated with the ionosphere would occupy a good-sized chapter. Wave propagation is a whole subject in itself—a fascinating one, too, and one that accounts for much of the charm of operating your own radio station. Here we must content ourselves with saying that the picture of wave travel, in simplest terms, is like Fig. 6-13. The waves radiated by your transmitting antenna travel up at some angle to the ionized layer, are bent downward in the layer, and come back to earth at the distant receiving station.

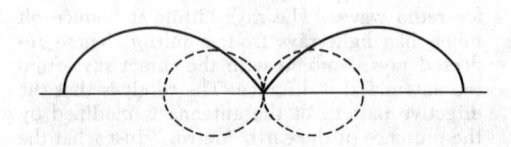

FIG. 6-12—Effect of the ground on radiation from a half-wave vertical antenna. In the absence of the ground, the pattern would be like the dashed line.

Transmission Lines

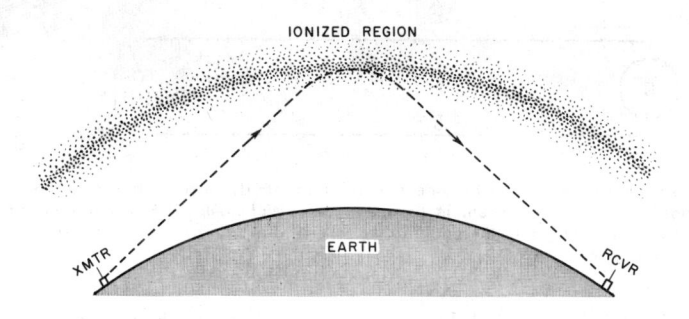

FIG. 6-13—A wave entering the ionosphere is bent back toward earth, when suitable conditions exist, to reach a distant point.

Wave Hops and Skip Distance

In this picture the signal got there in one **hop**. In many cases the returning signal hits the earth, is reflected upward again to the ionosphere, and comes down again still farther away. This can be repeated a number of times—enough times to carry the signal to the most distant parts of the earth, if you pick the right transmitting frequency and the right time of day. The time of day is important because the ability of the ionosphere to reflect signals depends on the sun.

The fact that a signal leaving the transmitting antenna at a low angle is more readily reflected than one going more-or-less directly upward leads to an interesting result. There are times (depending on the transmitting frequency again) when signals don't come back at short distances from the transmitter. This no-signal region around the transmitter is called the **skip zone**, because signals skip over it before coming back to earth. When there is "skip," you can work the longer distances with relative ease, but not the shorter ones. The skip zone may extend out for as much as 2000 miles from your station when you're using the very highest frequency that can be reflected at all. The **skip distance** is much shorter on lower frequencies such as 7 and 3.5 Mc. For much of the time, on these frequencies, there is no skip zone at all. As a rule, you can work the longest distances most easily on those frequencies for which the skip distance is greatest.

Transmission Lines

To radiate effectively, an antenna ought to be up in the air as high as it can be put. Also, it should not be close to houses, power lines and the like. You may not have an ideal spot, but even so you probably won't have to bring the antenna right into your operating room.* So in most cases the situation is this: The antenna is "out there" and the transmitter is "in here"; how is the r.f. power to get from the transmitter to the antenna?

The answer, of course, is a transmission line. Your 60-cycle power comes to you through a transmission line, too. However, there is a difference in the way r.f. lines and 60-cycle lines operate. The reason is the difference in wavelengths. One wavelength at 60 cycles is over 3000 *miles*. If we wanted to build a half-wave antenna for that frequency it would have to extend more than half way across the United States. So even though you may be 20 miles from a power station, you're only a very small fraction of a wavelength away. The time it takes for power to reach you is so short, compared with $1/60$ second (one cycle), that the standing-wave effects discussed earlier in this chapter are negligible.

But in transmitting power at a frequency of, say, 7 Mc., the time taken for the power to travel 50 feet isn't at all negligible compared with the duration of one cycle. This means that we can't look upon a transmission line as a simple electrical circuit, which we *can* do at 60 cycles. What is happening at the "far" or "output" end of the line may be quite different from what is happening at the "near" or "input" end at the same instant.

The "Infinite" Line

A useful concept in explaining transmission-line operation is the **infinite line**. This is an imaginary line consisting of two conductors, side by side and close together, extending so far that we can never reach the end.

If an r.f. voltage is applied to the input end of such a line, one terminal will be negative whenever the other is positive, and vice versa. This causes the current to flow in one direction in one wire and in the other direction in the second, as in Fig. 6-14. Because the currents flow in oppo-

---
*This has been done; moreover, it is quite possible to "work out" with an indoor antenna. However, it's better to put it outdoors if you can.

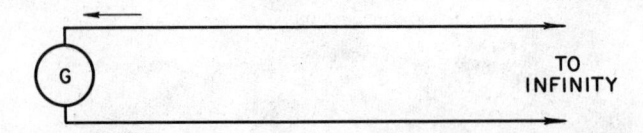

FIG. 6-14—An imaginary two-conductor line extending to infinity. Arrows show that the current in one wire flows in the opposite direction to current in the other; this *relationship* is true throughout the entire length of the line, although the actual currents periodically reverse direction as the polarity of the generator's voltage reverses each half cycle.

site directions, the electromagnetic fields set up by them are also opposite. The fields therefore cancel each other's effects, or nearly do so (there is always a *little* uncancelled field, because the two wires can't actually occupy the same spot). Since the fields cancel, there is no radiation from the line.

Thus all the energy put into the line travels away from the generator, following the line at almost the speed of light. And since the line is infinitely long, none of the energy ever comes back.

Characteristic Resistance

Probably the first question you'd ask at this point is this: If the generator voltage is known, how much current will flow in the line? From the discussion of the meaning of resistance in Chapter 2, you would be right to infer that such a line must act like a resistance, since energy is being taken continuously from the generator. But how many ohms?

This resistance, called the **characteristic resistance** of the line, has nothing to do with the actual resistance of the conductors. While it may seem odd, the fact is that it is a function of the inductance and capacitance per unit length of line. The resistance actually is determined by the line's *L/C* ratio. This ratio depends on the diameters of the conductors and the spacing between them. The smaller the conductor diameter and the wider the spacing, the higher the characteristic resistance. Practical values of resistance lie between about 150 and 800 ohms for a "two-wire" or **parallel-conductor** line as shown in Fig. 6-14.

It is important to realize that this characteristic resistance does not itself consume any power. The power is merely *following* the line on its way to infinity. The characteristic resistance is simply the ratio of voltage to current all along the line. Since the line is imaginary anyway, we can imagine further that the conductors have no actual resistance and there is no other energy loss along the line. Thus all the power put into the line is delivered to infinity, wherever that may be. This means that the characteristic resistance is "pure" resistance—no reactive effects at all.

Characteristic Impedance

But what if the conductors do have resistance of their own? Practically, of course, they must have. Also, the practical insulation between the two conductors is not perfect; there is some leakage between the two wires. This leakage is equivalent to a resistance (a high value) shunted across the two conductors. In the topsy-turvy world of transmission lines the presence of these two components of resistance gives rise to *reactance*. So if the line is a practical one having losses, the generator doesn't see a pure resistance but sees an impedance containing both resistance and reactance. This is called the **characteristic impedance** of the line.

Because things get complicated at this stage, we like to ignore the reactive part of the characteristic impedance, and do so by assuming that the line has no losses. As long as the losses per unit length are small we can get away with it. Fortunately, this is the case with lines used by amateurs at frequencies below 30 Mc. It is even a good-enough assumption in the lower v.h.f. range. When the losses are small the characteristic impedance is *very nearly* a pure resistance equal to the characteristic resistance. The term characteristic impedance is widely used to

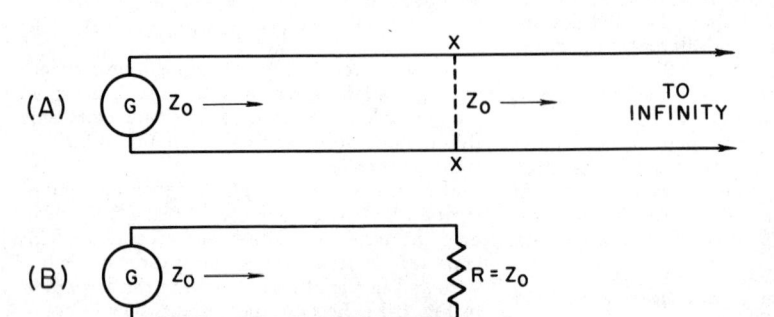

FIG. 6-15—An infinitely-long line can be simulated by terminating an actual line in its characteristic impedance.

Antenna Impedance

mean the characteristic resistance of a lossless line. We'll use it that way here, too.

The Terminated Line

An infinite line, even if we could have one, wouldn't be of any practical use. It happens, though, that a line can be tricked into *thinking* that it's infinitely long.

In Fig. 6-15A, suppose that the line is cut at XX. If the generator is moved up to this point it will still see the same characteristic impedance (which is commonly designated Z_0), since what is left of the line to the right of XX is still infinitely long. In the same way, the section of line to the left of XX "sees" the section to the right of XX as a resistance equal to the characteristic impedance. This is true anywhere along the line. It suggests the idea that the line section to the left of XX wouldn't know the difference if a resistor having the same value as the characteristic impedance were substituted for all the line to the right of XX.

This is actually so. If a line of any length is **terminated** in a resistance equal to its characteristic impedance the voltages and currents are just the same in that section as they would be if the line were infinitely long. If the line has no losses, all the power put into it at the generator end is delivered to the terminating resistance.

Matching

The terminating resistance doesn't have to be a resistor. It can be any device, such as an antenna, that uses up power and thus has an equivalent resistance. If the power-consuming device doesn't inherently have the right value of resistance to match the line, its resistance can be transformed by means of circuits (such as those described in Chapter 2) that will make it "look like" the proper value. Matching of this sort is done more often than not; only occasionally does the load have the right value of resistance, in itself, to match a practical line impedance.

One final point about a **matched line:** If the line has negligible losses, an ammeter inserted anywhere along its length will give the same reading. Also, a voltmeter connected across it at any point will give the same reading. There are no standing waves of current or voltage such as we find along an antenna, even though the line may be many times longer than the antenna. But this is true *only* when the line is terminated in its characteristic impedance.

Standing Waves on Lines

Now let's look at a line that *doesn't* simulate one that is infinitely long. The length of a matched line didn't matter, because all the power kept going in the same direction—to the load. If the line is not matched, its length becomes quite important.

To take an extreme case, suppose the line just stops, as in Fig. 6-16. The power goes out from

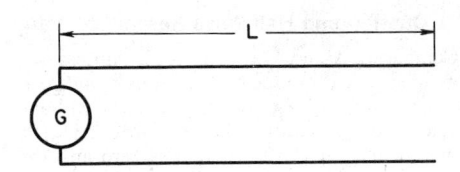

FIG. 6-16—A line with no termination—simply an open circuit.

the generator to the open end, at which point it has no path left to follow except to turn back and head toward the generator. This it does, just as in the case of the antenna discussed earlier. In coming back it sets up standing waves of voltage and current, just as it did along the antenna.

Here, too, the current and voltage distribute themselves along the line according to the wavelength. If the line length L is just one-quarter wavelength, the current and voltage distribution are as shown in Fig. 6-17A. If you will imagine the line to be unfolded so that the wires extend in opposite directions from the generator, you can see that this is the same voltage and current distribution as we found along a half-wave antenna, Fig. 6-3. The line, too, is resonant to the generator's frequency. The total length, for both wires, is still a half wavelength, although the line as a whole is only a quarter wave long.

Odd Lengths

If the line is less than a quarter wave, as in Fig. 6-17B, there is room only for the outer sections of the standing waves. The line is not resonant in this case. The generator sees it as a reactance, and in order to put maximum current into the line the reactance must be tuned out by adding reactance of the opposite kind. Inductive reactance is needed here for **loading** the line.

In Fig. 6-17C the line is more than a quarter wave long. Here we have not only the standing waves we had along the quarter wave line but the beginning of another set, too. This line is not resonant, either, and again it looks like a reactance to the generator. However, in this case its reactance must be tuned out by using capacitance for loading.

Finally, Fig. 6-17D shows a line a half wavelength long. Each wire is like a half-wave antenna. Since one terminal of the generator is always positive when the other is negative, and vice versa, the voltages and currents are always opposite in polarity along the wires, just as in the other cases. The half-wave line is also resonant at the applied frequency, since each wire will accommodate exactly a complete standing wave, no more and no less.

This could be continued on for still longer lines. In doing so we should find that the line is always resonant when its length is exactly a multiple of one-quarter wavelength. It is *not* resonant at any other lengths.

Quarter- and Half-Wave Resonance

Comparing A and D in Fig. 6-17, you can see that there is a difference even though both can be considered to be resonant. In A the voltage is zero at the generator, but the current has its highest value. In D the current is zero and the voltage has its highest value. Since the impedance seen by the generator is equal to voltage divided by current, the impedance at the input end of the line must be extremely low in A and extremely high in D. If there were no power lost in the line the impedance values would be zero and infinity, respectively. However, no line can be completely free from loss, so we don't have to worry about what might be meant by zero and infinity. Practically, the impedance is a very low resistance in A and a very high resistance in D.

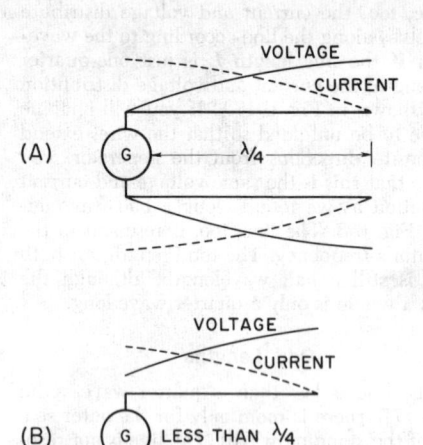

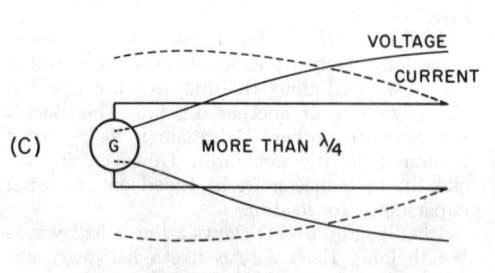

As you may remember from Chapter 2, the same descriptions were applied to series- and parallel-resonant *LC* circuits. A quarter-wave line open-circuited at the far end acts like a series-resonant circuit. A half-wave line open at the far end acts like a parallel-resonant circuit.

The Short-Circuited Line

Instead of being left open at the far end as in Fig. 6-16 the line could be short-circuited as in Fig. 6-18. Once again, energy traveling out from

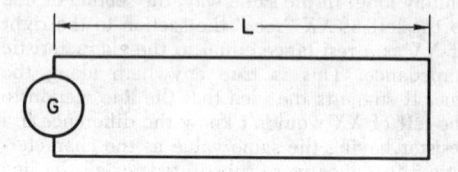

FIG. 6-18—Short-circuited line.

the generator must turn back when it reaches the short circuit. However, in this case there can be no voltage across the short circuit, although the current can be large. This is just the reverse of the open-circuited case of Fig. 6-16.

If you will look at Fig. 6-17D, you will see that just the same condition exists at the point ZZ, one quarter wavelength from the end of the open line. The voltage between conductors is zero (if there are no losses) at this point. This means that a short-circuit could be placed across the line at ZZ without disturbing the currents or voltages. Since it is a quarter wavelength from ZZ back to the input end of the line, this section of line also is resonant.

It is apparent from this that what the generator sees when looking into a quarter-wave short-circuited line is the same as what it sees when looking into a half-wave open-circuited line. That is, a quarter-wave short-circuited line is equivalent to a parallel-resonant circuit. The voltage and current distribution are as shown in Fig. 6-19A.

By carrying on this line of thought it is easy to demonstrate that a half-wave short-circuited line is equivalent to a series-resonant circuit.

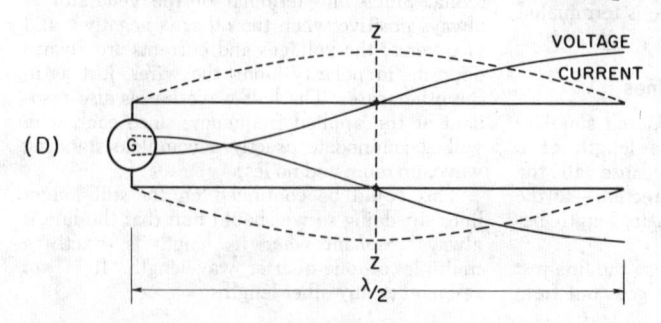

FIG. 6-17—Standing waves along open-circuited lines.

Transmission Lines

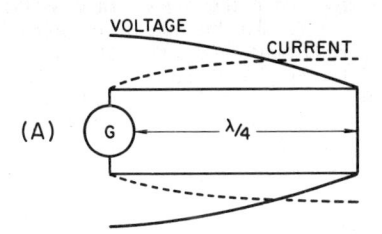

FIG. 6-19—Voltage and current distribution along resonant short-circuited lines.

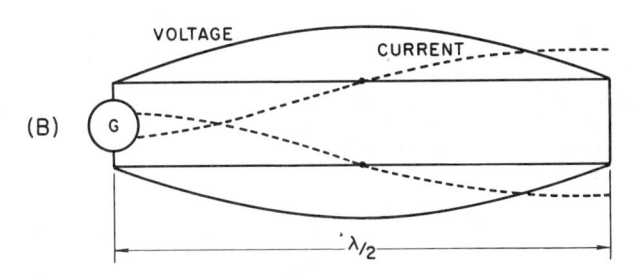

The current and voltage distribution are given in Fig. 6-19B. Lines having in-between lengths are not resonant, and will act like almost pure reactances. Table 6-I summarizes this.

Table 6-I
Transmission-Line Behavior

Length	Open-Circuited Line	Short-Circuited Line
Less than ¼ wavelength	Capacitive Reactance	Inductive Reactance
¼ wavelength	Series-resonant circuit	Parallel-resonant circuit
Between ¼ and ½ wavelength	Inductive Reactance	Capacitive Reactance
½ wavelength	Parallel-resonant circuit	Series-resonant circuit

The line behavior goes through the same series of changes with each added quarter wavelength.

Why Open- and Short-Circuited Lines?

Offhand, you might think that open- and short-circuited lines are about as useless, practically speaking, as an infinitely-long line. However, the fact is that they are quite useful.

In the first place, a resonant line can be substituted for a resonant circuit, and often is. The resonant line is especially useful at v.h.f. and u.h.f., where it may offer the only resonant-circuit structure that it is physically possible to use. Here is where the multiple resonance that goes with a series of quarter-wave sections often saves the day. A conventional LC circuit does not have this feature, and there is a limit to how large, physically, such a circuit can be made for a given frequency.

Second, nonresonant sections of line can be used in place of coils and capacitors, simply by adjusting the length to give a desired value of inductive or capacitive reactance. This is frequently done in antenna matching systems.

Finally, there are applications where multiple resonance in a line lets us do things like short-circuiting a harmonic of the transmitter while the fundamental frequency goes through unaffected. For example, a short-circuited line having a length of one-quarter wavelength at the fundamental frequency has a very high impedance—nearly an open circuit—and can be connected across another transmission line with little effect on the power flowing through it. But at the second harmonic it is a half wavelength long, and it will act as a short circuit across the other line at that harmonic (and all other even harmonics).

Mismatched Lines

You have seen that the power put into a matched line nearly all gets to the load at the output end. A small amount is used up by the losses in the line itself; this is converted into heat. We are assuming here, of course, that the line conductors are so close together that there is no radiation because of incomplete cancellation of the fields. If the spacing between the conductors is of the order of $\frac{1}{100}$ wavelength this is a good assumption, providing the currents and voltages in the line are **balanced**. Line balance means that the current and voltage in one wire are exactly duplicated in the other, except for reversed polarity.

But what if the load connected to the far end

of the line does not exactly match the line's characteristic impedance? A case like this falls somewhere between the perfectly-matched condition and the extremes of the open- and short-circuited lines. Some of the power reaching the end is absorbed by the load, but some of it also bounces back toward the input end. A **mismatch** is said to exist when the load resistance isn't the same as the line's characteristic impedance. The worse the mismatch, the greater the proportion of power reflected back.

Losses

The principal effect here, at least in transmitting, is that the line uses up a little of the power each time the power travels back and forth. But even though some of the power is handed back to the generator (the transmitter) we can still put the same total power *into* the line.

This is simply a matter of the coupling between the transmitter and line. The coupling that would deliver the transmitter's output to a matched line won't do it if the line isn't matched. But by changing the coupling as required, the transmitter can be loaded just as well. A little less power will reach the load than would get there if the load matched the line properly, because of the extra line loss. But the difference on this account is too small to cause any worry, if a low-loss line is used. Even with lines which, when matched, have fairly high losses, the *extra* loss caused by mismatching isn't much if you aren't mismatched by a factor of more than 3 or so.

On a perfectly-matched line there are no standing waves because no power is reflected from the load end. On open- or short-circuited lines there are large standing waves. Along such lines the voltage and current go to zero, or very close to it, at the nodes.

When a line is mismatched, but not open- or short-circuited, there are standing waves because some of the power is reflected. But only *some* of it. The reflected voltage and current can't completely balance out the **incident** voltage and current (the voltage and current traveling *to* the load) at the nodal points unless there is just as much coming back as is going out. Since this is not the case, there are no points of zero voltage and current along the line. Instead, there will be points of *minimum* current and points of *minimum* voltage. Likewise, there will be points where the voltage and current will be maximum.

Standing Waves on Mismatched Lines

If we went along a mismatched line measuring the amplitudes of the current and voltage, without paying any attention to polarity, we would find that both vary along the line. Fig. 6-20 is typical of what might be measured. The points of maximum and minimum are still one-quarter wavelength apart, as in the cases discussed before. The ratio of the current at B, a maximum point, to the current at A, a minimum point, is called the **standing-wave ratio**. Measurement of the maximum and minimum voltages would give the same ratio as measurement of current.

If very little power is reflected from the load —i.e., the line is nearly matched—there is relatively little variation in the current and voltage along the line, so the standing-wave ratio— usually abbreviated to s.w.r.—is low. The greater the mismatch the greater the reflected power and the larger the s.w.r.

S.W.R. and the Load

It happens that the standing-wave ratio can be measured more readily than the current or voltage, or even the load resistance. So it is customary to measure the s.w.r. in order to find out whether the line is matched. There is a very simple relationship between load resistance, the characteristic impedance of the line, and the s.w.r.:

$$S.W.R. = \frac{R}{Z_o} \text{ or } \frac{Z_o}{R}$$

where R stands for the load resistance and Z_o stands for the line's characteristic impedance. The reason for the choice in this formula is that it is customary to put the larger number on top,

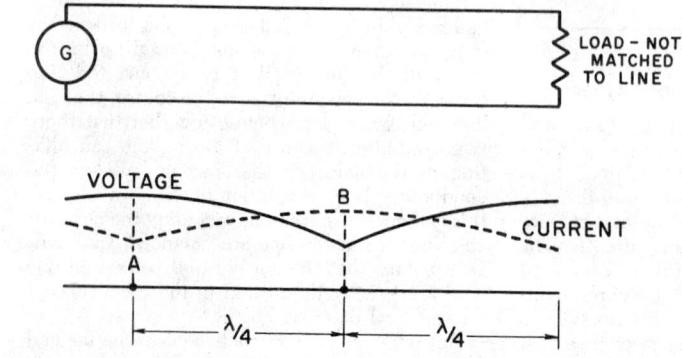

FIG. 6-20—The standing-wave ratio is the ratio of the current amplitude at B to that at A, or of the voltage amplitude at A to that at B.

Transmission Lines

so that the s.w.r. is expressed as, for example, 5 to 1, rather than 1 to 5.

Actually, you don't need to know R at all in making most adjustments of load resistance. If you're shooting for no reflected power—that is, an s.w.r. of 1 to 1, meaning that the maximum and minimum values are the same—you adjust for the smallest possible s.w.r. When you have it, you know you're right.

Fig. 6-20 shows the voltage high and the current low at the load. It could be the opposite. The drawing is for the case where the load resistance is larger than Z_0. The reverse would be true for a load resistance smaller than Z_0. The first case approaches the open-circuited line as R is made larger, and the second approaches the short-circuited line as R is made smaller.

With a mismatched load resistance, as in the cases discussed earlier, the generator sees a pure resistance when the line is some multiple of a quarter wave in length. Thus this same length indicates resonance. At all other lengths the generator will see reactance along with resistance. Table 6-I can be used to find the *kind* of reactance, if the short-circuit column is used for loads less than Z_0 and the open-circuit column is used for loads greater than Z_0.

Resistance Only!

Finally, a warning: To avoid confusing you with a lot of qualifications, in what was said above we have omitted one very important point. *The load has to be a pure resistance if any of this is to be true.*

Mostly, you will be working with loads that are "pure," or nearly so. You can't get an s.w.r. of 1 to 1 unless the load *is* a pure resistance; any reactance in it throws the whole thing off. So if you've been able to get the s.w.r. to 1 to 1 or close to it, you can take it for granted that the line behavior will be as described. If not, you can't find out where you stand without a much more detailed knowledge of transmission lines than we can give you in this book.

Practical Lines

Quite a few varieties of manufactured transmission lines are available. The ones that are of interest to amateurs are usually in stock at radio supply stores, since they are also used for television receivers. There are two general types. One is the parallel-conductor type we used for purposes of discussion in the foregoing part of this chapter. The other is the **coaxial line**. This also has two conductors, but one of them is a tube and the other is a wire centered in it.

The coaxial line, familiarly known as "coax" (pronounced with two syllables), obeys the same laws as the parallel-conductor line. All we have said so far applies to both types of line. However, the coax line has some distinctive features. The current is carried by the inner conductor and the *inside surface* of the tubular outer conductor. The *outside* surface is "cold" for r.f.,

if the line is properly used. In other words, the active part of the line is shielded from outside influences. This means, too, that there can be no radiation from the inside of the line.

Substantially all coaxial line in use by amateurs is the flexible type having a braided-wire tube for the outer conductor. Multistrand wire is often used for the inner conductor, although in some small-diameter lines a solid wire can be used without affecting the flexing. The insulation between the two conductors is a flexible solid plastic—polyethylene.

Velocity Factor

The presence of this solid insulation does two things: It increases the power loss, as compared with air insulation, and it reduces the speed at which power can go through the line. This means that the wavelength in coax cable is shorter, for the same frequency, than in air. The formula for wavelength given earlier has to be modified by a correction factor, called the **velocity factor**, on this account. For polyethylene-insulated solid-dielectric coax the velocity factor is 0.66. A line one-half wavelength long at 7.1 Mc., for example, would be 0.66 times 69.4 feet (a half wavelength in space), or 45.8 feet long.

Line Losses

If we should divide a line into sections of equal length and measure the power going in and coming out of each, we should find that there is the same *percentage* loss in each section. Suppose that 100 watts goes into the first section and 10 per cent of it is dissipated in heat in the line. Then 90 watts comes out to go into the second section. In the second section 10 per cent represents 9 watts, so now we have 81 watts going into the third section. This section loses 8.1 watts, and so on. This sort of power change is exactly what the decibel represents so nicely, so we can express line loss as so many decibels per unit length. The custom is to give the loss in decibels per 100 feet of line.

The loss becomes greater as we go higher in frequency. Losses in db. per 100 feet for the lines most used by amateurs are given in Table 6-II. These losses are for lines that are properly matched by the load. If there is a mismatch the loss will be higher. However, as we said earlier, the additional loss isn't usually serious unless the mismatch is 3 to 1—that is, an s.w.r. of 3 to 1—or more. Even then it is not considerable unless the line has high loss when matched.

Parallel-Conductor Line

The most common type of line is the parallel-wire TV lead-in, consisting of two wires separated by a web of polyethylene approximately ⅜ inch wide. It is sold under several trade names, and has a characteristic impedance of about 300 ohms. As shown by Table 6-II, its

Table 6-II
Transmission Lines

Type	Description	Characteristic Impedance, Ohms.	Velocity Factor	Matched Loss in Db. per 100 Feet							
				3.5 Mc.	7 Mc.	14 Mc.	21 Mc.	28 Mc.	50 Mc.	144 Mc.
RG-58/U	Small coaxial	53.5	0.66	0.68	1.0	1.5	1.9	2.2	3.1	5.7
RG-59/U	Small coaxial	73	0.66	0.64	0.9	1.3	1.6	1.8	2.4	4.2
RG-8/U	Medium coaxial	52	0.66	0.3	0.45	0.66	0.83	0.98	1.35	2.5
TV Twin Line, Standard	Parallel-cond., solid insulation	300	0.82	0.18	0.28	0.41	0.52	0.6	0.85	1.55
TV Ladder Line, 1-in. spacing	Parallel-cond. air-insulated with spacers	450	*	*	*	*	*	*	*	*

*Not known. Velocity factor app. 95 per cent. Losses very low in comparison with solid-insulation types.

losses are lower than the losses in coax. This is true of good-quality line, which you can be sure of getting only when you buy a well-known brand. Some of the "bargain" unbranded line is very poor, so it is best to steer clear of it.

The lowest-loss line available is the ladder type, consisting of parallel wires separated about an inch. The wires are held apart by small rods of polyethylene at intervals of a few inches. Thus most of the insulation is air, which has negligible loss.

There are many other types of line, both coaxial and parallel-wire, than those listed. Some have different characteristic impedances, and a few varieties have lower losses or greater power-handling ability. However, the types mentioned are easy to get, and are satisfactory for the majority of amateur installations of medium power.

Putting the Antenna and Line Together

The half-wave dipole is the basis for most amateur antenna designs. Different types of lines can be used to feed power to it. The line should just carry power to the antenna and not get into the radiating act itself. When this is so, and the dipole does all the radiating, one dipole is the same as another no matter how power may be fed to it. This obvious fact is too often overlooked. Amateurs frequently let themselves be dazzled by some trick name tacked on a dipole-plus-feeder combination, but names don't do the radiating.

The best place to feed a half-wave dipole is at the center. The dipole is a balanced antenna —that is, it is symmetrical about its center. To maintain this symmetry a balanced line—i.e., a parallel-conductor line—should be used. The dipole *can* be fed at one end, but this also upsets the symmetry of the system.*

If the impedance at the center of the antenna matches the characteristic impedance of the transmission line the two can simply be connected together and the line will operate without standing waves. One advantage of this matched operation is that the line length has very little effect on the coupling required between the line and the transmitter. Another is that the losses in the line are least, for a given length, when the line is properly matched. The line losses can either be very important or completely unimportant. They are quite important at v.h.f. even when the best possible job of matching is done. They are unimportant at the lower frequencies, even with a considerable mismatch. The only exception here is when a major error is made in selecting the proper type of line for the use to which it is to be put.

A matched antenna system is actually matched only for one frequency. At best, the system will stay matched over only a small band. As the 7-, 14- and 21-Mc. bands are narrow, in terms of percentage, an antenna that is matched at the center of one of these bands should work over the entire band without having the s.w.r. get too large at the band edges. But you can't do quite as well with antennas of this type on 3.5 and 28 Mc. Here it is best to cut the antenna for the section of the band that interests you most.

Matched Antenna Systems

Since the dipole has a center impedance of

---
*An exception to this is when *two* dipoles are fed from a parallel-conductor transmission line. An example is described shortly.

Antennas

about 70 ohms, it will match a line having a characteristic impedance of 70 ohms, or something close to it. (A small—i.e., 20 per cent or so—discrepancy doesn't cause any difficulty. When 70-ohm or 75-ohm line is mentioned it is to be understood that any impedance in that immediate vicinity is meant.) There is a polyethylene-insulated two-conductor 75-ohm line available for this purpose (Amphenol 214-023). The antenna is simply cut in the center and a connection made to each line conductor, as in Fig. 6-21A.

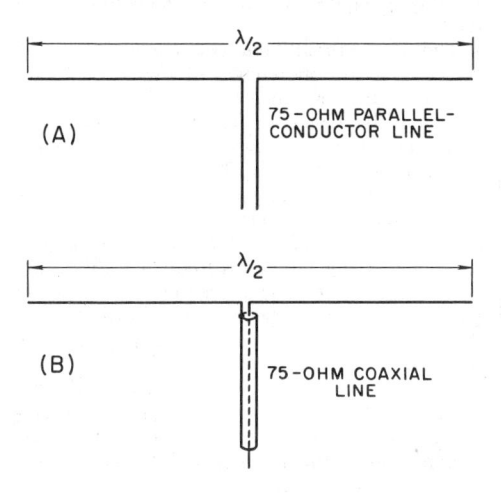

FIG. 6-21—Using 75-ohm line to match the center impedance of a half-wave antenna. The antenna length in feet is equal to 468 divided by the frequency in megacycles.

In spite of the fact that it is desirable to keep the system balanced, a good many amateurs use 75-ohm coax for the same purpose, as in Fig. 6-21B. This is not the best practice, although it will work. One side of the dipole is unavoidably connected to the *outside* of the outer conductor as well as to its inside. This makes the outside of the coax line a part of the antenna system. Thus the outside of the line radiates—but not in any predictable way, because everything depends on where and how the line is installed and how long it is. The principal thing to be said for this system is that the coax line is easy to get.

Very often, 52-ohm (a nominal value) line is used instead of 75-ohm. It is not matched as well by the antenna, but the mismatch is not serious. It has the same disadvantages as 75-ohm coax.

The Folded Dipole

The advantages of matched operation also are realized with the **folded dipole** shown in Fig. 6-22. The folded dipole has two half-wave conductors side by side. One is continuous, but the other is cut at the center for making connection

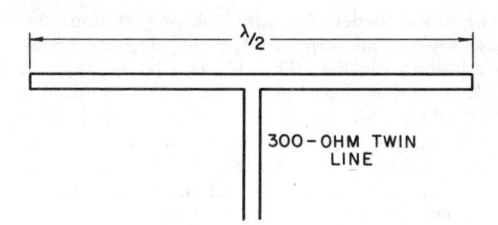

FIG. 6-22—The folded dipole. The antenna length is calculated in the same way as for a single-wire dipole.

to 300-ohm twin line. The two conductors are joined at their ends.

The wires radiate in parallel. In this respect, the pair is equivalent to a single half-wave dipole. But splitting the conductor into two parts has the effect of making the antenna impedance, as seen by the line, four times the impedance of a single-wire dipole. Thus at the point where the transmission line is connected the antenna impedance is approximately 300 ohms—just right for matching 300-ohm line.

Twin line can be used for the folded dipole itself, but ordinary TV line won't stand the mechanical stresses too well if the antenna is long. There is a special heavy-duty line available (Amphenol 214-022) which is better. TV ladder line also can be used for the dipole. The spacing between the dipole wires can be anything up to a few inches, so practically any construction that will keep the wires parallel can be used.

"Open-Wire" Feeders

Fig. 6-23 shows a half-wave dipole fed at the center through **open-wire** parallel-conductor line. This is line having mostly air insulation, such as the TV ladder line mentioned earlier. Here there is no attempt at matching the antenna to the line. Consequently there are fairly pronounced standing waves on the line. However, the high s.w.r. doesn't cause an undue power loss in open-wire line. The principal penalty is that more attention has to be paid to the coupling between the line and transmitter. The advantage is that the antenna can be made to take power at practically *any* frequency.

A transmission line operating with a high standing-wave ratio is often called a **tuned line**

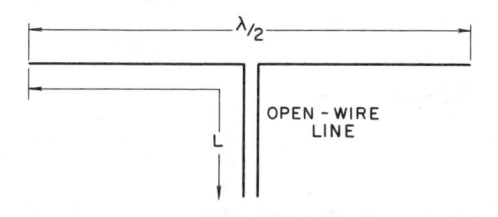

FIG. 6-23—Half-wave dipole fed with open-wire line.

or **tuned feeder.** Actually, the only tuning necessary is that required for coupling the transmitter to the line. The line can be any length. However, it does help simplify the transmitter coupling a bit if a resonant length is used. Such a length, as you have seen, will be some multiple of one-quarter wavelength. The line will "look like" a resistance at its input end in such a case, provided the antenna itself is resonant.

On the other hand, in this system the dipole doesn't have to be exactly resonant. Since there is no attempt at matching the characteristic impedance of the transmission line, the antenna doesn't *have* to look like a pure resistance, of just the right value, to the line. The over-all length of wire in the system, including both the dipole and the transmission line, is of more interest. It is this over-all length that determines whether or not the system as a whole is resonant. One line wire plus one side of the dipole (the length L in Fig. 6-23) should be a whole-number multiple of a quarter wavelength if you want the system to be resonant. The formula

$$\text{Length in feet} = \frac{234}{\text{Freq. in Mc.}}$$

will give the length of a quarter wave as accurately as is necessary.

Multiband Operation

As explained earlier, a matched antenna system is essentially a one-band system. There are ways of getting around this, but not with a simple dipole. One such scheme is shown in Chapter 15.

The simplest multiband antenna, and the most versatile, is the one shown in Fig. 6-24, using open-wire feeder. Since the amateur bands are harmonically related in frequency, we can take advantage of the fact that wires have harmonically-related resonances. The fundamental frequency of a center-fed wire is the one for which its length is a half wavelength. At twice the frequency each *side* of the antenna is a half wavelength long, so at this frequency the transmission line is feeding a pair of half-wave dipoles end-to-end. The current distribution is shown in Fig. 6-24, which also shows the other resonances up to the fourth multiple.

You should note a few especially interesting things in these drawings. In the second-harmonic case the polarity of the current is the same in both sides of the antenna. There is no reversal such as there was in a continuous wire of the same over-all length (Fig. 6-4). This difference comes about because we have, in effect, two half-wave antennas driven in push-pull, rather than a single antenna a full wavelength long.

There is a somewhat similar situation at the fourth harmonic. Here, too, the currents in the half-wave sections connected to the line have the same polarity. However, when we go out along either wire we find that the normal reversal occurs in the next half-wave section.

This type of current distribution occurs at all *even* multiples of the fundamental frequency. Note also that at the second harmonic the current is minimum where the feeder is connected. Although the voltage distribution isn't shown, the voltage is highest at these same points, just as in the cases discussed earlier. This means that the impedance is high at the connection point. If the antenna is resonant, it is a resistance rather than an impedance, and is of the order of several thousands of ohms. This same condition exists at all even multiples of the fundamental frequency.

Odd Harmonics

Now look at the drawing for the third harmonic. Here we have the normal current distribution for a wire three half-waves long (Fig. 6-4). The antenna current has its largest value right where the transmission line is connected. The voltage must be lowest at this point, so the impedance (or resonant resistance) of the an-

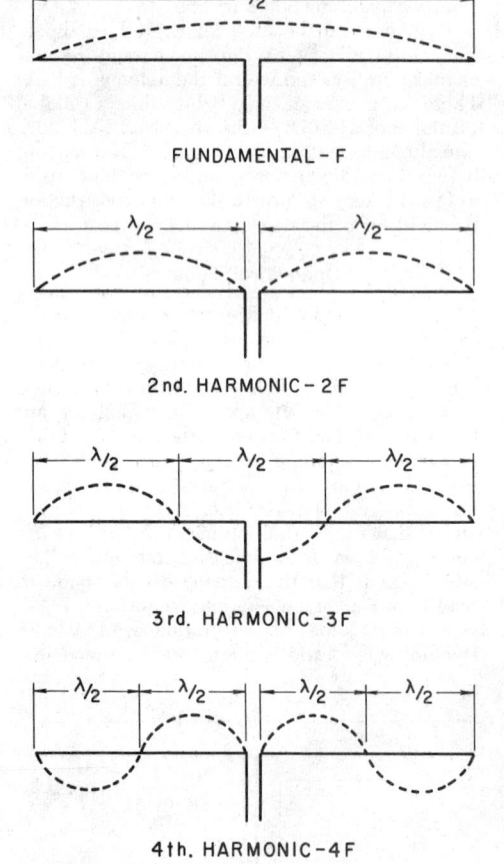

FIG. 6-24—Harmonic operation of a center-fed antenna. If the antenna is a half wavelength long at 7 Mc., for example, it will also be resonant in the 14-, 21- and 28-Mc. bands.

The Transmatch

tenna is low—more like the impedance at the fundamental.

Thus for all *odd* multiples of the fundamental, the current distribution is the same as in a simple continuous wire of the same over-all length, and the impedance at the feed point is low. The impedance goes up a little with each odd harmonic —to a little over 100 ohms at the third harmonic and to about 120 ohms at the fifth harmonic.

Because these figures do not differ too greatly from 70 ohms, it is possible to operate an antenna on its *odd* harmonics when it has been matched on its fundamental. The match is not as good as at the fundamental, but it is not so poor as to result in excessive line loss. Such operation does not really qualify the antenna for multiband work, because only a few bands—not a consecutive series—can be covered.

If the antenna is fed with 50- or 75-ohm line you should not try to operate it at *even* harmonics of the frequency for which it is matched. The line losses would be excessive because of the high s.w.r.

Transmitter-to-Line Coupling

Nowadays nearly all transmitter final tank circuits are designed for coupling into resistive loads of 50 to 75 ohms. A properly-matched coaxial line will "look like" such a resistance, and when a matched coax line is used there is no difficulty in making the final amplifier load up to the rated input. But if the load isn't properly matched, or some other type of line is used, you may have problems. The loading and tuning adjustments offered by the transmitter usually will give you some leeway—even if the matching at the antenna isn't perfect you may still be able to get the power input you want. Again, you may not.

You can get around troubles of this sort by using a special coupling circuit—a **transmatch** —between the output of the transmitter and the input end of the line. As we saw earlier, the input impedance of the line is not the same as the line's characteristic impedance unless the line is perfectly matched by the antenna. If the s.w.r. is greater than 1 to 1 the input impedance may differ widely from Z_0. If the line is connected directly to the transmitter, the latter may see a load that it can't handle. The transmatch takes the line input impedance and transforms it to what the transmitter wants.

It also does two other things. Practically all transmitter output circuits are single-ended— one side is grounded to the chassis, which is the right way to do it for coax line. What to do when a balanced line is used, as in Figs. 6-21A, 6-22 and 6-23? The transmatch easily handles this one; it provides the means for going from a balanced line to coax. In addition, it adds selectivity between the transmitter and the line— selectivity that often is badly needed. It is an unfortunate fact that most transmitters "put out" not only the frequency you want, but also harmonics of that frequency—and, in some cases, lower frequencies too, when lower frequencies are present in the stages leading up to the final amplifier. The transmatch is a circuit that, among other things, is tuned to your desired output frequency and so helps in keeping the unwanted frequencies from reaching the antenna.

Using the Transmatch

Fig. 6-25 shows how it is connected, and Fig. 6-26 is a typical circuit. It isn't the only circuit that can be used, but is probably as versatile as any. The circuit formed by L_1 and C_1 is tuned to

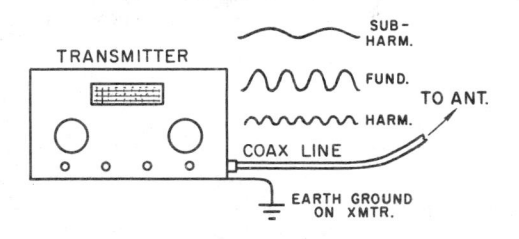

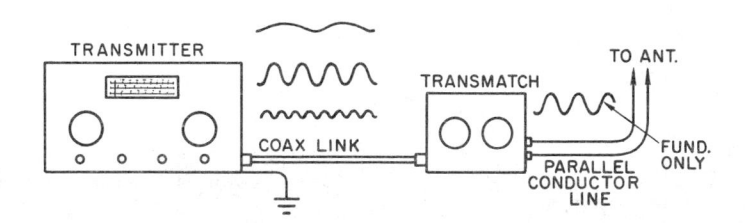

FIG. 6-25—The transmatch provides means for matching your transmitter's output impedance requirements, for going from a balanced transmission line to coax, and for filtering out frequencies that shouldn't be allowed to reach the antenna.

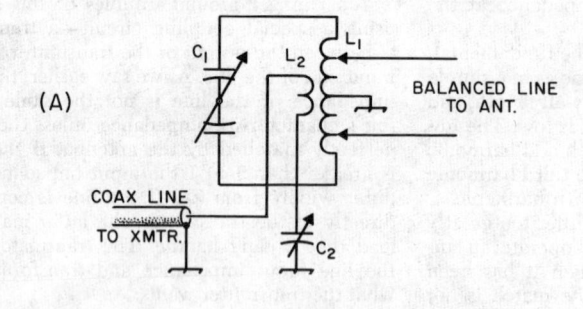

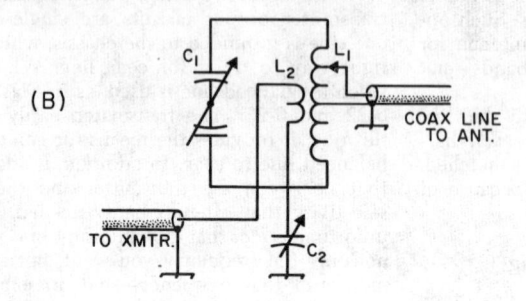

FIG. 6-26—A representative transmatch circuit, as applied to balanced lines (A) and coaxial lines (B). Matching adjustment procedure is the same for both cases.

your operating frequency. If the line is the parallel-conductor—balanced—type the wires are tapped on L_1 at equal numbers of turns from the center. The loading is adjusted by changing the positions of these taps. L_2 couples the power to L_1, and C_2 gives you a means for tuning this link circuit. A coax line goes from here to your transmitter's output terminal. Between these two adjustments you can transform a wide range of line input impedances into 50 or 70 ohms (whichever is the Z_0 of the coax line from the transmatch to the transmitter).

The method used for coupling to a coax line feeding the antenna is shown at B. It is very similar, the only difference being that the outer conductor of the line is connected to the center of the coil and only one tap is used. The coax link circuit to the transmitter remains the same. So does the method of adjustment.

The construction of a transmatch is described in a later chapter. The benefits of the circuit do have their price—you have to fix things so L_1C_1 can be tuned to each band you want to use. This usually means that L_1 is a plug-in coil. L_2 is generally made part of the same coil assembly, since it is advantageous to change it, too, for various bands. The same capacitors can be used for all bands, though, over at least the 3.5–30 Mc. range.

The adjustment of a transmatch is easy if you have a bridge such as the Monimatch described in a later chapter. Such a bridge is inexpensive and is an almost indispensable station accessory. However, you can arrive at a reasonably satisfactory adjustment simply by varying the tap positions, along with the settings of the two capacitors, while performing the normal tuning and loading operations on your transmitter. After a little cut-and-try you'll find the transmatch settings that let you load up the final amplifier to the input you want.

The Grounded Antenna

The antennas described earlier are usually hung more or less horizontally at a height of 25 feet or more above the earth. You may not have the room to put up a horizontal antenna a half

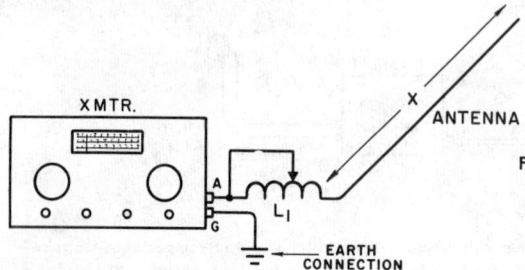

FIG. 6-27—A simple antenna for limited space.

The Transmatch

FIG. 6-28—Alternative to Fig. 6-27, using a transmatch for transmitter matching and filtering. The transmatch circuit can be the one shown in Fig. 6-26B, with the antenna tapped on L_1 of the figure in place of the coax line.

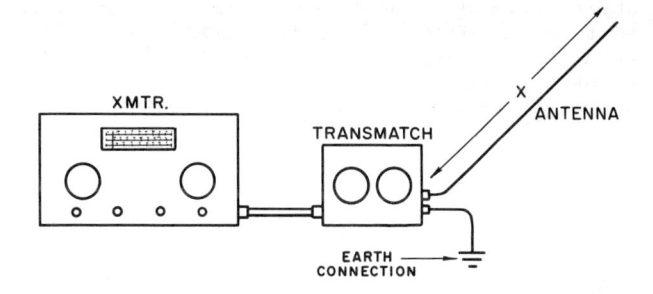

wave long at 80 meters; a length of some 130 feet is needed. If you can't, it's always possible to use a center-fed dipole having whatever length your location permits. Open-wire feeder should be used. Except for the antenna length, the system is like that of Fig. 6-23. It can be tuned up to take all the power your transmitter is capable of giving, if the length L in Fig. 6-23 is somewhere near 60 feet and you use a transmatch between the line and your transmitter. However, the length of one side of the antenna itself shouldn't be less than about an eighth wavelength—that is, 30 feet or so for 80-meter work.

Possibly the simplest scheme for getting on the air, when your antenna space is limited, is a wire "worked against ground." In this system you put out as long a wire as you can, making it as high as you can, and tune it to resonance at your operating frequency by means of a **loading coil**, L_1 in Fig. 6-27. If the transmitter has a coax output fitting, the A terminal in the drawing goes to the inner conductor. The outer sleeve should be connected to earth. Usually there is some piping in the house that you can hook onto for an earth connection. Cold water pipes usually are good, since they tie into the underground water distribution system. Or you can drive a regular ground rod (such as those made for TV grounding) into the earth. In either case you should keep the lead between the piping or rod and the transmitter short, since it will want to act as part of the antenna.

The length X in Fig. 6-27 can be up to one-quarter wavelength (60-odd feet at 80 meters) if the ground lead is only a few feet long. For longer ground leads, the lengths of the antenna and ground wires should be added together. The nearer this length to a quarter wavelength the less inductance you need at L_1. If the length is *over* a quarter wavelength you'll have to substitute a variable capacitor for L_1, in order to tune the system to resonance.

The antenna doesn't have to run in a straight line, although it is best to try to make it do so. Put the far end as high in the air as you can. Another scheme is to use metal tubing that can stand vertically, either self-supporting or with the help of insulated guys. This will radiate better than a slanting wire at lower height.

Antennas such as these aren't the best in the world. However, you may not have any other choice. And they *do* work—often better than you might expect.

The Receiving Antenna

In your amateur communication, the best antenna for receiving is the one you use for transmitting. All antennas work better in some directions than in others. These directions are the same whether the antenna is radiating a signal or picking one up from a distance. If you use your transmitting antenna for receiving, the stations you hear best also will be the ones that hear you best. With a different receiving antenna you might get strong signals from directions that your transmitting antenna covers only poorly. This can be frustrating, because you may spend a lot of time calling stations that don't hear you well enough to make a good contact—and vice versa.

Using the same antenna for both purposes does make it necessary to use some form of switching, since the antenna (or transmission line) has to be connected to the transmitter at one time and to the receiver at another. Manual switching is quite feasible, but a nuisance. Most amateurs use some form of automatic changeover. An antenna relay is probably the most common. Electronic switching also can be used, as described in Chapter 11. The electronic system is especially advantageous in break-in code work because it will follow keying. The regular-type relay does not do so well in this respect.

Because of these changeover problems a great many amateurs do use separate receiving and transmitting antennas. Then no switching is needed. Each antenna is permanently connected, one to the transmitter and one to the receiver. This system is useful for c.w. work on 80 and 40 meters. To send, you simply press the transmitter's key. When the key is open your receiver operates. The method works best at the lower frequencies because the amount of energy picked up by an antenna of given size increases with the *square* of the wavelength, so a receiving antenna just a few feet long will pick up over 60 times as much energy from an incoming 80-meter signal as it would from one on 10 meters.

Modern receivers have as much amplification as you can ever use on these two low-frequency bands. So you don't really need more than a few feet of wire as an antenna, if you're satisfied to rag-chew with stations within a radius of a few hundred miles of your station. For longer-distance work, though, you will be much better off to use your transmitting antenna for receiving. Changeover switching may be a bother, but the bother is more than outweighed by the results. On bands from 14 Mc. up, you shouldn't give serious consideration to any other method.

Beam Antennas

A flashlight bulb out in the open doesn't seem to shed much light. But put it into a properly-shaped reflector, as in the ordinary pocket flashlight, and it throws a bright beam. The bulb isn't giving any more light. The reflector is simply taking what light there is and intensifying it. It does this by focusing the rays into a narrow pencil. The price paid for the "gain" in this beam is *less* light in all other directions.

The radio rays from an antenna, too, can be focused into a beam. However, the type of reflector used in a flashlight has to be very much larger than the wavelength, in order to do any good. It is practical at radio wavelengths of the order of a few inches, but becomes too large for amateur use in bands below about 1000 Mc.

On the lower bands the focusing is done by combining the individual radiations from a number of dipoles. The waves from the dipoles are timed so that they add together when going in the desired direction, and tend to subtract (**interfere** with each other) in other directions. The subtraction means a *decrease* in intensity in those other directions. The total power must remain the same, since the antenna can't manufacture power itself. It can only rearrange it.

Types of Beams

A dipole in such a system is called an **antenna element**. It may or may not be exactly resonant, depending on the kind of system. A combination of antenna elements is called an **array**. There are two general types of arrays. In one, all the elements are connected to the transmitter through a system of transmission lines. This is called a **driven array**. However, it isn't *necessary* that all elements be driven directly. If an element is close to and more-or-less parallel with a second element that does have r.f. power in it, some of the power in this second element will be coupled into the first through the electromagnetic field. There is a resemblance here to inductive or capacitive coupling between ordinary circuits. Elements that get their power by this means are called **parasitic elements**, and arrays in which they are used are called **parasitic arrays**. One or more of the elements in a parasitic array has to be driven, of course; power has to be introduced into the system before any electromagnetic-field coupling can take place.

Besides multielement antenna arrays there are several other types of beam antennas in use. Principal among them are various forms of **long-wire antennas**. These work on much the same idea as the multielement types, but the "elements" are the half-wave sections of continuous wires operated on multiples of a fundamental wavelength.

The study of antennas is a whole field in itself.* It will suffice for our present purposes to become a little acquainted with a type of beam that is widely used on the amateur bands from 14 Mc. up—the Yagi antenna.

Yagi-type Beams

The Yagi antenna takes its name from the inventor of the directive system using parasitic elements. In its usual form, the antenna has a driven dipole, usually resonant, and one or more

---

*There are many books on the subject, among them *The A.R.R.L. Antenna Book*.

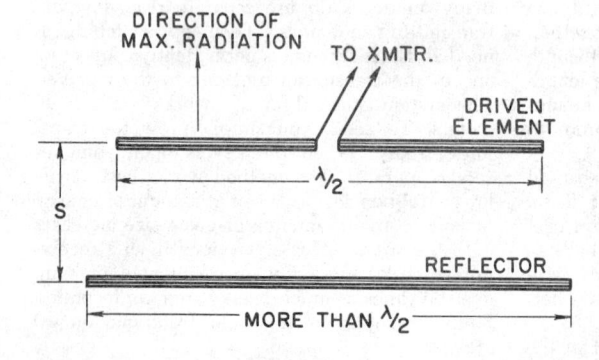

FIG. 6-29—A two-element parasitic beam. The lengths in wavelengths shown are to be understood to be electrical rather than physical lengths.

Beam Antennas

FIG. 6-30—The three-element parasitic beam. More parasitic elements—nearly always directors—can be added. The power gain in the favored direction is approximately proportional to the over-all length of the antenna measured in the direction of the arrow ($S_1 + S_2$, etc.), provided the elements are properly spaced and tuned.

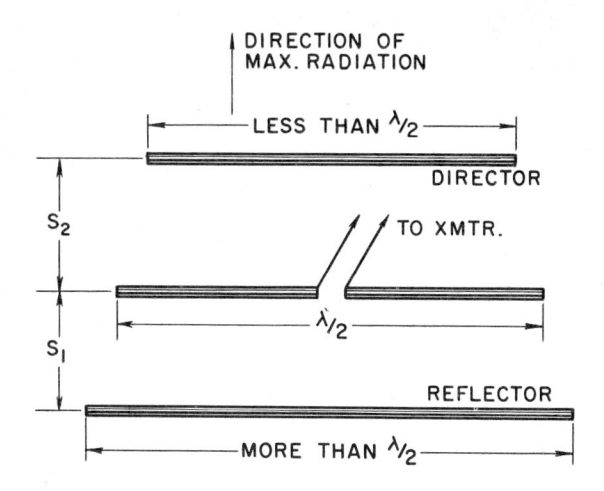

parasitic elements. The simplest arrangement is the **two-element beam** shown in Fig. 6-29. There is only one parasitic element. Usually it is a **reflector**, so-called because the energy it receives from the driven element bounces back to concentrate the radiation in the same direction that a reflector behind a flashlight lamp would concentrate it. This direction is shown in Fig. 6-29.

The power that a parasitic element picks up from the field of a driven element is practically all **reradiated**. Only a small fraction is lost in heating the resistance of the element itself—no more than in any dipole having the same ohmic resistance and carrying the same current. Thus, from a practical standpoint, all the power a parasitic element gets is used in enhancing the radiation of the system in one direction and tending to suppress it in others. What it does in this respect depends on how the parasitic element is tuned and on the spacing (in terms of wavelength) between elements.

A parasitic element acts as a reflector, at the spacings normally used, when it is tuned somewhat *lower* than the operating frequency. That is, it is made somewhat *longer* than an electrical half wavelength—about 5 per cent longer, ordinarily. The spacing S is usually about 0.15 wavelength. There are no magic figures here. Many values of S and reflector length will give good results. The two quantities are not independent; changing one will require changing the other for optimum results.

More than two elements can be used. When a third one is added it is usually a **director**, as shown in Fig. 6-30. This element helps the radiation along when it is placed in *front* of the driven element. To do this it has to be tuned *higher* than the operating frequency, at ordinary spacings. That is, it is made about 5 per cent *shorter* than an electrical half wavelength. The spacing, S_2, between the driven element and the director is from 0.1 to 0.2 wavelength in most antennas. S_1 is about the same as in the two-element beam.

We don't need to stop with three elements. The fourth and subsequently-added elements are practically always additional directors. Additional directors are usually a little shorter than the first one, and when a large number is used the spacing, for optimum results, tends to level off at about 0.2 wavelength. Antennas with many elements are practical only at very short wavelengths, because of the size of the structure required for a large number of elements. Three elements are commonly used at 14 Mc., three or four at 21 and 28 Mc., and four or more at 50 Mc. and above.

Beam Antenna Gain

The **gain** of a beam antenna is the ratio of the power radiated in the desired direction to the power radiated by some **reference antenna**, assuming the same power input to both. In amateur work it is understood that the reference antenna is a half-wave dipole. Such a dipole doesn't radiate equally well in all directions; its maximum radiation is along a line at right-angles to the direction of the antenna itself, as shown earlier in this chapter. So the reference dipole must aim, radiation-wise, in the same direction as the beam, in figuring the gain of the beam. Also, it must be at the same height above ground, and otherwise be installed under the same operating conditions as those for the beam.

Since gain is a power ratio, it is usually measured in decibels. Gains of up to 4 or 5 db. can be achieved in the two-element antenna, and up to 7 db. or so in the three-element. As a fraction of a decibel represents a power change that is not observable audibly (remember that one decibel is about the *least* detectable change) it doesn't pay to take too seriously attempts to squeeze out the last one-tenth of a decibel. Also, bear in mind that the same lengths and spacings will lead to the same results whether you make the antenna yourself or buy one ready-made. Once you've decided how many elements you'll use, it's sensible to forget about minor gain differences and concentrate on constructional features that will keep the antenna up in the air and operating throughout all kinds of weather.

Rotatable Beams

Of course, the gain of a beam antenna is useful only in the direction toward which the beam points. To be able to use the antenna all around the horizon you have to be able to rotate it. Yagi beams are practically always mounted so they can be rotated. The more solidly-constructed TV rotators can be used with good results for the smaller amateur antennas. Large ones, though, such as would be used on 14 Mc., usually require much heavier machinery for rotation.

Chapter 7

Workshop and Test Bench

Buy something ready-made and you take it for granted that it will perform. Put it together yourself and you savor that unforgettable moment when you first turn on the power, when the big question—will it or won't it work?—is about to be resolved. There's no satisfaction like that which comes from having produced a working piece of equipment with your own hands. And if it doesn't work at first try, you're a wiser amateur after you've found out why. You owe it to yourself to build as much as you can.

Doing a good job of building calls for certain mechanical facilities. Sometimes it takes ingenuity to provide them. Apartment-house dwellers rarely have space for workshops, but any available table space can be commandeered —temporarily, of course, and with suitable protection against damage from tools and hot soldering irons! It is better, though, to have a spot you can call your own, equipped with a bench or table large enough to let you work in some comfort.

A small shop-type bench, around 24 by 48 inches, has enough room for practically all the jobs you'll want to undertake, and even a 2 by 3 foot kitchen table will do. If you have the space, a real luxury is a bench about 3 by 8 feet, or even longer. It isn't hard to make one to fit your own needs if you have just a little aptitude for carpentry. If the bench can be wired with a.c. outlets, so much the better; you'll need them for soldering and testing. And you'll want plenty of light.

Given a working space that meets these general requirements, the next consideration is tools. It takes only a few to start out, but if you continue to do building you'll accumulate more as you progress.

Minimum Tools—Kit Building

Building from kits doesn't take many tools— probably no more than you'd need for general station use even if you did no construction work at all. The essential ones are:

7-inch screwdriver, ¼-inch blade (for 6-32 screws)
4-inch screwdriver (for 4-40 screws)
Pocket screwdriver with ⅛-inch blade (for knob setscrews, etc.)
6-inch long-nose pliers
Wire stripper and cutter
Pocket knife
Soldering iron, 60-watt

The wire stripper can be the inexpensive kind (such as the Miller type) but should be adjustable for the size of wire from which the insulation is to be stripped. Properly adjusted, it will strip the insulation off cleanly and will not nick the wire and weaken it.

In addition to the tools listed, a couple of socket wrenches will be handy. Two sizes—for ¼-inch and ⁵⁄₁₆-inch nuts—are most useful. You can find sets with a single handle and interchangeable wrenches in hardware and "dime" stores. They are inexpensive but quite adequate.

This selection of tools will suffice for kit building and ordinary servicing.

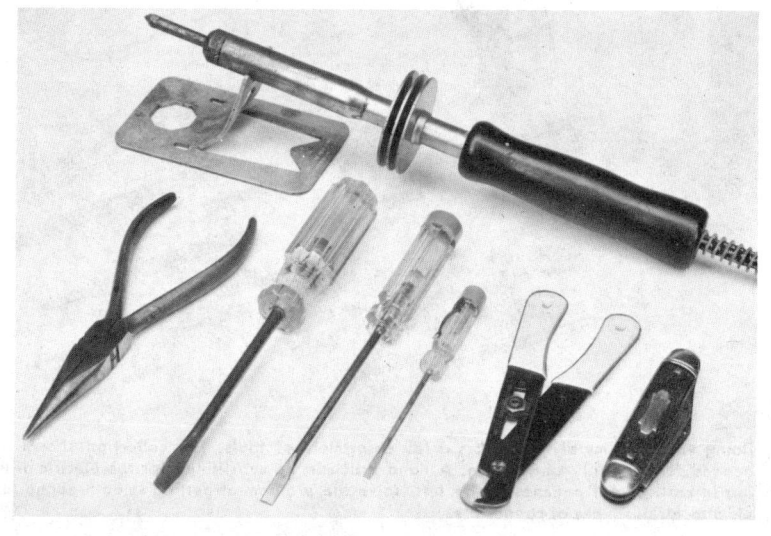

In fact, you can get similar screwdriver sets, and this is probably the cheapest way to get an assortment of both wrenches and screwdrivers.

A 60-watt soldering iron is large enough for ordinary wiring. As compared with higher-power irons, it is light in weight and easier to poke into tight spots such as those you run into in wiring miniature tube sockets. Also, it lessens the danger of overheating a connection. If you prefer a soldering gun, use one. However, for doing any considerable amount of wiring at one sitting most builders prefer the light iron; it stands up better and is less awkward to handle than the gun.

This is a pretty small list of tools, but it is actually sufficient for any assembly and wiring job, kit or not. In assembling a kit there is no layout work or metal cutting and drilling to do, and that simplifies things mightily.

Tools for Metal Fabrication

Most amateur equipment is built on metal—on chassis or boxes that come in a wide variety of sizes. Ninety-nine per cent of your metal work can be summed up in two words—cutting holes. Most of them are round, varying in size from $\frac{1}{8}$ inch to two inches or so. It takes a variety of tools to handle all the situations that may come up. There are two main categories—holes for component mounting screws, and holes for tube sockets. For the former you need a drill, preferably electric, capable of handling metal-cutting bits up to $\frac{1}{4}$ inch in size. The drill bits for this purpose come in numbered sizes, and the ones needed most are

No. 33 (to pass 4-40 screws)
No. 28 (to pass 6-32 screws)
No. 18 (to pass 8-32 screws)

A $\frac{1}{4}$-inch drill bit is also useful, both on its own account and because a $\frac{1}{4}$-inch hole can be reamed out to $\frac{3}{8}$ inch, which is the size hole required for the shafts of such items as volume controls, small variable capacitors, and phone jacks. A hand reamer, half-inch size (for metal, not wood) can be used for this job.

The best tool for cutting holes for tube sockets and similar components is the hand-operated

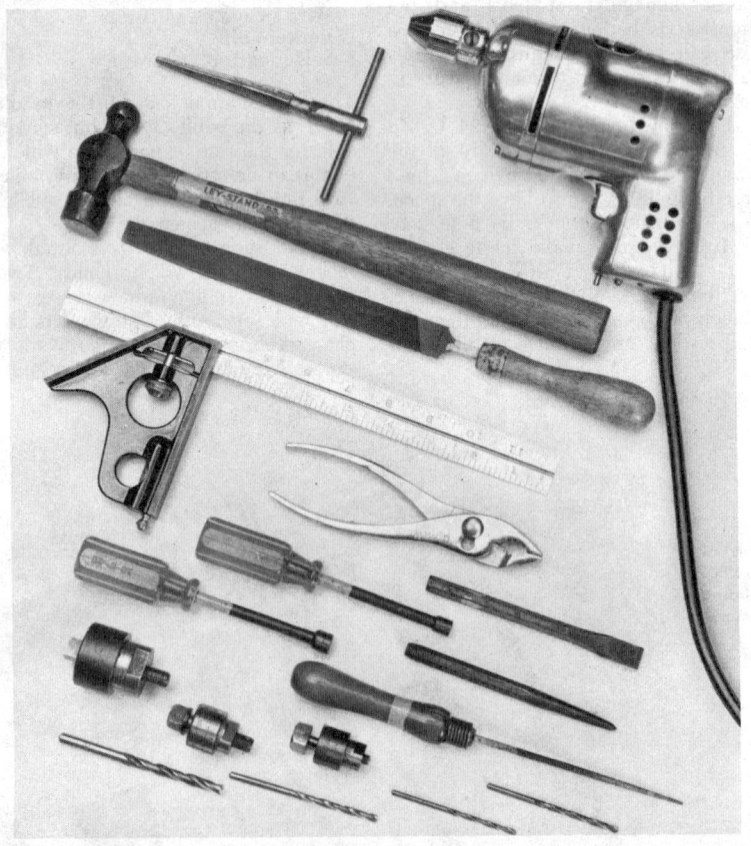

Doing your own metal work takes a fair assortment of tools. The collection shown here also includes those needed for assembly and wiring. A hand drill can be substituted for the electric drill shown, with some saving in cost. Socket punches, lower left, solve the problem of getting smooth-edged round holes for tube sockets and various types of connectors.

Workshop Practice

Before unwrapping the paper from your chassis, move the components around on it in trial positions until you get a satisfying layout. Keeping the wrapping on prevents finger marks and scratches.

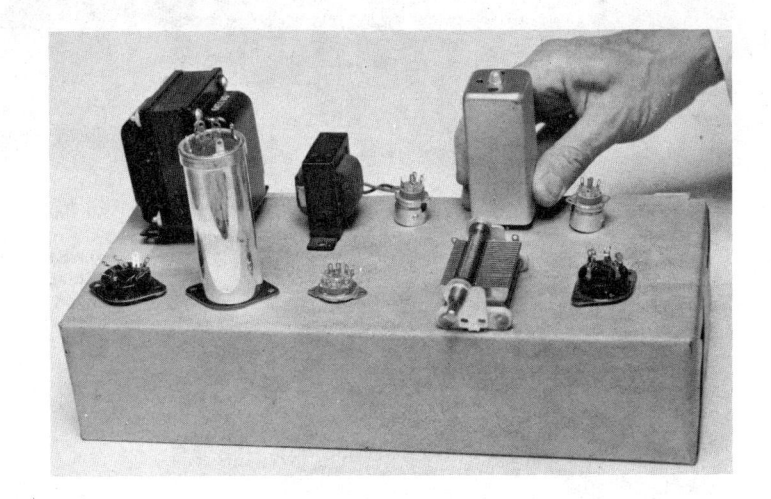

(Greenlee) socket punch. Three sizes will take care of practically all work:

5/8-inch punch (for 7-prong miniature sockets)
3/4-inch punch (for 9-prong miniature sockets)
1 1/8-inch punch (for octal sockets)

The 5/8-inch size also is just right for coaxial fittings of the UHF series, such as the SO-239.

The same tools can be used for working either steel or aluminum. The principal difference is that steel is much harder.

Holes larger than 1 1/8 inch are almost always for special components such as meters. Punches for the large holes are expensive, and for just occasional use probably won't earn their keep. It's more economical (although more work) to cut such holes by drilling a series of small holes around the inside of a circle of the right diameter, breaking or chiseling out the remaining metal, and then filing to smooth out the hole.

Every hole-cutting operation should be started by marking the right spot so the drill will go through where you want it. This takes a center punch. The kind you hit with a hammer is quite inexpensive. More costly, but convenient, is the "automatic" center punch, which has an internal spring-operated mechanism.

You'll also find a flat file or two useful, especially if you form small mechanical parts, such as brackets, from sheet metal.

Laying Out the Chassis

Whether the job before you is large or small, the first step in construction should be making a chassis drilling template. Take plenty of time here. Collect the components to be used, including "mechanical" items that don't appear on the circuit diagram such as tie-point strips and insulators that require mounting holes. Leave the paper wrapping on the chassis and try various arrangements of parts until you arrive at a layout that will let you get in everything without unnecessary crowding.

Don't forget to allow room for small items such as pigtail resistors and capacitors, and also make provision—at least mentally—for wiring them in place. These small parts tend to get crowded around such focal points as tube sockets, so you should have a fairly clear picture in advance of just how they will be installed, where those requiring grounding will connect to the chassis, and so on.

Watch out for "interferences" between top, side and bottom-mounted components—but by the same token, don't forget that you can take advantage of the fact that a component mounted wholly on one side will leave clear space on the other, except for mounting-screw holes. If you're building something from a printed description there usually will be photographs of the equipment to guide you.

Once you're satisfied with your layout, mark the exact spots where holes are to be drilled. These will correspond to the mounting holes in each component, of course, and the component can be used as a template for spotting the hole centers accurately. Components whose edges should be aligned with an edge of the chassis should have their mounting holes squared up. This means that you have to be able to draw lines parallel with the chassis edges. A combination square is handy for this, but is not a necessity. An ordinary ruler can be used; simply measure the proper distance from the edge along both adjoining sides of the chassis, mark the points and draw a line between them. Or you can use a sheet of letter paper as a square, lining one side up along an edge of the chassis.

Mark the *exact* spot for each hole with a pencil dot. Alongside it, jot down the size hole required. When this is done, put all the components on the template, setting their mounting holes over your marked points. This gives you a final check on the accuracy of your work, and if you've inadvertently mislocated some parts so they interfere with each other, you'll discover it before any damage is done.

Making the drilling template. The chassis wrapping can be used for this, and can be left on the chassis until all the drilling is finished. Then remove the paper, spray the chassis with clear lacquer, and it will retain that "new" look for a long time after the parts are mounted and the wiring is finished.

The template doesn't have to be made on the chassis wrapping paper, but can be worked out on a separate sheet of the proper size, if you want. One advantage of the separate template is that it can be discarded and a new one made readily if you get a better layout idea after one or more early trials. As we said before, this stage of construction shouldn't be hurried.

Drilling the Chassis

After the template has your final check and OK, get some wooden blocks a couple of inches square and a little thicker than the chassis is high. The height can be made up by stacking a number of pieces of ordinary board thickness, if necessary. The blocks provide a firm surface under the chassis top. Place the point of the punch exactly on your mark for a hole, with a block underneath. Hold the punch vertical, and give it a smart rap with the hammer—not so hard that there is a big dent in the metal around the punch mark, but hard enough so the mark will hold the drill point when you start drilling. It takes a more vigorous hammer stroke with steel chassis than with aluminum. It's a good idea to experiment a little with an old chassis, if you have one, to get the "feel" of it.

Finish your punching, then use the wood block as a foundation for drilling. (Don't drill a chassis without some such solid surface underneath, if you can possibly avoid it.) It's a good idea to drill "pilot" holes at all the punched points, using your smallest (No. 33) drill. Small holes can be drilled more accurately, and serve as a good guide for the larger drill sizes. If you want to save the template for any reason, remove it after drilling the pilot holes. Incidentally, if you haven't used the chassis wrapping for the template, a separate one can be fastened in place, for punching, with a few strips of Scotch tape.

Using Socket Punches

When laying out socket holes on the template, draw a pair of lines at right angles to each other through the center mark. These will let you use the gauge marks on the side of the socket punch for lining up the punch properly. Align both pairs of gauge marks with your pencil lines and the punch will be centered where you want it. See Fig. 7-1. Be sure to drill a large enough hole so the bolt in the punch can move freely; don't try to use the drilled hole for centering the punch. The punching from the chassis is dis-

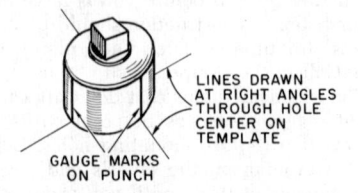

FIG. 7-1—How to use gauge marks on a socket punch for centering the punching where you want it.

Workshop Practice

torted when you pull the cutter through the metal, and will jam on the bolt unless you allow enough clearance. Keep the bolt lubricated with grease (vaseline will do) and be sure to clean off all the small metal chips after each use.

Punchings of this type come out with clean edges and the holes usually need no smoothing. Not so with the drilled holes. These have to be deburred. An easy way is to use a small cold chisel held with one side of the cutting edge flat on the chassis (Fig. 7-2). Light taps with the

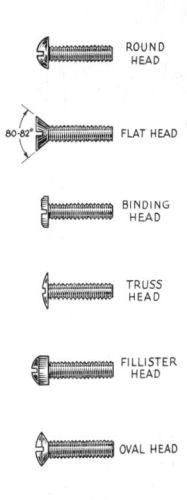

Machine screws are known by their heads. The flat-head and oval-head types require countersunk holes, but the others shown are intended for use on flat surfaces.

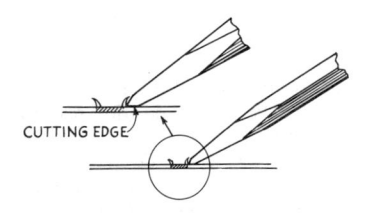

ROUND HEAD
FLAT HEAD
BINDING HEAD
TRUSS HEAD
FILLISTER HEAD
OVAL HEAD

FIG. 7-2—Using a cold chisel for cutting off burrs left after drilling.

hammer will cut off the burrs. A small amount of metal may be bent back into the hole; if so, it can be cleaned out by redrilling or filing with a small rat-tail file. Butt the edge of the chassis against something solid when using the chisel as a deburring tool.

Mounting the Components

Most small parts are mounted with either No. 4 or No. 6 machine screws; the size of mounting hole will indicate which to use. Miniature sockets usually have holes for No. 3 screws, but the metal is soft and can easily be reamed out to take a No. 4. The handle end of a small file can be used as a reamer. Larger components such as small power transformers usually mount with No. 8 screws. If you want your equipment to stay together solidly it's good practice to use star or lock washers under each nut.

In tightening screws use the right size of screwdriver. If the screwdriver is too small it will tend to gouge the edges of the screw slot.

Leave the heavy parts such as transformers and chokes off the chassis until you have the wiring completed to the point where these components have to be mounted. It's much easier to handle the chassis if it doesn't have heavy or bulky parts on it. Put on the sockets, tie points, and other small parts first, wire as much as you can, and then add the bigger ones as you come to them.

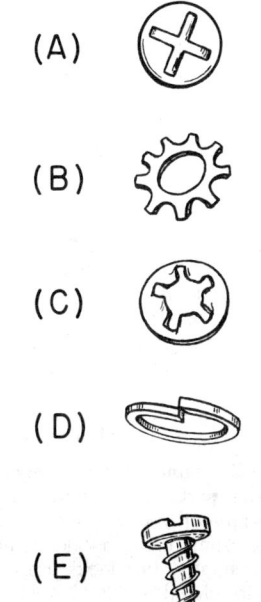

Wiring

Most builders find it advisable to put filament and heater wiring in first. Heater leads are best carried around the chassis in the folds at the edges, since this tends to reduce hum.

Leads in r.f. circuits should generally take the shortest route from point to point. However, keep them away from the chassis and other wiring, to reduce unwanted stray capacitance. Leads carrying only d.c. and 60-cycle a.c. can be run in any way that happens to be convenient. They can often be bunched together into a cable and laced together into a "harness"; this gives a neat job, especially if the cabled wires are run along the chassis edges.

In audio circuits the leads from plates and control grids should not be allowed to come too close to supply wires, especially those carrying a.c. The grid lead of a first audio stage often is a particularly touchy one for picking up hum. If you have any trouble on this score use a piece of shielded wire, grounding the shield to the same point at which the cathode resistor for the stage is grounded.

Plastic-covered hook-up wire—solid, not

The Phillips head (A) may be found on any type of screw; it requires a special screwdriver instead of the common flat-blade type used with slotted screw heads. Three types of lock washers are shown at B, C and D; the two former are most commonly used in radio equipment for holding screws up to No. 8 in diameter. E shows a sheet-metal screw, for assembling pieces of metal that are too thin for regular screw threads.

(A)
(B)
(C)
(D)
(E)

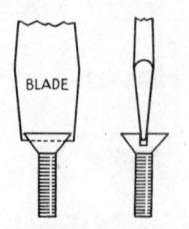

TOO LARGE TOO SMALL CORRECT

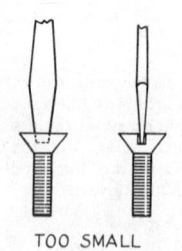

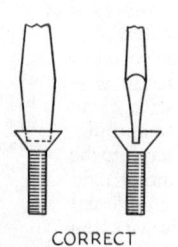

The wrong size of screwdriver blade won't turn the screw properly and probably will wreck the slot to boot.

stranded—is the easiest kind to strip, as it leaves a clean end. No. 22 is large enough for receiver wiring, including heaters, and for transmitters in stages of comparable power. And although you may often run across wiring instructions that tell you to wrap the wire around a soldering lug, tie point, or socket prong, *don't do it.* Just poke the wire through the hole so enough protrudes to take solder on both sides of the lug. A *good* soldered joint will have all the mechanical strength needed. Then later if you want to change something or have to replace a component you can unsolder it with no pain. Getting a "wrap-around" off under such circumstances will not only try your patience but probably will ruin the lug or prong.

Soldering

Everybody knows how to solder—so they think. But poor soldering is responsible for more trouble with radio equipment than any other single cause that can be named. Learn the *right* way to solder and you'll avoid much unnecessary grief.

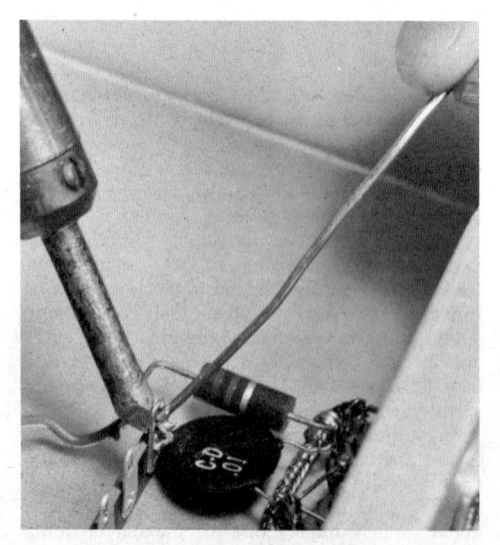

Apply the soldering tip to the joint, not to the solder. Let the joint get hot enough to cause the solder to run freely without actually having the solder touch the tip. You can't make a good soldered connection by using the tip to melt solder and letting it run over a cool joint.

The iron must be big enough for the job. As stated earlier, a 60-watt iron is just about optimum for practically all chassis wiring. It doesn't pay to use one too powerful because some components, such as small composition resistors, will be permanently damaged by excess heat. A heavier iron—100 watts or larger—will be needed for bigger jobs such as soldering No. 12 or No. 14 antenna wires, but you don't meet such large wire very often in a chassis.

Tinning the Iron

The tip of the iron must be clean and well tinned. A dirty tip with poor tinning simply won't make a good soldered joint. One of the first things you need to learn, therefore, is to tin the tip. Start by filing or sandpapering the tip until the metal all over the working end is bright. The end should be pyramid-shaped, pointed rather than stubby since you will have to get into tight regions around miniature tube sockets. Make sure that the entire tip is clean, by loosening the setscrew in the heater section of the iron and taking the tip out. Clean out the inside of the heater, too. Then replace the tip and make sure that the setscrew is tight. Let the iron heat up to working temperature.

With the iron hot, file it bright again and run on a little solder. Use the rosin-core type, preferably the 60-40 variety, although 50-50 can also be used. The metal will oxidize rapidly after you file, so you have to work fast at this point, or even do just a part of the tip at a time. With care, you will finish with a tip having a good coating of solder all over the point and extending a half inch or so down the cylinder. The tinning process can be helped along by using a little soldering paste (such as "Nokorode"). This will delay oxidation until you can get the solder to run on easily. But *don't* use soldering paste in actual wiring. If you use it in tinning, wipe the tip clean on a rag after the tinning job is done.

Making a Joint

Cleanliness is absolutely essential in soldering. The parts to be soldered *must* be clean and bright or you won't get a good joint. You may get something that looks like a good joint, occasionally, but looks are deceiving. Scrape the wire clean!

To make the joint, see that the parts to be

Workshop Practice

soldered are firmly in contact, then hold the hot iron on both parts (or on all, if there are two or more wires in a single lug or prong) until they are hot enough to make the solder flow freely when the solder is touched to the *parts,* not to the iron. Don't try to solder by touching the solder to the iron so it melts and runs over the parts. The iron is there primarily to heat the metal to be soldered so *that* metal will melt the solder.

After the solder has run into and all around the joint, take away the iron and let the joint cool. Don't let the wire or any part of the joint move while this cooling is going on. Especially, don't allow any movement when the solder cools to the mushy stage, which it passes through before becoming firm.

A good joint doesn't need a great deal of solder. Although you shouldn't try to be skimpy with solder, don't pile it up unnecessarily. The soldered joint should have a fairly bright surface, and the solder should taper off along the metal instead of drawing up into a bead.

Soldering paste should never be used in circuit wiring. It is almost impossible to remove it from the vicinity of the soldered joint, and it will collect dust and moisture with time. This can lead to noisy circuits. Rosin-core solder will leave a thin coating or drops of rosin on or around the joint, but these do no harm either electrically or mechanically.

Some Special Pointers

You have to have *enough* heat to make a good joint, but don't *overheat.* As soon as the solder flows the way it should, take the iron away. Small parts aren't helped by being overheated. Some, like semiconductor diodes and transistors, can be damaged even by normal soldering. In soldering the leads from these devices, hold the wire with pliers on the side away from the iron. This will conduct heat away from the device itself and prevent overheating it. Do the soldering just as quickly as you can, and don't take away the pliers until the joint has cooled to nearly room temperature.

Solder "takes" easily on bright copper and brass, or on leads and lugs that have been pretinned. Nickel-plated parts are a little harder to solder. The prongs on coil forms, plug bars, and many plug-in connectors are nickel-plated. It is a good idea to file the plating off the ends of such prongs and tin them by using the iron to apply a little solder before pushing the wires through the prongs. You can keep the hole from filling up with solder by giving the form a quick flick to throw the excess solder off while it is still molten. A good joint can be made between the pre-tinned prong and wire, with little or no fuss. After wiring a form or connector, always scrape the rosin off the prongs and wipe them with a rag wet with alcohol. You won't get good contact when the form is plugged into a socket unless the prongs are really clean.

Pre-tinning often is advantageous on the wire

Use pliers to conduct heat away from diodes and transistors while soldering. Grip the lead close to the body of the semiconductor and apply the iron only long enough to get a good soldered joint. As this is an awkward operation for only two hands, a heavy spring clip can be substituted for the pliers as a "heat sink." There are also special clips, made just for the purpose, on the market.

leads to small components such as capacitors and resistors. These leads are already tinned, but in time get dirty or an oxide coating forms. This sometimes makes soldering difficult. Fresh tinning will let you make your final joints faster and with more certainty.

Care must be used in soldering prongs or lugs on polystyrene parts such as coil forms. "Poly" softens at relatively low temperatures, and you may ruin a form in soldering to the prongs unless you keep the form itself cool. One way to do this is to make a "heat sink" from a scrap piece of aluminum. Drill holes in it at the same positions as the prongs, making them large enough for the prongs to pass through. Slip the metal sink over the prongs and hold it against the bottom of the form while soldering. It will absorb enough heat to prevent the form from softening. Bakelite and similar materials do not need this precaution since they cannot melt.

When several wires are to be soldered to one lug, use special care to ensure that *all* of them get soldered. The one on the bottom may not get hot enough to make a joint. Faults like this are hard to find when the equipment won't work. It isn't a bad idea to solder each wire separately, leaving enough room for succeeding wires to go through the hole in the lug or prong. Use just a little solder each time. Then when the last wire goes in place you can be confident that all of them are in good contact.

Keep a couple of old tubes handy and plug one into the socket when you're soldering prongs. Socket contacts usually fit loosely in the molding, and by the time you've wired all the prongs the contacts may be just enough off line to make it difficult to insert a tube, unless you seat a tube in the socket beforehand.

Finally, never forget that only *clean* metal can be soldered!

Taking Care of Tools

Good work can't be done with poorly cared-for tools. The first rule of tool care is to use a tool only for the job it was designed to do. A screwdriver, for example, makes a poor chisel, and attempting to use it as one makes it a poor screwdriver in short order.

The preceding section has described how to tin a soldering iron. Keep the tip clean and brightly tinned. You'll save time in the end by stopping work to re-tin whenever it's necessary. In using the iron, don't let solder pile up on the tip. Keep a clean rag handy and wipe off the excess solder regularly. Also, keep the entire tip clean. The part inside the heater usually gets a covering of scale after a time, and the tip should be periodically removed and scraped clean. The scale acts as a heat insulator and prevents the tip from heating properly. A tip should be replaced when its diameter has shrunk to the point where it can rattle around in the heater.

The tip of a screwdriver may occasionally require dressing. When you get a new screwdriver, inspect the tip. Observe that the end is *not* sharp, but is made flat so it fits snugly into the slot on the screw it is designed to drive, as in Fig. 7-3. If the tip gets nicked or worn with use, file or grind it back to the original shape, and test it by fitting it into a screw slot. It should not fit loosely, and the blade should go right down to the bottom of the slot so it will turn the screw firmly.

Knives, chisels, and the like can be kept in good condition by periodically touching up the

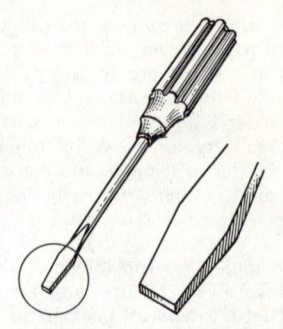

FIG. 7-3—A screwdriver should have a blunt end. Touch it up with a file when the edges get rounded with use.

edges on a carborundum stone. A small stone can also be used for touching up chassis punches that become dull, but use care in this operation. Never use the stone on the outside circumference of the punch. Look the punch over carefully and use the stone on the flattened part that angles up from the inside to the cutting edge. Keep the angle the same, by grinding parallel with the flat. And go easy, or you may wind up with the punch in worse shape than when you started.

Drills lose their cutting edges with time. Sharpening a drill is an art. We recommend getting a new drill bit in preference to trying to sharpen an old one. Buy only good drills. The "bargain" sets usually lose their edges on the first hole you drill, if the drill doesn't break off before it gets through the metal.

Files should be kept free from any accumulations of filings in the grooves. An inexpensive file brush is the best gadget for cleaning.

About Components

"Bread and butter" items—tube sockets, fixed resistors and capacitors, adjustable and variable resistors, and the like—make up the majority of parts used in any radio equipment. These are the components that you can find in any radio supply store or repair shop. All of them are made by more than one manufacturer, and components with identical ratings usually are mechanically interchangeable. You can freely substitute one make of part for another if you observe the required electrical ratings and values.

Small fixed resistors in ½-, 1- and 2-watt ratings are invariably color-coded to indicate the resistance value. The standard method uses three (sometimes four) colored bands starting from one end of the resistor. The band nearest the end gives the first figure of the resistance, the next is the second figure, and the third gives the number of zeros to be added after the first two figures. (A black third band indicates that no

Table 7-1
Standard Color Code
1—Brown
2—Red
3—Orange
4—Yellow
5—Green
6—Blue
7—Violet
8—Gray
9—White
0—Black

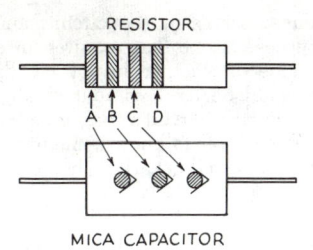

Color Codes

FIG. 7-4—Color coding of fixed resistors (½- to 2-watt size) and small mica capacitors. A—first band; B—second band; C—third band; D—tolerance band. B and D may not be present, indicating the widest tolerance (20 per cent) available.

Table 7-II

Standard Numbers for Fixed Resistor and Capacitor Values

A	B	C
10		
		11
	12	
		13
15		
		16
	18	
		20
22		
		24
	27	
		30
33		
		36
	39	
		43
47		
		51
	56	
		62
68		
		75
	82	
		91
100		

A—Available in 20, 10 and 5 per cent tolerance.
B—Available in 10 and 5 per cent tolerance.
C—Available only in 5 per cent tolerance.

This number system is based on overlapping tolerances expressed in percentage. Values may be increased or decreased in multiples of 10; e.g., 470, 6800, 27,000, etc.

zeros are to be added.) See Fig. 7-4. For example, a resistor having a red (2) first band, yellow (4) second band, and orange (3) third band has a resistance of 24,000 ohms. The fourth band, when present, gives the rated resistance tolerance. Silver indicates a tolerance of ±10 per cent; gold indicates ±5 per cent. If there is no fourth band the tolerance is ±20 per cent.

Fixed paper capacitors are sometimes color coded, too, using the same type of code. The unit in this case is the micromicrofarad, so if you have a tubular paper capacitor marked red (2), red (2), yellow (4) its value is 220,000 µµf. (0.22 µf.). Fixed mica capacitors are similarly coded, although dots rather than bands of color are used. Also, mica capacitors sometimes have additional colored dots to indicate voltage ratings and other characteristics.*

If you do any amount of construction you'll shortly find yourself recognizing the colors as readily as you recognize the corresponding numbers.

Tube socket prongs are numbered clockwise around the socket *looking at the bottom*. The standard numbering for the commonly-used types is given in Fig. 7-5. Note that on the miniature and novar sockets there is a wider gap between two of the pins than between the others. You start counting off in the clockwise direction from

---
*There are many variations of the capacitor color code but the one shown in Fig. 7-4 is most common on currently-produced components. More complete information can be found in *The Radio Amateur's Handbook*.

---

this wider gap. The octal socket has the pins evenly spaced, but there is a "key" in the socket. The No. 1 pin is just on the clockwise side of this key slot. Many sockets have the pin numbers stamped on the base, but since space is limited it is possible to confuse the pins and numbers. However, if you know the system you'll not have any trouble.

Volume controls and similar variable resistors, rotary switches, and phone jacks of various

BOTTOMS OF SOCKETS

FIG. 7-5—Standard pin numbering of common tube sockets.

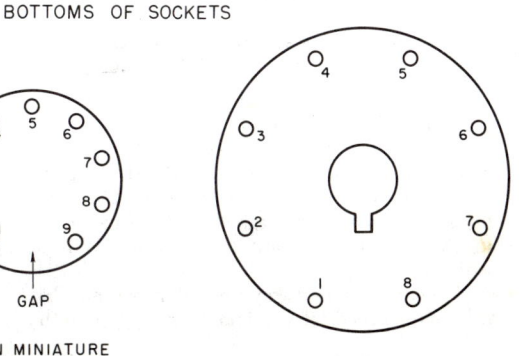

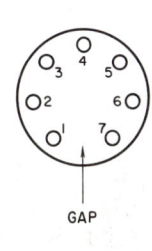

7-PIN MINIATURE
9-PIN MINIATURE AND NOVAR
OCTAL

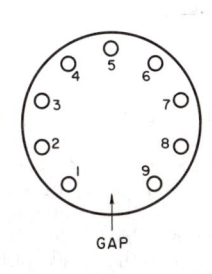

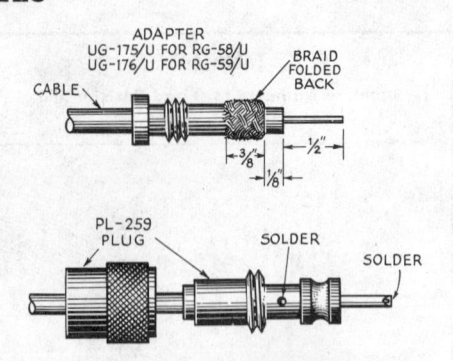

FIG. 7-6—Assembling PL-259 plug and adapter to small coaxial cable.

types all mount in the same size of panel hole—⅜ inch. Toggle switches take a larger mounting hole—15/32 inch, just slightly under ½ inch.

Coaxial Fittings

In practically all transmitters the r.f. output is taken from a standard coaxial chassis connector—the "UHF"-series type SO-239. The mating plug is the PL-259. These are military numbers; in some cases manufacturers have their own type numbers for the same components, but regardless of manufacture the numbers given above are sufficient for identification. There is a rather bewildering variety of r.f. connectors, each having its particular advantages, but the UHF series is used on most amateur equipment. The plugs in this series are designed primarily for use with cable approximately ½ inch in diameter, such as RG-8/U and RG-11/U. An adapter is needed for RG-58/U and RG-59/U cables, as shown in Fig. 7-6. To use the adapter and PL-259 plug, cut off the insulation as shown, leaving about ⅛ inch of the polyethylene dielectric protruding, then fold back about ⅜ inch of the braid. Screw the adapter into the body of the plug, solder at the spots indicated, and the job is done. But don't forget to put the outer part of the plug on the cable before doing the rest of the assembly!

With the larger cables, simply cut back the jacket so the braid will be exposed under the solder holes, trim off the insulation from the inner conductor just as shown in the upper drawing, assemble and solder.

Component Values and Substitutions

One of the rewarding things about building your own is that you can often save money by using parts salvaged from old equipment acquired in trades with other amateurs, or obtained from any of those sources that traditionally help fill the "junk box." Too often, though, the component you have is not exactly what is called for in the circuit you want to assemble. Will or won't it work?

You can go a long way toward answering that question correctly if you know what purpose the component serves in the circuit, and whether that purpose is one that requires an exact electrical value or whether any of a wide range of values could be used without making any difference in the circuit's performance. This is where it pays off to know something about the "theory" discussed in earlier chapters.

Bypasses

Take bypass capacitors, for example. The value of bypass capacitance is rarely "critical," in the sense that the exact value specified in the circuit has to be used. As you saw earlier in this book, a good bypass is one having low reactance compared with the impedance of the part of the circuit across which it is connected. "Low" usually can be taken to mean 10 per cent or less. Fig. 7-7 shows a sample r.f. plate circuit such as might be used in a transmitting buffer amplifier or frequency multiplier. The plate circuit, including the tuned tank, will usually have a total impedance of a few thousand ohms. A bypass reactance of 100 ohms would be sufficient, in most cases, so far as the plate circuit *alone* is concerned. The important thing here is what is on the *supply* side of the bypass. Perhaps the circuit designer didn't use the decoupling resistor, R. The r.f. impedance of the power supply circuit is generally a very much unknown quantity, so the safe thing to do in such a case is to use the largest bypass available in a suitable physical type. A ceramic capacitor of 0.01 µf.

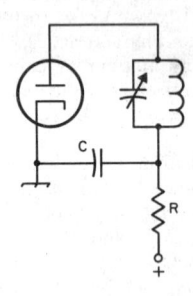

FIG. 7-7—Bypass capacitor and decoupling resistor. See text for discussion of range of values.

would be a logical choice, then, since it would offer the lowest reactance—a little under 5 ohms at 3.5 Mc.—conveniently available.

However, if the decoupling resistor R is included in the circuit, it is possible to proceed with less guesswork. Suppose R is 500 ohms. Then a reactance of 50 ohms or less would suffice, since this is 10 per cent or less of the lower of the two impedances to be bypassed. At 3.5 Mc. a capacitance of 0.001 µf. would do just as well, and possibly even better, than the 0.01 that isn't backed up by the decoupling resistor. You would have to consider whether the current flowing through R would have an appreciable effect on the voltage reaching the plate of the tube, of course, and possibly it would be better

Component Values

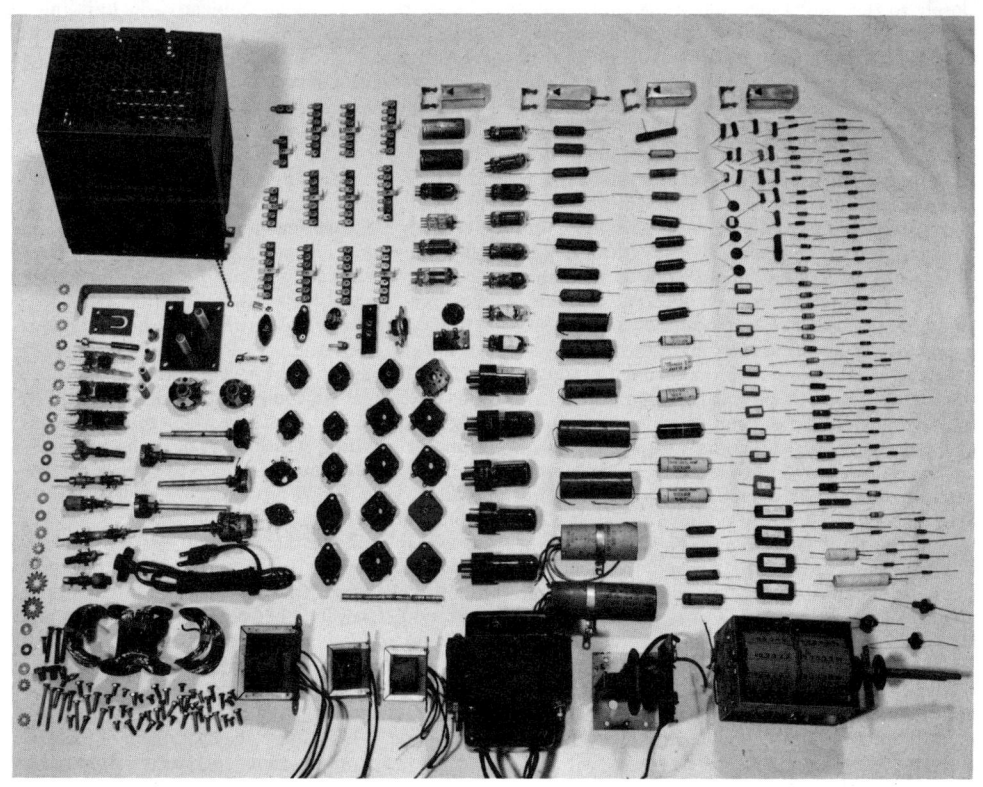

An old TV chassis gives you a magnificent start toward that prized possession of every constructor—a "junk box." Obtainable at practically any TV shop for a few dollars, many of the parts have an immediate money-saving application. This collection was salvaged from a single chassis by K4ATG.

to replace R by a small r.f. choke. A choke inductance of a few microhenrys would do as well as the 500-ohm resistor.

Capacitor Types

The same principles apply in audio circuits. Whether it is r.f. or audio, you'll rarely go wrong simply by substituting a *larger* value of bypass capacitance, provided you use a capacitor of the same type—ceramic, paper, electrolytic, and so on.

Aside from capacitance, these various types do differ in other important characteristics, some of which are not suitable in certain applications. Electrolytic capacitors, for instance, are not useful at radio frequencies, nor are they useful in purely a.c. circuits. Paper capacitors become inductive at radio frequencies, particularly in the larger sizes (over 0.01 μf.) and at the higher frequencies (14 Mc. and above), so they should not be substituted for disk ceramics. Mica and ceramic capacitors are generally interchangeable for bypassing, at least at frequencies below 30 Mc.

At v.h.f. you may not be able to get away with substituting a different type of capacitor for the one specified in the circuit, because at these frequencies the internal inductance of the capacitor, along with the inductance of the leads, becomes increasingly important. In v.h.f. equipment, better stick to the circuit designer's recommendations until you've gained enough experience to know what you can and can't do.

Resistance

Resistance values usually should be followed fairly closely, although there is often room for substitution here, too. A volume control in the grid circuit of a tube, for example, usually can be anything from 0.25 to 2 megohms without any very marked effect on the operation of an audio circuit. In most cases it is more important to use a control with an "audio taper," since this gives smoother control of volume than the "linear" type. Less liberty can be taken with cathode bias resistances, as a general rule, because these determine how the tube operates. However, you can go up or down in cathode resistance by 25 per cent or so without doing anything serious to the circuit operation, in most cases.

Tuned Circuits

In r.f. tuned circuits it pays to stick fairly

closely to the inductance and capacitance specified, even though you know that there is a wide range of *LC* combinations that will tune to the required frequency. But you don't always have to use the exact components mentioned. Variable capacitors of the same capacitance range, but of different manufacture, can be substituted. Sometimes this necessitates a minor change in the physical layout, because of different size, shape, or terminal arrangement, but this need not keep you from making the substitution.

The principal point about coils is that they should have the right inductance. The diameter, length, and number of turns can be varied to suit the form you have, just so long as you come out with the right value of inductance. Of course, you shouldn't use a physically large coil where the layout calls for a compact one, nor should you use a tiny coil to handle r.f. power when a big one is needed. However, if you can't get a slug-tuned form of exactly the diameter and length specified, one you do have or can get can be substituted if you change the winding so the coil can be tuned to the desired frequency. A smaller-diameter form will need more turns, and vice versa. If necessary, the wire size can be changed to the next nearest one that will let you get the turns on the form.

Don't be afraid to try substituting, but use your knowledge of circuit operation when doing it. If there is anything *really* critical in the circuit you're working on, the designer will have pointed it out, if the article describing the circuit was properly written.

Some General Remarks

There always have been, and always will be, "haywire artists" whose objective in building is to get something working no matter how it looks. However, most constructors are happier when their creations not only work but are presentable, too. Appearance and performance are not incompatible. Achieving both does take forethought, though. For example, a pleasing layout for panel controls doesn't follow automatically from a good r.f. circuit layout. You may have to spend some time moving parts around on the chassis, or making pencil sketches, before arriving at a plan that is desirable both electrically and mechanically. It's worth the effort.

Here are a few basic rules that should be kept in mind:

Make your r.f. wiring as short and direct as the placement of components will permit. Give wires that carry r.f. all the room you can. Don't bunch r.f. wiring with power or audio wiring.

In laying out and wiring screen-grid r.f. amplifiers, keep the r.f. grid wiring well separated from plate wiring, with as little wiring exposed as possible. Treat the grid and plate circuits as though they were deadly enemies, neither to be allowed to know that the other is in the neighborhood. Coupling between these two circuits, whether inductive or capacitive, can lead to all kinds of difficulty with instability.

Run r.f. ground (chassis ground, that is) leads directly to chassis, with the shortest possible lead length. The chassis itself, if made of material having good conductivity such as aluminum, has far less resistance and inductance than wire. If the layout permits, it is good practice to keep the grounds associated with a single stage close to each other, but not if you have to run long ground leads to do so.

You wouldn't be guilty of the pile-up at the left when a clean-cut layout like that at the right could be used. But, believe it or not, there are some who don't seem to appreciate the difference. They often have trouble making things work, too.

Wiring Practice

Lettering controls and connectors not only gives homemade equipment that "professional" look but serves a very practical purpose as well. The lettering here was done with panel decalcomanias that can be found in sets in radio supply houses.

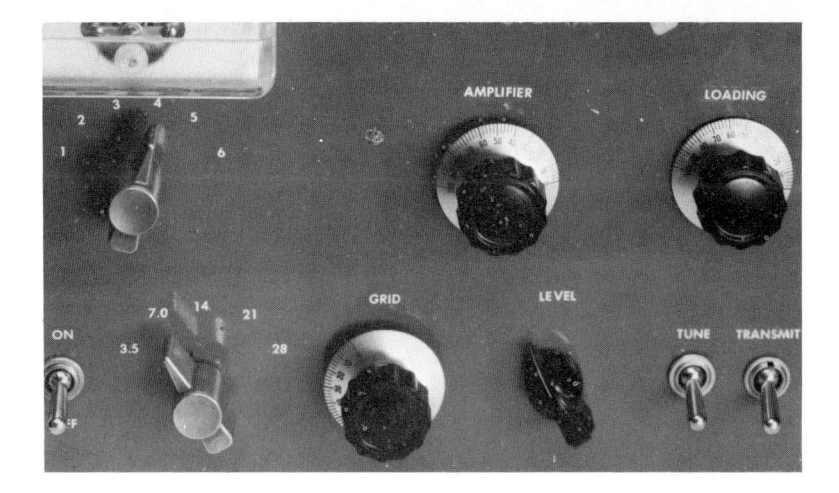

Don't put components in layers if you can help it. There may come a time when you have to do some trouble shooting. You don't want to have to dismantle half of the circuit just to get at a suspected component at the bottom of the pile.

Don't put controls so close together that you haven't finger room, or can't turn one knob without bumping into another and changing its setting.

When you've finished a piece of gear, make a diagram of it *exactly* the way it's wired. A written record is a time saver later. Your memory of what was done, and why, will be hazy after a few months.

Measure the d.c. voltages, and others if possible, at socket terminals and power distribution points. Jot them on the circuit diagram. If anything goes wrong later these voltages can be checked before you start taking things apart. Knowing what the voltages should be everywhere in the circuit gives you a head start in finding out what has gone wrong.

Label the controls and terminals on your equipment. Unless the unit is in use every day, you'll soon forget which terminals are which. Decalcomanias (such as "Technicals") sold by most radio stores are easy to apply, permanent, and give equipment a professional look.

On final tip: If you have an ohmmeter, test every resistor and capacitor before using it. Occasionally a defective or mislabeled one will be found, and it is easier to weed out the unsatisfactory ones before than after the set is ready for testing. Capacitors should be checked to make sure that they test "open"—extremely high resistance—although you may not be able to check the capacitance. In testing electrolytics, remember that the positive battery lead from the ohmmeter must go to the "plus" side of the capacitor. An ohmmeter will show an initial reading on capacitors of high value, but this is simply a charging current and will drop to zero or a very low value in a short time. Wait until the needle stops moving before deciding that the capacitor is good or bad.

On Kit Wiring

The real delights of construction are tasted only when you build something that is entirely the product of your own brain and hands. Designing your own takes knowledge and experience, and the experience of building from a published description inevitably adds to your store of knowledge. Your technical stature will grow with each project of this nature.

This cannot be said of building commercial kits. The principal experience you get here is practice in soldering. Not that there isn't a great deal to be said in favor of kits. They are economical; designing and building your own is not, either in time or money, unless you count the time as a worthwhile expenditure for the sake of learning, and unless you can acquire parts at something less than the price of new ones. Furthermore, since in the kit every part and wire is laid out for you, the set should work the minute you've made the last connection. All you have to do is to follow simple directions. Therein is the secret of successful kit building: follow the directions *literally*. Troubles with kits can practically always be traced to two things; *not* following directions, and poor soldering.

It isn't a bad rule, when building a kit, to avoid looking at the circuit diagram until the set is finished and tested. If you don't know what you're building you can do a better job, paradoxical though it may seem. The reason is that you won't be tempted to "improve" on the kit maker's procedure, and thus won't find yourself in blind alleys.

UNDERSTANDING

Trouble-Shooting

Radio equipment, like other machinery, occasionally will develop faults. Getting it back into good operating condition may take some doing. No discussion of trouble-shooting could anticipate *all* the things that might go wrong. The most anyone can do is to outline a few principles. The important factors in locating faults are adequate test equipment, knowledge of the circuit operation, and common sense. Locating the fault is the real problem, because once you've found it the remedy usually is obvious.

We have two different kinds of trouble-shooting to consider. One is the case where you finish building a piece of gear and can't make it work. The other is when a set has been performing properly right along until, more or less suddenly, trouble develops. In the second case the probable cause is a component that has failed, since you know the equipment *did* work. Almost anything could be wrong in the first instance.

Equipment

First, as to test equipment. The most useful item, all around, is the combination voltmeter, milliammeter and ohmmeter (v.o.m.). The instrument itself is discussed in Chapter 14. You need it to spot defects in resistors and capacitors, for measuring operating voltages, and for checking circuit continuity. A vacuum-tube voltmeter is even better for many tests, because its use has less effect on the circuit (see Chapter 14). However, it won't measure current directly. This is not a serious handicap, since current can be measured in terms of the voltage drop across a resistance.

The grid-dip meter is also useful, especially in testing newly-built equipment. It can also serve as a signal source for receiver alignment. Between the grid-dip meter and the v.o.m. or v.t.v.m. you can handle most of the problems you'll run into. For testing of a more advanced nature the oscilloscope is in a class by itself, although it doesn't supersede the other instruments. But to use a scope to good effect you have to know it well. We'll confine ourselves here to the simpler testing.

New Equipment

If you put together a circuit from a book or magazine article and it doesn't do what the author claimed, what to do? First, check your wiring to be sure it's right. *Really* check it. It isn't unusual to pass over a wrong connection time and again in checking; you simply have a blind spot for that error. If you can do so, get an amateur friend with some circuit experience to look it over, too; he may see right away the point you've missed.

Once you're sure the wiring is correct, check voltages. Measure the heater voltage on the tubes, and look at the heaters; with a little experience you'll be able to judge pretty well whether the cathode is showing the right color. If the heater voltage at the socket prongs is right but the cathode isn't up to temperature, something is wrong at the socket. The tube may not be making good contact, or you may have a "cold-soldered" joint on the prong. Measure the voltage on the prongs, not on the soldered connections.

Incidentally, measuring voltages in this way is good practice in *all* cases. If different voltages are measured on the conductors on either side of a soldered joint, something is wrong with that joint and it had better be resoldered. Cases like this crop up more often than you might think.

Having established that the tube heaters are operating properly, check the d.c. supply voltage and measure the voltages on plates, screens, and cathodes where cathode bias is used. The circuit from which you've worked often will have information to guide you here. If not, you must use your good judgment, based on knowledge of what the circuit is supposed to do. For example, screen voltage taken from the plate supply through a dropping resistor obviously will be lower than the supply voltage. (However, you have to be careful here—read the section on "Instrument Effects" in Chapter 14 before assuming that the measured voltage is too low. Usually, measurements made with a 20,000-ohms-per-volt meter or v.t.v.m. can be taken at face value, but beware of the readings of a 1000-ohms-per-volt meter in high-resistance circuits.) If no voltages have been specified with the circuit, your best plan is to assume that the values should be the normal ones for the type of tube and kind of service. The circuit discussions in earlier chapters can be used as a guide.

Oscillators can be checked by using your receiver to listen for the signal from them at the frequencies at which they should be working. Another method is to use a voltmeter to measure the d.c. voltage across the grid-leak resistor. If there isn't any, the oscillator isn't working. This measurement has to be made through an r.f.

Trouble Shooting

choke, as shown in Fig. 7-8A, to avoid short-circuiting the oscillator grid to chassis. Keep the lead marked A just as short as it can be made. If the oscillator already has a choke in its grid circuit and the grid leak is bypassed, as in 7-8B, the extra choke isn't necessary. Alternatively, a low-range milliammeter can be temporarily connected in series with the low end of the grid leak to measure the grid current. A 0-1 milliammeter will do for low-power oscillators, but a 0-10 instrument may be needed for more powerful circuits. If you get no grid voltage or grid current by this method, try tuning the oscillator circuit through its range; the circuits may simply be off tune. If there is still none, the circuit should be looked over carefully once more, for wiring mistakes or poor connections. If this too fails, use a grid-dip meter to check the circuit tuning, and adjust the tuned-circuit constants to bring the tuning into the right range.

Grid current in any stage of a transmitter can be checked by one of the methods shown in Fig. 7-8.

In a transmitter, if the plate current of an amplifier doesn't behave the way the circuit information said it should, the answer may be self-oscillation in the amplifier. If an amplifier has grid current with the excitation shut off (by pulling out the crystal or disconnecting the v.f.o.) *something* is certainly oscillating. This should be checked with the key closed, of course. The oscillation may be in the amplifier itself or in any preceding stage. Pull out the tube in the stage immediately preceding; if the oscillation continues it's in the stage you're working on. In final amplifiers, don't hold the key down any longer than is necessary to see if you have grid current. The tube usually will be badly overloaded when it isn't getting its normal excitation.

When an amplifier self-oscillates, it's usually either at a frequency near the intended operating frequency or at some v.h.f. frequency in the 100-150 Mc. region. In the former case, you may be able to hear the signal in your receiver by checking just as you would check your operating frequency. The v.h.f. parasitic oscillation is harder to spot, but if you can't find a signal near the right operating frequency it's safe to assume that a v.h.f. oscillation is responsible for the poor behavior. Practically all tubes need v.h.f. parasitic suppressors to be stable, and sometimes the suppressor that worked in the original circuit has to be modified in a copy. Here the only thing you can do is try various sizes of small coils and resistors, along the lines of the suppressors in the circuits of Chapter 10, until you find a combination that makes the amplifier stable.

If you discover an oscillation near your operating frequency, the remedy is to neutralize the stage. If the circuit already has neutralizing provisions, it's a matter of adjustment. The right setting of the neutralizing capacitor is the one that results in the least change in the amplifier's grid current when you tune its *plate* circuit through resonance. If there is no neutralizing cir-

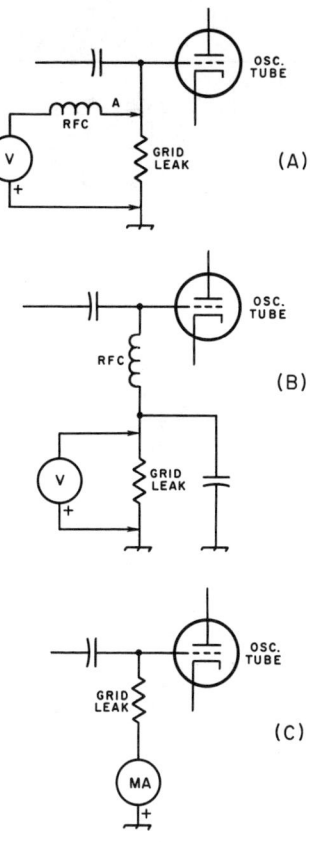

FIG. 7-8—Checking oscillator operation by measuring rectified grid voltage with a d.c. voltmeter, or grid current with a low-range d.c. milliammeter. The voltmeter resistance should be at least ten times the grid leak resistance in making such measurements.

cuit you'll have to put one in, in most cases. However, it is sometimes possible to stop self-oscillation by using resistance loading in the grid circuit. A resistor of a few thousand ohms (a carbon type, not wire-wound) across the tuned circuit connected to the grid may do it. Such loading cuts down the grid drive, though, and if you make the resistance low enough to stop the oscillation there may not be enough drive left to make the amplifier work properly. In such cases neutralization is the *only* answer. Circuits are given in Chapters 4 and 10.

Oscillators in receivers can be checked by the methods described above. In building converters, your principal problem will be getting the circuits to tune to the right frequencies, once you're sure the oscillator is working and the voltages on the converter are right. The grid-dip meter can be invaluable here.

Forethought

As a precaution in building any kind of equipment, test every component before you use it. A mismarked resistor or a faulty capacitor can

Table 7-III
Trouble-Shooting Chart

Symptom	Possible Cause	Procedure
Set is dead; tubes not lit.	Power-plug pulled out. Fuse blown.	Replace plug. Remove fuse and check for continuity. If blown, remove rectifier tube from power supply, replace fuse with new one. Then if tubes heat normally, shut off power and measure resistance from +B side of power supply filter to chassis. If zero or very low resistance, disconnect filter capacitors and measure each separately; a good electrolytic capacitor should show a resistance of 50,000 ohms or more. Measure resistance from filter choke winding to ground; this should show open circuit. If power supply filter tests OK, check +B on each stage individually, with other stages disconnected from line. Continue until defective component is located. If tubes do not light with rectifier out of its socket, remove all tubes and measure heater voltage. If zero or low, disconnect heater leads at transformer and check. If OK at transformer, look for short-circuit in heater wiring by checking individual circuits, disconnecting all except the one being checked. Normal voltage at all sockets with tubes out indicates a short in a tube.
Set is dead; tubes lit but less than normal brightness. Rectifier tube excessively hot.	Short-circuit in power supply.	Check as above.
Set seems dead, but all tubes appear to be normal.	Open resistor, preventing plate or screen voltage from reaching tube.	Measure plate and screen voltage on each stage, checking against voltage data if available. Measure voltage between cathode and chassis where cathode resistors are used; set r.f. (and i.f., if separate) gain control at maximum for this test. Cathode voltages should be only a few volts in a normal stage; zero cathode voltage indicates no plate current; excessively high voltage indicates open cathode resistor. If all such cathodes read high, vary the r.f. control; if no change, control is open. Measure voltages at all tube-socket prongs, using a high-resistance voltmeter. Measure voltages at both terminals of any resistors in series with plates or screens; if the same at both ends, tube in that stage may be defective. If voltage on tube side of resistor is abnormally low check values of resistors, disconnecting one terminal to make sure that other resistors in circuit are not affecting the measurement. Zero voltage at tube side of resistor while voltage at the other side is normal indicates open resistor or internal short in tube. Check which by removing tube from socket; voltage will show at both terminals of resistor if tube is defective.
Receiver does not bring in signals but tube hiss can be heard with gain controls at maximum.	Defective component in stage near front end of receiver.	Touch screwdriver blade to grid of each tube, working from audio output stage toward front end of receiver. A click or "pop" should be heard, becoming stronger as you move toward the front end. Tube which shows little or no noise with this test is probably in defective stage. Measure voltages as above. Check oscillator in h.f. oscillator-mixer or converter stages by measuring rectified voltage in grid leak (see Fig. 7-8). Lack of oscillation may be caused by defective tube, resistor or capacitor in oscillator circuit. Check these components separately. Check coils by measuring resistance; it should be very low if coil is OK. Measure antenna input coils, and check that antenna is connected.
Transmitter final amplifier plate current is high and does not respond to adjustment of tuning and loading controls. R.f. output is zero or very low.	No grid drive to amplifier, or very low grid drive.	Check band switch to make sure amplifier plate circuit is on proper band. Disconnect output cable; if plate current dips normally, load (antenna) circuit is shorted or badly off tune. Measure amplifier grid current (see Fig. 7-8). If none, or meter reads well below normal, adjust driver (buffer amplifier, doubler or oscillator, depending on circuit) for maximum possible grid current. If none, check oscillator tuning and rectified grid voltage (Fig. 7-8). Try another crystal, if crystal-controlled. Check oscillator plate and screen voltages; if zero or abnormally low, measure resistances used in circuit (disconnect if necessary to eliminate simultaneous measurement of other resistances) and check coils or r.f. chokes for continuity. If these components seem normal, try another tube. If oscillator is working, make similar checks on buffer stage.

Trouble Shooting

cause lots of your time to be wasted. The best rule is "don't trust—measure!"

And when you finish construction and have the gear working to your satisfaction, measure all d.c. voltages from socket prongs to chassis. Mark them on your diagram, or list them in notes that you will keep in a safe place. Record voltages out of the power supply, too—in fact, the more such measurements the better. If anything goes wrong later, you'll save time and effort in spotting the cause of the trouble.

Equipment Failures

As we said earlier, you have one thing going for you when a piece of equipment ceases to operate—you know that it once did work. When it ceases to do so the job is one of finding out what component went bad.

If you buy a receiver, transmitter or what-not, or put one together from a kit, it's good practice to record all voltages at socket prongs and power-supply outputs. The instruction book may have these figures in it, but they are simply representative of what you can expect. They aren't the voltages in *your* set, necessarily. Very often the equipment will work as it should with voltages differing by quite a percentage from those given. If you know what your voltages are at the start, you'll have specific values to check against if something goes wrong later.

No servicing instructions could possibly cover everything that could go wrong. New quirks are discovered every day. Again, your best asset is thorough understanding of the circuit and what each component is supposed to do.

Check the obvious things first. Don't take the set apart and spend fruitless hours on it because the power plug accidentally was pulled from the socket, or because an antenna connection fell off. When there is an actual failure, Table 7-III should help. It covers the principal causes of failure.

Your eyes and nose are good test equipment, too, and should be used to best advantage. Components that have "blown," such as resistors and capacitors, often will give themselves away by a burned or blistered appearance. A transformer or filter choke that has been badly overheated will have an unforgettable odor, as do other components in which insulation such as oiled paper or bakelite is an ingredient. Burned-out r.f. coils and chokes, too, usually will show external evidence of overheating.

When you discover which component failed, don't replace it until you're certain that you know *why* it failed. Components such as transformers and resistors rarely go bad on their own account, if the set has been designed with a reasonable safety factor. They fail because they have been asked to carry a current much beyond their safe rating, and the current was there because *something else* went wrong. So check for short circuits before replacing the component and turning on the power, otherwise you may have to do the same job all over again.

Success in servicing is pretty largely a matter of experience and knowing your circuit. Table 7-III is an elementary guide, but there is no magic road to all the answers. This is just as true of trouble shooting a piece of ham gear as it is true of repairing an oil burner, an automobile engine, or a television receiver.

Chapter 8

Building Receivers

No amateur expects that every QSO will be "100 per cent." Distant signals may fade out. Local man-made electrical noises may pop up at any time to drown out at least the weaker stations. There may be interference from a stronger signal on, or close to, the frequency of the station you're trying to receive. Some of these things can't be helped even with the best equipment so far invented. But there's a lot your receiver—and *you*—can do about some of them.

What features should a receiver have?

If you ask ten experienced hams that question you'll no doubt get ten different answers. The things each thinks vital will depend on the kind of operating he does—c.w., a.m. phone, s.s.b., DX chasing, traffic work, and so on. Receivers can be good in one department and not so good in others. But all good receivers do have one thing in common—they are easy to tune. Unless you can operate the controls comfortably and without strain, you can't get the most there is to be had from your receiver. The more advanced its design the more important it is that the set should be easy to tune.

The two main factors in ease of handling are the receiver's tuning rate and its stability.

Tuning Rate

The **tuning rate** of the receiver is the number of kilocycles it covers for each complete turn of the knob. Imagine for the moment that you have a receiver that can separate stations on channels 10 kc. apart. If the tuning rate is 20 kc. per turn of the knob, one turn will move you through two stations. But if the tuning rate is 300 kc. per turn, one turn will move you through 30 stations. It's harder to set the knob on one of those 30, by far, than to set it on one of the two you'd cover with the 20-kc. tuning rate. Fig. 8-1 should make this clear.

Sometimes tuning rate and bandspread are used as almost interchangeable terms. Bandspread really refers to the dial calibration—whether it's a long or short scale for a given number of kilocycles, and whether the *whole* scale is devoted to a comparatively small frequency range such as an amateur band.

Electrical bandspread and **mechanical bandspread** are names in common use. While either or both can have the spread-out dial scale associated with actual bandspread, the fact that there is a comparatively slow tuning rate is what is actually meant. In the electrical type the slow rate is attained by the design of the tuned circuits. In the mechanical type the tuning is slowed down by a high dial-drive ratio, with pulleys, gears, or the pinch-drive equivalent of gears. In practice, a combination of means may be used.

Stability

Good stability in a receiver means, primarily, that once you've set the tuning dial to the station you want the station "stays put." If the receiver isn't electrically stable the signal will gradually drift out of tune. Actually, it's the receiver that's doing the drifting, and you have to keep touching up the tuning to keep from losing the signal entirely.

(This assumes, of course, that the signal itself is stable. This isn't always the case, and sometimes receivers are accused of frequency drift when the blame rightly goes to the incoming signal. So before you condemn a receiver on this score, be sure you do your checking on a really stable signal. A simple crystal-controlled frequency standard such as the one shown in Chapter 14 is handy for this.)

There are other types of electrical instability, too. When your a.c. line voltage changes, the voltages in the set will also change. Receiver tuning can be quite sensitive to voltage. If a sudden load on the line, such as a refrigerator going

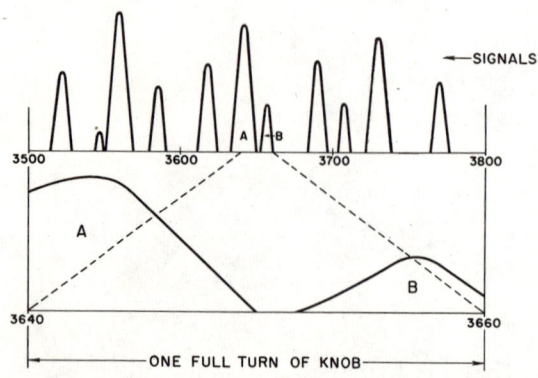

FIG. 8-1—How tuning rate affects ease of tuning. With fast tuning (top) you tune through many stations with one turn of the tuning knob, so setting exactly on the one you want is a rather critical operation. A slow tuning rate (bottom) gives a vernier effect and setting on the desired station is much easier. Although with the slow rate each signal seems broader, each comes in and goes out in the same number of kilocycles in both cases. That is, the selectivity is not affected by the tuning rate.

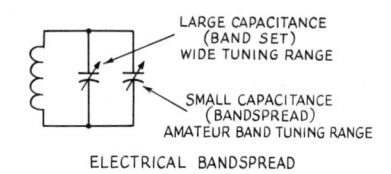

Receiver Construction 137

FIG. 8-2—Electrical and mechanical bandspread. These are really means of obtaining a slow tuning rate on a circuit that covers a wide frequency range. The band may or may not be spread over a large dial scale—in fact, it isn't likely to be spread out with mechanical bandspread. Other mechanical schemes than the pulley arrangement shown can be used.

on, detunes the signal you're listening to, your receiver is definitely a candidate for improvement in this respect.

A related effect is that the receiver tuning varies when the r.f. or i.f. gain controls are adjusted. (An audio gain control won't cause this because it doesn't vary any of the d.c. voltages on the tubes.) This, too, can be overcome by proper design.

The mechanical construction of the receiver is likewise important. Receiver tuning can be affected by mechanical shock and vibration. Try pounding the operating table or rapping the receiver cabinet. If the beat note on an incoming c.w. signal remains unchanged, you've got a receiver that is excellent mechanically. This is a rather rough test, but you certainly don't want to lose a signal because you accidentally bump the operating table when you move an arm or leg.

Not-So-Small Points

There are innumerable "little" things that make the difference between pleasure and pain in operating a receiver. For example, there is the matter of knob size. A large tuning knob in effect slows down the tuning rate, so a big knob is definitely an advantage.

There should be plenty of room for your hand, too—you don't want to be bumping into other controls when you tune. And the height of the knob above the table surface makes a big difference in comfort and fatigue.

You want a tuning system that has no **backlash**. There is nothing more annoying than to have the tuning apparently start off in the opposite direction to what you intend, then reverse itself and try to catch up. This is an extreme case of backlash, admittedly, but it does occur. The lesser case, where just nothing happens at first when you start turning the knob, is not enjoyable either.

Good design and construction will overcome these faults. Unfortunately, most of the cheaper receivers suffer from them in some degree, because it costs money to cure them—money that can't be put into a set to sell at a low price. There isn't much you can do to improve a ready-made set, ordinarily, especially in mechanical features. When you build your own it's different. Then you can keep matters fairly well under control, by following the principles set out in this chapter.

Sensitivity and Selectivity

Probably most amateurs think first of sensitivity and selectivity when "rating" a receiver. But all receiver features are pretty intimately related to each other. **Sensitivity**—the ability to bring in weak signals is the way most hams would define it—is rarely useful unless it is accompanied by good selectivity. If weak signals are always drowned out by strong ones on nearby frequencies, it doesn't do much good to have a sensitive receiver. But as you make the receiver more selective you find that the tuning becomes more critical; you no longer have wide tolerances in setting the dial.

Thus high **selectivity** puts a premium on stability and tuning rate, the very features we emphasized at the beginning. An unstable, hard-to-tune receiver with high selectivity can quickly drive you to some less nerve-wracking hobby.

In this day it is possible to get almost any desired degree of selectivity. As we have seen in Chapter 3, the amount that can be used depends on the kind of signal being received—that is, on the bandwidth of the signal.

A bandwidth of about 6 kilocycles is needed to receive amplitude-modulated phone signals. No phone signal *needs* to take up more bandwidth than this, and code signals can get along with far less.

You will want to be able to receive a.m. phone, so it's a good idea in building or assembling your first receiver to base your selectivity on that kind of reception. (You're used to selectivity of this sort from tuning an a.m. broadcast receiver.) Construction is easier than when you try for higher selectivity. The cost is generally less, too. Also, in getting experience with reception of other types of signals, such as c.w., you'll develop an appreciation of what higher selectivity can mean—probably more vividly than if you started out with the most advanced receiver.

Building Your Own

Building a complete receiver, even a fairly unpretentious one, is quite an undertaking—that is, if the receiver is to meet certain minimum standards of performance that present-day conditions make necessary. The only practical way of getting the adjacent-channel selectivity needed is to use the superhet principle. But as we saw in Chapter 3, the superhet will bring in all sorts of

UNDERSTANDING

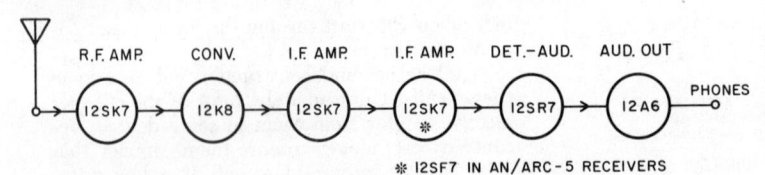

FIG. 8-3—Block diagram of the ARC-5 or "Command" receiver. It also has other names (see Table 8-1). The general layout is the same in all receivers in the range from 200 kc. to 9 Mc.

R.F. AMP. 12SK7 → CONV. 12K8 → I.F. AMP. 12SK7 → I.F. AMP. 12SK7* → DET.-AUD. 12SR7 → AUD. OUT 12A6 → PHONES

*12SF7 IN AN/ARC-5 RECEIVERS

signals that aren't even in amateur bands, if you give it the chance. Something has to be done about this, and that "something" is to use the double-super idea.

If you're thinking that this piles another complexity on the already-complex superhet, you're right. It happens, though, that an actual receiver can be assembled rather inexpensively and without a great deal of construction work. This fortunate circumstance comes about because of the ready availability of a receiver that was used in World War II—the **ARC-5** receiver, as it is commonly known.

The ARC-5 Receivers

Thousands of ARC-5 or "Command" receivers have been and are still being disposed of as military surplus. Depending on condition—which ranges from brand new to battered but operating—the prices have varied from under five dollars to fifteen or more. This is of course far below the cost of the parts that would be needed to make a comparable receiver. These sets were built for use on a 28-volt d.c. supply, and need a little modification before being usable with an ordinary power supply operating from the a.c. line. The modification is quite simple, as will be shown later.

The basic receiver is a 6-tube superhet having an r.f. amplifier, a converter, two i.f. stages, detector, and two audio stages. The layout is shown in block-diagram form in Fig. 8-3, and the receiver itself in Fig. 8-4. There are several tuning ranges, each covered by a separate receiver. The ones most commonly available are 6 to 9 Mc. and 3 to 6 Mc. Less common, but not too hard to find, are ones covering 200 to 500 kc., and 1.5 to 3 Mc. The one that covers 3 to 6 Mc. makes a good foundation for an amateur-band receiver.

Receivers covering a given tuning range do not all have the same type numbers. Different services had their own type designations. Also, there may be minor circuit differences between the various models. These are not serious. Table 8-I will help in identifying the receivers.

80-Meter Receiver as a Tunable I.F.

The 3-6 Mc. receiver can be used by itself for reception in the 3.5-4 Mc. amateur band. The

FIG. 8-4—A 3-6 Mc. receiver (BC-454) with tuning knob (not usually supplied with the receiver) and the adapter plate described in the text.

Command Receivers

		Table 8-I		
		"Command" Receivers		
Frequency Range	AN/ARC-5	(Signal Corps) SCR-274N	(Navy) ATA/ARA	Intermediate Frequency
190-550 kc.	R-23/ARC-5	BC-453	CBY-46129	85 kc.
520-1500 kc.	R-24/ARC-5	BC-946	CBY-46145	239 kc.
1.5-3.0 Mc.	R-25/ARC-5		CBY-46104	705 kc.
3.0-6.0 Mc.	R-26/ARC-5	BC-454	CBY-46105	1415 kc.
6.0-9.1 Mc.	R-27/ARC-5	BC-455	CBY-46106	2830 kc.

circuit must be modified a little to work with an a.c.-operated power supply, and the power supply itself must be built. Both points are covered subsequently in this chapter. For other amateur bands the 3-6 Mc. receiver can be used with a frequency converter, as shown in Fig. 8-5. By using a crystal-controlled oscillator in the converter, the stability is excellent on all frequencies. The actual tuning across each band is done with the 3-6 Mc. receiver, which thus becomes a **tunable intermediate-frequency amplifier**. The advantage of this arrangement (which is also used in some high-priced amateur-band receivers) is that you have the same band-spread and the same tuning rate, as well as the same frequency stability, on all bands. Also, a converter is quite simple to build, compared with building a complete receiver.

You don't have to have an ARC-5 receiver to take advantage of this idea. Any receiver that has good stability and a satisfactory tuning rate, and which is capable of tuning the 3.5-4 Mc. band, can be used. You may already have an inexpensive receiver that gives reasonably good results on 80 meters, even though its performance may be less than satisfactory on the higher-frequency bands. If so, it can be used with a converter. You simply keep it on 80 all the time, and convert the higher bands to the 80-meter range. The combination is a double super that will not only have the stability and bandspread of the 80-meter range but will also practically eliminate the images and other spurious signals that cause so much difficulty with low-priced receivers, especially on the 14-, 21- and 28-Mc. bands.

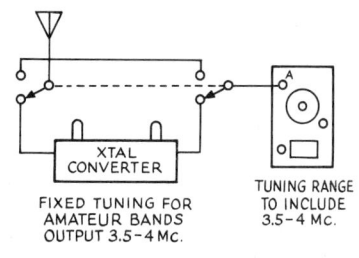

FIG. 8-5—Using a crystal-controlled converter for amateur bands from 7 Mc. up. The receiver with which the converter is used should cover at least the 3.5-4 Mc. range. For 28 Mc. reception a 2-Mc. range (for example, 3 to 5 Mc.) is desirable since the 28-Mc. band is nearly 2 Mc. wide. For 80-meter reception the converter is bypassed by the switching circuit as shown, the receiver then being used by itself.

There are several ways in which crystal-controlled converters can be built. A few are shown in this chapter. A converter can be built for just one band, if you wish, or it can be designed so that several amateur bands above 3.5-4 Mc. can be covered. Multiband design necessitates an assortment of crystals in the crystal-controlled oscillator, and either plug-in or switched coils in the circuits that are tuned to the frequency of the incoming signal. These circuits can be tuned over only a little more than a 2-to-1 frequency range, as a practical maximum, so it isn't possible to cover more than this with one set of coils. Many—probably most—receiver designs use a separate set of coils for each amateur band.

Modifying the Command Receiver

The ARC-5 receivers were designed for remote tuning, so some of the controls usually found on receivers are missing on these sets as they come from the dealer. There is no tuning knob, although there is a calibrated dial. The tuning capacitor and dial are driven through a spline mounted on the panel. Knobs that fit are available from a number of sources. There is no gain control on the receiver, and there is no b.f.o. switch. Provision has been made in the receiver for connecting both, through a connector in the recess in the front panel. It is also necessary to make provision for connecting a headset or loud speaker to the receiver, since in the original circuit the audio output is taken through a cable connector. A phone jack can be installed for this purpose.

Finally, you will need an external power

supply. The receiver uses tubes with 12.6-volt heaters and has them wired so that a 24-28 volt heater supply can be used. Here you have the alternatives of leaving the heater wiring unchanged and using a 25-volt filament transformer, or making a few simple changes in the heater wiring so a 12.6-volt filament supply can be used. (It would be possible, although not as economical, to make the same changes in the heater wiring and then substitute 6.3-volt tubes so a conventional heater supply would be usable.)

Heater Circuits

Probably most owners revamp the heater wiring so 12.6 volts will suffice. Fig. 8-6 shows how to do this. This drawing shows the set as it looks from underneath after the bottom plate has been removed. It is necessary to change the wiring on only three of the tube sockets. Although two capacitors must be dismounted in order to reach the wiring on two of the sockets, it isn't necessary to unsolder their leads. Leave the leads intact and remount the capacitors after changing the heater wiring.

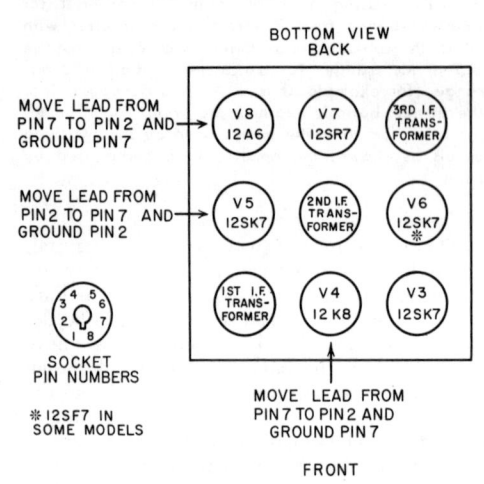

FIG. 8-6—This drawing shows the changes necessary for putting all tube heaters in parallel for use on a 12.6-volt supply. The metal-cased capacitors mounted on the side walls over V5 and V8 must be dismounted temporarily so the heater pin on the tube sockets can be reached. The leads to these capacitors need not be disconnected.

In removing leads on the tube sockets, heat the terminal just enough to melt the solder and then unwrap the lead. Don't cut any leads as they may not be long enough to reach to the new terminal after cutting. Pins to be grounded, as indicated in Fig. 8-6, are simply connected to the nearest chassis ground point through a short piece of wire.

Added Components

Connections for the b.f.o. switch, headphone jack, and gain control are shown in Fig. 8-7. The simplest way to mount the three components is to replace the switch panel adapter with a metal plate that covers the recess and uses the adapter mounting holes. This gives you a space approximately 2¼ by 1½ inches for mounting a jack, toggle switch, and gain control.

Fig. 8-4 shows a receiver with such a plate, and Fig. 8-8 gives dimensions and drilling information. The connections can be made to the internal socket by soldering flexible leads to the proper plugs, making them long enough so the three components can be wired before the plate is mounted. Be careful that none of your wires make contact with each other or with any of the socket pins, also that no solder runs between pins to short-circuit them.

Power connections should be made to the 3-pin connector on top of the receiver chassis at the rear. Fig. 8-9 shows how the power cable should be wired to this connector.

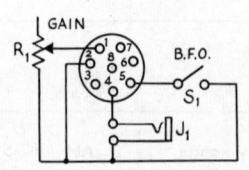

FIG. 8-7—Wiring of gain control, b.f.o. switch, and audio output jack. The b.f.o. is off when S_1 is closed and the gain is maximum when the arm of R_1 is at the bottom of the resistor.
J_1—Midget phone jack, single open or closed circuit.
R_1—25,000- or 50,000-ohm composition control, small (for example, IRC type Q or Mallory type U).
S_1—S.p.s.t. toggle switch.

Power Supply

With the filament circuit wired for 12.6 volts the current required for the six tubes is 0.9 amp. The total B current drain is about 50 milliamperes with a 250-volt d.c. supply. Thus only a relatively light power supply is needed. However, if a converter for the higher-frequency amateur bands is going to be used; it would be poor economy to build a supply that wouldn't take care of both the receiver and converter. Most converters will have at least two 6.3-volt

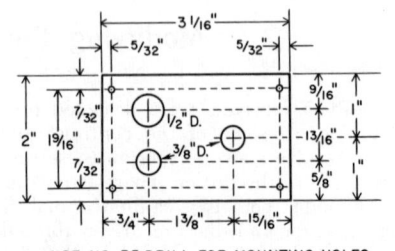

FIG. 8-8—Layout dimensions for mounting plate.

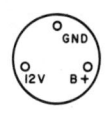

Command Receivers

FIG. 8-9—Power connector on top of chassis at rear. Connection cable to power supply may be soldered to the pins in this connector. Cover the exposed plus-B terminal with electrical tape so there will be no danger of accidental shock.

tubes and may have three, with a total heater-current requirement of about an ampere. The plate current will run 20 to 30 ma. in the average case. The additional drain should be provided for in the supply you build for the receiver.

Fig. 8-10 is a suitable power supply circuit. Of course, if you already have a supply that will do the job you don't need to build another one. You can use any plate voltage up to about 300, and if the supply happens to have more current capacity than you actually need it will make no difference—an underloaded supply will just run cool.

The power supply can be laid out in any fashion you please on a chassis large enough to contain all the parts. There are no special rules to follow, aside from using normal care in wiring so that no short-circuits can occur. Just don't mount the electrolytic capacitors where they can be "cooked" by the heat from the rectifier tube; their life can be shortened by running them at an unnecessarily high temperature. Finally, mount the rectifier tube so air can circulate around it.

S_2 in Fig. 8-10 is a B-plus on-off switch, not really a necessity but a distinct convenience if, as in testing, you want to turn off the plate voltage temporarily without having the heaters cool off. The 115-volt a.c. connection shown in the diagram can be a regular line cord. Notice that the extra 6.3-volt transformer is wired in series with the 6.3-volt winding on the power transformer to get 12.6 volts for the surplus receiver. Connect the side of this heater circuit marked A to the ground pin on the power socket of the receiver, along with the −B lead. If you build a converter, either let its heater wiring "float" or make sure that, if the converter filaments have one side connected to chassis, this A side is the one that connects to the converter chassis. The 6.3-volt winding on the power transformer should be the one used to supply heater power to a converter using 6.3-volt tubes, because it has higher current capacity than the separate 6.3-volt transformer.

The two 6.3-volt windings must be connected in the proper polarity to obtain 12.6 volts. The transformer leads will give you no clue as to this, so the thing to do is to connect two of the wires together, one from each winding, and then momentarily touch the remaining two together. If there is a spark, your wiring is right, and you'll have 12.6 volts across the open wires. If the polarity is wrong, touching the wires together will cause no spark. Another method is to make the trial connection between the windings and hook the two open wires to a tube heater. If the tube warms up, the polarity is right, but if the tube stays cold, reverse the connections of *one* of the windings.

Using the Receiver on 80 Meters

The 3 to 6 Mc. range of the receiver is much more than is needed for covering the 3.5 to 4 Mc. amateur band, but the additional coverage is useful on the 28- and 50-Mc. bands when the receiver is used with a converter. It is therefore recommended that the frequency range be left alone, although it *is* possible to modify it so the amateur band will cover practically the entire dial. However, the tuning rate is not too fast, and the selectivity of the receiver is such that tuning is not especially critical. The set is of

FIG. 8-10—Diagram of suggested power supply for ARC-5 receiver. See text for discussion of power requirements.

C_1, C_2—40-μf. electrolytic, 450 volts (dual 40-μf. unit may be used).
L_1—App. 8 henrys, 85 ma. (Stancor C-1709 or equivalent).
R_1—25,000 ohms, 5 watts, wirewound.
S_1, S_2—S.p.s.t. toggle.
T_1—App. 500 volts c.t., 90 ma., with 5-volt 2-amp. rectifier winding and 6.3-volt 3-amp. heater winding (Stancor PC-8404, PM-8404, or equivalent).
T_2—Filament transformer, 6.3 volts, 1.2 amp. (Stancor P-6134 or equivalent).

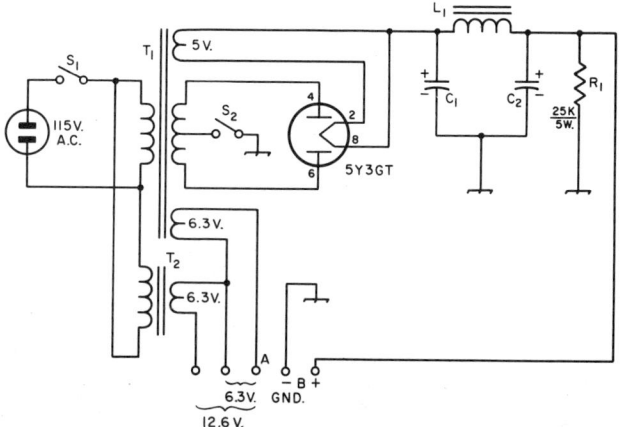

course not the last word in ham-band receivers, but its stability is good and the over-all performance is far better than you could expect from commercially-available receivers costing several times as much.

It is convenient to use a single-wire antenna with the receiver. The input circuit can be peaked with the trimmer operated by the small knob at the lower left corner. There is no external b.f.o. adjustment, but this is hardly necessary with a receiver of this selectivity (the bandwidth is about 4 kc. between the 6-db. down points). The b.f.o. can be brought into line with the i.f. amplifier, if necessary, by means of a screwdriver control which can be reached through a small hole on the right-hand side of the chassis near the back. You'll hear a swish as you turn this control (disconnect the antenna when making this adjustment). The point where the tone of this hissing noise sounds lowest is the right one.

After you become thoroughly familiar with the operation of the receiver you may want to touch up the tuning of the i.f. amplifier transformers. These are on top of the chassis in the compartment with the tubes. Unscrew the caps on top and the screwdriver controls are accessible. The simplest way to align the i.f. is to take the antenna off the receiver, turn on the b.f.o., and then adjust each trimmer for the lowest-pitch hiss. A metal screwdriver can be used, since the capacitor adjusting screws are grounded. Don't touch the b.f.o. tuning while making these adjustments. Only a small adjustment will be needed, unless the set happened to be very badly out of line when you got it. Start with the transformer at the rear and work toward the one at the front.

Greater I.F. Selectivity

After using the receiver for a while you may feel the need for greater selectivity—more ability to separate stations that are close together in frequency. A simple modification of the set will make a big improvement in this respect. It is done by making the second i.f. amplifier regenerative, which increases not only the selectivity but also the gain.

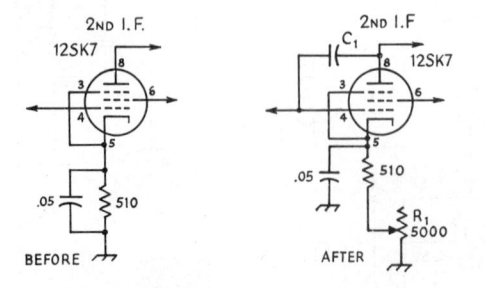

FIG. 8-11—Adding regeneration to the second i.f. stage for greater selectivity.
C_1—Very small capacitor—see text.
R_1—5000-ohm composition control.

You need only one additional component, a 5000-ohm control. This goes in the cathode circuit of the second i.f. tube to control its gain and thereby also control the regeneration in this stage. The original cathode circuit is shown at the left in Fig. 8-11. The tube is the one designated V_6 in Fig. 8-6. The modified circuit, at the right in Fig. 8-11, has a "gimmick" capacitor, C_1, connected between the plate and grid of the tube to provide the regenerative feedback. This capacitor is simply two ¾-inch pieces of insulated wire, one soldered to the plate pin and the other to the grid pin on the tube socket. The two wires are arranged side-by-side, but do not touch each other.

In order to get at the tube socket it will be necessary to dismount the metal-cased capacitor directly over it, but the lead to the capacitor need not be unsoldered.

Locate the cathode resistor, which is on the mounting board immediately to the rear of the tube socket. This resistor, 510 ohms (green-brown-brown), is the one nearest the tube socket. One side of the resistor is connected to the cathode terminal (Pin 5) and the other side goes to chassis ground. Unsolder the end of the resistor connected to ground. The simplest method of doing this is to hold the lead with long-nose pliers and heat the connection, then when the solder melts pull up gently on the lead and it will come out of the terminal.

The variable resistor, R_1, can be mounted anywhere you can find room for it, since it does not have to be close to the socket for V_6. Probably the most convenient method is to mount it on a bracket that is in turn mounted on the rear deck behind the tube compartment. If the bracket is formed so the shaft of the control is vertical, the shaft can be left long enough so its knob is about flush with the top of the receiver. This will make it easy to reach for adjustment.

Capacitor C_1 must be adjusted before remounting the metal-cased capacitor. Temporarily connect a wire between the metal case of the latter capacitor and the chassis, to put this capacitor back in the circuit. Then apply power to the receiver and set R_1 so that its resistance is nearly all in. Disconnect the antenna and turn on the b.f.o. Adjust the b.f.o. trimmer so that the background hiss has a high pitch. Then turn R_1 toward minimum resistance. If the capacitance of C_1 is large enough (try the two wires about $\frac{1}{32}$ inch apart at first) you will hear a squeal at some setting of R_1. This indicates that the i.f. stage is oscillating. If the stage just starts to oscillate with R_1 near minimum resistance no further adjustment is needed. If there is no oscillation with R_1 all cut out, increase the capacitance of C_1 by bringing the wires closer together. More likely, oscillations will start with R_1 at some fairly high-resistance setting, in which case the two wires should be spread apart. The object is to get the stage to go into oscillation with R_1 close to the minimum-resistance setting.

Converter Construction

Once this adjustment is completed the metal-cased capacitor can be remounted. Make certain that neither wire of C_1 can make accidental contact with any part of the circuit before you reassemble the set.

Now connect the antenna, reduce the gain with the regular gain control, shut off the b.f.o. and, with R_1 set just below the oscillating point, tune in a steady carrier of moderate strength. It will tune quite sharply. Set the tuning "on the nose," then turn on the b.f.o. and adjust its trimmer to give you the beat tone you like best for c.w. reception. As you tune through the band you'll find that signals will peak very definitely at this tone. The selectivity will be greatest when you keep the gain down and depend on the regeneration to build up the signal strength. If you operate the gain and regeneration controls properly, you'll get a very definite **single-signal** effect on c.w. signals, and will be able to separate phone signals much more readily than you could before adding regeneration to the i.f. amplifier.

Multiband Crystal-Controlled Converter

Once you have a satisfactory receiver for the 3.5-4-Mc. band—one that has good stability, adequate selectivity, and a slow-enough tuning rate—you're ready to make use of its good features for reception in other amateur bands. This can be done by means of a crystal-controlled converter, as discussed earlier in this chapter. The converter shown in Fig. 8-12 can be used, with an 80-meter receiver, for all bands from 7 Mc. to 50 Mc., if you wind all the coils and get the necessary crystals. This doesn't have to be done all at once. Make the coils and get the crystals for the bands you're particularly interested in. Later, if you want, the missing bands can be added.

The converter uses two tubes, one as an r.f. amplifier and the other as a combination mixer-oscillator. The r.f. and mixer stages are individually tuned. This eliminates the tracking problems usually associated with building a gang-tuned converter. Individual tuning also simplifies construction and design. The output circuit of the converter is untuned, so no adjustment of this circuit is necessary.

Circuit Details

As shown in Fig 8-13, the converter circuit uses a 6AU6 r.f. amplifier and a 6U8 (or 6U8A) mixer-oscillator, and the design is such that each r.f. coil assembly, L_1L_2, and mixer coil assembly, L_3L_4, will cover two bands. This means, of course, that only two sets of these coils are required to cover 7 through 28 Mc., a separate set being needed for 50 Mc.

The triode portion of the 6U8 operates as a crystal-controlled oscillator whose output is

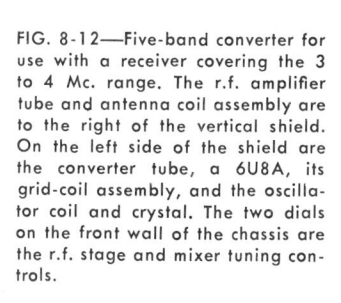

FIG. 8-12—Five-band converter for use with a receiver covering the 3 to 4 Mc. range. The r.f. amplifier tube and antenna coil assembly are to the right of the vertical shield. On the left side of the shield are the converter tube, a 6U8A, its grid-coil assembly, and the oscillator coil and crystal. The two dials on the front wall of the chassis are the r.f. stage and mixer tuning controls.

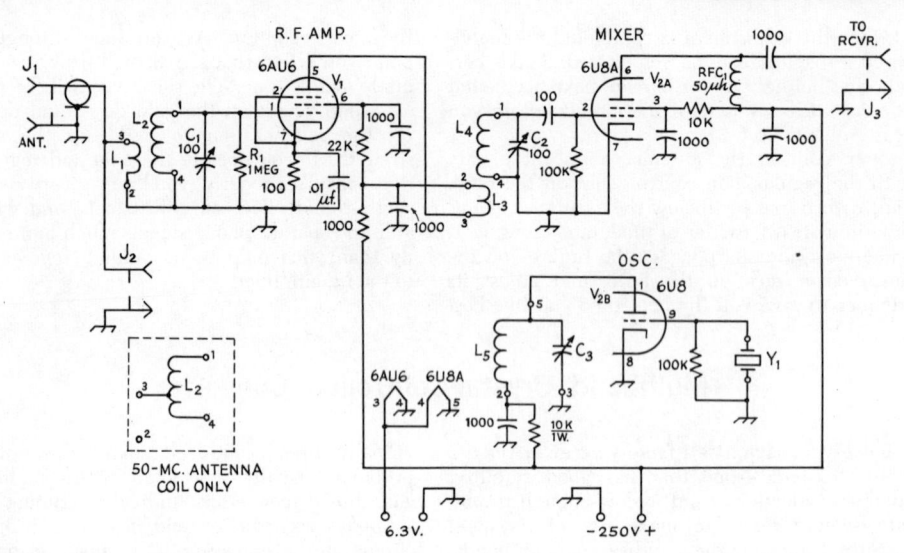

FIG. 8-13—Circuit diagram of the five-band crystal controlled converter. Unless otherwise indicated, capacitances are in µµf., resistances are in ohms, resistors are ½ watt.

C_1, C_2—100-µµf. variable capacitor (Hammarlund MAPC-100-B, Hammarlund MC-100-M, or E. F. Johnson 149-5).

C_3—Trimmer capacitor, one needed for each oscillator coil; 30 µµf. (Centralab type 827C, El Menco type 461, Allied 60 H 335) for all bands, except 7-Mc. when using a 3400-kc. crystal. C_3 for this 7-Mc. coil is a 100-µµf. mica fixed capacitor.

J_1—Coax chassis receptacle (Amphenol 83-1R).

J_2, J_3—Phono jack.

RFC_1—50-µh. r.f. choke (National R-33, Millen 34300-50).

Y_1—* 7 Mc.—3400 kc.
 14 Mc.—10,500 kc.
 *21 Mc.—17,500 kc.
 28 Mc.—24,500 kc.
 50 Mc.—46,500 kc. (International Crystal Co. type FA-9).

Coil Data

7 & 14 Mc.—L_1, L_3—4 turns No. 20, 1-inch diam., 16 t.p.i. (B & W 3015).
 L_2, L_4—15 turns No. 20, 1-inch diam., 16 t.p.i. (B & W 3015).
 *L_5, 7 Mc.—32 turns No. 24, 1-inch diam., 32 t.p.i. (B & W 3016).
 L_5, 14 Mc.—16 turns No. 24, 1-inch diam., 32 t.p.i. (B & W 3016).

L_1 and L_2 are made from a single length of B & W coil stock. The coils are separated by one turn, the 5th, which is cut and unwound from the support bars. L_3L_4 is exactly the same as L_1L_2.

21 & 28 Mc.—L_1, L_3—3 turns No. 20, 1-inch diam., 16 t.p.i. (B & W 3015).
 L_2, L_4—7 turns No. 20, ¾-inch diam., 16 t.p.i. (B & W 3011).
 *L_5, 21 Mc.—10 turns No. 20, 1-inch diam., 16 t.p.i. (B & W 3015).
 L_5, 28 Mc.—9 turns No. 20, ¾-inch diam., 16 t.p.i. (B & W 3011).

L_1 fits over L_2 and is positioned near the bottom of L_2. L_3L_4 is the same as L_1L_2.

50 Mc.—L_2—5 turns No. 20, ⅝-inch diam., 16 t.p.i. tapped 2 turns from ground end (B & W 3007).
 L_3—5 turns No. 20, ½-inch diam., 16 t.p.i. (B & W 3003).
 L_4—5 turns No. 18, ¾-inch diam., 8 t.p.i. (B & W 3010).
 L_3 is inserted inside L_4.
 L_5—4 turns No. 20, ½-inch diam., 16 t.p.i. (B & W 3003).

*Not actually required; see text.

coupled to the mixer via the capacitance existing between the pentode and triode elements within the tube. No additional mixer-oscillator coupling is required.

A short length of RG-58/U coaxial cable is used to couple the output of the converter to the receiver. For 7 Mc. and higher bands the center conductor at one end of the coax is connected to the antenna terminal of the receiver and the outer braid is connected to the receiver chassis. The other end is connected to a phono plug which goes into J_3. When 3.5-Mc. coverage is desired (the receiver by itself) the converter end of the coax should be plugged into J_2 and the r.f. coil, L_1L_2, should be removed from its socket. The purpose of R_1 is to provide a d.c. return for the grid of the 6AU6 when the coil is removed.

Construction

The converter shown in the pictures was mounted on a chassis large enough to provide space for a BC-454 and its power supply. However, the converter can be built on a separate chassis of its own by following the layout shown. The chassis size is unimportant so long as it is large enough to accommodate the components.

Converter Construction

FIG. 8-14—The input, mixer, and oscillator coil sockets run from right to left in this bottom view of the five-band converter. The 6AU6 socket is at the right and the 6U8A is at the left. The two variables, C_1 and C_2, are mounted on the front chassis wall in line with the associated coil sockets. The crystal mounting (Millen 33302) is near the oscillator coil socket. J_3 is beside it and J_2 is at the lower left. A piece of RG-58/U runs from J_2 to the antenna connector, J_1, on the rear wall of the chassis.

One as small as $2 \times 5 \times 7$ inches would provide more than enough room for the parts. If you build a separate power supply for the converter alone, it should furnish approximately 250 volts d.c. at 40 ma. and 6.3 volts a.c. at 1 amp. The circuit of Fig. 8-10 can be used by omitting the extra filament transformer, T_2, and substituting a power transformer with lower rating.

Note that a metal shield is used on top of the chassis to separate the r.f. and mixer circuits. The shield is made from a piece of aluminum measuring 3×5 inches.

Four-prong forms and sockets are used for the r.f. and mixer coils. A five-prong coil form and socket are used for the oscillator. The larger number of pins on the latter form helps avoid confusion when changing bands, although only three terminals actually are needed.

When installing the 6AU6 socket mount it so that Pin 1, the grid terminal, faces the L_4L_3 socket. The metal cylinder or shield in the center of the socket should be grounded to a lug mounted under one of the socket screws. When wiring, keep all lead lengths as short as possible.

The antenna terminal, J_1, is mounted on the rear chassis wall and is connected to J_2 by a short length of RG-58/U coaxial cable. Another short length of RG-58/U is used to connect J_2 to the socket terminal of L_1.

The Coils

Making the coils is a very simple process. Transparent polystyrene coil forms (Amphenol or Allied Radio 24-4P and 24-5P) are used to hold the coils, which are made from various sizes of B & W Miniductor stock. The coils are mounted inside the polystyrene forms and held in place by their own leads, the ends of which are soldered to the coil-form prongs.

The oscillator coils must be tuned to the crystal frequency. This is done by mounting a small capacitor along with the coil inside the form. With the exception of the one used in the 3.4-Mc. circuit all the capacitors are variable. The method of mounting the coil and capacitor is shown in the photograph of an assembly with a portion of the polystyrene form cut away to give a better view.

Before soldering the coil leads read what Chapter 7 has to say about soldering coil-form pins, particularly those on polystyrene forms.

Getting It To Work

The converter can be tested just as soon as

you have finished your first set of coils. Connect an antenna to J_1, connect the coax lead from the receiver to J_3, turn on the power supply, and let the tubes warm up. If you have a voltmeter you can check the voltages to make sure they are approximately correct. The voltages measured from chassis, using a 20,000-ohms-per-volt meter, should be approximately as follows: supply voltage, 250; plate of the 6AU6, 200 volts; screen, 125; 6U8 pentode plate, 250; screen, 150; 6U8 triode plate, 125. A variation of as much as 20 per cent either way will make little difference in the way the converter operates. Voltages measured through a high resistance, as at a screen grid, may be somewhat lower than given above if a voltmeter of lower internal resistance, such as 1000 ohms per volt, is used.

FIG. 8-15—An oscillator tuned-circuit assembly with part of the polystyrene coil form removed to give a better view of the coil and capacitor mounting.

Next, check to see if the crystal-controlled oscillator is working. This is simple if the receiver you plan to use with the converter is one that has continuous coverage of the high-frequency spectrum up to 30 Mc. or so, as many of them do. Set the receiver to the frequency of the crystal you have plugged in and tune around a bit to see if you can hear the signal from the oscillator. You won't be able to miss it if the oscillator is working; it will be very strong. The crystal for the 50-Mc. band, 46.5 Mc., may be outside the range of such a receiver, but you will probably be able to hear the oscillator by tuning the receiver around 22.5 or 23 Mc.

This method won't work with an ARC-5 receiver because of its limited tuning range, so other methods of checking must be used. One way is to connect the positive lead from your voltmeter (it must be a high-resistance meter) to the chassis and the negative lead to the grid of the 6U8 triode. Use an r.f. choke (2.5 or 1 mh.) in series with the lead from the voltmeter to the grid. If the oscillator is working the meter will show a reading of a few volts. If there is no reading, adjust C_3 so that the oscillator starts. The best adjustment of C_3 is the one that gives the maximum voltage.

Another method is to measure the positive voltage at the junction of the 10,000-ohm, 1-watt resistor and L_5. When C_3 is tuned through its range the voltage will be found to vary. The oscillator is working when the voltage is maximum.

After the oscillator is known to be working, set the receiver to 3.5 Mc. and adjust C_1 and C_2 for maximum background noise. Peak the antenna trimmer, if any, on the receiver. Tune in an amateur signal and readjust C_1, C_2 and the receiver's antenna trimmer for maximum signal strength. These controls will have to be re-peaked when you tune over a considerable frequency range. However, a single setting of them should allow you to cover the c.w. portion of any band without the necessity for readjustment, provided you peak the tuning for the center of such a range. On the 28- and 50-Mc. bands a range of about 200 kc. can be covered without re-peaking. Readjusting these controls actually is not a hardship in ordinary operating, since you generally are working with stations within a few kilocycles of your own transmitting frequency. Thus the controls do not have to be touched unless you move to an entirely different part of the band.

When you first put a set of coils into operation it's a good idea to vary C_3 to find the setting that gives best signal strength. Once set, however, this capacitor will not need any further touching up.

With the crystal oscillator frequencies specified in Fig. 8-13, the low-frequency edge of each amateur band will be at 3.5 Mc., with one exception. The exception is the 7-Mc. band, the low-frequency edge of which starts at 3.6 Mc. To make the band start at 3.5 Mc. the oscillator frequency also would have to be at 3.5 Mc., and all you would hear at the low end of the 7-Mc. band would be the crystal oscillator.

If you don't mind tuning the receiver "backwards," two crystals and two of the oscillator coils can be eliminated. The oscillator coil information for the 3400- and 17,500-kc. crystals is given for those who prefer to have their receivers tune in the same direction on all bands. If you choose to use the 10,500-kc. crystal and oscillator coil to cover both 7 and 14 Mc., both bands will start at 3.5 Mc. but you'll tune toward 3.2 Mc. for the 7-Mc. band and toward 3.85 Mc. for 14 Mc. A similar method of tuning is required for 21 and 28 Mc. when using the 24,500-kc. crystal and oscillator coil for both bands.

Converter Construction

FIG. 8-16—A three-band crystal-controlled converter using a single set of coils. It has its own power supply. The controls on the chassis front, from the left, are r.f. tuning, mixer tuning, band switch, and on-off switch. On top of the chassis, the 6CB6 is at the left, the 6U8A is in the center, and the crystals are at the right. The power transformer, T_1, is behind the 6U8A.

Another Converter Arrangement

Fig. 8-16 shows another converter that can be used with a receiver that has good stability and a satisfactorily-slow tuning rate in the region of the 3.5-Mc. amateur band. Its circuit is basically the same as that of the plug-in coil converter shown in the preceding pages. However, there are differences in detail. You can take the converter out of operation and connect the antenna directly to the receiver by means of a switch. Also, crystals for the various bands are selected by switching rather than plugging them into the crystal socket. This converter covers three popular DX bands—14, 21 and 28 Mc.—with a single set of coils, so it isn't necessary to change coils when changing bands. Finally, it has its own power supply, and so is a completely self-contained piece of equipment.

As shown by the circuit, Fig. 8-17, the r.f. amplifier in this case is a 6CB6. This is a "hotter" tube than the 6AU6 used in the other converter, especially at low plate voltage. The tube is especially useful with an inexpensive power supply such as the one shown. The mixer-oscillator tube is a 6U8A. If you compare this circuit with Fig. 8-13 you will see some minor variations in resistor values; these are the result of the fact that the plate supply voltage is in the neighborhood of 100–125 volts instead of the 250 specified for the other circuit.

The antenna tuned circuit (grid circuit of the r.f. amplifier), $C_1 L_2$, is designed for continuous coverage from 14 Mc. to 29.7 Mc. The output of the r.f. stage is fed to the mixer grid circuit, $C_2 L_4$, which is similar to the r.f. stage grid circuit in that it covers the three bands by tuning C_2. A double-pole, three-position switch, S_2, is used for band changing in the crystal oscillator, which uses the triode portion of the 6U8A. One section of the switch selects the proper crystal and the other section changes the capacitor across L_5, the tank coil of the oscillator, for the three bands.

A three-pole, two-position switch, S_1, is used for on-off switching. In the "out" position the antenna is connected directly through the converter, from J_1 to J_2. This makes it possible to use the receiver in the normal way, without removing the converter connections. When S_1 is set to the "in" position the power transformer is turned on, the antenna is connected to L_1 in the r.f. amplifier, and the plate circuit of the mixer is connected to J_2.

The power supply uses a small transformer and selenium rectifier in a simple half-wave circuit. The filter uses no choke, but has a 1000-ohm resistor connected between two electrolytic filter capacitors. This type of filter gives all the smoothing needed for the low current taken by the tubes in the converter. In wiring the rectifier make sure that the correct polarity is used. The output terminal usually is marked "cathode," "+," or has a red dot; this terminal should be connected to the filter side of the circuit.

Construction Details

The converter shown in the photos is built on a 2 × 5 × 7-inch aluminum chassis. Most of the construction is simple, although a few precautions should be taken.

Note in the bottom view, Fig. 8-18, how the coils are mounted. All three should be installed with their axes at right angles to each other. L_2 and L_4 are so mounted in order to minimize coupling between the grid and plate circuits of the 6CB6; otherwise the r.f. stage might oscillate. A metal shield, 1¾ by 3½ inches, is installed between these two circuits to reduce unwanted coupling still further. The shield should be mounted so that it crosses the 6CB6 socket between Pins 2-3 and 5-6. Solder the metal cylinder in the center of the socket to the shield.

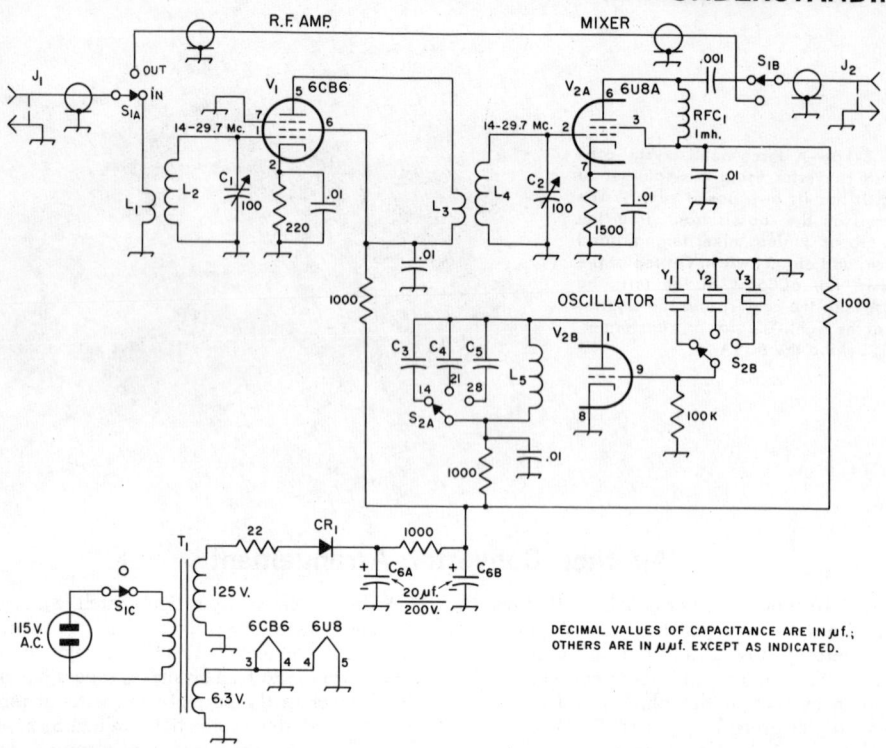

FIG. 8-17—Circuit diagram of the three-band crystal-controlled converter. Resistances are in ohms; resistors are ½ watt. Capacitors not specified below are disk ceramic.

C_1, C_2—100-$\mu\mu$f. variable capacitor (Hammarlund MAPC-100-B).
C_3—150-$\mu\mu$f. mica (14 Mc.).
C_4—47-$\mu\mu$f. mica (21 Mc.).
C_5—15-$\mu\mu$f. mica (28 Mc.).
C_6—20-20-μf. 250-volt electrolytic.
CR_1—65-ma. 130-volt selenium rectifier (Federal 1002A).
J_1, J_2—Phono jacks.
L_1–L_4, inc,—See Fig. 8-19.
L_5—9 turns, same coil stock as L_1–L_4.

RFC_1—1 mh. (National R-50, Millen 34300-1000).
S_1—Phenolic rotary, 1 section, 3 poles, 3 positions, two positions used (Centralab PA-1007).
S_2—Phenolic rotary, 1 section, 2 poles, 3 positions, (Centralab 1472).
T_1—Power transformer; minimum ratings: 125 volts, 15 ma.; 6.3 volts, 0.75 amp. (Knight 61G411, Triad R2C).
Y_1, Y_2, Y_3—14 Mc.: 10,500 kc.; 21 Mc.: 17,500 kc.; 28 Mc.: 24,500 kc.; see text (International Crystal Type FA-9).

The object is to shield Pin 1, the grid, from Pin 5, the plate. Thin sheet copper (flashing copper) or any metal that can be soldered can be used for the shield.

Lengths of RG-58/U are used to connect J_1 and J_2 to the terminals on S_{1A} and S_{1B}. The shielding on the cable is useful for minimizing stray pickup. These cables run along the front fold of the chassis, as close inspection of the bottom view will show. The outer conductor of the cable should be soldered to the ground lugs on J_1 and J_2, and to grounded soldering lugs installed at S_1.

Making the Coils

Fig. 8-19 shows the details of the coils. If you are careful in cutting the coils from the original piece of coil stock you'll only need one 3-inch length of stock. A simple method of cutting through the coil support bars is to heat a razor blade and slice through the insulation. If you attempt to cut the bars with a saw you will deform the coil.

The r.f. and mixer coils are mounted on terminal strips as shown in Fig. 8-19. L_5 is mounted between the pole terminal on S_{2A} and Pin 1 of the 6U8A.

Testing

This converter is given a preliminary checking the same way as the five-band converter described previously. That is, first make sure that the oscillator is working. If you can't get evidence of oscillation, check the oscillator plate voltage (Pin 1 on the 6U8A socket). It should be somewhere near 100 volts. If the oscillator doesn't work even with plate voltage present, carefully check your wiring for errors. Also, make sure you have the correct value of capacitor across L_5.

Converter Construction

FIG. 8-18 — Bottom view of the three-band converter. The input circuit, L_2C_1, is at the right-hand side in this view. It is separated from the mixer circuit, $L_3L_4C_2$, by the shield partition. The switch at the left front corner is S_1. The power-supply components are mounted along the rear (top) edge of the chassis. J_1 and J_2 are at the upper right on the back wall of the chassis. L_5 is end-on in this view, just below the socket for the 6U8A.

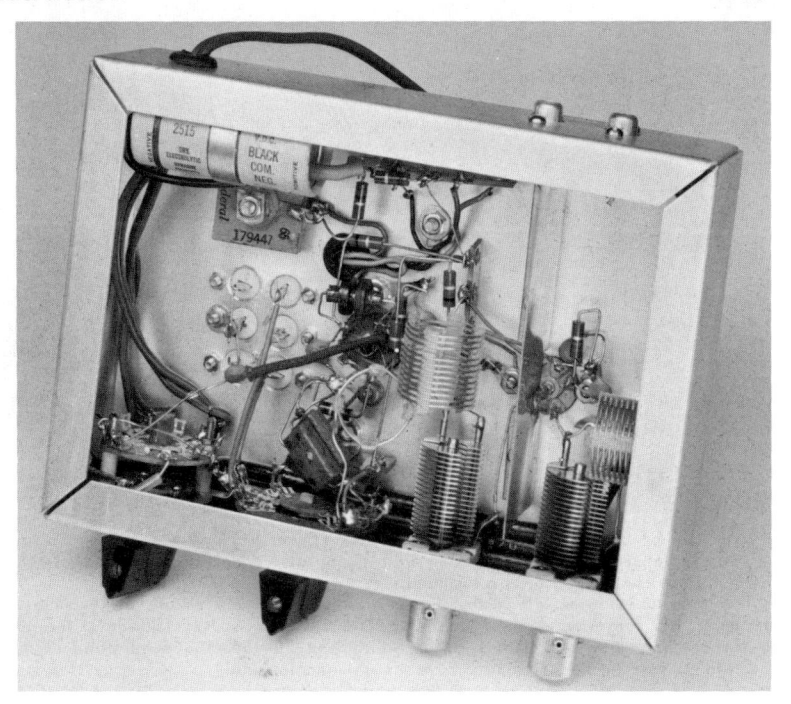

Next, the converter should be connected to the receiver with a length of coaxial line. Either RG-58/U or RG-59/U is suitable. Although the length is not critical, it is advisable to use no greater run of cable than is necessary. The end that connects to the converter should be fitted with a phono plug to go into J_2. The other end should have its braid connected to the ground post on the receiver and the inner conductor connected to the antenna post. If the receiver has "doublet" antenna connections, tie one of these to ground and connect the inner conductor to the other. The coaxial cable is essential for connecting the converter and receiver together because an unshielded wire here would pick up 3.5-Mc. signals which would interfere with your reception of signals on the desired band.

Connect your receiving antenna to J_1. If the antenna is just a single wire, solder its end in the prong of a phono plug so it can make contact with the inner part of J_1. You won't need a ground connection, particularly. If the antenna uses a transmission line, one side of the line should connect to the inside terminal of J_1 and the other side should connect to the converter chassis (outside terminal of J_1).

With the antenna connected, set your receiver to approximately 3750 kc. and switch the converter for 14-Mc. reception. Set C_1 and C_2 near maximum capacitance—plates about two-thirds meshed—and listen for an amateur signal. You may have to try several settings of C_1 and C_2, because the tuning of these controls is quite sharp. Once you locate a signal, peak the two capacitors for maximum signal strength and make a note of the capacitor settings so you can return to the same point when changing bands. Follow the same procedure for 21 and 28 Mc. C_1 and C_2 will be very near minimum capacitance for 28 Mc. and only slightly higher for 21 Mc. Don't be discouraged if you don't get the correct settings at the first try; as we said, the tuning of the two capacitors is rather critical. If your receiver has an antenna trimmer, it should be peaked for each band.

A single setting of the controls is usually sufficient to cover the c.w. portion, about 200 kc., of any of the three bands. You'll probably have to re-peak the controls when going from one end of a band to the other.

FIG. 8-19—Details of coil construction. Coils are made from a single length of coil stock, either B & W Miniductor 3011 or Illumitronic Air Dux 616T, ¾-inch diameter, 16 turns No. 20 wire per inch.

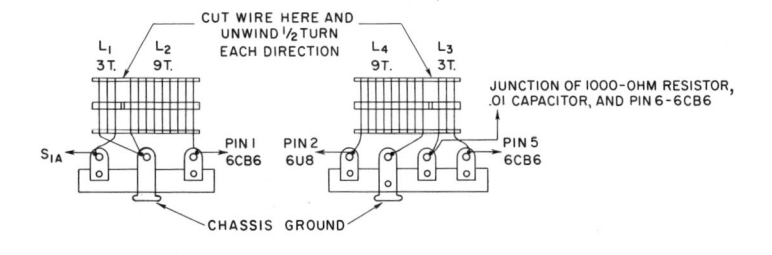

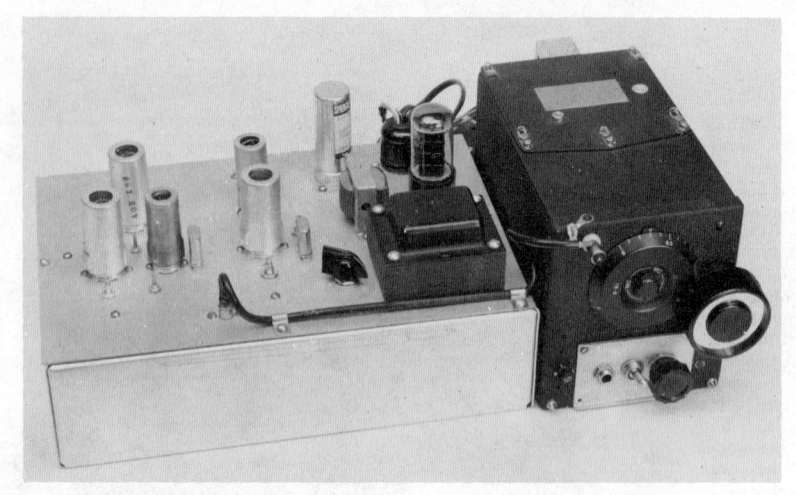

FIG. 8-20—A receiving setup for 50 and 144 Mc. It uses a 6 to 9 Mc. "Command" receiver as a tunable intermediate-frequency amplifier, working with a separate crystal-controlled converter for each band. A power supply for the complete receiver is included.

The three tubes at the left on the chassis are in the 144-Mc. circuit. The two tubes in the center are part of the 50-Mc. converter. A short length of RG-58/U coax is used to connect the outputs of the converters to the antenna terminal of the receiver, which is a BC-455. The outer braid of the coax is grounded to the receiver case at the screw just above the antenna terminal.

50- and 144-MC. Reception

The crystal-controlled converter and lower-frequency receiver combination we have recommended for reception in the 7- to 30-Mc. range become doubly desirable in the v.h.f. bands. In fact, this is the method used by most advanced v.h.f. operators. An ordinary superhet receiver with a tunable "front end" having satisfactory stability is pretty hard to build. Letting a crystal oscillator take over makes the design problems much easier. A crystal-controlled converter with a stable low-frequency receiver will give you a type of reception that is quite comparable with what you would expect from a receiver for the lower-frequency amateur bands.

The design of converters for 50 and 144 Mc. can be based on using a "Command" receiver as the tunable intermediate-frequency amplifier. However, the use of the converters is not restricted to these receivers. The converters to be described can be used with a general-coverage communications receiver, if you have one and are satisfied with the way it works in the 6- to 10-Mc. range. If you don't have such a receiver, probably the most inexpensive way to get started is to use the ARC-5 set that covers 6 to 9 Mc. You will need to make the preliminary modifications described earlier, of course (Fig. 8-7).

The dual converter setup shown in Fig. 8-20 includes a power supply that will handle the converters along with a Command receiver. With this power supply you don't need to re-wire the heater circuits for 12.6 volts since the supply includes a 25-volt filament transformer. If you prefer, though, the power supply circuit in Fig. 8-10 can be substituted, in which case you will need to change the heater wiring as shown in Fig. 8-6.

V.H.F. Circuits

V.h.f. circuits are more "touchy" than those for lower frequencies. All sorts of effects that can be ignored at, say, 7 Mc. start to govern how the set will work at v.h.f. Among these are such things as stray capacitances and tube capacitances. Above 50 Mc. these may be the major capacitances in the circuits. The inductance in even an inch or two of wire may be equally important.

This isn't hard to understand when you consider that in order to double the resonant frequency of a tuned circuit you have to cut *both* the inductance and capacitance in half. It works out this way: If we want a tuned circuit at 144 Mc. to have the same L/C ratio as one at 7 Mc., both the inductance and capacitance have to be reduced in the ratio 144/7, or about 20 to 1. If the 7-Mc. circuit used a tuning capacitance of 35 $\mu\mu f.$, including all the stray capacitances in the circuit, the comparable capacitance at 144 Mc. would be about 1.5 $\mu\mu f.$ Unfortunately, this is much smaller than the grid-to-cathode capacitance built into the tube with which the circuit will be used. The tube and stray capacitances are fairly constant no matter what the frequency, so it is obvious that circuits have to become more **high-C** as we go up in frequency.

Inductance presents a similar problem. Since relatively more capacitance *has* to be used in the tuned circuits, the inductance has to be correspondingly smaller. If care is not used, all the

V.H.F. Reception

FIG. 8-21 — Circuit diagram of the 50-Mc. converter. Unless otherwise indicated, resistances are in ohms; resistors are ½-watt composition. All 0.001-μf. capacitors are disk ceramic.

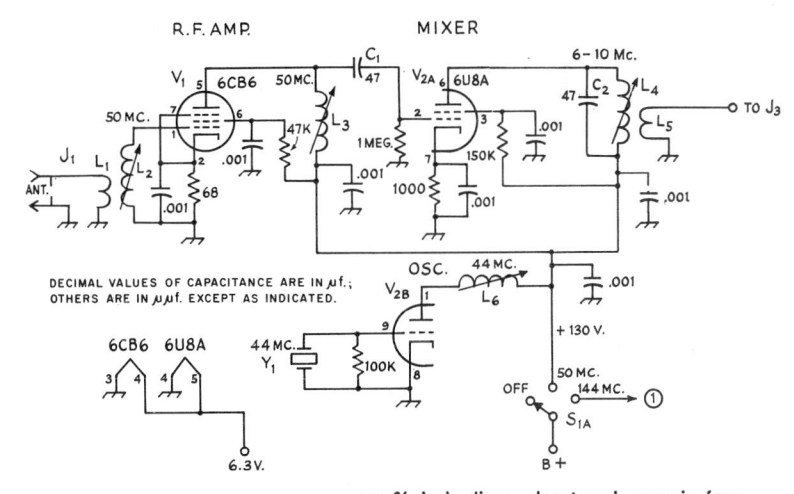

C₁, C₂—47-μμf. mica.
J₁—Phono jack.
L₁—2 turns No. 30 enam., wound at ground end of L₂.
L₂, L₃, L₆—0.9–1.6 μh.; 7 turns No. 28 enam., close-wound on ⅜-inch diam. slug-tuned ceramic form, iron core (Miller 4403).
L₄—6.7–15 μh.; 26 turns No. 30 enam., close-wound on ⅜-inch diam. slug-tuned ceramic form, iron core (Miller 4406).
L₅—8 turns No. 30 enam., close-spaced, wound over ground end of L₄.
S₁—Rotary, 2 poles, 3 positions (Centralab 1472).
Y₁—44-Mc. overtone crystal (International Crystal FA-9).

inductance will be in connecting leads instead of where it can be used to best advantage—in a coil. So lead lengths in r.f. circuits have to be reduced to a minimum. At 144 Mc. even an inch of lead may be equivalent to a fair-sized coil at 7 Mc. This is important not only in the tuned circuits but also in leads used in bypassing tube elements and other components that are supposed to be at ground potential at r.f.

It all adds up to the fact that v.h.f. equipment demands specialized methods of layout and construction. It would be nice to be able to continue making plug-in coils for converters such as the five-band one described earlier in this chapter, but it wouldn't be very practical. The 50-Mc. band is about the limit for that type of construction, and even on that band it is possible to get a considerable improvement by building a converter especially for the band. Consequently, the converter assembly shown here actually consists of two *separate* circuits, one for 50 Mc. and one for 144 Mc. However, there is no objection to using the same receiver as a tunable i.f. for both.

The Intermediate Frequency

There is nothing sacred about a particular intermediate-frequency range for v.h.f. converters. The intermediate frequency chosen should not be too low, because of possible image response. Selectivity, for equally good tuned circuits, is a percentage-type thing; that is, the response depends on the *percentage* departure from the resonant frequency, not the actual number of kilocycles off resonance. So if we are to maintain the same ratio between desired-signal response and image response at two different signal frequencies, we should increase the intermediate frequency in proportion to the signal frequency.

In other words, if we have a certain signal-to-image ratio at 7 Mc. with an intermediate frequency of 500 kilocycles, we would get the same image ratio, with equally good signal-frequency circuits and the same number of them, only by increasing the intermediate frequency in the ratio 144/7. This would call for an i.f. of 10.3 *megacycles*.

You can see now why it is preferable to use the 6 to 9 Mc. ARC-5 receiver with v.h.f. converters. The image response will be less than it would be if you used the 3 to 6 Mc. receiver. The 6-to-9 receiver has a higher internal i.f. than the 3-to-6 receiver (see Table 8-I) and therefore does not separate signals on adjacent channels as well. However, this is not a serious handicap and may even have some advantages at v.h.f., since it makes tuning less critical.

If you use a regular communications receiver with these converters you should still work in the 7-Mc. range with it. A general-coverage communications receiver should be used, rather than the amateur-band-only type, because the 50- and 144-Mc. bands are both 4 megacycles wide. You would be able to tune only a small segment of either band with a purely-amateur-band receiver.

Converter Circuits

The two converters, one for 50 and the other for 144 Mc., are arranged side-by-side on the chassis shown in Fig. 8-20. The circuits are shown separately in Figs. 8-21 and 8-22, since the two are quite independent. You need only

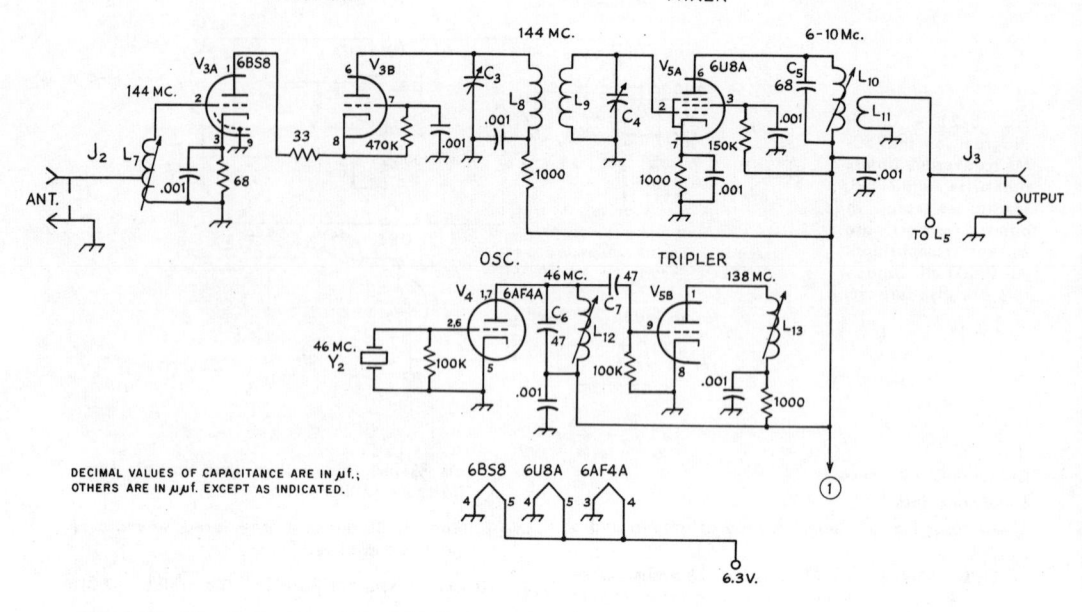

FIG. 8-22—Circuit diagram of the 144-Mc. converter. Unless otherwise indicated, resistances are in ohms; resistors are ½-watt composition. All 0.001-µf. capacitors are disk ceramic.

C_3, C_4—1.6-µµf. tubular trimmer (Centralab 829-6).
C_5—68-µµf. mica.
C_6—47-µµf. mica.
C_7—47-µµf. mica.
J_2, J_3—Phono jacks.
L_7—0.17–0.27 µh.; 5 turns No. 24 enam., close-wound on 3/16-inch diam. slug-tuned ceramic form, iron core (Miller 4301), tapped one turn from grounded end.
L_8, L_9—3 turns No. 20, ½-inch diam., turns spaced to cover ⅜ inch.
L_{10}—5–9 µh.; 30 turns No. 30 enam., close-spaced on ¼-inch diam. slug-tuned ceramic form, iron core (Miller 4505).
L_{11}—8 turns No. 30 enam., close-spaced, wound over ground end of L_{10}.
L_{12}—0.64–0.95 µh.; 12 turns No. 28 enam., close-wound on 3/16-inch diam. slug-tuned ceramic form, iron core (Miller 4304).
L_{13}—0.27–0.41 µh.; 7 turns No. 28 enam., close-wound on 3/16-inch diam. slug-tuned ceramic form, iron core (Miller 4302).
S_1—See Fig. 1.
Y_1—46-Mc. overtone crystal (International Crystal FA-9).

build one, if you are interested in working in just one of these two bands.

The circuit of the 6-meter converter, Fig. 8-21, is essentially the same as the circuits used in the lower-frequency converters described earlier. It has a pentode r.f. amplifier, a pentode mixer, and a triode crystal-controlled oscillator. The latter two are in a combination tube, a 6U8A. The r.f. amplifier is a 6CB6. Its grid coil, L_2, is tuned to the 50-Mc. band by the input capacitance of the 6CB6, plus the stray capacitance associated with the grid-circuit wiring. The inductance of L_2 can be adjusted to the proper value by means of the slug in the coil form. The amplifier grid circuit is coupled to the antenna through a small coil, L_1, wound around L_2 at its grounded end.

The r.f. amplifier output is capacitively coupled to the grid of the mixer tube by C_1. L_3 tunes the amplifier plate circuit, the capacitance here being the output capacitance of the 6CB6, the input capacitance of the 6U8A, and the stray capacitances across the coil, a total of 8 or 10 µµf. With a 44-Mc. crystal in the oscillator the mixer output frequency is 6 to 10 Mc. for signal frequencies of 50 to 54 Mc. A simple triode crystal oscillator circuit is used, the plate circuit being tuned to 44 Mc. by means of L_6, which is adjusted to resonate at approximately 44 Mc. with the tube capacitance. The oscillator voltage is introduced into the mixer grid circuit by coupling within the 6U8A, no external coupling being needed.

If a 6 to 9 Mc. Command receiver is used as the tunable i.f. the whole 50–54 Mc. band cannot be covered with a 44-Mc. crystal in the oscillator, since the band is 4 Mc. wide. In many cases this abbreviated coverage will not be a handicap since the frequencies between 53 and 54 Mc. are not as heavily occupied as those nearer the lower edge. The upper part of the band can be reached by using a crystal at 45 Mc., which will give a range of 51 to 54 Mc. with the ARC-5 set. If a general-coverage communications receiver is used as an i.f. the 44-Mc. crystal will permit covering the entire band.

V.H.F. Reception

The mixer output circuit in this converter is tuned to the 6–10 Mc. range by means of $C_2 L_4$. A coupling link, L_5, feeds the i.f. output to the receiver. This type of coupling is particularly suitable for working into a regular communications receiver since the antenna circuits of these receivers are generally of the low-impedance type. For the Command sets, it will suffice to couple from the mixer plate to the receiver's antenna post through a blocking capacitor of about 50 $\mu\mu f$. With either type of coupling the line from the converter to the receiver should be coax (RG-58/U or RG-59/U). This line should not be more than a foot or so long in the case of capacitive coupling, but with the coupling link, L_5, the connecting line may be any convenient length. Shielded wire (coax) has to be used. Otherwise you'll be getting stray pickup of signals in the 6–10 Mc. region. Even with only a foot or two of exposed wire these can be strong enough to interfere seriously with your 50-Mc. reception.

S_1 is the control switch for the converters. In the "off" position the plus-B supply is disconnected from both converters, but the heaters of both are kept hot if the main power-supply switch is closed. In the second position plate voltage is applied to the 50-Mc. converter, and in the third position to the 144-Mc. converter.

144-Mc. Converter

The circuit of the 144-Mc. converter differs quite a bit from the circuits so far discussed. In one respect it resembles the 50-Mc. circuit—the mixer is the pentode section of a 6U8A and the mixer output circuit is tuned to the 6–10 Mc. range. However, at 144 Mc., it isn't practical to use the triode section of the 6U8A as a simple crystal oscillator. For a 6-to-10 Mc. i.f. output the oscillator frequency has to be near the 144-Mc. band; a frequency of 138 Mc. is used in this converter. Reliable crystals for such a high frequency aren't to be found in the catalogs, so it becomes necessary to use frequency multiplication in order to reach the desired frequency. This takes an extra tube and stage. In Fig. 8-22 a separate triode oscillator with a 6AF4A tube is used, operating at 46 Mc.—one-third the frequency wanted. The triode section of the 6U8A is used as a frequency tripler to get output on

FIG. 8-23—Grouped along the left in this view are the components of the 144-Mc. converter. Note how the interstage coils, L_8 and L_9, are mounted on opposing tie points. The 50-Mc. components are mounted across the middle of the plate. The switch to the left of the power transformer at the upper right corner is S_1. At the lower right corner is J_4, with the 0.01-μf. ceramic bypass capacitors mounted on the socket prongs. The line cord goes through a grommet in the chassis just to the right of J_4.

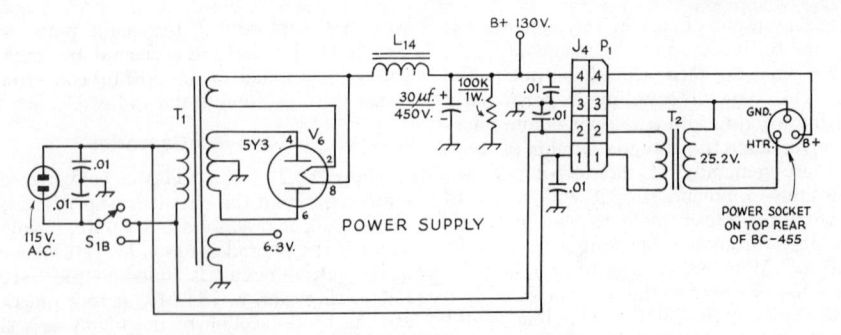

FIG. 8-24—Circuit diagram of the power supply and connections to the BC-455 (see also Fig. 8-9). All 0.01-μf. capacitors are disk ceramic.

J$_4$—Octal socket.
L$_{14}$—Filter choke, 8.5 hy., 50 ma. (Knight 62G136).
P$_1$—Octal plug (Amphenol 86-PM8).
S$_1$—See Fig. 8-21.

T$_1$—Power transformer, 480 volts center-tapped, 40 ma., 5 volts, 2 amp.; 6.3 volts, 2 amp. (Knight 62G034).
T$_2$—Filament transformer, 25.2 volts, 1 amp. (Knight 61G421).

138 Mc., which is then coupled to the mixer grid circuit internally in the 6U8A. The oscillator tank circuit is C_6L_{12}, tuned to approximately 46 Mc. This drives the grid of V_{5B} through the coupling capacitor, C_7. The plate circuit of V_{5B} is tuned to 138 Mc. by L_{13}, which resonates at that frequency with the tube capacitance.

The r.f. amplifier in Fig. 8-22 uses a double-triode circuit known as the **cascode**. The plate of the first triode section, V_{3A}, is connected to the cathode of the section, V_{3B}, through the 33-ohm stabilizing resistor. The grid of V_{3B} is grounded for r.f. through the 0.001-μf. bypass capacitor. Thus the second section is a **cathode-driven** or **grounded-grid** amplifier with essentially direct coupling to the plate of the preceding stage, and no tuned circuit is needed at this point. Note that the plate-to-cathode circuit of V_{3B} is in series with the plate-to-cathode circuit of V_{3A}, for direct current. That is, the same plate current flows through both triodes and the plate-supply voltage divides across the tubes.

The coupling between V_{3B} and V_{5A} uses two separate tuned circuits inductively coupled. With proper adjustments of C_3 and C_4, the tuning capacitors, and also the coupling between the coils L_8 and L_9, the circuit can be made to give quite uniform energy transfer across the 144–148 Mc. band. The advantage of this circuit, as compared with a simple tuned circuit, is that it discriminates to a much greater extent against signals *outside* the desired frequency range. This helps reduce images and other responses that cause spurious signals.

Initial Alignment

The preliminary checking and alignment of these converters should follow the same sequence as in the case of the lower-frequency converters described earlier. First, check the power-supply voltages. The tubes should light normally and the d.c. voltage at the output side of the supply should be approximately 130.

Next, make sure that the crystal-controlled oscillator is working. Measure the d.c. voltage across the 0.1-megohm oscillator grid leak, using a high-resistance voltmeter with an r.f. choke in series, as described earlier. As an alternative, you can measure the current through the oscillator grid resistor by lifting the lower end of the resistor from the chassis and inserting a low-range (0–1 ma. or less) d.c. meter in series. Bypass the bottom of the grid leak to ground with a 0.001-μf. ceramic capacitor, using short leads, when measuring grid current. Adjust the slug in the oscillator plate coil (L_6, Fig. 8-21, or L_{12}, Fig. 8-22) for maximum grid current.

The crystals are overtone types and under some conditions may want to oscillate at their fundamental frequencies, approximately one-third the rated frequencies. The frequency can be checked by coupling an absorption wavemeter (see Chapter 14) to the plate coil. There will be a dip in the d.c. grid voltage or grid current when the wavemeter is tuned through the frequency at which the crystal is oscillating.

The easiest way to peak up the r.f. circuits and the i.f. output circuit is with a grid-dip meter, also described in Chapter 14. Adjust the r.f. circuits so they resonate at about the center of the band. The i.f. output circuit can be adjusted to approximately 8 Mc. Later, after you've used the converters for actual reception of incoming signals, you may want to peak the circuits for optimum performance in some particular part of the band, but these mid-band settings will suffice at the start.

If you don't have a grid-dip meter, the tuning can be adjusted by connecting an antenna to the converter and using incoming signals for alignment. Set the communications receiver somewhere in the 6 to 7 Mc. range and adjust its antenna trimmer, along with the tuning of the converter's i.f. output circuit, for maximum noise. Tune around until you pick up a signal and then

Receiver Construction

adjust the r.f. circuits for maximum response. If the communications receiver is the BC-455, turn on the beat oscillator and keep the r.f. gain at a low level so the change in signal strength with each converter tuning adjustment will be easy to observe. With other types of communications receivers, turn off the a.g.c. and use the b.f.o. and manual gain the same way, if the receiver has no S meter. If it does have an S meter, use the meter as a signal-strength indicator for alignment.

Some General Remarks on Building Receivers

The converter-receiver combinations featured in this chapter represent, in terms of amateur-band performance, about as good a value as you can hope to get for your receiver dollar. We recommended your following these designs, in the early stages of your amateur career, in preference to attempting the construction of a complete receiver. With increasing experience you'll want additional features, no doubt. You may have to build a receiver from the ground up in order to get exactly what you want in all respects. The construction of advanced receivers is beyond the scope of this book, but you can find plenty of information in this field in *The Radio Amateur's Handbook* and in articles in *QST*.

An ARC-5 receiver and crystal-controlled converters will get you accustomed to receiving equipment that is stable both electrically and mechanically. Later, if you build a complete receiver of your own, you'll want to concentrate on stability, along with the ease of tuning discussed earlier. These are the really basic receiver requirements, so a few remarks about them are in order, in case you're tempted to go farther in the receiver-construction field.

Frequency Drift

Slow drift in frequency is usually caused by heating in the receiver. The heat is generated in the tubes, power transformer, and resistors. It takes a good deal of time—sometimes several hours—before the heat distributes itself throughout the set and all the components settle down to a stable temperature. While the components in the tuned circuits are warming up, the materials of which they are made expand, causing their electrical values to change slightly. In other words, the heating actually tunes the receiver over a small frequency range.

There are several ways of combating this frequency change. An obvious one is to put the critical components as far as possible from the sources of heat—away from tubes, transformers, and resistors that get hot. An ideal way would be to take the critical tuned circuits out of the receiver entirely. This can be done, actually, but is a little cumbersome; most people like to have all the receiver controls on one panel. Another scheme is to remove the main source of heat—the power transformer, rectifier and remainder of the power supply—from the receiver and make a separate unit out of it. This was quite common at one time, and is still good practice if you build your own. However, in the interests of convenience most commercially made receivers today are complete with power supply in one cabinet.

Heat Reduction

If you're using, or planning to buy, a manufactured receiver there isn't a great deal you can do about the heat problem. But if you're going to build your own, here are some ways of keeping the temperature rise down:

1) Use no more power from the line than you actually need. The biggest power consumer, aside from the power supply itself, usually is the audio amplifier. Don't build in a high-power audio output stage; use an audio tube that takes only a small amount of heater and plate power.

2) Don't skimp on chassis size. Compactness —which means that components are crowded together—may sometimes be necessary, but usually isn't. The same number of input watts spread over twice the chassis area will cut the temperature rise by a large margin.

3) Give the set, especially the high-temperature components, plenty of ventilation. Don't put transformers, tubes, and power resistors below chassis. The heat they give off will be trapped there. Put them on top so the heat can rise out of the receiver. You're better off if you don't use a cabinet, since cabinets trap heat too. If you *do* use one, punch holes in the top, and also in the sides just above the top-of-chassis level. This will give you a good air draft so the heat can get out.

4) Make sure that the tuned-circuit components, especially those in oscillators, are not close to the big producers of heat such as the power transformer, rectifier, and audio tubes. It may help, too, to use a heat shield between these high-temperature elements and other parts of the circuit. A heat shield is nothing more than a metal plate or baffle big enough so that the circuit element you want to protect can't "see" the source of heat. This cuts off the direct heat radiation in the same way that it cuts off light. The shield should be made of bright metal because that kind of surface is the best reflector of heat as well as light.

Temperature Compensation

Manufactured receivers often use **temperature-compensating** elements—usually capacitors

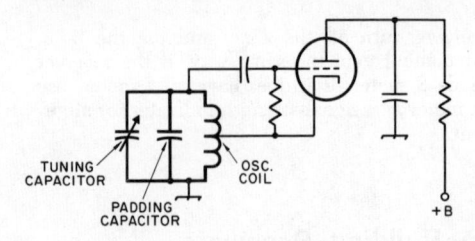

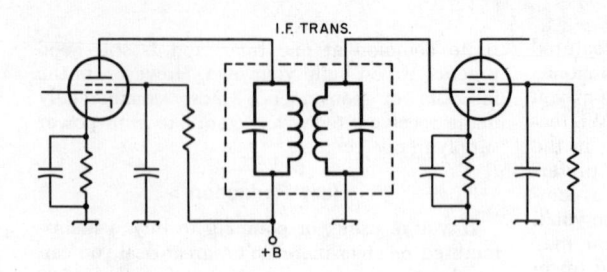

FIG. 8-25—The circuits that are most susceptible to heating effects, causing frequency drift, are shown heavy in these drawings. I.f. transformers and oscillator tuned circuits need the most care in design and installation. Auxiliary components such as resistors and bypasses, shown light, have much less effect on the tuning.

—in the vulnerable tuned circuits. As the temperature rises, the change in value in these compensators is in the opposite direction to the change that normally accompanies heating. Thus under ideal conditions the "positive" change in one part of the circuit is just balanced by the "negative" change in another, resulting in a net change of zero. The design of circuits using such compensation is rather tricky, and to do a good job requires a fair amount of equipment and extensive measurements. You can go part way in your own home-built gear by using **zero-temperature-coefficient** capacitors in tuned circuits where fixed capacitors are applicable. These do not compensate for changes in other circuit components, but at least show little or no shift in value by themselves, as the temperature changes.

Actually, there are many parts of the receiver that aren't particularly affected by heat. The really critical circuits are the oscillators and those where the selectivity is quite high, as in the i.f. amplifier. Even in these circuits it is only the components actually in the tuned circuits that have much effect on the receiver tuning and thus on the observed frequency drift. These components include oscillator coils, tuning capacitors, and i.f. transformers. Coupling and bypass capacitors, as well as resistors, usually are not critical as to value. As a result, a small change in the values of such components, such as might result from heating, has little or no effect on frequency drift. The really sensitive circuit elements are shown in heavy outline in Fig. 8-25, in typical oscillator and i.f. amplifier circuits. Those shown light will have only a minor effect—although it should not be forgotten that resistors are *sources* of heat and should be kept away from the other components.

Voltage Stability

The voltages in a receiver circuit are never 100 per cent stable. There are two principal causes of changes in plate voltage. One is the voltage of the a.c. line. It will vary with the time of day, the number of appliances in operation, and so on. These changes are reflected in corresponding changes in the voltages in the receiver. The second cause is to be found in the operation of the receiver itself: Adjusting the r.f. or i.f. gain of the receiver changes the current through the tubes where gain is controlled, and this change in current represents a change in the load on the power supply. Since the voltage regulation of the supply is not perfect, the change in load is accompanied by a change in output voltage.

Voltage changes of this nature affect the frequency of oscillators in the receiver, but are of little consequence otherwise. The effect is most pronounced in c.w. or s.s.b. reception. A change in oscillator frequency will show up as a shift in the c.w. beat note or will cause an s.s.b. signal to go "Donald Duck." Good design in the oscillator circuits—use of high-C tuned circuits, for example—will help a great deal. The simplest remedy is to use regulated voltage on the oscillators. **Voltage-regulator** (**VR**) tubes are widely used for this purpose.

If the oscillator uses a screen-grid tube it is far more important to use regulated voltage on the screen than on the plate. Also, if the oscillator is connected to a mixer or converter tube, the voltage on the screen of the latter should be regulated (Fig. 8-26). This is because any change in the load that the oscillator "sees" will affect the oscillator frequency, and a change in the voltage on a mixer screen will change the load on the oscillator.

Mechanical Stability

A receiver that "jiggles" the signal when there is a little vibration is a receiver that will get more cussing out than praise. Again, it is the oscillator circuits that require the attention. They *have* to be solidly built—no flimsy con-

Receiver Construction

struction, no floppy wiring. The higher the frequency the more important this becomes.

Variable tuning capacitors rate careful selection when the oscillator operates at a high frequency. Pick out one with wide spacing between the plates. Also, it is preferable to use one with rather small, relatively heavy plates. Use a double-bearing capacitor, because bearings at both ends of the shaft will hold the rotor rigidly in line with the stator. Finally, don't handicap a good capacitor; anchor it firmly to the chassis. It should be mounted on the same support that is used for the oscillator coil and other components, so it can't move with respect to any other oscillator components or wiring.

Coils, too, should be solidly built and mounted. Small diameter coils on acetate bars ("Miniductor" type) are good when firmly supported. If you wind your own on slug-tuned forms, "dope" the wire so it can't move. Duco or a similar cement can be used. Pick out a form in which the slug fits tightly.

Crystal-controlled oscillators are less susceptible to drift and vibration than the tuned-circuit kind, since the frequency of a crystal is a pretty stable thing. However, even a crystal isn't *perfectly* stable, and the oscillator frequency *can* be affected by the same things that affect the frequency of tuned-circuit oscillators. So use a reasonable amount of care in constructing crystal oscillators, too.

Circuit Values

After a while, one gets a "feel" for the right values to pick for circuits of various kinds operating at various frequencies. This problem of circuit values worries many beginners. Sometimes the search for exactly the component value specified in a circuit is justified, but many times it isn't. You can judge adequately on this point only when you have some idea of the *purpose* of the component.

As a general rule, tuned circuits in receivers use as much inductance as the shunt capacitances permit. (Oscillator circuits are sometimes an exception, since a high-C oscillator circuit is often desirable for the sake of stability.) The reason for this is that a high *L/C* ratio causes the circuit to develop a high resonant voltage across its terminals—in effect, the signal voltage is amplified more in a **low-C** circuit. This is only true, incidentally, when the grid of the tube does not take power from the circuit. While this is generally the case in receiving amplifiers at frequencies below 20 Mc. or so, at higher frequencies the tube does consume power, and places a load on the circuit. Thus as the frequency goes up a low-C circuit is not always the most effective. At v.h.f., though, the minimum circuit capacitance (see discussion earlier in this chapter on v.h.f. reception) often makes it impossible to get too much inductance into the circuit.

The minimum circuit capacitance includes the input capacitance of the tube—that is, the effec-

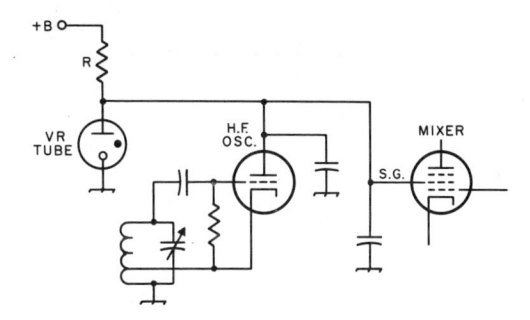

FIG. 8-26—The oscillator and mixer are most in need of voltage regulation if a receiver is to be stable in frequency with inevitable changes in voltage in the receiver. The VR tube is one that regulates at approximately 105 volts, in most receivers. The value of resistor R can be calculated for a particular case by the method given in Chapter 12.

tive capacitance from grid to cathode—along with the minimum capacitance of the tuning capacitor and stray capacitances added by the wiring. The tube and stray capacitance together add up to at least 10 $\mu\mu$f. in the average case. You can find the tuning capacitor's minimum in the data supplied by the manufacturer of the capacitor. It will run 5 to 10 $\mu\mu$f. in the capacitor sizes used in amateur-built receivers. Another stray is distributed capacitance in the coil, but this is negligible with the small "air-wound" coils often used at the higher frequencies (14 Mc. and up). On the other hand, a slug-tuned coil form will add several micromicrofarads across the coil, just because of the capacitance between the slug and the winding.

Finally, there is the possibility that the circuit may be **trimmed** by a small capacitor for the sake of exact adjustment of the tuning range. All these capacitances are shown in Fig. 8-27. This is quite a formidable array of capacitances, so you can appreciate why it is something of a battle to get the total capacitance down at v.h.f. In the frequency range up to 30 Mc. about the best you can hope for is a minimum (not including any trimming capacitance) of 15 to 20 $\mu\mu$f. At v.h.f. you still have C_D, C_S, and C_I no matter what you do, but in these bands the tuning and trimmer capacitors very often are left out entirely, the circuits being tuned by varying the inductance. The signal-frequency circuits (r.f. amplifiers and mixers) often can be **fixed-tuned** at these frequencies. That is, the tuning can be adjusted once for a band or a large portion of it and then left alone. This is possible because the circuits are broad-tuning as a result of the tube loading.

Bypass and Other Values

Bypassing has only one object—to provide a current path that is *much* more attractive than the route that is being bypassed. It would be simple if capacitors had capacitance and noth-

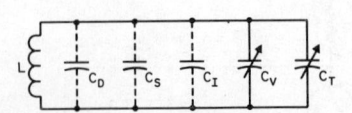

FIG. 8-27—Each of the capacitances shown contributes to the minimum capacitance in a circuit and thus sets an upper limit to the frequency to which the circuit can be tuned with a given coil.

C_D—Distributed capacitance of coil plus capacitance of coil to chassis, including capacitance to tuning slug if used.

C_S—Capacitance of wiring, sockets, etc.

C_I—Input (grid to chassis) capacitance of vacuum tube used with the circuit. If the circuit is connected to the plate of the tube, C_I is the plate to chassis capacitance.

C_V—Minimum capacitance of tuning capacitor.

C_T—Actual "in-use" capacitance of trimmer if used.

The first three, shown with dashed connections, are unavoidable and are only partly under the control of the constructor.

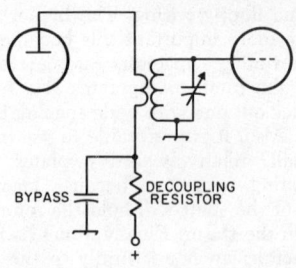

FIG. 8-28—A decoupling resistor helps out considerably in making a bypass effective.

ing else. Then the largest capacitance always would be the best bypass. But capacitors do have inductance and, in some cases, absorb energy as well, and these things affect the efficiency of a bypass.

The higher the frequency the less bypass capacitance you need to do a job. This is because the reactance of a capacitor goes down as the frequency goes up. Thus if a bypass capacitance of 0.001 µf. is satisfactory at 10 Mc., we would need ten times that much capacitance (0.01 µf.) to do the same job at 1 Mc., and only one-tenth as much (100 µµf.) at 100 Mc. It is well to keep this in mind in considering substitutions. In actual practice, circuit designers usually pick the largest capacitance that can be obtained in a component that has low-enough *inductance* to be useful at the frequency under consideration. Meeting this requirement at v.h.f. usually means that the capacitor must be physically as small as possible and have very short leads.

Bypassing can always be helped out by making it harder for the current to choose the unwanted path. In receiver circuits this can be done by connecting a decoupling resistor in series. This is shown in Fig. 8-28 in a typical amplifier plate circuit. If this resistor is, say, 1000 ohms and the reactance of the bypass capacitor is 10 ohms, it is easy to see that the current has 100 times more incentive to go through the capacitor than through the resistor. Decoupling resistors can be used only where the direct current through them is low. Otherwise there would be an excessive loss of plate-supply voltage in the resistor. The current is usually low enough in receiver circuits.

Chapter 9

Accessories for Your Receiver

Receiver circuits designed for construction by beginners are simplified by leaving out features that, although useful, are not actually necessary. This is a good approach, because it lets the builder concentrate on essentials, with no side issues to cause confusion and possible trouble. After the receiver has been finished and the operator has had time to become familiar with how it works, other circuits that will improve reception in one way or another can be added. Many of these additions can be in the form of external accessories, such as the ones shown in this chapter. Often they are useful with the more elaborate receivers, too.

Small outboard gadgets usually have only one or two tubes, at most, and need very little a.c. and d.c. power for their operation. Provided you don't try to use too many of them at once, it is usually possible to take the power from the receiver's accessory socket.

Amplitude Limiting at Audio Frequencies

For a.m. phone reception most receivers include automatic control of the r.f. and i.f. gain. This keeps the audio output in the speaker or phones reasonably close to a constant level no matter how the strength of the incoming signals varies. These automatic gain control (a.g.c.) systems respond to the amplitude level of the incoming phone carrier. When you use the b.f.o., as in c.w. or s.s.b. reception, the b.f.o. signal is the "carrier." Since the b.f.o. should be far stronger than any signal it beats with, the a.g.c. system simply would reduce the gain to a low value and keep it there. Most receivers, therefore, are wired so the a.g.c. is switched off with the same control that turns on the b.f.o. (This is not true of all receivers; some have special circuits that permit the a.g.c. to operate properly with the b.f.o. on.)

True automatic gain control on c.w. signals is possible, with proper receiver design. From a practical standpoint, though, it isn't necessary. For code reception, it is just about as satisfactory—many operators think it more so—to limit the maximum amplitude that a signal can have in the headphones. Any signals weaker than the predetermined level will give their normal response, but when one gets over the limiting or clipping level it hits a ceiling; it can't get any

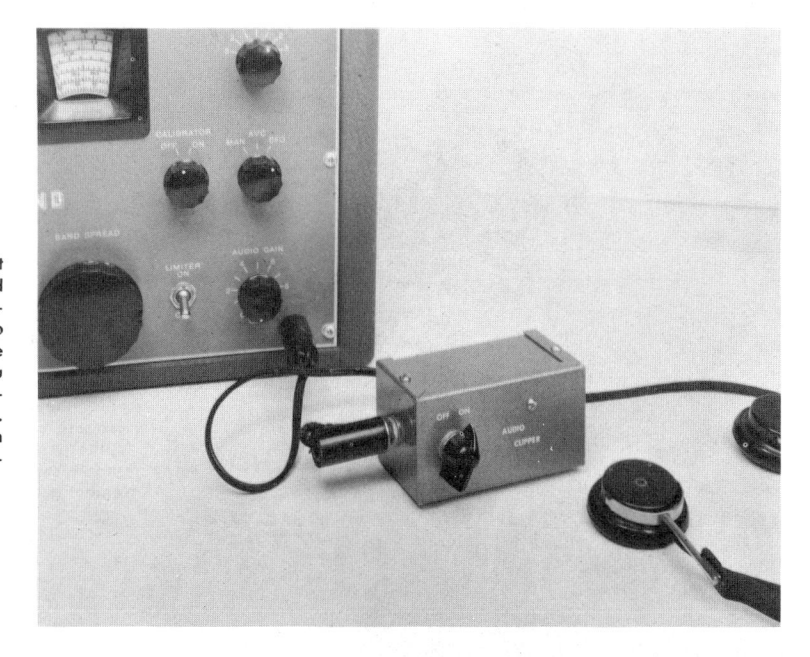

FIG. 9-1—For what it costs in time and money, an "ear-saver" simple audio limiter is one of the best investments you can make for operating comfort in c.w. work. It's useful in other types of reception, too.

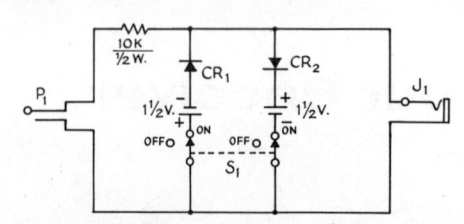

FIG. 9-2—Circuit diagram of the audio limiter.
CR₁, CR₂—1N34 or similar germanium diode.
J₁—Open-circuit headphone jack.
P₁—Headphone plug.
S₁—2-pole, 2 position rotary switch (Centralab 1464). Photo shows a 2-pole 3-position switch, which was used because it was available conveniently. A d.p.d.t. toggle switch also can be used.

stronger. This saves your ears a lot of pounding. A bonus is the fact that peaky noise, such as the kind you get from passing cars, is clipped at the same maximum level. Because of the nature of noise of this type, clipping it gives a very noticeable improvement in the signal-to-noise ratio. The same thing is true of key clicks, "pops" on the a.c. line, and other noises of the **impulse** type.

There are innumerable clipper or limiter circuits, but about the simplest is the diode type shown in Fig. 9-2. The two diodes are biased so that they are nonconducting until the signal voltage applied to them exceeds the battery bias. When this happens the diodes suddenly show a low resistance to the excess signal (or noise) voltage. In effect, the diodes become a short-circuit across the headphones. Thus the voltage at the phones is not able to go above the bias voltage.

The circuit of Fig. 9-2 uses semiconductor diodes CR_1 and CR_2, back-biased by 1½-volt penlite cells. Two diodes are needed in order to clip both the positive and negative peaks of the signal or noise.

The limiter circuit as shown works best with high-impedance headphones, 2000 ohms or more. Many of the headphones that were available on the surplus market are the low-impedance type and the clipper will not work well with these unless a step-down transformer is used. A tube-to-line output transformer having 500-ohm output is suitable. The headset should be connected to the 500-ohm winding. The high-impedance winding can be connected to a phone plug to go into J_1.

Construction and Wiring

The clipper shown in Figs. 9-1 and 9-3 was constructed in a 2¼ × 2¼ × 4-inch Minibox. However, any convenient size box or chassis can be used as neither the dimensions nor layout are critical. The penlite cells are held in place by a small bracket, secured by a screw and nut that also holds a two-terminal tie point. If desired, battery holders can be used for the cells.

The leads from CR_1 and CR_2 are soldered directly to the penlite case (negative) and tip (positive), respectively. If you happen to get the type of cell that has a foil wrapping, take it out of the wrapping so you can solder directly to the case. The contact between the cell case and the disk at the bottom of the wrapping is poor unless there is pressure on the bottom. Wrap the bare cell case with tape, to avoid short circuits.

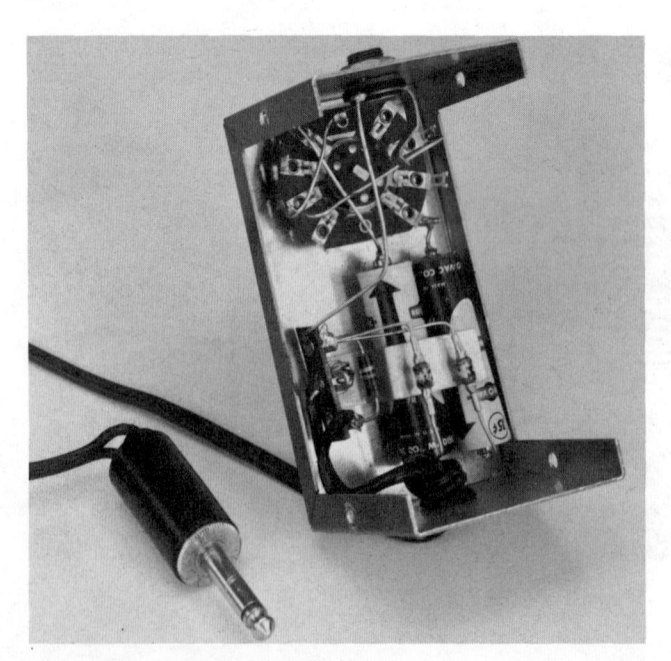

FIG. 9-3—The simplicity of the audio limiter is apparent from this interior view. The headphone jack is on the end at the top. The two flashlight cells are clamped to the case by the small bracket, above which the two diodes and the series resistor can be seen. The cord to the phone plug (ordinary lamp cord is suitable) can be any convenient length.

Audio Selectivity

FIG. 9-4—Selective audio filter ("Audiofil") with built-in amplifier and power supply. It can be used for c.w. reception with any receiver. The receiver does not have to be modified. The two filter inductors are at the right in this view. The power transformer is on the left side of the chassis.

How To Use It

To use the limiter, simply plug it into the receiver phone jack and plug your headphones into J_1. Set your receiver gain controls as you normally operate them, and then tune across the band with limiter switched off. Next, switch in the limiter and go across the band again. You should immediately note the difference, since the stronger signals will be held to one level and any noisy clicks should be suppressed to the point where they are hardly noticeable.

In this simple limiter circuit there is no control over the maximum headphone volume, because the peak signal voltage is set by the bias voltage (1.5 volts in this case) on the clipper diodes. As it happens, this represents a signal level that is satisfactory to most operators. The fact that the clipper bias is not adjustable simply means that the over-all signal level has to be fitted to the clipper, for optimum results. The optimum condition usually is one where clipping just begins on the moderately strong signals, leaving the weaker ones and the usual hiss-type noise background unaffected; this gives good protection to the ears without making the general run of signals take on the somewhat "thin" sound that accompanies heavy clipping. Generally, it is best to run the audio gain fairly wide open and adjust for the clipping level by means of the r.f.-i.f. gain controls on the receiver.

S.s.b. reception is handled the same way as c.w. For a.m. phone, use the receiver's a.g.c. system as you normally would, and adjust the audio gain to set the desired signal just below the clipping level, so that noise peaks above the signal will be clipped off. As it stands, the clipper is only usable with headphones, so speaker reception is "out" unless the circuit can be installed somewhere between audio stages in the receiver itself.

If you want to eliminate the bias cells and are content to operate with a somewhat lower signal level, silicon diodes can be substituted for the germanium diodes specified in Fig. 9-2. Silicon diodes do not conduct until a "forward" voltage of 0.5 to 0.6 volt is applied. If diodes of this type are used without battery bias, the clipping level will be of the order of ½ volt, peak. Any small general-purpose silicon diode may be used.

Audio Selectivity for Code Reception

In copying c.w. signals with the receiver's b.f.o. on, the signal you're listening to generates just one audio tone. You can vary the tone by varying the receiver's tuning or by adjusting the b.f.o., of course, but there is always just *one* beat note. It follows that the audio amplifier in the receiver only has to amplify this one audio tone. In fact, it helps a great deal if it *doesn't* amplify other tones, because that means that the signal you want will be amplified more than the signals you don't want. In other words, an audio amplifier that builds up one tone more than others will give you selectivity for code reception— selectivity that isn't built into the receiver itself. This "audio selectivity" not only can be used to help reception with a receiver that isn't very selective, but often will be a welcome addition to a receiver that *does* have a fair amount of selectivity of its own.

The chief disadvantage of audio selectivity is that it isn't very practicable to vary the tone conveniently while operating. That is, you have to build the circuit for a selected tone and then stick with it. This isn't actually much of a hardship; the audio selectivity need not be so great that the tone becomes monotonous. Even a mod-

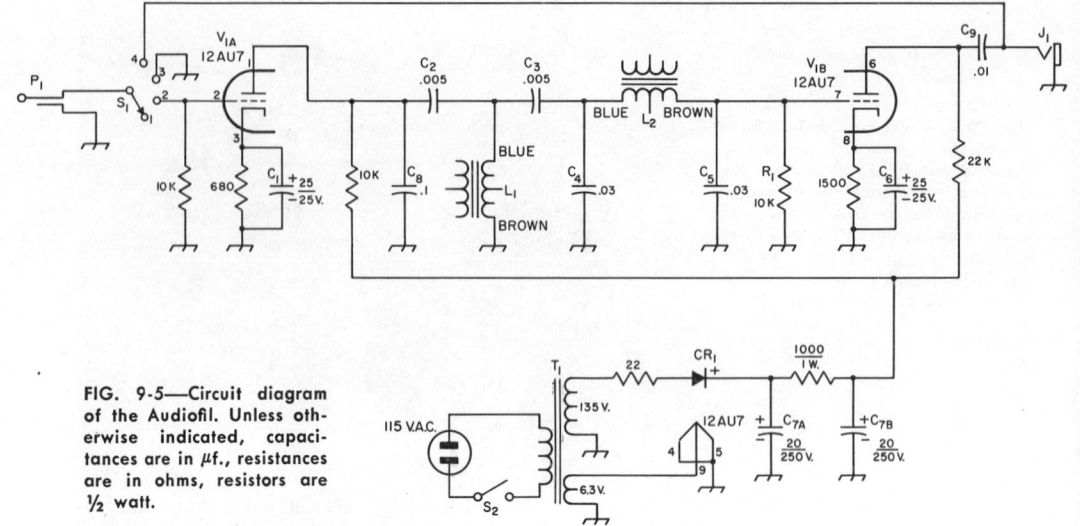

FIG. 9-5—Circuit diagram of the Audiofil. Unless otherwise indicated, capacitances are in μf., resistances are in ohms, resistors are ½ watt.

C_1, C_6—25-μf. 25-volt electrolytic.
C_2, C_3—0.005-μf. mica, 20 per cent tolerance.
C_4, C_5—0.03 μf. paper, 20 per cent tolerance (Mallory type GEM-413, or Sprague type 4TM-S3).
C_7—Dual 20-μf. 250-volt electrolytic.
C_8—0.1-μf. 400-volt paper (Mallory GEM-401 or Sprague 4TMP10).
C_9—0.01-μf. ceramic.
CR_1—Selenium rectifier, 130 volts, 65 ma. (Federal type 1002A or 1386, or equivalent).
J_1—Open-circuit phone jack.
L_1, L_2—4.5 henrys (approximate) total primary winding of Triad S-53X universal output transformer. See text.
P_1—Headphone plug.
S_1—Single-pole, 4-position with a.c. line switch (S_2) attached (Centralab type 1465).
T_1—125 volts at 50 ma., 6.3 volts at 2 amp. (Stancor PA8421 or equivalent).

erately selective circuit will go a long way toward reducing interference from undesired signals. It will also cut out a lot of background noise, thus improving the signal-to-noise ratio.

As there is usually some loss of signal voltage through selective circuits, it is good practice to add some amplification to any selective audio device. Fig. 9-5 is a circuit of this type. Dubbed the "Audiofil," it uses a tuned audio filter circuit made from inexpensive components. It also has a dual-triode tube for providing amplification and optimum coupling into and out of the circuit. Since there is only one tube, the power requirements are small—25 ma. at 125 volts for the plates and 0.3 amp. at 6.3 volts for the heater. It may be possible to take this power from your receiver's accessory socket. However, to make the Audiofil completely independent of the receiver a simple power supply has been included.

Switch S_1 lets the operator use the Audiofil or not, as he pleases, and also makes provision for turning off the power. The switch is one having a separate a.c. switch attached. It comes with four contacts. In the first position the a.c. is off; in the other three positions it is on. Position 2 connects the filter circuit; position 4 cuts out the filter circuit. The third position is grounded; it is not needed particularly, and grounding it reduces stray coupling between the active contacts, 2 and 4. If your receiver is used as a source of power, a simple two-position switch, single pole, will suffice.

Construction Details

A 2 × 5 × 7-inch aluminum chassis was used for the Audiofil shown in Figs. 9-4 and 9-6. However, any chassis large enough to accommodate the components can be used. When mounting L_1 and L_2 on the chassis, set their cores at right angles to each other. Also, mount these coils on the side of the chassis away from the power transformer. This minimizes chances of hum pickup.

An inductance of approximately 4.5 henrys is required at L_1 and L_2. The primary winding of a Triad S-53X audio output transformer has this inductance. (The primary center tap and the secondary are not used.) In order to increase the Qs of the chokes, their iron mounting frames should be removed. This is easily done by bending back the small tabs and then slipping the frames off the cores. Cardboard strips replace the frames, to provide a nonmetallic clamp for the cores.

Using the Audiofil

When you finish the wiring, check it carefully against the circuit diagram, then plug the line cord into an a.c. outlet, turn on S_1 and allow a minute or two for the tube to warm up. Plug P_1 into the headphone jack in your receiver, and plug your headset into J_1. With the switch in

A WWV Converter

FIG. 9-6 — Below-chassis view of the Audiofil. The power-supply components are in the left-hand lower section. Construction is not critical, and with the exception of keeping the power transformer away from the filter inductors to reduce hum pickup, any convenient layout may be used.

position 2 (filter in) you should notice that the background noise tends to peak at an audio tone in the middle-frequency range—actually about 700 cycles, slightly higher than the 600-cycle audio tone broadcast by WWV, if you have no other way of checking audio frequencies. Tune in a c.w. signal and as you go through it you will find that it too will peak noticeably at this same frequency. If there are no other signals at different tones audible, switch out the filter (S_1 to position 4) and you should get a good demonstration of the usefulness of audio selectivity in bringing up the signal you want while cutting down those

you don't. Of course, the filter can't separate two signals that are so close in frequency that they both have about the same beat note, but that is true of other kinds of selectivity, too.

The Audiofil has a nominal pass band of 500 to 900 cycles; that is, tones above 900 and below 500 will be reduced considerably in amplitude. The pass band is wide enough so that your receiver tuning is not made critical, but the selectivity is such that you'll be able to pull through many signals that otherwise would be hard to copy through interference created by other signals near the same frequency.

WWV Converter

Nearly every amateur needs to use the standard-frequency transmissions from WWV or WWVH for checking a crystal calibrator (see Chapter 14), getting correct time, or for other services provided by these stations. The frequencies used for these transmissions are outside the tuning range of an amateur-band receiver. If your receiver cannot tune to them, a simple converter such as is shown in Figs. 9-7 and 9-9 will let you bring in either the 5- or 10-Mc. ones. These two, by and large, are the easiest frequencies to receive.

The 5-Mc. transmissions are superimposed on the amateur 3.5-Mc. band, and the 10-Mc. transmissions are transferred to the 7-Mc. band. The oscillator in the converter is crystal-controlled. Only one crystal is needed.

The converter uses a single tube, a 6BA7, as shown in Fig. 9-8. Heater and plate power can be taken from the accessory socket that is found on most receivers. A slug-tuned coil, L_1, serves for both WWV frequencies; a small fixed tuning capacitor is used for 10 Mc. and a larger fixed-variable combination is switched in for 5 Mc.

The crystal oscillator frequency is 8500 kc., in round figures. (Any frequency between 8500 and 8650 kc. will keep the converted signal inside the amateur band limits.) The fundamental frequency is used to convert the 5-Mc. signal to 3.5 Mc., and the second harmonic on 17 Mc. converts the 10-Mc signal to 7 Mc. The crystal oscillator uses capacitive feedback for fundamental operation. A tank circuit, L_2C_4, connected in the oscillator anode circuit, picks off

FIG. 9-7—Built in a 2 × 4 × 4-inch box, the WWV Converter takes up very little room alongside the receiver. The 5- and 10-Mc. standard frequency transmissions are converted to the 3.5- and 7-Mc. amateur bands, respectively.

the second harmonic. The converted output is choke-capacitor coupled to the receiver's antenna input circuit.

The assembly shown in the photographs is in a small aluminum box with removable sides. Stood on edge, it takes up very little operating-table space. This is only one of many forms in which it could be built, since there is little in the layout that requires particular care. The leads in the r.f. circuits shouldn't be unduly long, but with this in mind you can use any layout plan that appeals to you. The principal thing, from an operating viewpoint, is that the frequency-selector switch, S_2, and the on-off switch, S_1, should be easily accessible. One pole of S_1 is used to transfer the receiving antenna from the converter directly to the receiver. The second pole turns the B plus on when the converter is in

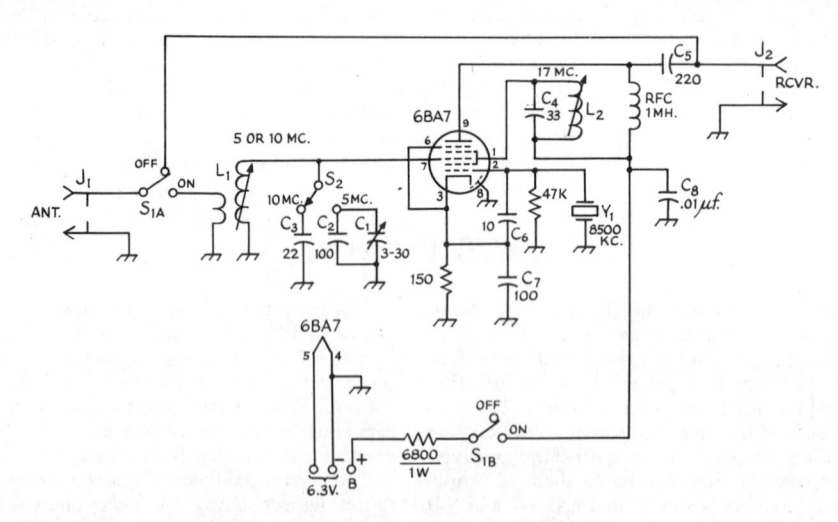

FIG. 9-8—Circuit diagram of the WWV converter. Except where indicated otherwise, capacitances are in $\mu\mu f$., resistors are ½ watt. "B" voltages up to 250 volts can be used.

C_1—3-30-$\mu\mu f$. compression trimmer.
C_2–C_7, inc.—Mica.
C_8—Ceramic disk.
J_1, J_2—Phono jack.
L_1—Approx. 8 μh., slug-tuned. (North Hills 120-C, Miller 4406, or equivalent.) Link is 2 turns No. 28 insulated wire wound around grounded end of coil.
L_2—Approx. 2.5 μh., slug-tuned (North Hills 120-A, Miller 4404, or equivalent).
S_1—D.p.d.t. toggle.
S_2—Rotary, 1 pole, 2 positions (Centralab 1460).
Y_1—8500 to 8650 kc.; see text.

An Automatic Noise Limiter

FIG. 9-9—Inside the WWV converter. The tube and crystal sit on top of a small aluminum shelf mounted on one side plate of the box. S_1 and S_2 are mounted on the front edge. L_1 is mounted on top, with the slug adjusting screw available for convenient alignment. J_1, J_2 and L_2 are on the rear edge. The power cable also comes out the rear. Other components are wired to the crystal and tube sockets.

use and turns it off when the antenna goes directly to the receiver.

Only a few preliminary adjustments are needed, once the wiring is finished. With power applied and S_1 in the "on" position, set S_2 for 10 Mc. and set your receiver to 7000 kc. If the frequency of Y_1 is not exactly 8500 kc., the proper 7-Mc. frequency will be the difference between 10 Mc. and twice the crystal frequency. For example, if the crystal frequency is 8575 kc., its second harmonic is 17,150 kc. and the difference frequency is 17,150 − 10,000 = 7150 kc. Turn the adjusting screw in L_1 for maximum noise or response to any signals that may be picked up near 10 Mc. If it's the right time of day for good signals from WWV you'll hear that station and can peak L_1 on it. Then adjust the slug in L_2 for maximum signal. Next, set your receiver to 3500 kc.—that is, to the difference between the crystal fundamental and 5000 kc.—and set S_2 in the 5-Mc. position. Then adjust C_1 for maximum signal from WWV. After this there's nothing to be done except to put the cover on the box and switch in the converter whenever you need it.

Automatic Noise Limiter and A.G.C. for the ARC-5

The "automatic" noise limiter is a limiting or clipping device resembling in principle the c.w. limiter described earlier in this chapter. However, its bias is not fixed but is regulated by the average strength of the incoming signal. It is useful on a.m. signals where the carrier amplitude can be used to set the limiter bias at the proper level. This is done at the detector circuit in the receiver, where the modulated carrier is rectified and converted into a direct current.

Although many manufactured receivers, even low-priced ones, have noise limiters of this general type built in, the military surplus receivers do not. The circuit shown here was built especially for use with the BC-455 receiver (this receiver is the basis of several setups described in Chapter 8) or other of the ARC-5 series. Combined with it is an improvement in the a.g.c. system of the receiver to make it more satisfactory for a.m. phone. The noise limiter is especially useful when the receiver is used with a v.h.f. converter. On the v.h.f. bands automobile ignition systems cause potent interference when your station is near a highway, and limiters of the type shown can do a good job on ignition noise.

Improving the A.G.C.

A better a.g.c. system can be incorporated into the BC-455 by the addition of two resistors and a capacitor. Fig. 9-10 shows the original and modified circuits. This diagram also shows the noise limiter circuit. The original wiring is shown by light lines. All the components with the wiring shown in heavier lines are additions. If you happen to have access to a complete diagram of the BC-455, you'll find that the circuit component designations in the upper drawing are the same as those in the instruction-book circuit of the BC-455.

The first step in making the modification is to remove the bottom plate of the receiver. Locate the socket for the detector tube. Some models of the receiver used a 12SR7 while others had a 12SQ7. However, the base connections for both tubes are identical. Pin 5 is connected to the chassis; remove this lead and connect a 100-$\mu\mu$f. mica capacitor, C_1, between Pins 4 and 5. Con-

FIG. 9-11—The a.g.c.-b.f.o. switch connections. Grounding either the a.g.c. or b.f.o. turns it off; in other words, with the switch thrown to "a.g.c." as shown, the a.g.c. is off and the b.f.o. is on. Throwing the switch to the other side turns off the b.f.o. and turns on the a.g.c. The headphone and gain-control wiring shown in Chapter 8 are not affected by this switching arrangement.

nect a 0.47-megohm resistor between Pin 5 and chassis. Another 0.47-megohm resistor should be connected between Pin 5 and C_{15A}. C_{15} is a potted capacitor consisting of three sections, 0.05 μf. each, located directly below the 12A6 audio tube. The terminal you want is the one closest to the front of the BC-455; leave the old (blue) lead connected when you solder on the new lead. Next, find R_{11}, a 0.1-megohm, ½-watt resistor (brown-black-yellow) mounted in a block of four resistors located on the same side of the chassis as C_{15}, close to the 12SK7 socket. Remove R_{11} from the circuit by heating the mounting points and gently pulling up on the leads.

For c.w. reception the a.g.c. should be turned off. This is made possible by using an s.p.d.t. toggle as a combined a.g.c.-b.f.o. switch as shown in Fig. 9-11. The arm of the switch should be connected to chassis ground. One switch terminal goes to Pin 5 of the connector in the front compartment of the BC-455. The other switch terminal goes to Pin 3 of the connector. Connect an insulated wire between the base of Pin 3 and the junction of C_{15A} and the 0.47-megohm resistor, R_2. You'll have to remove the plug-in coil assembly that is immediately to the rear of the connector in order to get at the base of Pin 3. Take out the two screws, one on either side of the BC-455, that hold the coil assembly in place. Once the screws are removed, the coils can be lifted out.

While you have the coils out, install the audio gain control, R_5 in Fig. 9-10. There is room for mounting the gain control, if you remove the 3-μf. potted capacitor mounted on the front panel of the receiver. This capacitor is held in place by two screws. Unsolder and remove the lead that goes from the capacitor to Pin 1 on the connector. Next, mount R_5 in the space formerly occupied by the capacitor. Connect a lead from one end terminal of R_5 to chassis. The remaining two leads to R_5 can be installed when the noise limiter is wired into the set. Replace the plug-in coil assembly, making sure that none of the wir-

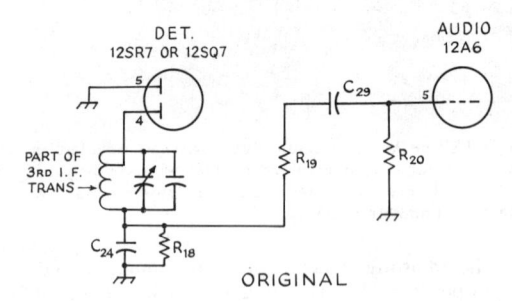

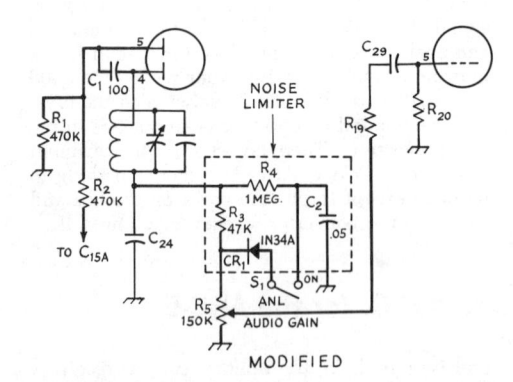

FIG. 9-10—Circuit diagram showing the original and modified circuits for the a.g.c. and noise-limiter additions. Wiring in heavy lines indicates added components. All resistors are ½ watt.
C_1—100-$\mu\mu$f. mica.
C_2—0.05-μf. (or 0.047-μf.) 200-volt paper.
CR_1—1N34A germanium diode.
R_1, R_2—0.47 megohm.
R_3—47,000 ohms.
R_4—1 megohm.
R_5—0.15-megohm potentiometer, audio taper.
S_1—S.p.s.t. toggle switch.

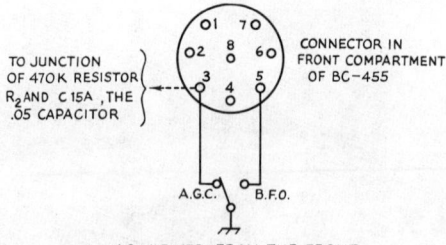

An S Meter

FIG. 9-12—The noise limiter components mounted on an insulating board. The resistor at the left is R_4. In color-coded diodes, the end toward which the colored bands are grouped is the cathode. Other types have a bar or dot to indicate the cathode.

ing to the base terminals of the connector is shorted to the coil box.

The Noise Limiter

The simplest method of installing the noise limiter is to mount all the components, with the exception of R_5 and S_1, on a small insulating board. The assembly is shown in Fig. 9-12. A piece of bakelite or plastic can be used for the board. The board shown in the photograph was cut from a plastic saucer.

An s.p.s.t. switch, S_1, is used to switch the noise limiter in or out as needed. There isn't enough room to mount the switch on the front panel of the BC-455, so the best spot is on the side. The switch can be installed between C_{15} and the next potted capacitor toward the front. This has the BC-455 designation C_{30}, and is a 15-μf. audio bypass capacitor. It is necessary to remove the bottom screw holding C_{30} and swing the capacitor more toward the front of the receiver in order to get enough room for S_1.

Next, locate R_{18}, R_{19}, and C_{24} (BC-455 circuit numbers). The two resistors are on a block just to the left of and below the 12SR7 socket as you view the bottom of the set with the panel to the left. The first resistor is R_{18}, 0.51 megohm (green-brown-yellow), and the second is R_{19}, 0.1 megohm (brown-black-yellow). Remove R_{18} from the receiver. One side of R_{18} was grounded and the other side was connected to R_{19} and also, through a short lead, to C_{24}. Lift this short lead at the end where it is attached to R_{18}. Next, connect an insulated wire from R_{19} to the arm of the audio gain control, R_5. Dress this lead across the chassis and then down the side to the front, running it under the potted capacitors along the side. You can now install the board holding the noise limiter components in place.

The lead from the junction of R_3 and R_4, Fig. 9-10, should be connected to the end of the short lead you lifted from R_{18}. The lead from C_2 to chassis ground can be connected to the ground terminal that formerly held R_{18}. An insulated lead from the junction of R_3 and CR_1 should be dressed along the side of the chassis to the remaining terminal on the audio gain control. The remaining two leads from the limiter board should be connected to the two terminals on S_1.

Although the limiter assembly shown in Fig. 9-12 was designed especially for the ARC-5 receivers, it is small enough to fit in almost any receiver that is in need of such a device and which uses a conventional diode detector. Simply substitute R_3-R_5 of the limiter circuit for the detector diode load resistor. In some receivers the load resistor may also be the volume control, in which case the control probably can be used as R_5 without changing the resistance value.

This type of limiter is not effective in c.w. reception because the b.f.o. voltage rectified by the detector diode usually is far larger than the incoming signals. This biases the limiter diode to such an extent that substantially no limiting takes place. The limiter should be turned off for c.w. reception, but may be left on continuously while receiving a.m. phone signals.

S-Meter Circuit

In a.m. phone work an **S meter** (signal strength indicator) is not only an almost indispensable conversation piece, but it has its practical uses, too. It is helpful, for example, in receiver alignment, permitting you to adjust i.f. and r.f. circuit tuning "on the nose." It is also useful for making comparative checks in tuning and pointing v.h.f. beams, helping others in tests, and similar projects. Bear in mind that it gives *relative* indications—"measurements" would be too strong a word—and is in no sense an absolute measuring device. In fact, the S meters of no two models of commercially-built receivers will give the same readings, and it would be hard to find two of the *same* model that would read exactly the same on a given signal.

The S-meter circuit shown in Fig. 9-13 can be used with any receiver, requiring only a single connection to the receiver's circuits. This is a connection to the a.g.c. line, which provides the control voltage for the meter circuit. The circuit is essentially a bridge, with the triode tube acting as one of the arms. The negative d.c. voltage

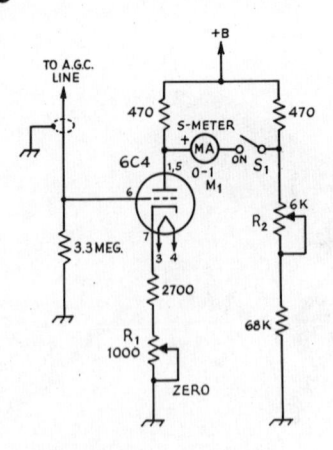

FIG. 9-13—S-meter circuit diagram. All fixed resistors are ½ watt.
M₁—0–1-ma. milliammeter.
R₁—1000-ohm control, linear taper.
R₂—6000-ohm control, linear taper
S₁—S.p.s.t. toggle.

from the a.g.c. line changes the plate current through the tube, thus causing the meter to read. This circuit has the advantages that the meter reads "up"—that is, increasing signal strength causes the current through the meter to increase —and that the pointer cannot be driven off scale at the maximum end. The current from the B supply is only of the order of 5 ma., and the 6C4 heater takes 0.15 ampere at 6.3 volts. A B-supply voltage of 100 to 150 volts can be used.

Since the voltages and currents in the circuit are all d.c., there are no special precautions to be observed in building and wiring. The meter and the two controls can be mounted in a sloping-front meter case, if you like. The leads to the receiver can be any convenient length. The one to the a.g.c. line should be shielded, to prevent introducing hum into the receiver. The shield should be grounded to the receiver chassis.

R_1 and R_2 require only a preliminary adjustment, after which they need not be touched. Before connecting the a.g.c. line take the 6C4 out of its socket and apply the B voltage. Close S_1 and adjust R_2 for a full-scale reading on the meter. Then replace the 6C4, let it come up to temperature, and adjust R_1 for zero reading. Connect the a.g.c. line and the meter will read on incoming signals, and possibly give a small reading on the receiver noise with the gain full on. You can calibrate the meter in S units if you want; generally, a half-scale reading is considered to be S9. The lower S numbers can be marked off proportionately. With a 6C4 or other medium-μ triode (such as one section of a 12AU7) the total range of the meter will be 60 to 70 db.

Using the S Meter with the BC-455

This meter circuit can be used with the BC-455 or other receivers of the ARC-5 series. In addition to the modification to improve the a.g.c. action (Fig. 9-10) it is desirable to raise the screen voltage on the controlled tubes, especially if a relatively low-voltage (130-volt) supply such as that shown in Fig. 8-24 is used. This is easily done. Toward the rear of the receiver are two wire-wound enamel-covered resistors mounted vertically. There is a black lead to the terminal near the chassis on the resistor that is on the same side of the receiver as C_{15}. All that it is necessary to do is to disconnect this lead from the resistor.

The connection to the a.g.c. line should be made at the point marked "To C_{15A}" in the lower drawing in Fig. 9-10. The power supply of Fig. 8-24 has enough spare capacity to take care of the 6C4's heater and plate power.

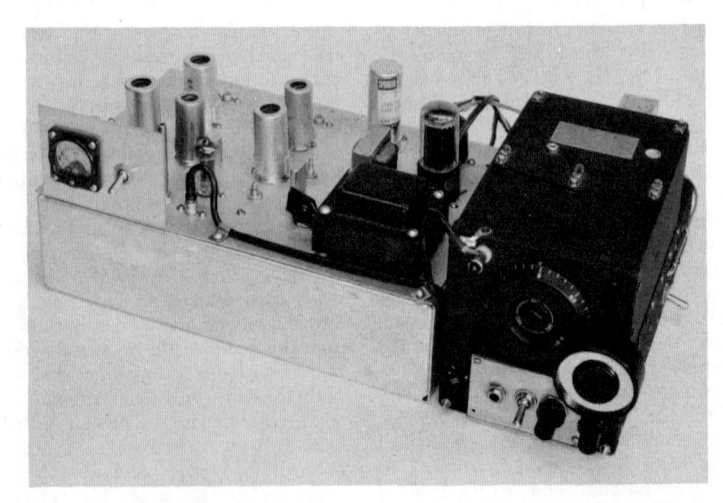

FIG. 9-14—S meter installed on the 50–144-Mc. converter described in Chapter 8. The meter can be mounted in any desired way on this and other receiving setups, since only d.c. voltages and currents are used.

Chapter 10

Building Transmitters

Many elements go into the making of a good transmitter, but most amateurs would agree that a most important one is stability. A signal that can't be held in tune by the receiving operator isn't going to get many contacts. Of course, you want to get the maximum possible power from the outfit, consistent with the amount of power put into it; the more r.f. power radiated, the bigger the signal. But if the signal does not stay put on one frequency, well enough to be copied on a highly-selective receiver, all the attention you may put into other details will be wasted.

At this point it would be a good idea to go back to the latter part of Chapter 8 and read what is said there about oscillator stability. All the effects described there, and the remedies suggested, apply just as much to the oscillator in a transmitter as to the oscillators in receivers. However, it takes even more care to stabilize a transmitting oscillator, simply because it is associated with higher-power amplifier stages than is the case in a receiver. These stages can "kick back" on the oscillator in various ways.

Generally, too, we try to get more power out of a transmitting oscillator than out of a receiver oscillator. This tends to exaggerate the effects of temperature, of voltage changes, and so on. Therefore it's good practice to operate the oscillator at a very low power level and build up the power by amplifiers, even if it takes more stages. It is well worth the sacrifice of some simplicity.

The crystal-controlled oscillator offers the easiest way to get satisfactory stability. It would be an exaggeration to say that a crystal oscillator is foolproof in this respect. However, it does avoid many of the problems that have to be solved in constructing a good self-excited variable-frequency oscillator. That is the reason why Novice licensees are required by the amateur regulations to use crystal control. The disadvantage is that you can't move around in the band at will, unless you have a whole selection of crystals. However, this is by no means a fatal defect, especially in the Novice bands.

Crystal Frequencies

When using harmonics of a crystal frequency you want to be sure that the harmonic falls in the band it should. Bear in mind that *all* of your signal, including key clicks in c.w. and sidebands in phone, must be inside the limits of the band or sub-band assigned for that type of operation. Table 10-I gives the fundamental crystal frequencies that will keep you inside the various band limits. The frequencies shown for phone

Table 10-I
Crystal Frequencies for Various Output Frequencies

Output Frequency	General Class Licensees	
	"80-Meter" Crystals	"40-Meter" Crystals
80-meter c.w.	3501-3999 kc.	
40-meter c.w.	3501-3649 kc.	7002-7298 kc.
20-meter c.w.	3501-3586 kc.	7002-7173 kc.
15-meter c.w.	3501-3574 kc.	7002-7148 kc.
10-meter c.w.	3501-3711.5 kc.	7002-7423 kc.
75-meter phone	3804-3996 kc.	
40-meter phone	3603-3647 kc.	7205-7295 kc.
20-meter phone	3552-3585 kc.	7104-7171 kc.
10-meter phone	3564-3711 kc.	7128-7422 kc.
	Novice Class Licensees	
80-meter c.w.	3701-3749 kc.	7152-7198 kc.
40-meter c.w.	3576-3599 kc.	7035-7081 kc.
15-meter c.w.	3518-3539 kc.	

These figures include allowances explained in the text.

include a 3000-cycle allowance for sideband width at the output frequency. All frequencies include an allowance for the manufacturer's tolerance, which is approximately 1 kc. in the 3.5-Mc. band, 2 kc. in the 7-Mc. band, and so on. The frequencies have been rounded off to the nearest kilocycle (on the safe side) because that is the way crystal frequencies usually are specified. The table is thus useful for ordering crystals by mail.

If you're tempted to crowd the edge of a band, stop and think for a moment. Besides the manufacturer's **tolerance** and the bandwidth allowance, the frequency at which a crystal actually operates depends on the circuit you use it in, how much voltage you put on it, the circuit tuning

169

conditions, and the crystal temperature. It's dangerous to assume that your actual frequency is exactly what is marked on the crystal holder—it may be a kilocycle or two off. Unless you can measure frequency very accurately, it is best to choose your crystal frequencies a couple of kilocycles, at least, inside those given in Table 10-I. Remember that the responsibility for staying inside the band is entirely yours, in the eyes of the FCC monitors.

Tubes for Crystal Oscillators

In any crystal oscillator some of the power fed back from the plate circuit is used in keeping the crystal in mechanical vibration. In vibrating, the crystal develops the r.f. voltage that drives the oscillator tube. If the tube is one that will give a large output with a small driving voltage, we can look for a relatively large amount of power output without danger of overheating or damaging the crystal. Also, for the electron-coupled circuits (see Chapter 4), we want a tube with a screen grid that does the best possible job of shielding the plate from the control grid and cathode.

The best tubes for this purpose, if a moderate amount of power is wanted, are those made for use as video amplifiers in television receivers. The 6AG7 has long been a standby for this application. Newer types with equivalent performance are the 6CL6 and 12BY7. Audio pentodes such as the 6AQ5 and comparable types are also frequently used. All of these, when operated with 100 to 150 volts on the screen grid and 200 to 300 volts on the plate, can give outputs of a few watts on the fundamental frequency of the crystal. This type of operation is not dangerous to the crystal. When the plate tank circuit is tuned to a crystal harmonic the output is less; about half as much on the second harmonic, and one-third or a quarter as much on the third harmonic. These may seem to be rather small amounts of power, but the fact is that they are quite sufficient for driving even a fairly good-sized beam tetrode amplifier. There is no real need to have more.

Crystals

Besides the fact that crystals can handle only a small amount of r.f. power, there are some limitations on the frequencies for which they can be manufactured. Two kinds of crystals are available—**fundamental** and **overtone** types. The former oscillate on a frequency which is determined by the thickness of the quartz plate. This is called the fundamental frequency of the crystal.

Fundamental-type crystals are not usually made for frequencies higher than 10 Mc. or so. They have to be very thin at much higher frequencies, and are mechanically weak. For this reason, the design of transmitters for the amateur bands up to and including the 28-Mc. band commonly is based on using fundamental crystals operating at frequencies no higher than the 7-Mc. band. Frequency multiplication is needed if output is to be obtained in the 14-, 21- and 28-Mc. bands from such crystals.

Overtone crystals have higher frequencies marked on their holders—frequencies that often are in the v.h.f. part of the spectrum. In suitable circuits, they vibrate in a complex way at a frequency that is a close approximation to an odd-numbered harmonic of the fundamental frequency determined by the thickness of the crystal. The term overtone is used in preference to harmonic because a crystal having a fundamental frequency of, say, 10 Mc. will not have its **third overtone** at exactly 30 Mc. The actual frequency will be slightly different. The third and fifth overtones are the ones most used.

Almost any crystal, whether fabricated for overtone operation or not, can be made to operate on its third overtone. Many of them will also oscillate on the fifth overtone. However, for best overtone operation the crystal must be specially processed. Even so such crystals are less **active** than the fundamental types—that is, getting them to oscillate reliably requires more care in adjustment of the circuit.

Overtone crystals help to eliminate frequency-multiplying stages in a v.h.f. transmitter. Also, starting out at the highest possible frequency reduces the number of spurious frequencies that may give trouble. For example, if you want to work on 144 Mc. you can start out with a 48-Mc. overtone crystal oscillator and triple directly to 144 Mc. in a frequency-multiplier stage. The frequency multiplier probably will develop a small amount of output on the second (96 Mc.) and fourth (192 Mc.) harmonics of 48 Mc., too. However, the selective circuits associated with the tripler and the final amplifier will pretty effectively suppress these frequencies.

Suppose, though, that you start out with a 6-Mc. fundamental crystal, as many 144-Mc. transmitters do. This requires a total frequency multiplication of 24 times, and the outputs of the multipliers will contain *some* energy at 6-Mc. intervals all along the way and beyond. Among these will be birdies 6 Mc. either side of the final output frequency—that is, at 138 Mc. and 150 Mc. To get completely rid of them may take more selectivity than an ordinary transmitter can supply. Problems of this sort can be avoided by using the highest possible crystal frequency.

Crystal Frequencies for the V.H.F. Bands

Table 10-II is prepared on the same basis as Table 10-I, but for the 50- and 144-Mc. v.h.f. bands. That is, the crystal frequencies listed are between limits that you should observe in ordering crystals for multiplication into these bands. The figures include allowances for the manufacturer's tolerance and for the band occupied by a phone signal, in the case of phone. They do *not* include allowances for variations in frequency resulting from the oscillator operating

Crystals

Table 10-II

Crystal Frequencies for 50 and 144 Mc.

Output Frequency	6-Mc. Crystals	8-Mc. Crystals	12-Mc. Crystals	24-Mc. Crystals
General Class Licensees			
6-meter c.w.	6252-6748 kc.	8336-8997 kc.	12,501-13,499 kc.	25,003-26,997 kc.
6-meter phone	6265-6747 kc.	8353-8997 kc.	12,527-13,498 kc.	25,054-26,996 kc.
2-meter c.w.	6002-6165 kc.	8003-8220 kc.	12,002-12,332 kc.	24,003-24,664 kc.
2-meter phone	6002-6160 kc.	8003-8214 kc.	12,002-12,323 kc.	24,003-24,647 kc.
Technician Class Licensees			
6-meter c.w. and phone same as General Class. 2-meter c.w. and phone	6044-6123 kc.	8058-8164 kc.	12,084-12,248 kc.	24,170-24,497 kc.
Novice Class Licensees			
2-meter c.w. and phone	6044-6123 kc.	8058-8164 kc.	12,084-12,248 kc.	24,170-24,497 kc.

conditions. Play safe and stay well inside the limits given when ordering by mail.

Of course, if you have a good frequency standard against which you can check your actual frequency (see Chapter 14) you can work anywhere within the FCC assignments, just so long as you keep all your sidebands inside the assigned band limits. It's a good idea to have such a standard whether or not you try to crowd the band edges with your actual operating frequency.

Equipment in This Chapter

The transmitting circuits and equipment shown in this chapter were chosen primarily to satisfy the requirements of the Novice regulations. In most cases, though, the circuits look ahead a bit—to the day when the Novice will become a General Class licensee. Thus other bands than those the Novice can use may be provided for. Also, the equipment may be capable of running a higher power input than Novices are permitted. If you are a Novice licensee, you simply stay inside the Novice assignment and keep your power input down to 75 watts. Then on the happy day when the "General" arrives, you branch out into wider fields wherever they attract you.

Money is an important factor in ham radio just as it is in most other human activities. Partly because of economics—but from preference, too, in a very great many cases—most amateur transmitters at frequencies up to 30 Mc. run no more than around 150 watts input. You'll be right in there with the majority, in power, if you use transmitters of the type shown here.

In the constructional examples described, there is considerable emphasis on saving money by using salvaged parts. There is a wealth of useful material available in obsolete TV receivers and in military surplus. These transmitters make liberal use of such components.

The General Class licensee usually wants freedom in choosing an operating frequency within the band. This is done by substituting a variable-frequency oscillator for the crystal oscillator. A v.f.o. can be added to practically any transmitter shown, and a v.f.o. design is given later.

Finally, the transmitters can be put on amplitude-modulation phone. This conversion is covered in Chapter 13.

On the 50- and 144 Mc. bands there is a great deal of low-power operation, nearly all of it on phone. The transmitter for these two bands is one that the Novice or Technician will find quite practical, even though its power is small compared with the transmitters for the lower frequencies.

A Low-Cost Transmitter

The two-stage transmitter shown in Fig. 10-1 will operate from a power supply such as you might find in an obsolete television receiver. You can often pick up old TV sets in service shops—sets that will, in most cases, be junked because it isn't worthwhile to try to restore them to operating condition. A few inquiries should unearth a chassis that you can get for a few dollars—maybe even just for the asking. Look around for one that has a power transformer in good condi-

FIG. 10-1—This crystal-controlled transmitter uses receiving tubes and will operate from a power supply of the type usually found in television receivers. It has two stages, as shown in Fig. 10-2. Along the front of the chassis, from the left, are the crystal, oscillator cathode switch, oscillator tuning control, meter switch, meter, and the amplifier plate-tank tuning and loading controls.

tion. Most transformer-type receivers will give you 300 to 350 volts at about 200 milliamperes —plenty for a small transmitter—plus heater current for as many tubes as you're likely to need.

The man from whom you get the chassis probably can tell you whether the transformer is good. If he doesn't know, look it over carefully for damage. If it is burned out, your sense of smell probably will warn you—burned transformer insulation has an unmistakable odor! If the transformer is good, the chances are excellent that the rest of the power supply is in good shape, too. Although an old TV chassis is a rather bulky thing to have around, you can save some trouble by using its power supply just as it comes, without remounting and rewiring the parts. However, if you prefer a neater power supply on its own chassis, it isn't a very ambitious undertaking to transfer the essential components to a new and smaller chassis, using the same circuit.

The transmitter circuit of Fig. 10-2 has been designed around tube types that you're likely to find in an old TV chassis, too. Although a 6K6-GT and 6BG6GAs are shown, other types intended for the same purposes can be substituted. The 6K6GT is an audio power amplifier, primarily, and any similar power pentode such as the 6AQ5, 6V6GT, or 6AG7 will work satisfactorily. The 6BG6G is a TV horizontal-deflection tube; you can substitute the 6BQ6 or 6DQ6 for it. The ratings of these latter tubes are somewhat lower, particularly in the case of the 6BQ6, so if you use a pair of this type the transmitter input should be kept down to 50 watts or less.

You may be able to salvage some of the other parts, such as resistors, sockets, and ceramic bypass capacitors, from the TV receiver chassis. In fact, one of these chassis gives you a first-rate start toward that treasure-house of every ham builder—the junk box. The one thing you won't have any use for, at least in the ham station, is the picture tube. Disposing of picture tubes is a real safety problem, so avoid taking it when you get the chassis, if you possibly can.

The Transmitter Circuit

The transmitter operates on five bands, 3.5 through 28 Mc. The power input to the amplifier stage will depend on the plate voltage available, and will range up to 60 watts or so with a 300- to 350-volt supply.

The crystal oscillator circuit is one that has been discussed in Chapter 4 (Fig. 4-13), and will give output at harmonics of the crystal as well as at the fundamental frequency. A plug-in coil, L_2, is used in its plate circuit to select the desired harmonic. For 3.5-Mc. fundamental-frequency output L_2 is not used; the plate tank circuit then consists of L_1 and C_3. L_1 is left in the circuit all the time, the various coils used at L_2 being adjusted so that with L_1 and L_2 in parallel the resultant inductance has the right value for the desired output frequency. The feedback in the oscillator portion of the circuit is adjusted by means of C_1, a small trimmer capacitor.

The amplifier has two tubes in parallel. Two are used in order to take advantage of the amount of plate power that can be obtained from a TV receiver power supply. However, if you don't mind cutting the power input in half you can use just one tube. In that case, ignore the wiring to the second tube. When two are used, the corresponding elements are connected together at the sockets. The plates, which are top-cap connections on these sweep tubes, are not directly connected; each has a **parasitic suppressor** right at its cap, before the parallel connection is made.

The amplifier operates straight-through on all bands except 28 Mc., where it is used as a doubler. For working in the 3.5-Mc. band you need

A Low-Cost Transmitter

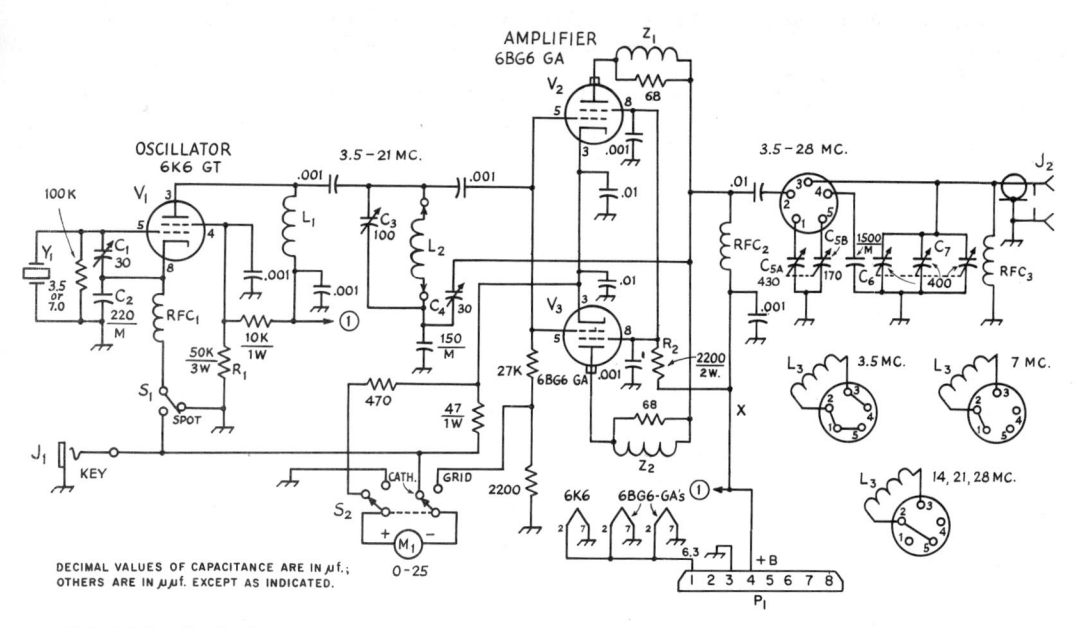

FIG. 10-2—Circuit diagram of the inexpensive transmitter. Unless otherwise indicated, resistances are in ohms, resistors are ½ watt. Fixed capacitors marked M are mica; others are disk ceramic.

C₁, C₄—3-30-μμf. mica trimmer.
C₂—220-μμf. mica.
C₃—100-μμf. variable (Hammarlund HF-100).
C₅—Two-section receiving variable, approx. 170 μμf. and 430 μμf. (Allied Radio 61-H-065 or Philmore 9045).
C₆—1500-μμf. mica.
C₇—Three-section receiving variable, approx. 400-μμf. per section (Allied Radio 60-H-726 or Philmore 9047).
J₁—Open-circuit phone jack.
J₂—Coax chassis receptacle, SO-239 or phono jack.
L₁—25-μh. r.f. choke (Millen 34300-25).
L₂, L₃—See coil table.

Z₁, Z₃—3 turns No. 14 wound on a 68-ohm 1-watt resistor (parasitic suppressors).
M₁—0-25 milliammeter (Shurite Model 950 or 550).
P₁—Four-prong plug, cable mounting (Amphenol 86PM4).
R₁—50,000 ohms, 3 watts (three 150,000-ohm 1-watt resistors in parallel).
R₂—See text.
RFC₁, RFC₂, RFC₃—1-mh. r.f. choke (Millen 34300-1000).
S₁—S.p.d.t. toggle.
S₂—D.p.d.t. toggle.
Y₁—3.5-or 7-Mc. crystal, as required.

3.5-Mc. crystals. Crystals in this same band also can be used (provided their harmonics fall within the proper higher-band limits, of course) for 7- and 14-Mc. work. The oscillator doubles frequency in the first case and quadruples in the latter. These and the other bands can be covered with 7-Mc. crystals, doubling in the oscillator for 14-Mc. amplifier output, tripling for 21-Mc. output, and doubling for 28-Mc. output, where the amplifier also doubles. These combinations are shown in a table.

The amplifier plate tank is a pi network designed for working into 50- to 75-ohm loads. The tank capacitor, C₅, is a two-section variable of a type found in many small broadcast receivers. The one used here has a maximum capacitance of 430 μμf. in one section and 170 μμf. in the other. The loading capacitor, C₇, is a three-section broadcast-receiver type variable having a capacitance of about 400 μμf. per section. Any capacitor that has a total of 1000 to 1200 μμf. with the sections parallelled will do. An additional 1500 μμf. is connected in parallel with the loading capacitor for 3.5 Mc. This is a fixed mica capacitor.

The amplifier is neutralized by the capacitive-bridge method. C₄ is the neutralizing capacitor, while the remaining arm of the neutralizing bridge is the 150 μμf. mica capacitor connected from the junction of L₂, C₃ and C₄ to ground.

Metering and Keying

The transmitter has an inexpensive d.c. milliammeter, 0-25 range, for checking grid and cathode currents. For reading grid current it is connected across a 2200-ohm resistor that is in series with the grid-leak resistor. Since the meter resistance is low compared with 2200 ohms, practically all the current flows through the meter. Thus the range in this case is 0-25 ma. For reading cathode current a 470-ohm resistor is connected in series with the meter, and the combination is connected across a 47-ohm resistor in the amplifier cathode circuit. This multiplies the scale by 10, making the meter read 250 ma. full scale. The 47-ohm resistor also gives the amplifier tubes a little **protective bias** in case the r.f. excitation should fail.

The amplifier is keyed in its cathode circuit. The oscillator cathode also can be keyed along with the amplifier, for break-in operation. Alternatively, if the arm of S_1 is grounded (**spotting** position of S_1), the oscillator runs continuously and only the amplifier is keyed. Sometimes the keying is better, especially on the higher bands such as 14 Mc. and above, when only the amplifier is keyed. There is usually less tendency to **chirp** if the oscillator runs all the time. Listen critically to your keying, using your receiver with the antenna disconnected, and then decide which is better. If the oscillator is keyed, the setting of C_1 will have a marked effect on chirpiness, and the setting of C_3 probably also will affect this keying characteristic.

Power Supply

If you follow the suggestion of using the power supply as it comes on the TV chassis, remove all the tubes from the chassis except the rectifier, which no doubt will be a 5U4G. Three output leads will be needed. One, connected to the chassis, carries one side of the 6.3-volt heater supply and the negative side of the high-voltage d.c. supply. The second is the "hot" side of the heater wiring; you can pick this up at one of the tube sockets. The third is B-plus, which should be taken from the positive terminal of the last filter capacitor in the set. You may have to trace the power-supply wiring from the rectifier through the filter choke to find this point.

In the circuit as shown here, these three connections are made through an octal socket and a corresponding octal plug which is on a cable from the transmitter. Take the existing wiring off an octal socket in the TV set and use three of the prongs for this purpose.

If you prefer to strip the power-supply parts from the TV chassis and build up a separate supply, Fig. 10-3 gives a suitable circuit. The values used in the filter, $C_8 C_9 L_4$, can be almost anything you can salvage from the TV set, as they aren't critical. You can readily identify the electrolytic capacitors in the TV filter since they are nearly always marked with the capacitance and working-voltage rating. Generally they will be of the metal-can type that mounts on the chassis. Capacitance values range from 8 to 40 μf. or more, and frequently two or more capacitors will be in the same can. If so, the metal can is usually the negative terminal for all units in the can. Divide the available capacitance nearly equally between C_8 and C_9, but if the division doesn't come out even, use the larger value at C_9. Be sure that you use only those capacitors having at least a 450-volt working rating. In the case of the filter choke, use what you find in the receiver.

In the circuit shown in Fig. 10-3, S_4 is for turning off the high voltage (as when receiving) without turning off the heater power for the tubes in the transmitter.

Construction

A $3 \times 8 \times 12$-inch aluminum chassis is used for the transmitter. Follow the general arrangement shown in the top and bottom views, Figs. 10-4 and 10-5. C_3 must be insulated from the chassis; fiber washers are used for this purpose. Be sure to allow sufficient room between C_5 and C_7 so the two rotors won't strike each other when set near minimum capacitance.

The top shield, which is necessary for TVI reduction, is made from a section of Reynolds' "do-it-yourself" perforated aluminum stock. The "fence" that runs around the top of the chassis is made from two sections of the same stock, each 2 inches wide and 21 inches long. The perforated stock comes in a 36×36-inch piece, so it is impossible to get a single length long enough to go around the entire chassis. The completed fence is 1¾ inches high, with a ¼-inch lip which is secured to the chassis top with machine screws and nuts. The two sections are each formed into

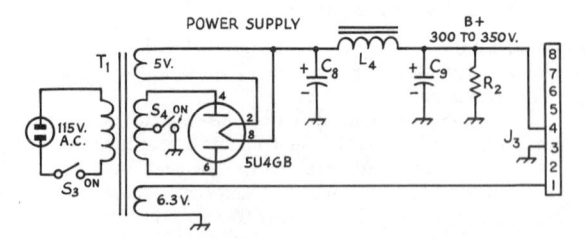

FIG. 10-3—Power supply circuit. If parts are salvaged from an old TV receiver, use the components available; see text. Suitable values for a supply built from new parts are:

C_8, C_9—8 μf. or more (electrolytic), 450 volts working.

L_4—Approximately 2 henrys, 200 ma.

J_3—Octal tube socket.

R_2—50,000 ohms, 10 watts.

S_3, S_4—S.p.s.t. toggle.

T_1—Power transformer, 675 volts center-tapped, 200 ma.; 5 volts, 3 amp.; 6.3 volts, 5 amp. (Stancor P-5059 or equivalent).

Crystal and Coil Combinations for the Low-Cost Transmitter

Amplifier Output Band	Crystal	L_2	L_3
3.5 Mc.	3.5 Mc.	None	C
7 Mc.	3.5 Mc.	A	D
14 Mc.	3.5 Mc.	B	E
7 Mc.	7 Mc.	A	D
14 Mc.	7 Mc.	B	E
21 Mc.	7 Mc.	B	F
28 Mc.	7 Mc.	B	G

A Low-Cost Transmitter

FIG. 10-4—The transmitter with the top screen removed. L_2 is between the oscillator tube (at the right) and the amplifier tubes. The amplifier coil is at the left. Along the back (facing side) of the chassis from the left are the output jack, power cable, and key jack.

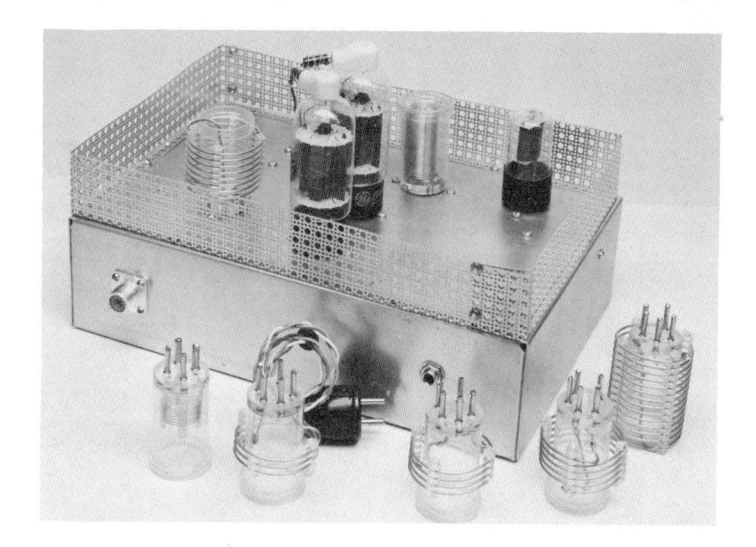

an L shape measuring 8 × 12 inches, the remaining inch being used at two of the corners for an overlap to fasten the two sections together with screws and nuts.

The sides of the shield are made from two pieces of perforated stock measuring 6½ × 20½ inches before folding. The side dimensions of the two pieces after folding are 7¾ and 11¾ inches; the extra inch is used for the overlap to connect the two pieces together. A one-inch flange is folded in around the top so that the overall height is 5½ inches. The top is made from a piece of stock 7¾ by 11¾ inches and is secured to the sides with machine screws and nuts. When the completed cover is slid down inside the fence and flush with chassis, the overlap is sufficient to prevent harmonic leakage, provided care has been used in folding the stock to insure a snug fit. No screws are needed to hold the cover down. This makes coil changing simple because the cover can be removed and replaced quite easily.

The cable used to connect the transmitter to the power supply can be made any length, depending on where you install the power supply.

Making the Coils

The plug-in coils are made from commercially-wound coil stock, the oscillator coils being mounted inside the plug-in coil forms and the amplifier coils on the outside. The Air Dux coil stock specified has exactly the right diameter to fit over the forms.

When cutting the oscillator coils from the original stock, allow three extra turns on the 20–15-meter coil and five extra turns on the 40-meter coil. When these extra turns are unwound you'll have sufficient lead length to reach through the prongs on the forms.

When making the amplifier coils, put jumpers between the prongs as shown in Fig. 10-2.

Tuning

The first step in testing is to neutralize the final amplifier. The lead that feeds the plates and screens of the 6BG6Gs should be disconnected at point X in Fig. 10-2 so that the only voltage on these tubes is the heater voltage. Plug in a 7-Mc. crystal and the 7-Mc. grid and plate coils. Turn on the power and let the oscillator tube warm up. Set S_2 to read grid current. Next, close the key and adjust C_3 for a grid-current reading of 4 to 5 ma. Set C_7 at maximum capacitance (plates fully meshed) and then tune C_5 through its range. At one point (where the amplifier tank circuit is resonant at the crystal frequency) you should notice a dip in the meter reading. Next, carefully

---

Plug-In Coil Data for the Low-Cost Transmitter

L_2—(A) 7 Mc.—29½ turns No. 20, 16 turns per inch, ¾-inch diam.
    (B) 14-21 Mc.—7½ turns same (B & W Miniductor 3011 or Illumitronic Air Dux 616T).

L_3—(C) 3.5 Mc.—13 turns No. 14, 6 turns per inch, 1¾-inch diam.
    (D) 7 Mc.—8 turns same.
    (E) 14 Mc.—5 turns same.
    (F) 21 Mc.—3½ turns same.
    (G) 28 Mc.—2½ turns same.
    (Illumitronic Air Dux 1406T).

Note: A single length of Illumitronic 616T or B & W 3011 will suffice for the 7- and 14-21-Mc. oscillator coils. One length of Air Dux 1406T is sufficient for all the amplifier coils. The L_2 coils are mounted in four-prong plug-in coil forms, 2 required (Amphenol or Allied Radio 24-4P), and the L_3 coils in five-prong forms, 5 required (Amphenol or Allied Radio 24-5P).

FIG. 10-5—The oscillator components are grouped around the 6K6 socket at the upper left. Below the toggle switch, next to the meter, is the coil socket for L_2. The two 6BG6 sockets are at the center of the chassis. The two-section variable capacitor at the upper right center is C_5. C_7 is to its right. The coil socket for L_5 is below C_5. To the right of the socket are C_6 and RFC_3.

adjust the neutralizing capacitor C_4 so that the *least* amount of change occurs in the meter reading when C_5 is tuned through resonance. When you find this point, the amplifier will be neutralized. The plate and screen leads may now be reconnected—*remembering to turn off the power first.*

A dummy load should be used for testing the amplifier. A good load for this purpose is a 60-watt lamp bulb. Connect a lead from the inner terminal of J_2 to the center contact on the base of the bulb and another lead from the chassis to the threaded part of the lamp base.

Now turn on the power and set both C_5 and C_7 at maximum capacitance, plates fully meshed. Set S_2 to read grid current and close the key. Tune C_3 for a grid-current reading of 2 to 4 ma. Don't hold the key down very long, because the final tank will be off resonance and the amplifier will take excessive plate current. This could cause permanent damage to the tubes.

Next, set S_2 to read cathode current and close the key again. Tune C_5 for a dip in the meter reading; this will indicate that the final is tuned to resonance. Start decreasing the capacitance of C_7 while keeping the amplifier in resonance (at the dip in the meter reading) by adjusting C_5.

The lamp should start to light and should get brighter as you adjust C_5 and C_7. The 6BG6Gs are good for 100 ma. per tube, so you can load the amplifier to about 200 ma. However, don't increase the loading beyond the point that results in maximum brightness of the lamp. When you get to that point, try readjusting C_3 to see if you can get more output. When you've reached the limit of output, note the amplifier cathode-current reading. This is the value of current you should adjust for when the amplifier is feeding an actual antenna. The settings of C_5 and C_7 probably will be different with a real antenna system. This simply means that the antenna load is not the same as the load the bulb gives you. The bulb test is to show you what the transmitter is capable of doing, and to familiarize you with the proper tuning procedure. The procedure itself should be followed with any kind of load on the transmitter.

The tuning method is the same on all bands. One point to watch out for is that the 14- and 21-Mc. bands are both covered on one coil with C_3. Thus there are two settings of this capacitor that will result in maximum final-amplifier grid current. The one nearest maximum capacitance is 14 Mc. and the one near minimum capacitance

A 100-Watt Transmitter

is 21 Mc. Be sure you use the right setting for the band you want to work on. Try out all the bands on the lamp dummy to get familiar with the settings, and check your output with a wavemeter if you have one. Chapter 14 tells how to make an inexpensive one.

The feedback capacitor, C_1, in the oscillator circuit should be adjusted with the transmitter on 21 Mc. Adjust C_1 for a grid-current reading of no more than 4 ma. on this band. This adjustment need not be changed for other bands with crystals of ordinary activity.

You'll find antenna systems suitable for use with this transmitter in Chapter 6. Some types of antennas will require a feeder-matching circuit or "transmatch" between the transmitter and the feeder. Details on making such circuits are given in Chapter 11.

One final point: The plate current that the amplifier tubes will take depends more on the screen voltage than on the plate voltage. To get the tubes to take 100 ma. each it is necessary to have 300 volts on the screens. You may have to tinker with the value of R_2 to get this voltage. With the transmitter operating into the lamp dummy load, measure the voltage at the 6BG6G screens with the plate tank controls adjusted for maximum lamp brightness. *Don't touch any of the wiring when making this measurement.* If the voltage is low, try less resistance at R_2. If it is appreciably higher than 300 volts, make R_2 larger until you get about 300. The right value of R_2 will depend on the voltage, under full load, available from the power supply. Since this will depend on the components used, it will be necessary to check the voltage and change R_2 accordingly, if you want to get the maximum output from the set consistent with operating the tubes within safe ratings.

A 100-Watt Transmitter

A somewhat more advanced transmitter circuit than the one just described is shown in Fig. 10-6 and the accompanying photographs. Although more powerful, it is a comparatively inexpensive set to build, since the amplifier tubes can be obtained very cheaply from surplus dealers and the power transformer can be salvaged from an old TV receiver. If you're a Novice licensee, you can run the transmitter at 75 watts input until you move up to the General Class.

Circuit Details

The transmitter can be operated on any band from 3.5 Mc. through 28 Mc. at inputs up to 100 watts. It uses a 6AG7 crystal oscillator driving a pair of 1625s. Either 80- or 40-meter crystals are used, depending on the band.

The plate circuit of the oscillator is tuned by the combination of C_3 and L_1L_2, Fig. 10-7. The correct inductance for each band is selected by using S_1 to short out part of L_2. The oscillator can be operated either straight-through, doubling, or tripling, depending on the crystal used. An 80-meter crystal is used for 3.5-Mc. operation, and the same crystal will provide more than enough excitation for 40 meters with the oscillator working as a doubler. However, the

FIG. 10-6—A 100-watt transmitter for five bands, using salvaged TV power transformer and surplus 1625 amplifier tubes. Across the bottom on the front of the chassis are the crystal, amplifier grid (or oscillator plate) tuning capacitor, oscillator band switch, and power switch. On the box panel, from the left, are the tune-operate-spot switch, meter switch, and amplifier plate tuning and loading controls. The amplifier band switch is at the upper right.

At the rear of the chassis are L_8, the 5U4 and 6DE4s (see Fig. 10-7) and the power transformer. The key jack, not visible in this view, is mounted on the rear chassis wall.

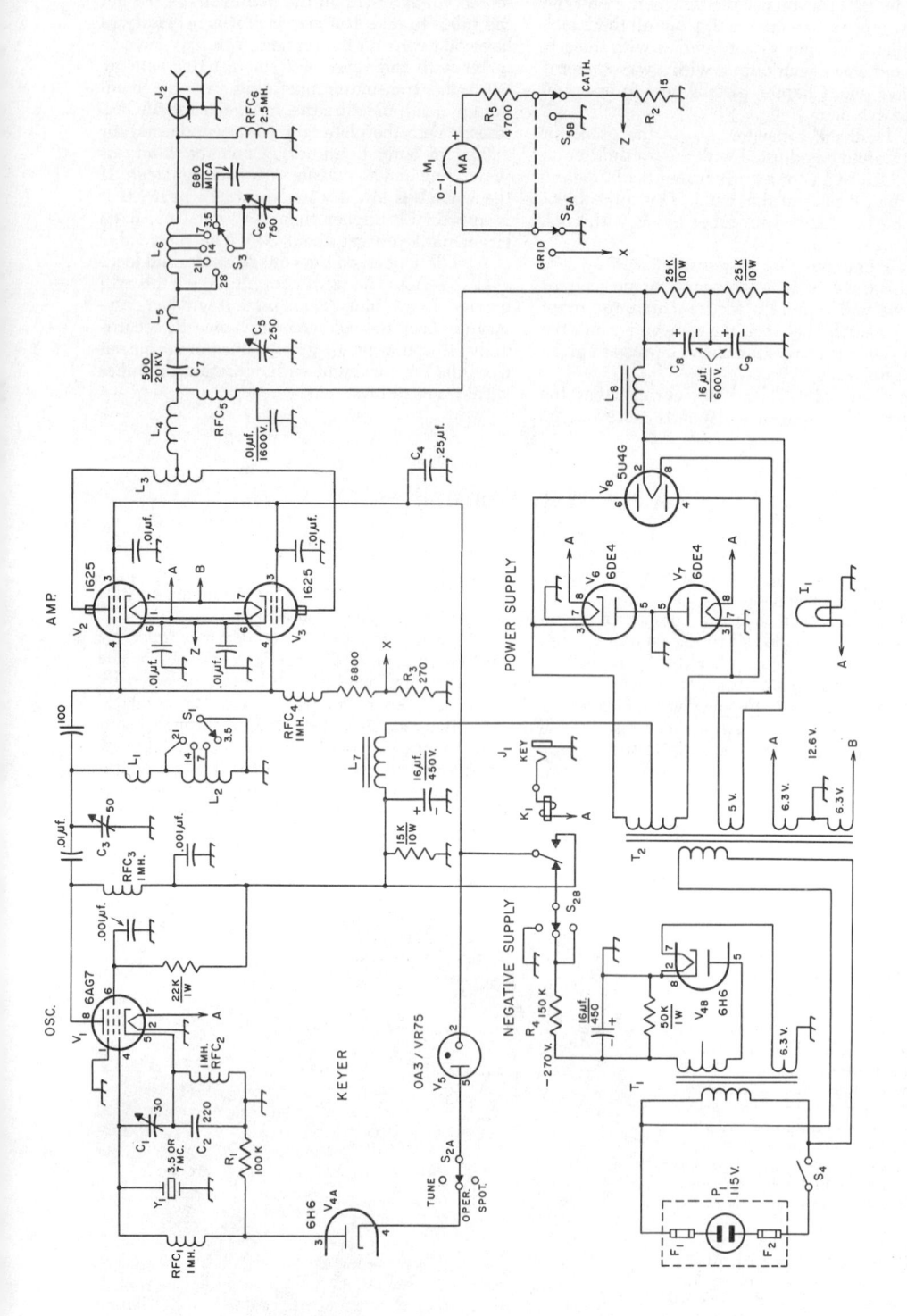

FIG. 10-7.—Circuit diagram of the 100-watt transmitter. Unless otherwise specified, capacitances are in µµf., resistances are in ohms, resistors are ½ watt. Capacitors with polarity marked are electrolytic.

A 100-Watt Transmitter

C_1—3–30-$\mu\mu$f. mica trimmer.
C_2—220-$\mu\mu$f. mica.
C_3—50-$\mu\mu$f. variable (Hammarlund HF-50).
C_4—0.25-μf. paper, 400 volts.
C_5—250-$\mu\mu$f. variable (Hammarlund MC-250-M).
C_6—Approx. 750-$\mu\mu$f. variable, dual-section broadcast type, with stators connected in parallel (Allied 60 H 725).
C_7—500-$\mu\mu$f., 20,000-volt, TV "doorknob."
C_8, C_9—16-μf., 600-volt electrolytic.
F_1, F_2—5 amp., type 3AG.
I_1—Dial lamp, 6 volts, type 47.
J_1—Phone jack, open circuit.
J_2—Coax chassis receptacle, SO-239.
K_1—Keying relay, s.p.d.t., 6-volt a.c. coil (Potter Brumfield KA5A).
L_1, L_2, L_5, L_6—See coil table.
L_3—18 turns No. 22 enam., wound on a 1-watt resistor (any value over 1000 ohms) as a form, tapped at center.
L_4—12 turns No. 22 enam. on same type form as L_3.
L_7—8.5-hy. 50-ma. filter choke (Allied 62 G 136).
L_8—Filter choke, current-carrying capability over 200 ma. (see text).
M_1—0–1 d.c. milliammeter.
P_1—Fuse-type line plug.
R_1—0.1 megohm, ½ watt.
R_2—15 ohms, ½ watt.
R_3—270 ohms, ½ watt.
R_4—0.15 megohm, 1 watt (see text).
R_5—See text.
RFC_1–RFC_4, inc.—1 mh. (National R-50, Millen 34300-1000).
RFC_5—1 mh. (Millen 34107) or 120 μh. (Raypar RL-102).
RFC_6—2.5 mh. r.f. choke.
S_1—Phenolic rotary, 1 pole, 11 positions (4 used), 1 section (Centralab 1001).
S_2—Phenolic rotary, 2 poles, 3 positions, 1 section (Centralab 1473).
S_3—Ceramic rotary, 1 pole, 6 positions (5 used), 1 section (Centralab 2501).
S_4—S.p.s.t. toggle.
S_5—Phenolic rotary, 2 poles, 3 positions (two used), 1 section (Centralab 1473).
T_1—250 volts, center-tapped, 25 ma.; 6.3 volts, 1 amp.; h.v. center tap not used (Allied 62 G 008).
T_2—App. 700 volts center-tapped, 200–300 ma.; 6.3 volts, 5 amp.; 6.3 volts, 1 amp.; 5 volts, 3 amp. (from old TV receiver; see text).
Y_1—Crystal, 3.5- or 7-Mc. band as required.

grid drive to the amplifier on 14 Mc. is not great enough when the oscillator is operated as a quadrupler from a 3.5-Mc. crystal. Adequate 14-Mc. excitation is obtained with the oscillator working as doubler from a 7-Mc. crystal, and also on 21 Mc. working as a tripler; the 7-Mc. crystal, of course, also can be used for 40-meter work with the oscillator working straight through. For 28-Mc. work, a 7-Mc. crystal is used with the oscillator doubling to 14 Mc., and the amplifier also works as doubler.

The parallel 1625s are operated as a straight-through amplifier on all bands except 28 Mc. The amplifier tank circuit is a pi network designed primarily to work into 50- or 70-ohm loads. C_5 is the plate tuning capacitor. The variable loading capacitor, C_6, is a two-gang broadcast type consisting of two sections of approximately 375 $\mu\mu$f. each. These two sections are connected in parallel to provide 750 $\mu\mu$f. at maximum. In addition, a 680-$\mu\mu$f. mica fixed capacitor is switched into the circuit on the 80-meter band. This, with C_6, provides the approximately 1450-$\mu\mu$f. capacitance required for 50-ohm loads at 80 meters.

L_3 and L_4, in the plate leads of the 1625s, are for suppressing parasitic oscillations. The 1625 plates are parallel fed. Either of the two chokes specified in the parts list for RFC_5 will work satisfactorily. RFC_6 serves as a safety precaution in the event that C_7, the plate blocking capacitor, should break down, in which case the d.c. plate voltage would be shorted to ground through RFC_6 rather than appearing on the antenna circuit.

Note that the cathodes of the 1625s are individually bypassed, as are also the screens. The bypass capacitors, 0.01 μf., should be installed right at each tube socket, using the shortest possible path to chassis.

Keying System

A feature of the circuit is the **differential keying** system. A keyed oscillator usually has either clicks or chirps. The chirp is a slight change in the oscillator frequency as it is keyed. It can be eliminated, or much reduced, in a two-stage transmitter by letting the oscillator run continuously during a transmission and doing the keying in the following stage. However, break-in operation is not possible when this is done, and it is necessary to have a send-receive switch to turn off the oscillator at the end of each transmission. In "differential" keying, the oscillator does not actually run all the time but is turned on just before the power is applied to the amplifier and turned off shortly after power is taken off the amplifier. Only the amplifier keying is heard on the air, and this keying can be shaped as desired, to eliminate key clicks, without affecting the oscillator frequency.

In the system used here, the keying circuit can be adjusted so that the oscillator "hangs on," even at slow speeds. In fact, it can be set to turn

FIG. 10-8—Arrangement of parts inside the 100-watt transmitter. The loading capacitor, C_6, in the upper left-hand corner, is mounted at the rear of the box near J_2, the output connector. A panel-bearing assembly (Allied 60 H 385) and a shaft coupler are connected to the rotor of C_6. (Some of the broadcast-type capacitors have ⅜-inch diameter rotor shafts. A ⅜-to-¼-inch shaft coupler (Allied 60 H 362) can be used in such an installation.)

The 28-Mc. coil, upper center in this view, is connected between the stator of C_5 and one end of L_6. Two steatite standoffs are used to support L_6 from the side of the box. When placing the 1625 sockets be sure to allow sufficient clearance for installing and removing the tubes.

off as much as a second or two after the key is opened, thus eliminating the need for manual switching even though the oscillator runs continuously while the operator is sending at ordinary keying speed. A "Tune-Operate-Spot" switch, S_2, is included in the circuit: in the "Spot" position, the oscillator is turned on and, if desired, can be left on continuously while keying is done entirely in the amplifier stage. The system as shown also eliminates key clicks by shaping the keyed signal.

The differential keying circuit uses a 6-volt a.c. relay, single-pole, double-throw. In the key-up position a negative voltage is applied to the screens of the 1625s through the contacts of K_1, thus cutting off the plate current. The same voltage is applied to the 0A3/VR75, causing it to conduct and thereby applying negative voltage as a bias to the 6AG7 grid through one diode of a 6H6 (the other half of the 6H6 is used as a rectifier for the negative supply). In this condition the 6AG7 plate current is cut off and the circuit does not oscillate. When the key is closed, a positive voltage is applied to the circuit, through the contacts on K_1. The positive voltage does not reach the grid of the oscillator because it cannot get through the 6H6, and since the negative biasing voltage has been removed the oscillator comes on. Meanwhile, the positive voltage has gradually been overcoming the negative charge left on C_4, so the amplifier screens come up to the normal positive operating voltage relatively slowly. The slow rise in voltage—actually, it takes place in just a small fraction of a second—eliminates the click on closing the key.

On opening the key the negative voltage is again applied to the 1625 screens through K_1, but must overcome the positive charge left on C_4 before cutting off the amplifier. This slow change in voltage eliminates the click on "break," and also delays the application of negative bias to the oscillator, so the oscillator holds on for a while after the key is opened.

Control and Metering

The first position of S_2 grounds the amplifier screens and turns on the oscillator. This allows tune-up at reduced amplifier plate current. The second position of S_2 permits the oscillator and amplifier to be keyed together, as described above. In the third position the oscillator is turned on but the amplifier is off until the key is closed. This position can be used for spotting your frequency. It also lets the oscillator run continuously, if plain amplifier keying is preferred.

The metering circuit uses a 0–1 milliammeter as a low-range voltmeter that can be switched, by S_5, across appropriate shunts to read either grid or cathode current of the 1625 amplifier. The internal resistance of the meter used is approximately 50 ohms, and with the series and shunt values listed in Fig. 10-7 the full-scale readings are approximately 20 ma. for grid current and 300 ma. for cathode current. Some of the lower-priced meters have internal resistances of 1000 ohms, and if such a meter is used the 4700-ohm series resistor, R_5, should be changed

A 100-Watt Transmitter

to 3900 ohms. The total of the meter resistance and R_5 should be approximately 5000 ohms, whatever the actual meter resistance.

Power Supply

To obtain as much voltage as possible from the TV-type transformer a bridge rectifier circuit is used in the power supply. A pair of 6DE4 half-wave rectifiers and a full-wave 5U4G are used in the bridge.

The two 6.3-volt a.c. windings on the power transformer are connected in series to provide the 12.6 volts required for the heaters of the 1625s.

The power supply has two output voltages, approximately 600 and 300 volts. The actual voltages will depend on the particular type of power transformer used, but will be in this vicinity. Choke-input filters are used in both the high- and low-voltage legs. Practically any TV power-supply choke will be usable for L_8, providing its current rating is at least 200 ma. or more. Two 16-μf. 600-volt electrolytic capacitors, C_8 and C_9, are connected in series to provide 8 μf. at 1200 volts for the filter capacitance in the high-voltage side of the power supply. These two capacitors are shunted by two 25,000-ohm, 10-watt resistors, to help equalize the voltage drops across the two capacitors and to serve as a bleeder.

A 15,000-ohm, 10-watt resistor is used as a bleeder in the low-voltage leg of the supply. The filter in this side of the supply consists of an 8.5-hy. choke (L_7) and a 16-μf. 450-volt electrolytic capacitor.

One of the diode sections of the 6H6, V_{4B}, is used as a half-wave rectifier in the negative-voltage supply for the keying system. The secondary of T_1 has two windings, one at 250 volts, center-tapped, and the other at 6.3 volts. The center tap on the high-voltage winding is not used. When installing the electrolytic filter capacitor be sure that its positive side is connected to chassis.

Construction

A 17 × 12 × 3-inch aluminum chassis is used as the base and the r.f. components are housed in a 12 × 7 × 6-inch aluminum box (Premier AC-1276). In laying out and mounting components inside the box be sure to allow clearance for the ½-inch lip around the bottom. It is a good idea to follow the general layout shown in the photographs.

L_6 is a length of Air Dux pi-network coil stock that comes mounted on a piece of plastic. Two steatite standoffs, ½ × 1 inch, are used to support the coil on the side of the box. The 28-Mc. coil, L_5, is connected between one end of L_6 and the stator of C_5. The output terminal, J_2, is mounted on the rear of the box just behind C_6. The front-view photograph shows the layout of the panel controls.

Underneath the chassis, L_2 is mounted on a 1-inch cone insulator by cementing the coil support bars to the insulator with Duco cement. L_1 is supported between the rotor of C_3 and the 21-Mc. switch terminal. The keying relay should be mounted on a rubber grommet to reduce the relay noise: the grommet size is ¼-inch diameter and a ¼-inch hole is required.

The power-supply components are mounted at the rear of the chassis. The layout shown in the photographs can be followed if desired but the arrangement of parts is not critical. However, when mounting the rectifier sockets be sure to allow clearance for the tube envelopes when the tubes are inserted in the sockets.

In order to obtain the 12.6 volts required for the 1625 heaters, the two 6.3-volt windings on T_2 are connected in series. In the TV-type transformers there is usually one winding of 6.3 volts at 5 or more amperes and another winding of the same voltage at a little more than 1 ampere rating. Connect the two windings in series and check the voltage at the outside ends with an a.c. voltmeter. If you don't have such a voltmeter, use the 1625 heaters instead: if the heaters light up the connections are correct: if they stay dark, reverse one of the 6.3-volt windings. You will get 6.3 volts between either lead A or B (Fig. 10-7), and chassis. The 6.3-volt winding with larger current rating (heaviest leads) will be your lead A. This is the one that will handle the heater current for the 6AG7, 6DE4s, the pilot light, and the keying relay. Current for the 6H6 heater can be taken from the 6.3-volt winding on T_1.

F_1 and F_2 mount in a fuse-type line plug.

Testing and Adjustment

Before applying power to the transmitter, carefully check your wiring for errors. If an ohmmeter is available measure the resistance between the low-voltage B-plus line and chassis. Before using the ohmmeter make sure the filter capacitor is discharged, by shorting the positive side of the electrolytic to chassis. Connect the negative lead of your test instrument to chassis

---

Coil Data for the 100-watt Transmitter

L_1—13 turns No. 20, ½-inch diam., 16 turns per inch (B & W Miniductor 3003).

L_2—46 turns No. 24, 1-inch diam.; 14-Mc. tap 5 turns from junction of $L_1 L_2$, 7-Mc. tap 17 turns from junction of $L_1 L_2$ (B & W Miniductor 3016).

L_5—4 turns No. 16, 1-inch diam., 1 inch long.

L_6—28 turns No. 16, 1½ inch diam., 12 turns per inch; 21-Mc. tap 1½ turns from junction of $L_5 L_6$, 14-Mc. tap 5½ turns from junction of $L_5 L_6$, 7-Mc. tap 17½ turns from junction of $L_5 L_6$ (Pi Air Dux 1212A with 4 turns removed).

FIG. 10-9—Below chassis view of the 100-watt transmitter. The oscillator components are in the upper left corner. The amplifier grid circuit components, C_3, L_1, L_2, and S_1, are at the top center. Directly below are the sockets for the 1625s. Lower left, on the chassis side is T_1, the negative-supply transformer. T_2, the power transformer, is visible in the lower right-hand corner. Just above it, on the chassis wall, is L_7.

and make your test with the positive lead. The resistance should be approximately 15,000 ohms, the value of the bleeder in the low-voltage supply. On the high-voltage side the resistance will be about 50,000 ohms. These tests will show whether there are any shorts or opens in the B-plus circuits.

A 100-watt lamp can be connected to J_2 to be used as a dummy load. Put S_2 in the "Tune" position, thus grounding the screens of the 1625s. Plug in a 3.5-Mc. crystal and set S_1 and S_3 for the 3.5-Mc. band. Turn on the power and allow about two minutes for the tubes to warm up. With the meter switch in the grid-current position you should get a reading when C_3 is tuned. Adjust C_3 so that the grid current is about 7 ma. Next, set C_6 near maximum capacitance (plates fully meshed), set S_5 to read cathode current, and adjust C_5 for a dip in the reading. The current will be small because the screens of the amplifier are grounded. Then turn S_2 to "Operate," close the key, and adjust C_5 and C_6 to cause the lamp dummy to light.

Reducing the capacitance of C_6 will increase the lamp brightness, if C_5 is readjusted for the dip in plate current each time the setting of C_6 is changed. Eventually a point will be reached where decreasing the capacitance of C_6 further will give no more output, although the plate current may continue to increase. The best loading adjustment is the one that gives maximum output with the least plate current. The plate current itself should be in the neighborhood of 175 to 200 ma. at this point.

Go through the same procedure on the other bands to familiarize yourself with the controls. On each band, S_1 and S_3 should be set to the same frequency, with the exception of 28 Mc.; on this band, set S_1 to 14 Mc. and S_3 to 28 Mc. Remember, also, that 7-Mc. crystals should be used for 14, 21 and 28 Mc. When testing the setup on 21 Mc., adjust C_1, the oscillator feedback capacitor, so that the grid current is no more than 7 ma. This adjustment should be made with C_3 at the position that gives maximum drive.

Typical voltage readings, with a TV power transformer having a 700-volt center-tapped winding, are: high voltage, 570 volts with the amplifier plate current 170 ma.; low voltage, 260 (screens of 1625s and plate of 6AG7), 6AG7 screen, 150 volts. Your transformer may give slightly higher or lower voltages, depending on the type. However, practically all TV power transformers will be suitable.

If you're a Novice licensee you must not run more than 75 watts input. If you know what your plate voltage is, Ohm's Law will give the plate current you can run on the amplifier to get 75 watts input. For example, if the plate voltage is 570 and the amplifier plate current is 130 ma. the input is 74 watts. If you can't measure the plate voltage, the simplest thing to do is to assume that it will be 600 volts, a round figure that is not likely to be exceeded with a power supply of the type described here when the amplifier is loaded to a plate current of 125 ma. Thus a plate current of 125 ma. should be on the safe side, for an estimated 75 watts input.

To load the amplifier to the Novice limit, first set the output capacitance, C_6, to maximum and "dip" the final by adjusting C_5. Note the current reading and, if you can measure the plate voltage, calculate your input. You'll probably find it to be considerably less than 75 watts. Decrease the capacitance of C_6 a little and redip the final, then repeat until the input is brought up to the 75-watt point. Alternatively, follow the same tuning process until the dipped plate current is 125 ma., as suggested above.

The grid voltage and current for 1625s at normal operating conditions are −45 volts and 7 ma. The 6AG7 gives more than enough output, being capable of developing over 100 volts bias

A 150-Watt Amplifier

on the 1625 grids on 80, 40, and 20, and about 70 volts on 15 meters. However, the drive on 15 should be adjusted as outlined above, to minimize crystal heating. If the grid current is too high on the lower bands it can be reduced by adjusting C_3.

With the component values specified for R_4 and C_4, the oscillator tends to stay on for two or three seconds after opening the key. You can adjust the circuit so that the oscillator stays on for longer or shorter times by changing the value of R_4. A higher value, such as 220K or 330K, holds the oscillator on for longer periods. This also tends to soften the keying.

A 150-Watt Amplifier

Many Novices start out with transmitters running 20 to 40 watts input, often constructed from kits. It isn't a bad way to break in. However, after a little experience on the air—and particularly when the General Class license is within sight—the thought of higher power is inevitable. The amplifier shown in Fig. 10-10 represents a practical way to get up with the majority of hams at fairly low cost. The amplifier is band-switching, covers all bands from 3.5 to 28 Mc., and can be driven by practically any low-power transmitter already in your possession. For example, it makes a good companion piece for the low-power transmitter shown earlier in this chapter. You will need only one tube in the amplifier stage of that set for driving the amplifier described here, and that tube can be any of the TV sweep tubes from the 6BQ6 up.

The amplifier tubes are 1625s, used because they are very cheap in surplus. You can use 807s instead, if you wish; the two types are identical except for heater voltage and basing. The 807 has a 6.3-volt heater and a standard 5-prong base, while the 1625 takes 12.6 volts and has a large 7-prong base. In either case the amplifier can be run at 150 watts input if the plate supply can deliver 750 volts at approximately 220 ma., 200 of which is for the plates, and about 20 for the screens. Since the rated plate current for the two tubes is 200 ma., the power input is reduced if lower plate voltage is used. However, provided the screen voltage is adjusted to 300 volts under full load, the amplifier can be operated at a plate current of 200 ma. with any plate-supply voltage from 500 to 750.

The power supply shown with the amplifier is an "economy" supply using a receiving-type power transformer with a bridge rectifier. It is similar to the one in the 100-watt transmitter described earlier, and uses a salvaged TV-receiver transformer and choke. If you're lucky enough to get a really husky TV transformer, you can realize about 700 volts at 200 ma. from such a supply. Everything depends on the voltage available from the transformer; in some cases you may get no more than 600 volts or so. The difference in input is not great, although you may not reach the maximum power the amplifier is capable of handling. Of course, other types of supplies may be used; just follow standard design practices as outlined in Chapter 12 if you want to use regular transmitting components.

The Circuit

Two parallel-connected 1625s are used in the amplifier circuit, shown in Fig. 10-12. Another 1625 is a **clamp tube** to prevent excessive plate current in the amplifier tubes when there is no r.f. grid excitation.

The grid circuit of the amplifier is untuned. This makes it necessary to supply somewhat more driving power than would be needed with

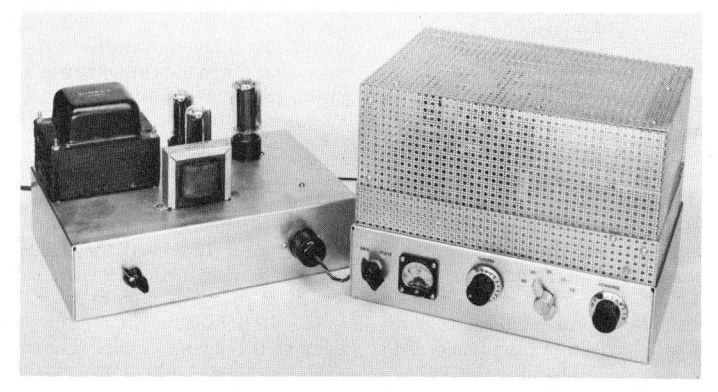

FIG. 10-10—The five-band amplifier and its power supply. The switch on the left front of the power-supply chassis is S_3. On the amplifier, from the left, the controls are the meter switch, plate tuning, band switch, and loading control.

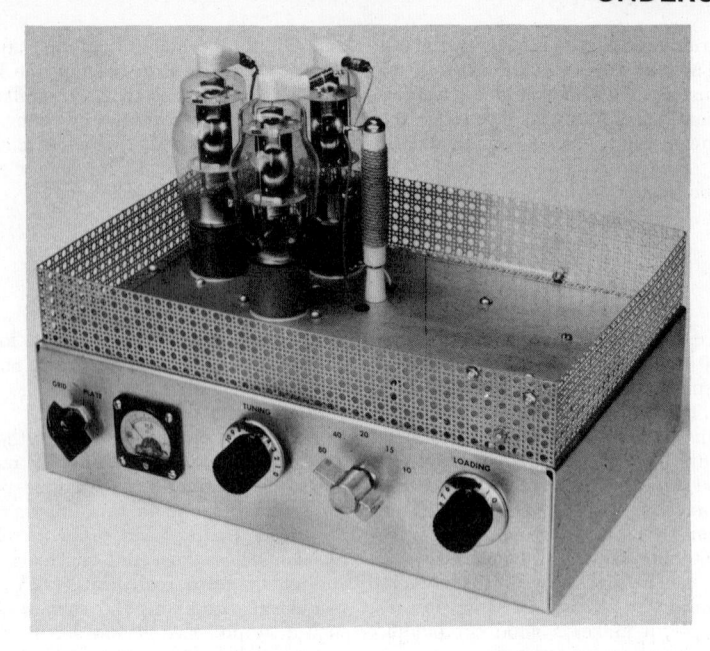

FIG. 10-11—A view of the amplifier with the cover removed. The two 1625s nearest the upright plate choke, RFC₂, are the amplifiers. The tube to the left is the clamp tube. Note that the parasitic suppressors are mounted right at the plate caps.

a tuned grid circuit, but eliminates an extra control. Nearly any Novice transmitter will furnish more than enough drive on all bands.

When there is no drive for the amplifier there is no bias on the grid of the clamp tube, V_3, so its plate current tends to be large. This causes a large voltage drop in the screen resistor, R_3, so the voltage at the screens of V_1 and V_2 drops to about 100 volts. The low screen voltage holds the amplifier tubes' plate current to a low-enough value so they idle at well below their rated plate dissipation. When excitation is applied, grid current flowing through R_1 develops enough negative grid bias to cut off the plate current of V_3, so the amplifier screen voltage rises to its normal operating value of about 300 volts.

A band-switching pi-network tank circuit covering 3.5 to 30 Mc. is used in the amplifier. This network is designed for 50- to 75-ohm loads. One pole of the band switch, S_1, shorts out sections of L_4 for various bands. The other pole adds capacitance (C_6) across C_5 on 3.5 Mc. The output loading capacitor, C_8, is a three-section variable having a capacitance of about 400 µµf. per section. All three sections are connected in parallel. On 3.5 Mc. a 1500-µµf. mica capacitor, C_7, is connected across C_8 by means of S_1. RFC_3 is a safety precaution to short-circuit the d.c. to ground in the event the plate blocking capacitor, C_4, should short out.

A 0-1 milliammeter, connected as a voltmeter, is used to measure either the plate or grid current. It does this by measuring the voltage drop across shunts of appropriate resistance—R_2 in the grid circuit and R_4 in the plate. The full-scale current is 20 ma. when the meter is connected across R_2, and 300 ma. when connected across R_4 in the plate lead.

Power Supply

The bridge rectifier circuit, Fig. 10-14, uses a pair of 6DE4s and a 5U4G. The filter choke, L_5, is approximately 2 henrys. Two 30-µf. 500-volt electrolytic capacitors are connected in series to provide a working voltage of 1000 volts. The +B voltage can be turned on and off with S_{3B}. Another section, S_{3A}, of the same switch is used to turn the supply on and off. P_2 is a fused plug; F_1 and F_2 protect the supply in the event of overload.

Nearly all TV transformers have at least two 6.3-volt filament windings. These can be connected in series to provide the required 12.6 volts for the 1625 heaters. The 6.3 volts required for the 6DE4s is taken from one of the two 6.3 windings; use the winding with the heavier current rating (heavier wire) for this purpose if there is a choice.

Construction

The amplifier is built on a 3 × 8 × 12-inch aluminum chassis, with the power supply as a separate unit. If desired, the builder can combine both units on a single chassis, but a larger one would, of course, be required.

The bottom-view photograph will show you most of the layout details. The three 1625s are mounted at one side of the chassis and the rest of the room is taken up by the tank circuit components. An Air Dux 1212D6 coil assembly is used for L_4. This assembly is supported on the

A 150-Watt Amplifier

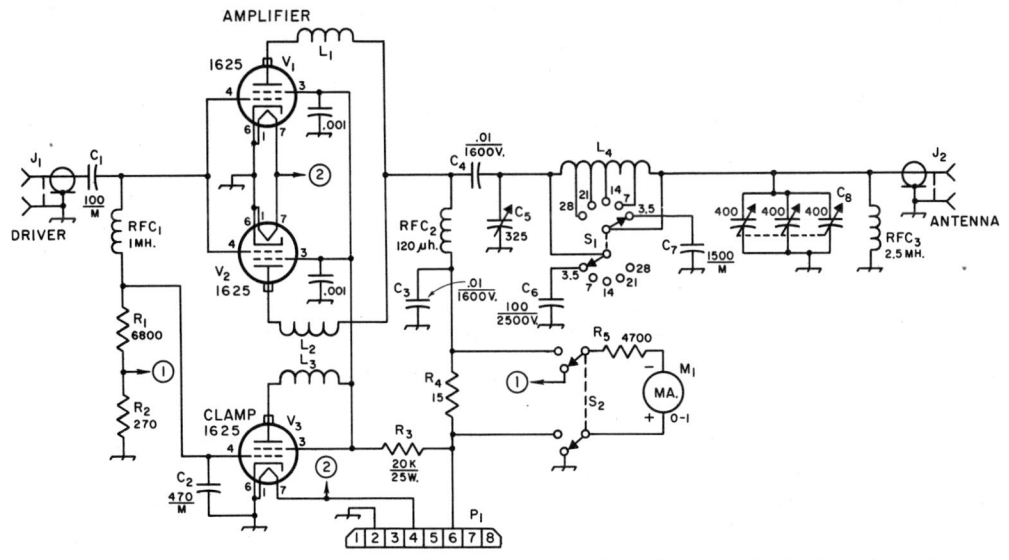

FIG. 10-12—Circuit diagram of the 1625 amplifier. Unless otherwise indicated, decimal values of capacitance are in $\mu f.$, others are in $\mu\mu f.$; M = mica. Resistances are in ohms.

C_1—100-$\mu\mu f.$ mica.
C_2—470-$\mu\mu f.$ mica.
C_3, C_4—0.01-$\mu f.$ 1600-volt disk ceramic.
C_5—325-$\mu\mu f.$ variable (Hammarlund MC-325-M).
C_6—100-$\mu\mu f.$ mica, 2500 volts.
C_7—1500-$\mu\mu f.$ mica.
C_8—Three-section receiving variable, approx. 400-$\mu\mu f.$ per section (Allied Radio 60-H-726 or Philmore 9047).
J_1, J_2—Coax chassis receptacle, type SO-239.
L_1, L_2, L_3—10 turns No. 18 enam., close-wound on a 1-watt resistor, 1000 ohms or more.
L_4—19 turns No. 14, 1½-inch diam., 9 turns spaced 12 turns per inch, 10 turns spaced 6 turns per inch (Illumitronic Air Dux 1212D6). The end of the coil with wide spaced turns is connected to C_5. 7-Mc. tap: 12 turns from the C_5 end of the coil. 14-Mc. tap: 6 turns from the C_5 end of the coil. 21-Mc. tap: 4 turns from the C_5 end of the coil. 28-Mc. tap: 2 turns from the C_5 end of the coil.
M_1—0-1 milliammeter, 1⅝ inch square, D'Arsonval movement.
P_1—Octal plug (Amphenol 86PM8).
R_1—6800 ohms, 1 watt.
R_2—270 ohms, ½ watt.
R_3—20,000 ohms, 25 watts.
R_4—15 ohms, ½ watt.
R_5—4700 ohms, ½ watt.
RFC$_1$—1 mh. (Millen 34300-1000, National R-50).
RFC$_2$—120 $\mu h.$, 500 ma. (Raypar No. RL-101).
RFC$_3$—2.5 mh. (Millen 34103, National R50).
S_1—Ceramic rotary, 1 section, 2 poles, 5 positions (Centralab PA-2003).

chassis by two 1¼-inch high isolantite standoff insulators. The tap leads for the various bands are brought forward to S_1, which is mounted on the chassis front between C_5 and C_8.

As it comes from the dealer, L_4 has more turns than are needed. Remove 17 turns from the close-wound end, leaving a total of 19 turns. To prevent shorting, the turns adjacent to the 40-meter tap point should be bent in toward the center of the coil. The remaining taps are on portions of the coil where the turns are not so close together.

The coils L_1, L_2 and L_3 in the plate leads of the three tubes are for suppressing v.h.f. parasitic oscillations. These coils should be mounted directly at the plate caps. The forms on which they are wound are one-watt resistors; any resistance value over 1000 ohms is suitable.

The top cover is made from Reynolds do-it-yourself perforated aluminum. The construction is exactly the same as in the low-power transmitter described earlier in this chapter.

Testing and Tune-Up

Connect the power supply and your exciter to the amplifier. You can use a short length of coax cable—either the 50- or 70-ohm type is suitable—between the exciter and amplifier, but keep the length of coax as short as possible. You should use a dummy load on the amplifier for initial testing. A 100-watt lamp bulb will be suitable. Turn on the power supply, but leave the +B off (second position of S_3) and let the heaters warm up. Switch the exciter and the amplifier to 80 meters and turn on the exciter. With S_2 set to read grid current, tune the exciter for a reading of 8 ma. Next, set S_2 to read plate current, set C_8 at maximum capacitance, plates fully meshed, and turn on the +B voltage. Resonate the amplifier by tuning C_5 for a dip in plate

FIG. 10-13—Bottom view of the amplifier. The clamp tube socket is at the far left in this view, with the two amplifier sockets to the right. The L_4 assembly is mounted on two standoffs, the assembly being positioned between C_5 on the left and C_8 at the right. On the rear wall of the chassis J_1 is at the left side and J_2 at the right-hand side, in this view.

current reading. You can now start to load the amplifier by decreasing the capacitance of C_8. Retune C_5 for the resonance dip as you continue to increase the plate current. The lamp should increase in brightness as the amplifier loading is increased. Maximum rated plate current for the tubes is 200 ma., or two-thirds of full-scale reading on the meter.

The same procedure should be followed in checking the other bands.

To get maximum output from the 1625s the screen voltage should be about 300 when the amplifier is operating fully loaded. The value given for R_3, 20,000 ohms, is about right when the plate voltage is 700 or more. With 600 volts on the plates, a 15,000-ohm 25-watt resistor will give about 300 volts on the screens. A variation of plus or minus 20 volts will not materially affect the operation of the amplifier.

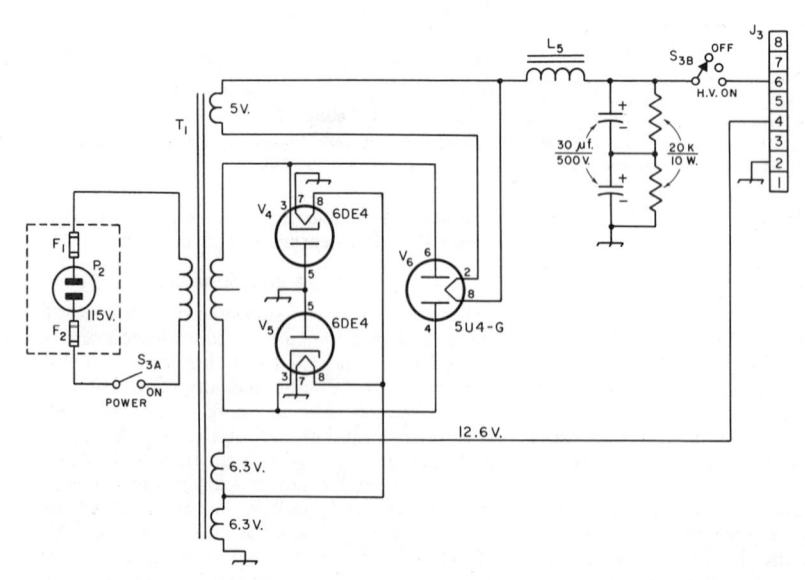

FIG. 10-14—Circuit diagram of the amplifier power supply.

F_1, F_2—3-amp. type 3AG.
J_3—Octal socket.
L_5—Approx. 2 hy., taken from TV set.
P_2—Line plug, fuse-in-plug type.

S_3—Single-pole, four-position with a.c. switch on back (Centralab 1465).
T_1—Power transformer taken from TV set; see text.

ARC-5 Transmitters 187

FIG. 10-15—ARC-5 transmitter converted to crystal control for Novice use. The crystal oscillator, the separate small unit at the left, is connected to the oscillator-tube socket in the transmitter through a short length of cable. A small notch should be cut in the transmitter cover to provide clearance for the cable when the cover is installed.

The power transformer, rectifier, and choke are mounted on top of the power-supply chassis at the rear, and the control switches are mounted on the wall as shown. Remaining components are underneath.

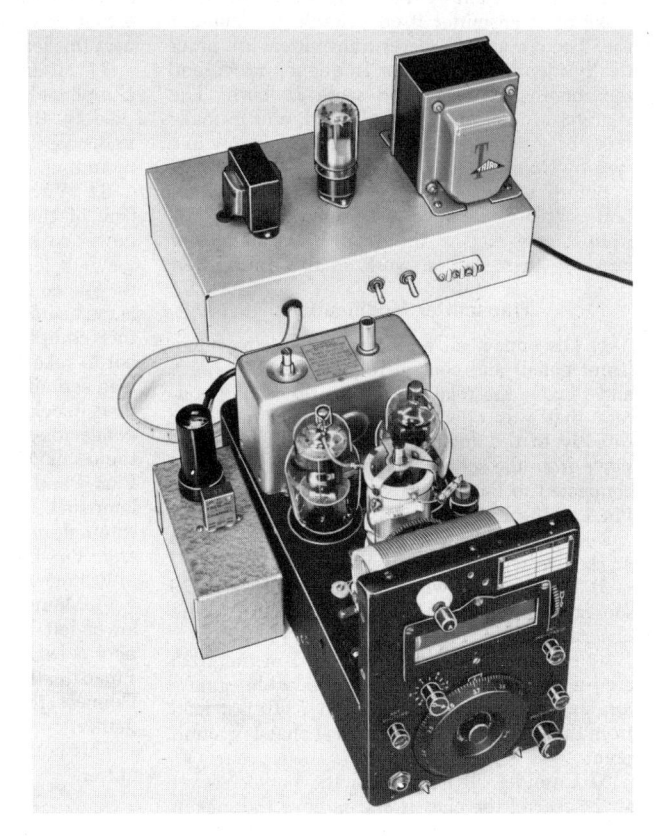

The ARC-5 Transmitters

If you have looked through Chapter 8 you've seen that a good deal of the receiving equipment described there is designed around surplus "Command" receivers, part of what is generally called the "ARC-5" series. There are useful transmitters in the same series, too, and they are plentiful and cheap. Prices for two transmitters of special interest to Novices—the ones covering 3 to 4 Mc. and 7 to 9.1 Mc. respectively—are about the same as for the receivers. The 3 to 4 Mc. transmitters commonly available are the BC-457 and the T19 (ARC-5). Corresponding numbers for the 7 to 9.1 Mc. transmitters are BC-459 and T22 (ARC-5).

These transmitters are all alike in basic circuit. They have two stages, a variable-frequency oscillator and a straight-through amplifier. The oscillator tube is a 1626 and the amplifier uses a pair of 1625s in parallel. These tubes have 12.6-volt heaters. The sets also include a magic-eye tube (1629) which is of no interest in amateur use, having been included for the purpose of calibration checking.

The v.f.o. in the ARC-5 transmitter has very good stability, and many amateurs use the sets for everyday communication. The transmitters have also been used extensively as v.f.o.s for driving bigger amplifiers. However, a Novice has to use crystal control. The transmitter can very readily be converted to crystal by the method described below. The modification does not in the least affect the usefulness of the v.f.o. in the set; the v.f.o. can be restored to normal operation in a matter of minutes.

The principal disadvantage of the ARC-5 transmitter is the fact that it is built for one band only. Thus if you get the 3 to 4 Mc. transmitter you can only work on the 3.5-Mc. amateur band; similarly, the 7 to 9.1 Mc. outfit is good only for 7-Mc. operation.

In the modification to be described, the Novice requirement for crystal control is met by using a separate crystal-controlled oscillator. The output of the external oscillator is fed into the transmitter through a plug that fits into the 1626 oscillator socket. The 1626 is not used. The transmitter modifications are such that to restore the transmitter to v.f.o. operation it is necessary only to remove the external oscillator plug and put the 1626 back in its socket. No wiring changes are needed to go from crystal control to v.f.o.

In addition to the external oscillator, a power

supply is required for the oscillator and transmitter, and certain wiring changes are needed to make the transmitter itself suitable for amateur use. The 80- and 40-meter transmitters are practically identical except for frequency range and the modifications are the same in both. The changes are not at all complicated, and to make them easy to follow they are outlined below in a step-by-step sequence.

Before starting, remove the top cover and bottom plate. Remove the tubes and crystals from their sockets so there will be no danger of breaking them as you work on the transmitter.

Transmitter Modifications

1) Disconnect all wiring to the antenna relay (front panel) and control relay (side of chassis), and remove the relays. Connect the spring contact that was operated by the antenna relay directly to the antenna post. Unsolder and remove from the set all remaining wires that were connected to the relays, except the one going to Pin 4 on the 1626 oscillator socket.

2) Remove the wire-wound resistor mounted on the rear wall of the transmitter.

3) Unsolder the wire from Pin 7 of the 1629 socket and move it to Pin 2. Ground Pin 7 to the chassis.

4) Unsolder the wires from Pin 1 of the 1625 closest to the drive shaft for the variable capacitors and solder the wires to Pin 7. Run a lead from the same Pin 1 to the nearest chassis ground point.

5) Unsolder all leads from the power socket at the rear of the chassis and remove the socket. The socket can be pried off with a screwdriver.

6) Unsolder the end of the 20-ohm resistor (red-black-black) that is connected to Pin 4 on the oscillator socket and connect it to Pin 6 of the crystal socket. There is also a lead on Pin 4 that was connected to the keying relay: connect this lead to the nearest chassis ground point.

7) Mount an octal male chassis connector (Amphenol 86-RCP8) in the hole formerly occupied by the power socket. Install a solder lug under one of the nuts holding the connector mounting.

8) Wire the connector as shown in Fig. 10-18. One of the leads unsoldered from the original power socket is red with a white tracer. This is the +B lead for the 1625s. The yellow lead is the screen lead for the 1625s and the white lead is the heater lead. Although the manuals covering this equipment specify these colors, it's safer not to take them for granted: check where each lead actually goes before connecting it to the new power socket. The lead from Pin 1 on the power connector, J_1, to Pin 6 on the crystal socket is the oscillator plate-voltage lead. The leads from Pins 7 and 8 on J_1 to Pins 1 and 6 on the oscillator socket are new leads to carry power to the external crystal-controlled oscillator. The lead from Pin 4 on J_1 to Pin 2 on the 1629 (resonance indicator) socket is the 12-volt heater lead.

9) Mount a closed-circuit phone jack at the lower left-hand corner of the front panel. Connect a lead from the tip contact of the jack to Pin 6 (cathode) of either of the 1625 sockets. Connect the remaining two jack terminals together.

This completes the modification of the transmitter.

Crystal-Controlled Oscillator

The external crystal-controlled oscillator circuit, shown in Fig. 10-16, uses a 6AG7 in the grid-plate oscillator circuit. Either 80- or 40-

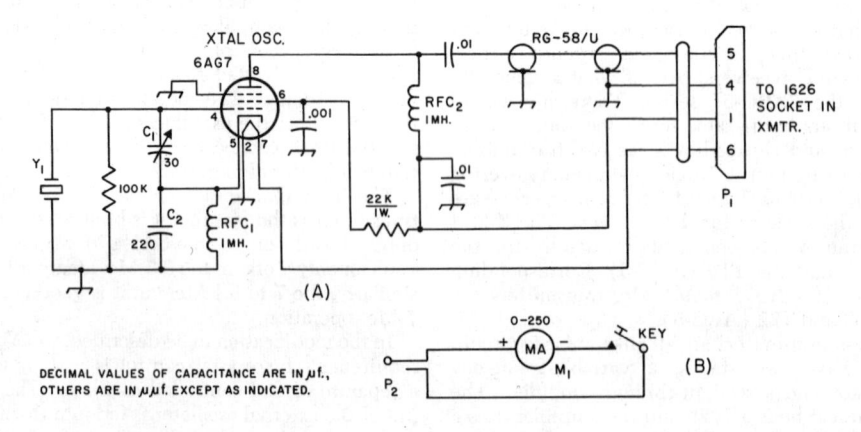

FIG. 10-16—(A) Circuit diagram of external crystal-controlled oscillator. Unless otherwise specified, resistances are in ohms, resistors are ½ watt. The 0.01- and 0.001-µf. capacitors are disk ceramic. (B) Method of connecting the milliammeter in series with the key.

C_1—3–30-µµf. trimmer.
C_2—220-µµf. fixed mica.
M_1—0–250 d.c. milliammeter.
P_1—Octal plug, male (Amphenol 86-PM8).
P_2—Phone plug.
RFC_1, RFC_2—1-mh. r.f. chokes.
Y_1—3.5- or 7-Mc. Novice-band crystal, as required.

ARC-5 Transmitters

FIG. 10-17—Bottom view of the crystal oscillator showing the arrangement of components. Terminal strips are used for the cable connections and also as a support for C_1, the feedback capacitor.

meter crystals are required, depending on the band in use.

Output from the oscillator is fed to the transmitter through an 8-inch length of RG-58 coax cable. The cable is terminated in an octal plug, P_1, which goes into the oscillator tube socket in the transmitter. Power for the external oscillator is obtained through this socket.

The crystal-controlled oscillator is built in and on a $4 \times 2 \times 2\frac{3}{4}$-inch aluminum box. The tube and crystal sockets are mounted on top and the remaining components are inside. Layout of parts is not particularly critical but the general arrangement shown in Fig. 10-17 should be followed to insure good results.

The 1625s are keyed and the crystal oscillator runs continuously during transmissions. It is thus necessary to turn the oscillator off during standby periods. This method is used in preference to keying the oscillator and amplifier simultaneously because keying the oscillator is likely to make the signal chirpy.

Power Supply

The power supply, shown in Fig. 10-18, uses a 5U4G rectifier and a capacitor-input filter. The power transformer, T_1, is a type made by several manufacturers. To obtain 12.6 volts for the heaters, a 6.3-volt filament transformer is connected in series with the 6.3-volt winding on T_1. This setup also will provide 6.3 volts for the heater of the 6AG7.

To turn off the plate voltages on the transmitter during stand-by periods, the center tap of T_1 is opened. This can be done either by S_2 or by an external switch connected in parallel with S_2 through a two-terminal strip mounted on the power-supply chassis. The remotely-mounted switch can be installed in any convenient location at the operating position. A single-pole, single-throw switch can be used for this purpose. If desired, a multipole switch can be used to perform simultaneously this and other functions, such as controlling an antenna-changeover relay.

The high-voltage and heater leads are brought out in a cable to a female octal plug, P_3, that connects to J_1 on the transmitter. The length of the cable will, of course, depend on where you want to install the power supply. Some amateurs prefer to have the supply on the floor under the operating desk rather than have it take up room at the operating position.

The supply shown in Fig. 10-15 was constructed on a $3 \times 6 \times 10$-inch chassis. The layout is not critical, nor are there any special precautions to take during construction other than to observe polarity in wiring the electrolytic capacitors. Also see that the power leads are properly insulated.

When wiring P_3 don't connect the B-plus lines to Pins 2 or 3, the amplifier plates and screens, at first. It is more convenient to test the oscillator without plate and screen voltages on the amplifier.

When the supply is completed, measure between chassis and the 12.6-volt lead with an a.c. voltmeter to see if the two 6.3-volt windings are connected correctly. If the voltage is zero, reverse one of the windings. If you don't have an a.c. meter, watch the heaters in the 1625s. They will light up if you have the windings properly connected. Leave the B plus off, by opening S_2, for this check.

Next, set the slider on the bleeder resistor, R_1, at about one-quarter of the total resistor length, measured from the $+B$ end of the bleeder. Be sure to turn off the power when making this adjustment. With the tap set about one-quarter of the way from the $+B$ end of the bleeder the oscillator plate and amplifier screen voltages will be approximately 250 volts.

Testing the Transmitter

To try out the transmitter you'll need a key and meter connected as shown in Fig. 10-16. When P_2 is plugged into the jack in the transmitter the meter will measure the cathode current of the 1625s. The cathode current is the sum of the plate, screen and control-grid currents. Some amateurs prefer to install the meter in the plate lead so it reads plate current only. This can be done by opening the $+B$ line at the point marked X in Fig. 10-18, and inserting the meter in series with the line. However, unless you have more than one meter, don't install it in the power supply in this way until after you have made the tests described below.

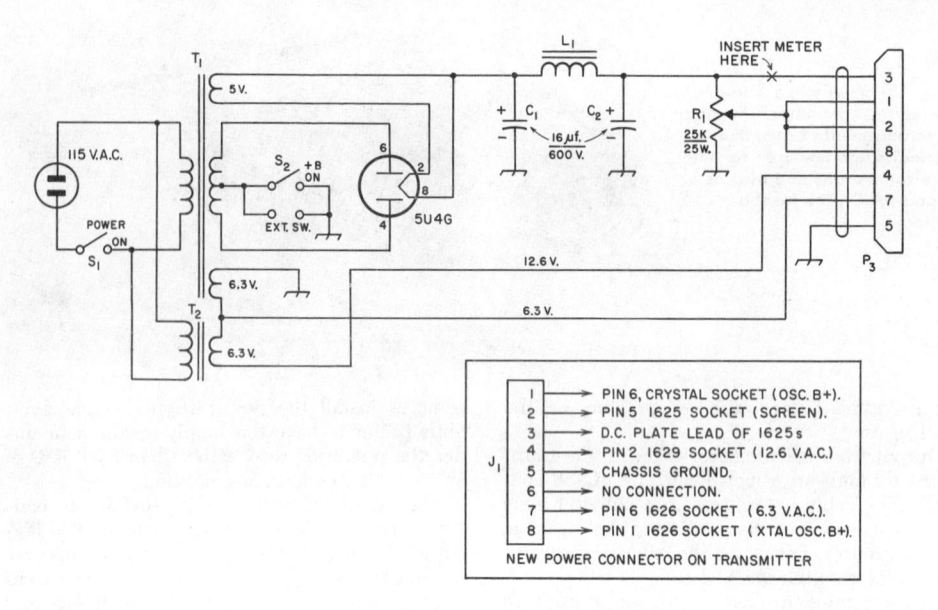

FIG. 10-18—Circuit diagram of power supply.

C₁, C₂—16-μf., 600-volt electrolytic (Sprague TVA-1965, Aerovox PRS).

J₁—Octal chassis-mounting connector, male (Amphenol 86-RCP8).

L₁—1- to 2-hy., 200-ma. filter choke, TV replacement type (Stancor C-2325 or C-2327, or equivalent).

P₃—Octal cable connector, female (Amphenol 78-PF8).

R₁—25,000 ohms, 25 watts, with slider.

S₁, S₂—Single-pole, single-throw toggle switch.

T₁—Power transformer, 800 volts center-tapped, 200 ma.; 5 volts, 3 amp.; 6.3 volts, 6 amp. (Knight 61G414, Triad R-21A, or equivalent).

T₂—Filament transformer, 6.3 volts, 3 amp. (Triad F-16X, Knight 62-G-031, or equivalent).

Insert the external oscillator plug, P_1, into the 1626 socket and connect P_3 to the transmitter. Plug P_2 into the key jack. With S_2 open, turn on the power and allow a minute or two for the tubes to warm up. Next, close the center-tap connection, S_2, on the power transformer. Set the transmitter dial to the same frequency as that of the crystal you are using and close the key. You should get a slight indication of grid current on the meter. (There is no plate or screen current because you don't have screen or plate voltages on the amplifier until you wire Pins 2 and 3 of P_3.) If you don't get a reading, adjust C_1 to the point where you do.

The next step is to peak the amplifier grid circuit—that is, the 1626 v.f.o. tank—for maximum grid-current reading. The v.f.o. trimmer capacitor is in an aluminum box on the top of the chassis at the rear. There is a ½-inch diameter hole in the side of the box: loosen the small screw visible through this hole, thus unlocking the rotor shaft of the trimmer capacitor. Move the rotor-arm shaft in either direction, observing the meter reading as you do, and find the position that gives the highest reading. This should be something more than 10 ma.

You are now ready to connect the plate and screen voltage leads to P_3. Be sure to turn off the power supply before making the connections!

The first test should be with a dummy load; a 115-volt, 60-watt light bulb can be used for this purpose. The lamp should be connected between the antenna terminal and chassis ground. However, to make the lamp take power it may be necessary to add capacitance in parallel with it. A receiving-type variable capacitor having 250 μμf. or more maximum capacitance will be adequate for the job.

Turn on the power and allow the tubes to warm up, but leave the key open. Set the antenna coupling control on the transmitter to 7 or 8, and set the variable capacitor connected across your dummy load to about maximum capacitance. Next, close the key and adjust the antenna inductance control for an increase in cathode current. Turn the frequency control for a dip in current reading. You'll probably find that the indicated frequency differs from that of the crystal you are using, but don't worry about it.

Adjust the three transmitter controls, antenna inductance, antenna coupling, and frequency, along with the variable capacitor across the lamp load, until the lamp lights up to apparently full brilliance. The cathode current should be between 150 and 200 ma. With the transmitter fully loaded, adjust C_1 in the crystal oscillator so that the lamp brilliance just starts to decrease. This

V.H.F. Transmitters

is the optimum capacitance for C_1 and it can be left at this setting, no further adjustments being required.

If you have a d.c. voltmeter, check the voltages. Using the power supply of Fig. 10-18, the plate voltage on the 1625s is approximately 400 with the amplifier fully loaded. With the plate voltage on the oscillator and screen voltage on the 1625s adjusted to 250 volts (bleeder tap), the oscillator screen voltage is 160 volts. The oscillator takes approximately 30 ma. and the 1625 amplifier screens about 10 ma. when the amplifier is fully loaded.

Getting on the Air

To put the transmitter on the air it is necessary only to connect an antenna wire to the antenna post and connect a ground lead from the transmitter chassis to a water-pipe ground or to a metal stake driven in the ground. Since the transmitter has a variable antenna coil that can be used to tune the antenna to resonance, almost any length of antenna can be used. However, for best results the minimum length should not be less than about ⅛ wavelength for the band in use. This is approximately 33 feet for 80 meters and 16 feet for 40 meters. You'll do better if you can make the antenna longer—and be sure to get the far end as high as possible.

An output indicator will prove to be a handy device for knowing when power is actually going into the antenna. For this purpose you can use a 6.3-volt, 150-ma. dial lamp. Connect two leads, each about one foot long, to the shell and base of the bulb, respectively. Clip one lead to the antenna post and the other lead on the antenna wire two feet from antenna post. A small amount of power will go through the bulb and will provide a visual indication of output. Follow the same tuning procedure as outlined above for the dummy antenna. If the bulb gets so bright that it is in danger of burning out, move the leads closer together to reduce the pickup.

You may find that certain antenna lengths won't work—that is, the amplifier won't load—no matter where you set the antenna coupling and inductance. In such a case, connect a variable capacitor—like the one used with the lamp dummy—between the antenna post and the transmitter chassis. Adjust the capacitor and antenna inductance for maximum brilliance of your output indicator: this will be the best setting for the control.

V.H.F. Transmitters

In Chapter 8 we had a few things to say about the special features of v.h.f. circuits. The remarks about capacitance and inductance apply to transmitters, too. The operating principles of oscillators, frequency multipliers, and amplifiers are the same as for those used on lower frequencies, but we can't afford to take any liberties with stray capacitances and lead lengths. This also goes for the tubes: physically small tubes, with short internal leads, are necessary if we are to get reasonably efficient performance above 50 Mc. Components such as variable capacitors and coils likewise must be small.

In fact, above 144 Mc. it is often desirable—and frequently downright necessary—to dispense with conventional tuned circuits entirely. Transmission-line circuits are used instead. However, we won't touch on these in this book, since they don't have to be used in transmitters such as the one shown here. You're bound to meet them later in your amateur career, though, if you elect to specialize in v.h.f. operating.

The Novice or beginning Technician will do well to start out on low power with equipment using familiar coils, capacitors and tube structures. Much of the equipment used in everyday rag-chewing on the 50- and 144-Mc. bands is of this type, whether home-built or of commercial manufacture. Many of the popular manufactured sets have only a couple of watts of r.f. output, and they are giving excellent results in so-called "local" work. The transmitter shown here is in this general class. It operates with an input of 15 watts to the modulated final stage—possibly a little higher power than the "gooney boxes" (a nickname for a popular type of v.h.f. transceiver).

As we explained earlier, it is desirable to use a reasonably high crystal frequency in order to avoid spurious frequencies as much as possible. As an economical compromise this transmitter uses crystals in the 8-Mc. region. Military surplus crystals of this type usually are quite cheap. The fact is that most of them could be operated on the third overtone in the neighborhood of 24 Mc. However, overtone operation is somewhat tricky. In the beginning, at least, you want something that will be as reliable as possible. Later you can experiment with overtone circuits if you wish, using the same crystals. The circuit changes would not be extensive at all. At that stage, too, you can use regular 24-Mc. crystals, which are all of the overtone type. Bear in mind that part of the fun of the game is changing things and trying new circuits—you don't have to stick with your first selection forever, especially when you do your own building!

The transmitter shown in Fig. 10-19 is bandswitching, covering both 50 and 144 Mc. It has three stages—oscillator, doubler, and final amplifier—on 50 Mc., and four stages—oscillator, doubler, tripler, and final amplifier—on 144 Mc. The oscillator and doubler tubes are 5763s; the tripler and amplifier tubes are 7558s. The latter

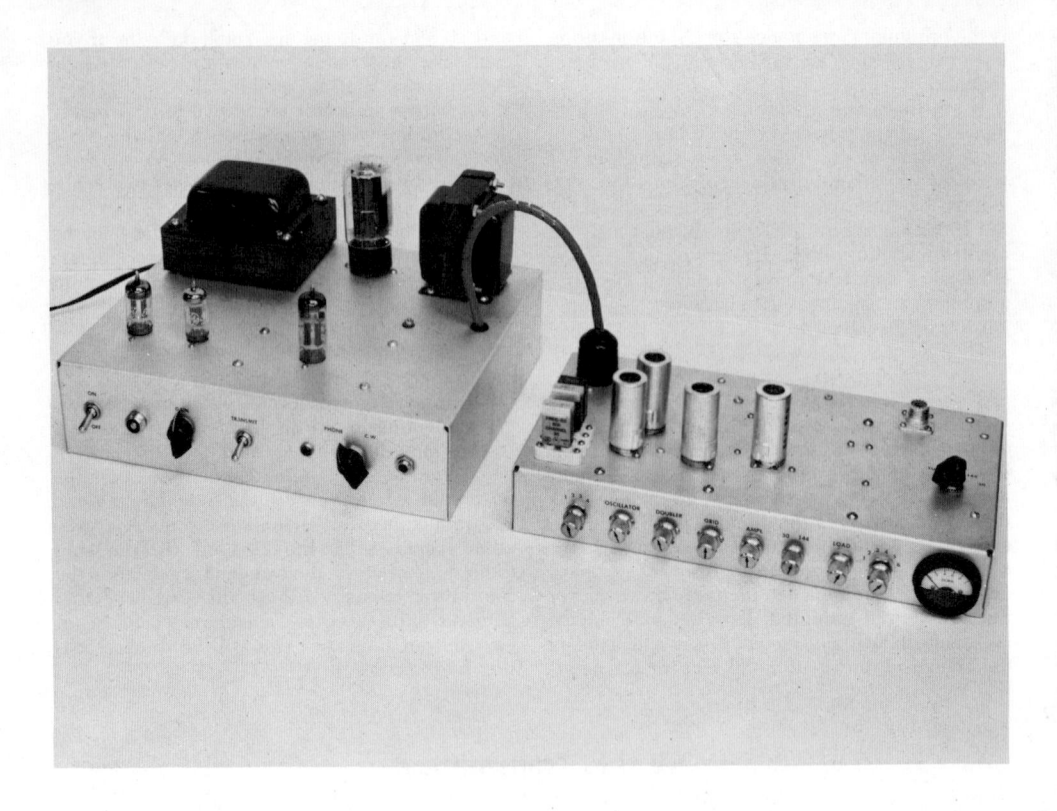

FIG. 10-19—Low-power 50–144 Mc. transmitter for the Novice or Technician. This is a complete phone transmitter, including plate modulator and power supply. The controls on the front of the modulator and power supply chassis, at the left, are the power switch, microphone jack, audio gain control, transmit-standby switch, phone-c.w. switch, and key jack. Components and controls on the r.f. chassis, right, are identified in the layout drawings.

is like the 5763, but is especially adapted for v.h.f.

R.F. Circuit

The r.f. line-up of the two-band transmitter is shown in Fig. 10-20. The oscillator, V_1, a 5763, uses 8-Mc. crystals in the grid-plate oscillator circuit, tripling in the plate circuit for both 50- and 144-Mc. operation. The plate tank, L_1C_1, of the oscillator covers 24 to 27 Mc. Output from V_1 is used to drive a 5763 doubler, V_2. On 50 Mc., output from the doubler is fed directly to V_4, the 7558 amplifier, through S_{2A}. For 144-Mc. work the output from V_2 is used to drive V_3, a 7558 tripler stage.

When operating on 50 Mc. the screen of V_3 is grounded by means of S_{3A}. Thus this stage is out of operation on 50 Mc. Although C_5 and the output capacitance of the tube are added to the capacitance of the 50-Mc. doubler circuit, C_3L_2, the minimum capacitance is low enough so that this circuit is capable of tuning to 54 Mc., the top limit of the band.

For 144-Mc. operation, S_{2A} feeds the output of V_2 to the grid of V_3. C_5L_3, together with the input capacitance of the 7558 final, becomes the 144-Mc. grid circuit of the amplifier. On this band S_3 disconnects the tripler screen from ground. S_3 also serves as tune-up control by grounding the screens of V_3 and V_4, as required, to prevent damage to the tubes if their circuits are left off resonance. V_1 and V_2 are protected by cathode bias.

The tank circuit of the amplifier consists of L_4, L_6, and C_7. When the circuit is tuned to 50 Mc. L_6 is the tank coil. This coil acts as an r.f. choke when the circuit is used on 144 Mc., where L_4 is the tank coil. The proper output links, L_5 and L_7, are selected by S_{2B}. C_8, a 50-$\mu\mu$f. variable capacitor, is the loading control.

A 0-1-ma. milliammeter connected as a low-range voltmeter is used for checking the operation of the various circuits. Current is determined by measuring the voltage drop across resistors in series with the circuits in which the current is to be measured. Regular ranges are as follows: modulator and amplifier plate current, 100 ma. each; tripler and amplifier grid current, 5 ma. each. Another position of the meter switch, S_4, is used to connect M_1 as an r.f. voltmeter across the output coax connector, thus providing a visual indication when power is actually going

V.H.F. Transmitters

Modulator and Power Supply

The speech-amplifier and modulator, Fig. 10-22, utilizes a 12AX7 dual triode, V_5, as a two-stage resistance-coupled amplifier, followed by a 6C4 driver, V_6. Output from the driver is transformer-coupled through T_1 to the grids of V_7, a 12BH7 operated with its two sections in push-pull. Either crystal or ceramic microphones can be used. Output power from the modulator is enough for fully modulating the 15 watts input to the r.f. amplifier.

The tripler screen is also modulated, along with the plate and screen of the amplifier. This increases the drive to the final amplifier on modulation peaks, with a resulting improvement in the modulation characteristic, and simplifies the phone-c.w. switching.

RFC_6, between the microphone jack J_3 and the grid of V_{5A}, is for preventing feedback troubles because of r.f. pickup on the microphone leads.

The power-supply components were selected to provide a B-plus voltage of 250, as this is the maximum rating for the 7558 when operated as a plate-modulated r.f. amplifier. A choke-input filter, consisting of L_8 and C_{9B}, is used.

S_6 is a double pole, single-throw toggle switch with one section serving as the transmit-standby control and the other section, S_{6B}, controlling 115 volts a.c. for an external antenna relay. The transmit-standby function is accomplished by opening and closing the center tap of T_3.

The phone-c.w. switch, S_5, is used to short out the modulation transformer and transfer the screens of the tripler and amplifier to the keying line. A single-pole double-throw 6-volt a.c. relay is used to key the screens of the tripler and amplifier tubes. In the key-up position the screens of the two tubes are grounded. When the key is closed K_1 is energized and screen voltage is applied to the two stages.

Construction

The r.f. section and power supply-modulator are separate units, both using fairly large chassis to make construction easier. A $2 \times 7 \times 13$-inch aluminum chassis is used for the r.f. unit, with a 6×8-inch piece of flashing copper mounted on the under side of the chassis. All r.f. grounds are made to the copper. The copper sheet is used because short, direct ground connections can be soldered to it. This helps prevent ground currents from wandering all over the chassis.

Figs. 10-23 and 10-24 give the important dimensions for mounting components on the front and top of the r.f. chassis. After making the socket holes in the chassis lay the copper sheet against the chassis top and mark off the socket holes on the copper, or else fasten the copper sheet to the chassis in the proper position and cut all the holes simultaneously in both. This will help you insure correct alignment of the two pieces.

The grid and plate terminals of the 7558 should be shielded from each other to prevent external feedback when the tube is operated as a straight-through amplifier. A shield shaped like a right angle is used for this purpose. It is made from a piece of aluminum measuring $2 \times 6\frac{1}{2}$ inches. It is $1\frac{3}{4}$ inches high, with a $\frac{1}{4}$-inch wide lip for securing it to the chassis, and is 3 inches long on one side and $3\frac{1}{2}$ on the other. It is secured to the chassis with four screws and nuts. The shield crosses the socket of V_4 between Pins 2 and 3 on one side and between Pins 8 and 9 on the other. C_7 and the rear section, S_{2B}, of the band switch are mounted on the shield.

The rotor shafts of C_1, C_3 and C_5 should not touch the chassis where they come through the front wall. This means that particular care should be taken when installing the capacitors because there isn't much space to spare between the rotor shaft and mounting holes. Connect the rotor soldering lugs to the copper, using short leads. All r.f. ground connections should be made to the copper plate, keeping the leads as short and direct as possible.

The r.f. coils, L_1 through L_7, are all of the air-wound type. One end of L_1 is supported by the stator of C_1. The other end goes to a lug on a terminal strip mounted between the socket for V_1 and the edge of the copper sheet. L_2 is installed between the stator of C_3 and a tie point mounted alongside the socket for V_3. L_3 is mounted between the stator of C_5 and one side of C_6; one of the unused terminals on S_{2A} serves as a tie point for the junction of L_3 and C_6.

In the amplifier tank circuit, L_4 is connected between the plate pin of the socket for V_4 and the stator of C_7. The 50-Mc. coil, L_6, has one end connected to a tie point on a strip mounted near the rear edge of the copper. The other lead from L_6 is soldered to the center of L_4. The 144-Mc. link, L_5, is mounted inside L_4 and is connected at one end to the grounded rotor terminal of C_7 and at the other end to a switch terminal on S_{2B}. Spaghetti sleeving is placed over L_5 to make sure that the two coils are adequately insulated from each other. The 50-Mc. link, L_7, is connected between the chassis and a tie point on the same terminal strip that supports L_6. The link is oriented so that it is coupled to the bottom (cold end) of L_6.

All r.f. chokes should be mounted as close to the coils as possible (although preferably not inductively coupled to them), keeping the leads short. Also, the grid resistors should be connected to the grid pins on the tube sockets with the shortest possible leads. All bypass capacitors should be connected close to the tube terminals or coils they are bypassing, using short lead lengths.

Shielded wire is used for the connections from J_2 to the tube heaters, for the screen leads to S_3, and for the B-plus leads to the terminal strips that hold L_1 and L_6. The shield wire is used to

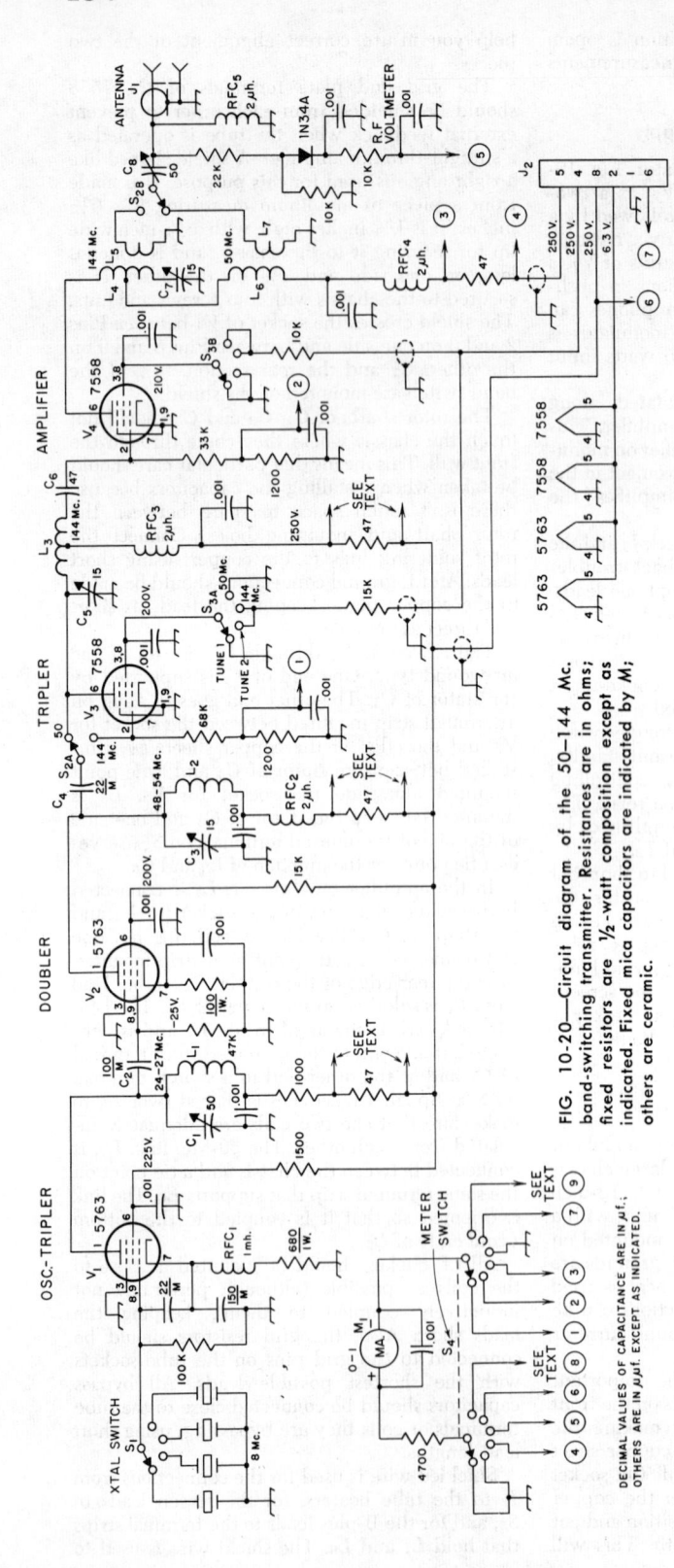

FIG. 10-20—Circuit diagram of the 50-144 Mc. band-switching transmitter. Resistances are in ohms; fixed resistors are $\frac{1}{2}$-watt composition except as indicated. Fixed mica capacitors are indicated by M; others are ceramic.

DECIMAL VALUES OF CAPACITANCE ARE IN μf.; OTHERS ARE IN $\mu\mu f$. EXCEPT AS INDICATED.

C$_1$, C$_8$—50-$\mu\mu f$. variable (Hammarlund MAPC-50-B).
C$_2$—100-$\mu\mu f$. mica.
C$_3$, C$_5$, C$_7$—15-$\mu\mu f$. variable (Hammarlund MAPC-15-B).
C$_4$—22-$\mu\mu f$. mica.
C$_6$—47-$\mu\mu f$. ceramic.
J$_1$—Coaxial connector, chassis mounting.
J$_2$—Octal connector, male, chassis mounting (Amphenol 86-CP8).
L$_1$—9 turns No. 20, $\frac{5}{8}$-inch diam., $\frac{9}{16}$ inch long, 16 turns per inch (B & W Miniductor 3007).
L$_2$—5 turns No. 16, $\frac{1}{2}$-inch diam., approx. $\frac{1}{2}$-inch long (see text).
L$_3$—3 turns No. 16, $\frac{1}{2}$-inch diam., approx. $\frac{1}{2}$ inch long, tapped at center.
L$_4$—5 turns No. 14 enam., $\frac{1}{2}$-inch diam., turns spaced to $\frac{5}{8}$ inch. Enamel should be scraped from the area at center of coil (2$\frac{1}{2}$ turns) for soldering lead from L$_6$.
L$_5$—2 turns No. 14 enam., $\frac{3}{8}$-inch diam., spaced $\frac{3}{8}$ inch (see text).
L$_6$—4$\frac{3}{4}$ turns No. 14 enam., $\frac{1}{2}$-inch diam., close-wound.
L$_7$—4 turns No. 20 insulated wire, $\frac{1}{2}$-inch diam., close-wound.
M$_1$—0-1 d.c. milliammeter.
RFC$_1$—1 mh. (National R-50).
RFC$_2$–RFC$_5$, inc.—2 μh. (National R-60).
S$_1$—Phenolic rotary, 1 section, 1 pole, 4 positions (Centralab PA-1001).
S$_2$—Ceramic rotary, 2 sections, 2 poles, 2 positions (two Centralab PA-1 sections and one PA-302 shaft assembly).
S$_3$—Phenolic or ceramic rotary, 1 section, 2 poles, 4 positions (Centralab PA-1003 or 2003).
S$_4$—Ceramic rotary, 1 section, 2 poles, 6 positions (Centralab PA-2003).

V.H.F. Transmitters

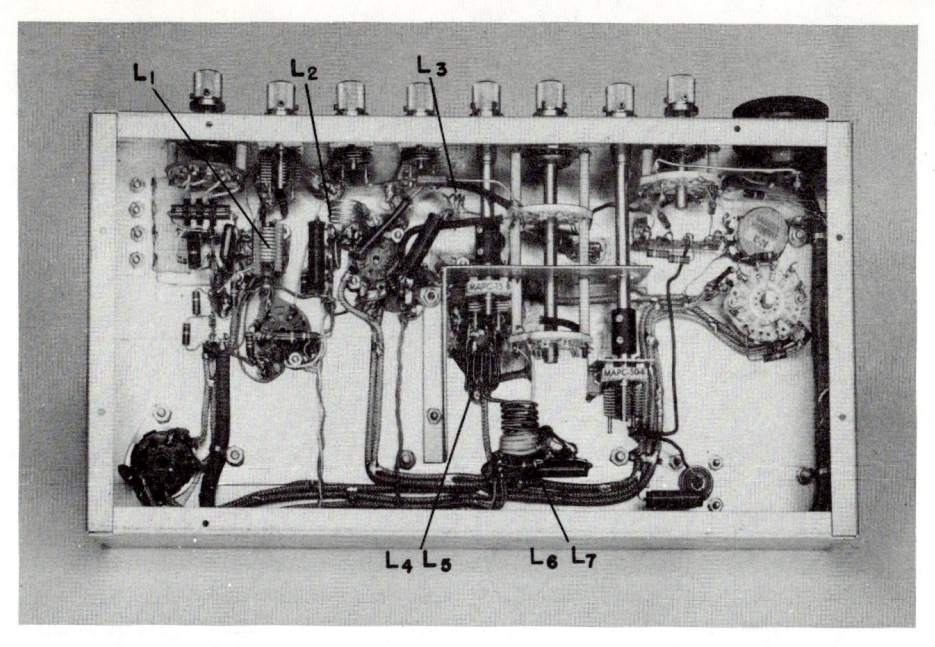

FIG. 10-21—The final amplifier is at the right in this bottom view of the r.f. chassis. Near the lower right-hand corner of the copper plate is C_8, which is mounted on the aluminum bracket. An insulated coupler is used to connect C_8 to the panel bearing. Next to the coupling is S_{2B} and to the left of S_{2B} is C_7. The 144-Mc. coil, L_4, is below C_7. L_6 is near the edge of the copper plate.

minimize r.f. coupling through the power-supply leads and, with a bottom plate on the chassis, helps confine harmonics within the chassis.

A miniature d.c. milliammeter was used in the transmitter shown in the photographs. If a larger meter is used it can be mounted on a small panel fastened to the top of the chassis with angle brackets. The length of leads to the meter is unimportant. However, the meter terminals are "hot" with the d.c. voltage when reading amplifier plate current, so they should be protected if an external mounting is used. They can be covered with electrical tape.

Construction of the power supply and modulator is not critical. The general layout shown in the photographs can be followed. The power-supply components are mounted along the rear of the chassis and the modulator is near the front. The keying relay has a single mounting screw, and a rubber grommet should be used when installing the relay to minimize its mechanical noise while keying.

The cable which connects the two chassis together can be any length, depending on the individual operating arrangement.

In wiring the modulation transformer, T_2, only the black and slate-colored leads from the secondary are connected to S_5. The remaining ones should be taped to prevent accidental short circuits and then tucked out of the way.

Using the Meter

Before applying power to the units, carefully check all wiring for errors. Then put the transmit- standby switch in the standby position and turn on the power switch.

In order to facilitate testing, all the important voltages are shown in Fig. 10-20. The plate and screen voltages can be measured with a regular test meter, if you have one. If not, you can use the milliammeter, M_1, as a voltmeter by setting the meter switch, S_4, to the last (open) position. Then ground the negative side of the meter through connection 9, and connect a 510,000-ohm ½-watt resistor from point 8 to a test prod. This converts the milliammeter into a voltmeter with a full-scale reading of 500 volts. Be sure to use insulated wire for the test lead and cover the resistor with tape or spaghetti in order to prevent accidental shock. (This meter cannot be used to check the negative d.c. voltages at the grids of V_1 and V_2 because its resistance is too low. A vacuum-tube voltmeter is the best instrument for this purpose.)

Don't be concerned if the voltages you measure differ slightly from those given. Variations are to be expected because of component tolerances, and a difference of 10 per cent or so won't affect the over-all performance.

The open position of S_4 also can be used for measuring the plate currents of V_1, V_2 and V_3. Each of these tubes has a 47-ohm resistor in its d.c. plate lead. With S_4 in the open position, connect clip leads to points 8 and 9, and clip the other ends across the 47-ohm resistor in the circuit to be measured, with point 9 connected to the plate side of the resistor in each case. (Be sure the power is off while these connections are

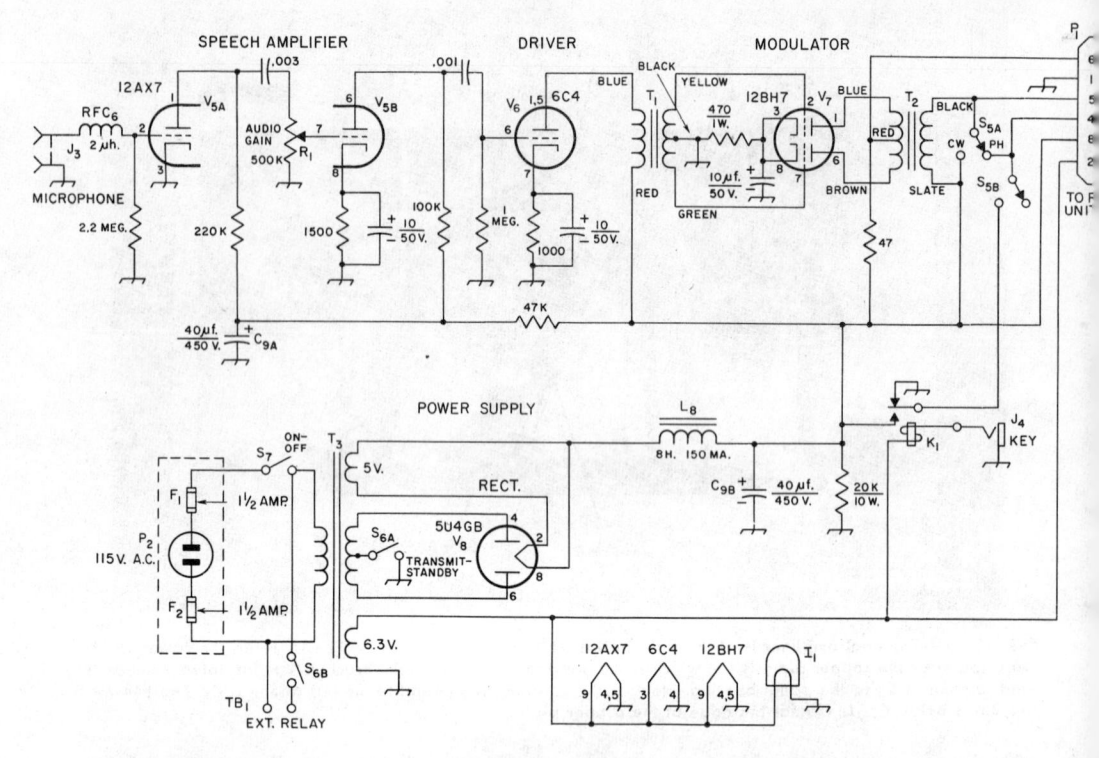

FIG. 10-22—Circuit diagram of power supply and modulator. Unless specified otherwise, capacitances are in $\mu f.$, resistances are in ohms, resistors are ½ watt. Capacitors with polarity marked are electrolytic.

C_9—Dual electrolytic, 40 $\mu f.$ per section, 450 volts.
F_1, F_2—1.5 amp., type 3AG.
I_1—No. 47 pilot lamp.
J_3—Microphone connector, chassis mounting (Amphenol 75-PC1M).
J_4—Open-circuit phone jack.
K_1—S.p.d.t., 6-volt a.c. coil (Potter-Brumfield KA5AY).
L_8—8 henrys, 150 ma. (Thordarson 20C54).
P_1—Octal plug, cable mounting, female (Amphenol 78-PF8).
P_2—A.c. plug, fuse mounting type.
R_1—0.5-megohm composition control, audio taper.

RFC_6—2 $\mu h.$ (National R-60).
S_5—Phenolic rotary, 1 section, 2 poles, 2 positions (Centralab PA-1003).
S_6—D.p.s.t. toggle.
S_7—S.p.s.t. toggle.
T_1—Driver, 5.2 to 1, primary to ½ secondary (Thordarson 20D76).
T_2—Modulation, 10 watts, 10,000 ohms plate-to-plate to 4000-ohm load (Thordarson 21M68).
T_3—Power, 700 volts center-tapped, 200 ma.; 5 volts, 3 amp.; 6.3 volts, 6 amp. (Thordarson 22R07).
TB_1—Terminal strip, 2 screw terminals.

being made or shifted!) The meter has a full-scale range of 100 ma. in this case. V_1 and V_2 each take a plate current of approximately 30 ma. The plate current of the tripler, V_3, is about 40 ma. These currents do not have to be measured in the course of normal tuning procedure, so a check of this type need be made only when the transmitter is first built, or in case maintenance is required after, for example, a component failure.

Testing and Adjustment

For 50-Mc. tune-up put S_3 in the "Tune 1" position, set S_4 to read amplifier grid current, and adjust C_1, C_3, and C_5 for maximum grid current, which should be between 2 and 4 ma. If you cannot get grid current, make sure the oscillator is working by listening for the signal in a receiver tuned either to the crystal frequency or to its third harmonic. If the oscillator isn't working you've got a bad crystal or a wiring error. A possible reason for insufficient drive is that C_3L_2 isn't tuning to the 50-Mc range. Check the setting of C_3 that gives the most grid current; if the plates are fully open the circuit may not quite reach 50 Mc. In that case reduce the inductance of L_2 slightly by spreading the turns. On the other hand, if the plates of C_3 are fully meshed as you approach maximum grid current, the coil turns must be squeezed together to lower the frequency enough to give you adequate tuning range.

V.H.F. Transmitters

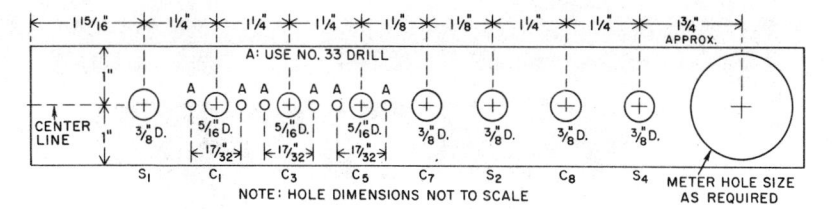

FIG. 10-23—Hole sizes and placement of controls on front of r.f. chassis.

Once you have a grid current of 2 ma. or more you're ready to test the amplifier. You'll need a dummy load; four 6-volt 150-ma. dial lamps connected in parallel will be suitable. Set S_3 to the 50-Mc. position and turn on the "transmit" switch, S_6. The meter switch should be set for reading amplifier plate current. Adjust C_7 so that the plate current "dips," indicating that the final tank is in resonance. Off-resonance plate current may go as high as 90 ma., while the plate current at resonance will depend on the setting of C_8. Set the meter to read output and adjust C_7 and C_8 for maximum indication. Then switch back to read plate current, which should not exceed 70 ma. The best setting of the controls is the one that shows maximum output with minimum plate current—minimum being some value close to 70 ma., but in no case higher than is necessary for getting the largest possible reading on the r.f. voltmeter.

The tune-up procedure on 144 Mc. is similar, with a few additions. Set S_3 to the first tune-up position and set the meter switch to read tripler grid current. Adjust C_1 and C_3 for maximum tripler grid current, which should be 2 to 4 ma. If you find that you cannot get enough grid current, you may have to adjust L_2 as outlined in the 50-Mc. tune-up procedure. Advance S_3 to the "Tune 2" position and switch the meter to read amplifier grid current. Adjust C_5 for maximum grid current, and also repeak C_1 and C_3. If you find that you cannot get enough grid current you probably will have to decrease or increase the inductance of L_3 by spreading or squeezing the turns. The amplifier tune-up procedure is similar to that described for 50 Mc.

On either band the output reading on the r.f. voltmeter will depend on the antenna and transmission line characteristics. The actual reading does not matter a great deal; the important thing is to tune for maximum.

The transmitter is designed to work into a 50- or 70-ohm load, so your antenna system should be such that the s.w.r. on the transmission line is not over 2 to 1.

The plate current of the modulator, without speech input, should be approximately 25 ma. Because of the nature of speech waveforms, the

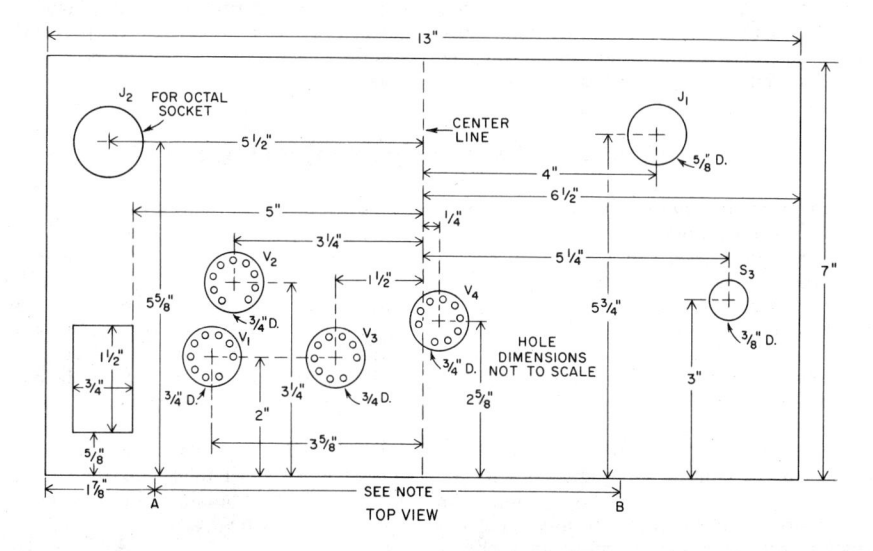

FIG. 10-24—Layout drawing of top of r.f. chassis, showing orientation of tube sockets. This is a top view; sockets should be mounted so the pins as seen from the top of the socket match this drawing. Note: Copper plate, 6 × 8 inches, butts against front wall of chassis between points A and B.

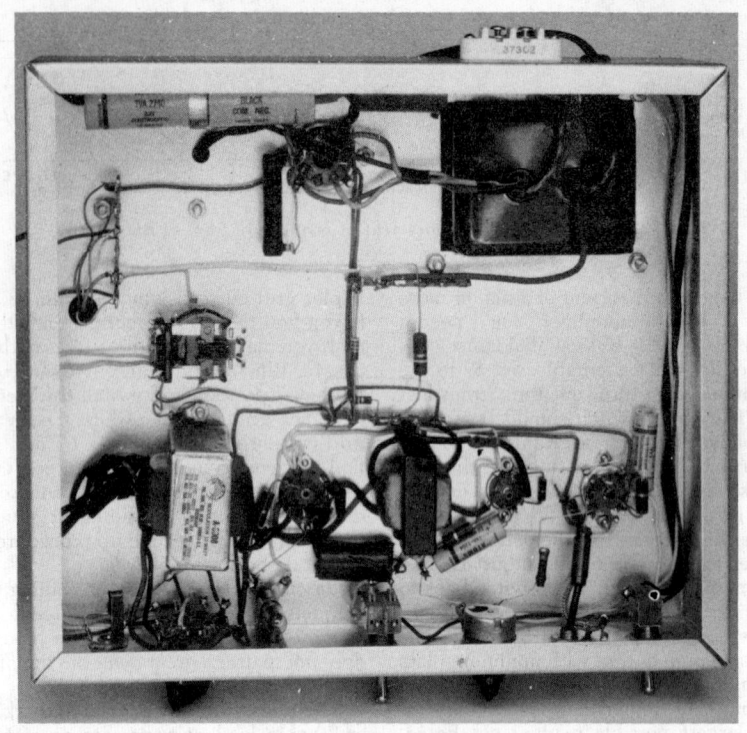

FIG. 10-25—In this bottom view of the modulator and power-supply chassis the 12AX7 speech amplifier socket is at the lower right; to the left are the 6C4 socket, driver transformer, 12BH7 socket, and modulation transformer. Immediately above the modulation transformer is the keying relay. (A double-pole relay is shown but only one pole is required.) Power-supply components are mounted along the rear (top) edge of the chassis.

plate current just "kicks" slightly when you are modulating the transmitter 100 per cent on voice peaks. Beware of any large swings in the modulator plate current as you talk—these mean overmodulation and distortion. And don't assume that you can use the r.f. output meter to indicate modulation—the pointer will be rock steady when you're modulating properly. If it flickers, you're hitting the microphone too hard. Keep your volume within proper limits and you'll have a good-sounding 6- and 2-meter phone signal.

The Variable-Frequency Oscillator

We have remarked before that crystal control is not a fool-proof answer to all problems of transmitter frequency stability. You can verify this for yourself by listening, especially on the 14-Mc. and higher bands. Many of the simpler crystal-controlled rigs fall far short of deserving the "T9X" reports that are handed out so indiscriminately. Nevertheless, the emanations from transmitters using variable-frequency oscillators often are worse.

This doesn't have to be so. It is possible to make a v.f.o. transmitter with signal quality indistinguishable from *good* crystal control. It is done by care in construction, by intelligent coordination with the rest of the transmitter, and—not least—by careful and critical checking of the signal in your own station. Reports from distant stations are not to be relied upon. For one thing, the other operator may not want to hurt your feelings. Or he may not even recognize some of the faults that make the difference between a really good signal and one that is below par. Some hams can't hear a chirp even when it hits them between the ears. So check your signal quality yourself. Use your receiver to do it by taking off the antenna and setting the controls as described in Chapter 14, in the section on checking frequency.

Once again, we recommend that you reread the discussion in Chapter 8 on oscillator stability. You can't be over-familiar with those points when you build or use a v.f.o.

What Circuit?

It may surprise you to know that the kind of circuit you use—Hartley, Colpitts, or whatnot—

The Variable-Frequency Oscillator

is far less important than *how* you use it. The electrical effects we are concerned with principally are two: a change in the oscillator frequency when any of the voltages on the oscillator happen to change, and a slow change in frequency—drift—when the temperature of some component changes. These will occur in any circuit. The methods used to minimize them are the same with all circuits, whatever their names may happen to be.

However, you do have to settle on an actual circuit. Fig. 10-26 shows one that is inherently good. It is used—with occasional modifications in detail—in probably the majority of v.f.o.s in current operation. It is basically a "hot-cathode" Colpitts-type circuit, with a tuned tank having a high C-to-L ratio. Fig. 10-26A shows the triode version, while the pentode electron-coupled adaptation is shown at B. Everything to the left of the tube in A would also be used in B. The difference between the two is in the provision for doing something with the extra grids in the pentode version, plus a difference in the methods of coupling the oscillator to a following amplifier.

The electron-coupled pentode circuit helps to isolate the following amplifier from the oscillator. This is an aid to stability. With the triode, output coupling is generally from the cathode, as shown. A buffer amplifier (discussed later) should be used with the triode amplifier. The combination of the triode oscillator and buffer usually gives better isolation than the electron-coupled pentode by itself. However, it's again as much a matter of *how* it's done as it is the actual circuit.

Which Band?

In principle, you could use an oscillator operating at your final amplifier frequency, whether that frequency be 3.5 or 30 Mc., or even higher. In practice, it isn't a good idea to try it. It becomes progressively harder to get good stability the higher we go in frequency, even on a percentage basis. But, as we have seen in Chapter 3, the selectivity of receivers is *not* on a percentage basis but on an actual *number* of cycles or kilocycles. For the receiving operator to hold your signal in tune, you have to have just as much "cycle stability" at 30 or 50—or even 144—Mc. as you do at 3.5 Mc. This isn't easy. However, it is *easier* to get adequate stability on the higher bands by putting your oscillator on a low band.

Experienced amateurs who build their own v.f.o.s rarely attempt to put the oscillator on a higher frequency than the 3.5–4 Mc. band. Half that frequency—1.75–2 Mc.—is often used. From there on, the transmitter design is just the same as it would be if you were using crystal control with crystals in those same frequency ranges. As a beginner, you will do well to choose one or the other of these frequency ranges. If you want to experiment, the suggested values in Fig. 10-26 will give you a starting point. But don't be afraid to try others. Varying the capacitance ratio of C_4 to C_3 may be fruitful; in general, try to use as much capacitance as you can at C_4 while using enough at C_3 to give you the same *total* capacitance in the two. You will need about 500 μμf. total, for the 3.5–4 Mc. range, whatever the C_3/C_4 ratio. Remember that the two are in series, when you figure out what the total capacitance is (see Chapter 2). And whatever you try, *listen* to the signal, critically, whenever you make a change.

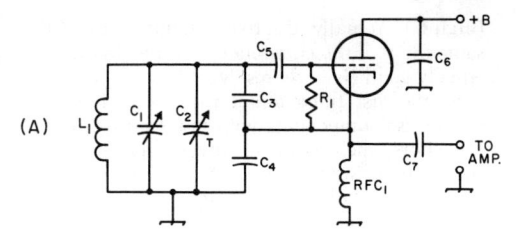

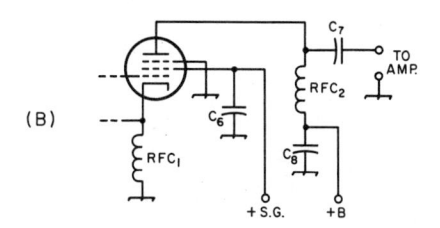

FIG. 10-26—Triode and pentode v.f.o. circuits. The two are identical in the actual oscillator circuit, the principal difference being in the way the oscillator is coupled to a following amplifier.

C_1—Tuning capacitor; capacitance range as required for covering the desired frequency range. A capacitance change of approximately 160 μμf. is required to cover 3.5–4 Mc. if the minimum capacitance is 520 μμf.

C_2—Trimmer for adjusting calibration; 25-μμf. variable.

C_3, C_4—Resultant series capacitance, 500 μμf. (two 0.001-μf. fixed capacitors, or other combinations having the same total; see text).

C_5—100 to 470 μμf.; not critical.

C_6, C_8—0.01-μf. ceramic; not critical.

C_7—470 μμf. to 0.001 μf.; not critical.

L_1—3.1 μh. for capacitance and tuning range as given above. Typical coil: 12 turns, 1 inch diameter, 16 turns per inch, No. 20. Small adjustment probably required for exact bandspread desired.

R_1—10,000 to 47,000 ohms; not critical.

RFC_1, RFC_2—1 to 2.5 mh.; not critical.

Note: For 1.75–2 Mc. range, double values given for C_1-C_4, inclusive, and L_1. Other values unchanged.

The Tuned Tank

The tuned circuit in an oscillator should have a high *operating* Q. There are two ways to get it. One is to use loose coupling between the tube and the circuit. The other is to use a tuned circuit with a large ratio of capacitance to inductance (**high C**). Actually, the two amount to about the same thing, if the components in the circuit have equally good Qs in themselves.

So the first thing about the circuit is that it should use components with low losses—high-Q capacitors, such as air or silver-mica dielectric, and a low-resistance coil. A good coil will be wound with fairly heavy wire, perferably on a low-loss ceramic form. **Air-wound** coils—actually they use strips of plastic for support—have even lower losses than ones wound on ceramic forms. They are more susceptible to vibration, though, and have to be mounted carefully on that account.

In fact, protecting the circuit against mechanical vibration and heating effects may be even more important than extreme attention to reducing the electrical losses. Mechanical vibration of any kind will cause minute changes in the electrical constants of the components. This causes the frequency to be modulated correspondingly. If the hum from a power transformer vibrates the coil or capacitor, the output of the oscillator will sound as though it came from a poorly-filtered plate supply. Or if the table on which the oscillator is sitting happens to get a bump, the frequency may be "wobbulated" in a most annoying way. Effects of this sort can be avoided by rigid construction and by isolating the oscillator as much as possible from sources of vibration. In other words, don't let components hang by their leads; give them a firm mounting. *Two* points of support are better than one. The moral is that you should use a double-bearing tuning capacitor—one having the rotor supported at both ends. Finally, don't ask for trouble by putting the oscillator next to a power transformer.

Heat

Heat—or, rather, a *change* in temperature—is the next enemy. We generally think of it as heat because the equipment temperature rises after the power is turned on. Keep the tuned circuit away from tubes and resistors that get hot. The v.f.o. design shown a little later offers a good method of isolating the tuned circuit from heat sources—in this case, the oscillator tube.

Nowadays no one thinks of getting much power from an oscillator, but there was a time when they did. One of the lessons learned in that era is just as valid today: The tube must be operated with a power input well below what it is capable of handling. The reason is simple. Excess power input means excess heating of the tube elements. The more heat in the tube, the greater the dimensional changes in the elements. Result—the internal capacitances shift. And since these capacitances are unavoidably a part of the oscillator circuit when the tube is connected to its tuned tank, the frequency shifts accordingly. The heating-up process is fairly slow, so the frequency drifts—maybe for minutes only, if conditions are favorable, but more likely for an hour or more.

One of the best ways of combating the effects of tube heating is to make the tube "see" large values of shunt capacitance in the tuned circuit. The bigger these shunt capacitances, the smaller the *percentage* change in circuit capacitance when the tube heats up. This is because the tube changes are primarily a function of the actual tube temperature, and thus are the same (for the same power input) regardless of the circuit constants. This is another potent argument for using a high-C circuit.

Frequency drift never really stops if the oscillator is turned on and off periodically—as most oscillators are, either during receiving periods or with keying. The ideal situation would be one in which the oscillator runs throughout the entire operating period, never being turned off until the station closes down. In most stations this can't be done because the oscillator would interfere with reception. (It isn't impossible to eliminate this interference, although in this book we can't go into the somewhat elaborate methods needed. It's something for the more advanced amateur.)

The Tube

Most oscillators today use receiving-type tubes. They operate with low plate voltage, and in general design are quite similar to the oscillators used in receivers. The type of tube isn't terrifically important, although it should be one having fairly high transconductance. More important is the fact that the anode input should be held below one watt with these small tubes. The word **anode** was used instead of "plate" intentionally, by the way. In the electron-coupled circuits using screen-grid tubes the anode is the screen grid rather than the actual plate. The plate input often will run higher—two to three watts, perhaps. However, all of the input, whatever electrode it may go to, helps contribute to the total heating. The lower the input the better, if drift is to be kept to a minimum.

The tube should be protected against mechanical vibration. A separate chassis for the entire oscillator (*not* including power transformers!) is always a good idea. When the oscillator is part of

The Variable-Frequency Oscillator

a transmitter or exciter the separate chassis easily can be shock-mounted, using rubber grommets to soak up vibration. Most tubes are reasonably non-microphonic, but occasionally you may run into one that wants to "ring" on the slightest provocation. The best remedy in such a case is to replace it.

Output Coupling

You can ruin a good oscillator by the way you take power from it.

Consider the simple triode oscillator. You

HEAVY LOADING CAUSES INSTABILITY

might couple to it by using a small link coil wound around the oscillator tank coil, for instance. Or you might couple to a following stage by taking output from the plate of the tube through a small capacitor. One way is just as good—or bad—as the other. No matter how loose your coupling may be, the mere fact that you have placed a load on the oscillator will affect its frequency stability. For if you take anything *out*, the coupling also serves to bring something *in*. That "something" is an effect on the oscillator frequency. Any variation in the load will change the frequency, to a degree dependent upon the tightness of the coupling as well as on the load itself.

The cardinal rule, then, is that the load on the oscillator should be very light and should be as nearly constant as it is possible for a load to be. A **buffer amplifier** is the best answer to the load problem. An ideal buffer is one that operates with practically no driving power, that offers the best possible isolation between its output and input circuits, and that runs continuously. A screen-grid Class A amplifier is the closest approach to the ideal buffer, provided it uses a tube—such as a receiving r.f. amplifier pentode—that has very low grid-to-plate capacitance.

The buffer amplifier will give maximum power output when its plate circuit is tuned to the same frequency as the oscillator. Unfortunately, even the best-screened amplifier will "kick back" on the oscillator frequency when its plate circuit is tuned through resonance. This can be avoided by using an untuned plate circuit, as in Fig. 10-27. In such a case the amplifier plate is merely fed through an r.f. choke. This is an inefficient way of operating the tube, but it is well worth

doing if you don't really need high output. Untuned coupling from the buffer to the following stage will go a long way toward giving the oscillator the complete isolation it should have, especially when the buffer is operated as a Class A amplifier.

You can use a dual tube in the v.f.o., one section as the oscillator and the second as the buffer. A number of triode/pentode tubes can be used in this way, the triode being the oscillator. The 6U8A is a typical example. Another good combination is a dual triode, with one section as the oscillator and the second as a cathode-follower buffer. This buffer circuit is shown in Fig. 10-27. Whether the buffer is a pentode or triode, it should be loosely coupled to the oscillator. With "hot-cathode" oscillator circuits, such as are used almost universally, it is convenient to drive the buffer from the oscillator cathode. The r.f. voltage at the oscillator cathode has just about the right amplitude for feeding a Class A buffer.

Keying

The foregoing discussion may have suggested the thought that keying an oscillator or its following buffer isn't the best way in the world to go about getting a stable v.f.o. If so, you've got the right idea. Let them both run all the time, if you possibly can. If you get too much interference from the v.f.o. while receiving, at least let it run continuously while you're transmitting. You

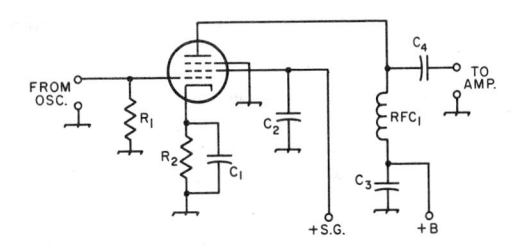

UNTUNED BUFFER

CATHODE FOLLOWER

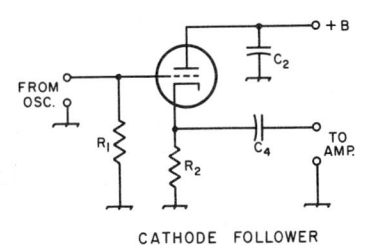

FIG. 10-27—Buffer-amplifier circuits.

C_1, C_2, C_3—0.01-μf. ceramic; not critical.

C_4—470 $\mu\mu$f. to 0.001 μf.; not critical.

R_1—47,000 ohms; not critical.

R_2—1000 ohms or less; use value required for cathode-bias resistor for type tube chosen when used as Class A amplifier.

RFC_1—1 to 2.5 mh.; not critical.

won't be able to work break-in, but most amateurs don't use real break-in anyhow. So-called break-in is principally a way of avoiding having to throw a send-receive switch. Actual breaking into a transmission by the other operator—*real* break-in—is seldom successful in general operating. (It's an art practiced quite regularly by traffic handlers and others who work the same group of stations habitually, though.)

Why shouldn't you key the oscillator? The answer is simple. An oscillator can't be keyed without a change in frequency. We saw in Chapter 8 that all oscillators will react to a voltage change by shifting frequency—slightly in a well-designed oscillator, very noticeably in a poorly-designed one. The circuit design can overcome this to a very worth-while extent, and we can reduce the problem to vanishing by using regulated d.c. voltage on the oscillator anode. Eliminate it, that is, if the oscillator runs continuously. Regulated plate supply or no, keying the oscillator changes the voltage on one or more of the tube elements. The voltage on the plate, grid or cathode, or some combination of them, *has* to change when the oscillator goes from the non-oscillating state to the oscillating state, and vice versa. If there were no voltage change there would be no keying.

Chirps and Clicks

If the oscillator is made to change state very quickly, the frequency shift—chirp—may take place in such a short time that no ear can detect it. Unfortunately, such a rapid make and rapid break is accompanied by **key clicks**. Thus we can have a chirpless signal if we're content to have a clicky one—but those who have to put up with the clicks will hardly be equally contented. The remedy for clicks is to slow up the make and break—"soften" the keying by suitable **shaping**.

This is done by using circuits that cause the voltages on the oscillator to change relatively slowly when the key is closed and opened. The chirp now becomes painfully evident.

But suppose we let the oscillator and buffer run, and key in a stage which—no matter how we tune it or whether we turn it on or off—has no effect on the oscillator frequency. Then we can do what we please in the way of softening the keying. That way we can have both clickless and chirpless keying.

"Yoops"

The voltage-change chirp is one kind. There is another—a slower change in frequency that begins when the key is closed and continues throughout a dot or dash, or for a longer period if the key is held down. This is really a drift caused by heating. It may be because the input to the tube is greater with key down than up, or it may actually take place in one of the circuit components because they are carrying an r.f. current when the key is closed but not when it is open. The result is a "yooping" sound to the keyed signal. (It's quite typical of many keyed crystal oscillators, because the crystal heats when oscillating.) The remedies are of course the same as for ordinary frequency drift.

However, if the oscillator is not keyed it will stabilize more quickly, so after a short warm-up period the drift will stop. This, too, is an argument for letting the oscillator— and buffer—run continuously.

Differential Keying

Nevertheless, you may want actual break-in. If the oscillator is designed for minimizing the slow heating change, so the frequency stays the same throughout a dot or dash, you can use the differential keying system incorporated in one of the transmitters in this chapter. This keys the oscillator as rapidly as possible, with no attempt at softening the keying *at this point*. The softening is done in a later stage, so any clicks generated by the oscillator don't get through to the output. The rapid chirp on make and break is over so quickly that it isn't heard. Besides, the final output is low at the instant of make or break, because of the subsequent shaping.

Aside from building a v.f.o. that can run continuously without interfering with your reception, this is the only approach that will give you complete freedom to shape your keying characteristic as you want it—without chirps.

Isolation

If we seem to have put great emphasis on v.f.o. stability, it's because it deserves stressing. The stability of your signal—or lack of it—is its most self-evident characteristic. Aside from keying problems, as such, there are other things that demand attention.

On the surface, some of them may seem to be associated with keying. For example, you may have a chirp or yoop even though you've got a good oscillator design and are using a recommended keying scheme. There are at least two possible explanations. One is that your keying may turn a lot of power on and off—enough so

A Practical V.F.O.

that the line voltage drops appreciably when the key is down. Although the oscillator plate voltage may be regulated, the heater voltage of the oscillator tube isn't. The change in heater power with keying can affect the oscillator frequency.

The other is that a portion of the high-power r.f. is getting into the oscillator circuit to shift the oscillator tube's operating conditions. If the oscillator and amplifiers use the same power supply this can easily happen. (It's pretty hard to keep r.f. off filament and heater leads, for instance, when they are all connected to the same filament transformer.) Good bypassing and filtering of supply leads may clean it up, but often the best remedy is to give the oscillator its own power supply. The power requirements are small, and such a supply can be built at very little cost.

It should be taken for granted that shielding the oscillator is an essential. Leaving the circuits exposed in the vicinity of a similarly-exposed amplifier, even though the transmitter as a whole is shielded, is simply a way of asking for trouble.

Phone and the V.F.O.

When you're using amplitude-modulated phone you may not be aware of instability that would show up very quickly with keying. Here we have no keying chirp to worry about, of course. But drift can still be with us. Likewise, amplifiers can react on the oscillator frequency. Both the effects just mentioned can be observed on many phone signals, when you know how to look for them.

Two distinct kinds of oscillator instability are typical of a.m. phone. The first is a simple shift of frequency when the transmitter is modulated. It is usually the result of a heavier load on the power line when a Class B modulator is developing audio-frequency power. With each spoken syllable the oscillator frequency jumps a little. You can hear it on an incoming signal if you switch on your receiver's b.f.o. The "jitter" in the beat tone with modulation is easily recognizable.

The second is nastier in its over-all effect because it can easily cause the transmitted signal to be broader than it should. The oscillator itself is modulated, but the modulation is frequency modulation rather than amplitude modulation. This is purely a result of insufficient isolation between the modulated stage and the v.f.o. The principal reason is usually lack of an adequate buffer stage or stages, but the stray coupling effects mentioned above can cause it, too.

A bad case of oscillator f.m. will have easily-observable effects. The speech will have a good deal of distortion, and the signal will splatter outside the channel width that a good signal would occupy. However, there are other malpractices that can lead to these same results—for example, overmodulation. There are ways of detecting f.m. unmistakably, but they require a good deal of skill and experience in handling a receiver. If the carrier shifts with modulation, as described in the paragraph above, the chances are that there is oscillator f.m., too. Cleaning up one usually will also clean up the other.

A Practical V.F.O.

The v.f.o. shown in Fig. 10-28 illustrates the practical application of the principles just discussed. The oscillator circuit is basically that of Fig. 10-26A combined with the cathode-follower buffer of Fig. 10-27. The oscillator covers the 80-meter band, the 3.5–4 Mc. range being spread over practically all of the tuning dial.

The use of large capacitances (0.001 µf.) at C_5 and C_6, Fig. 10-30, almost completely swamps out the effect of changes in tube capacitances. The series capacitor, C_4, permits using a larger coil than would be possible if L_1 and C_1, the tuning capacitor, were connected directly across $C_5 C_6$. C_2 is parallelled with C_1 so that the change in capacitance as C_1 is varied is just sufficient to give complete bandspread. C_3 is a small trimmer for setting the calibration exactly on the dial scale.

Considerable attention has been paid to the mounting of the oscillator components. The complete oscillator is built in a 4 × 5 × 6-inch aluminum utility box. The oscillator tube is mounted horizontally on the back of this box, together with the power and output connectors and the 25-µµf. frequency-setting capacitor, C_3. The advantage of this mounting arrangement is that the heat developed by the tube does not have as much effect upon the circuit components as it would if the tube were mounted inside the box. The tuning capacitor should be a high-quality, two-bearing type.

Maximum rigidity of the oscillator circuit is attained by use of a special aluminum bracket formed from one of the original box covers, as shown in Fig. 10-31. The box cover material is soft aluminum and can easily be bent with the aid of wood blocks and a vise. A hardwood block and hammer can be used to make the bends square and sharp. The bracket is bolted securely to the front and back of the oscillator box; thus it not only supports the circuit components but also aids considerably in stiffening the box itself. To aid in fitting the variable capacitor, the holes in the bracket for the mounting feet are slotted, and in assembly the shaft nut is first tightened to the side of the box and then the 6-32 screws for the feet are tightened. Special clamps to hold the coil are cut from thin Lucite sheet in strips ¼ inch wide and 2½ inches long. Holes are drilled at the ends of the strips so that they can be bolted

FIG. 10-28—A stable v.f.o. designed and built by W2YM. The panel is supported on metal pillars to provide space for the dial mechanism, since the tuning capacitor is firmly mounted on the front side of the shield box. Note the heavy ($\frac{1}{8}$ inch thick) top and bottom plates for stiffening the aluminum box.

to $\frac{1}{2}$-inch standoff insulators and, in assembly, the coil is clamped between the two Lucite strips to provide a sturdy coil mounting. The strips can be cemented to the coil if desired.

The silver-mica capacitors must be mounted so that there is no possibility of any motion. Again, half-inch stand-off insulators are used as tie points, as can be seen in the top view. For further stiffening, top and bottom covers are cut from $\frac{1}{8}$-inch aluminum panel stock and fastened to the box with liberal use of self-tapping screws.

Although the particular arrangement shown uses a National MCN dial mounted on a small panel and bolted to the v.f.o. box with $1\frac{1}{4}$-inch metal bushings, any dial and panel arrangement can be used. The v.f.o. box is sturdy enough so that it could be completely supported from the panel.

Feedthrough-type bypass capacitors are provided for making power-supply connections to

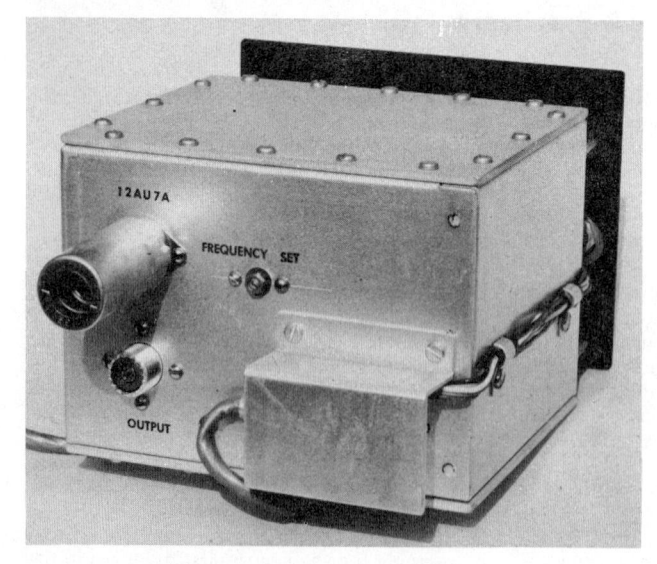

FIG. 10-29—External connections, brought out through feedthrough-type capacitors, are covered with a small aluminum "awning" to prevent accidental shock. Wiring to the "zero-operate" switch and dial lamp is cabled and clamped to the side of the box.

A Practical V.F.O.

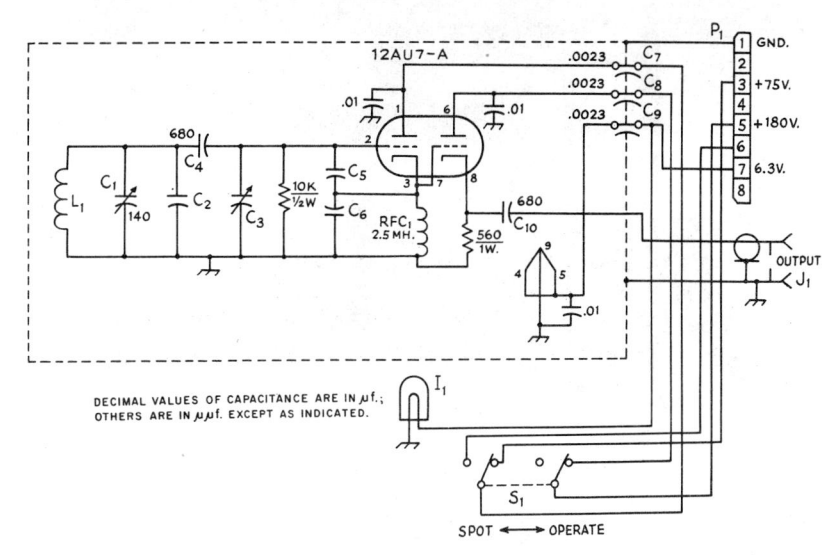

FIG. 10-30—The v.f.o. circuit. 0.01-µf. capacitors are disk ceramic. Components outside dashed line are external to the case.

C_1—140-µµf. variable (Hammarlund MC-140-S).
C_2—Approximately 100 µµf.; see text.
C_3—25-µµf. air padder (Hammarlund APC-25).
C_4—680 µµf., silver mica.
C_5, C_6—1000 µµf., silver mica.
C_7, C_8, C_9—Feedthrough type (Centralab FT-2300).

I_1—6.3-volt pilot lamp.
J_1—Coaxial connector, chassis-mounting type.
L_1—14 turns No. 20, 16 turns per inch, 1-inch diameter (B & W 3015).
P_1—8-prong (octal) cable connector, male.
S_1—D.p.d.t. toggle.

the v.f.o. These not only provide handy terminals, but also reduce the radiation from the v.f.o. The small aluminum bracket that covers these connections is used to minimize shock hazard.

The tolerances of the three capacitors connected between coil and grid, grid and cathode, and cathode and ground are such that some experimentation with the value of C_2 may be necessary. Combinations have been found where the value of C_2 is as low as 56 µµf. or as high as 120 µµf. C_3 varies the frequency about 35 kc.

If the v.f.o. is installed in a well-ventilated location away from heat sources, no temperature compensation should be found necessary. The fixed capacitors, C_2, C_4, C_5 and C_6, in the tuned circuit should be silver mica, for maximum temperature stability.

Voltage Amplifier

The output from the cathode follower of Fig.

FIG. 10-31—Mounting bracket for the v.f.o. tuned-circuit capacitor and inductor.

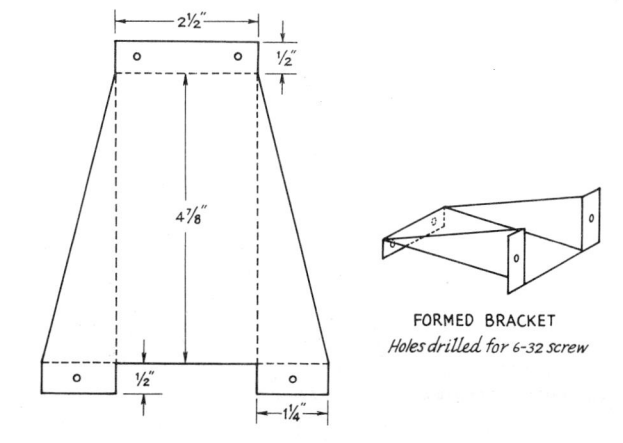

FORMED BRACKET
Holes drilled for 6-32 screw

FIG. 10-32—The interior layout of the v.f.o. is very simple. The tuning capacitor and coil are mounted on the central bracket, which also forms a bridge between the front and back of the box for further stiffening. The small padder for adjusting the frequency calibration is on the rear wall of the box.

10-30 is only 1½ to 2 volts, and it may be necessary to amplify this output to a level of 10 to 30 volts, depending upon the transmitter to be used. Almost any receiving-type pentode can be used for this auxiliary amplifier. The amplifier stage makes a convenient place to key the transmitter, so a 6DT6A is recommended as a combined amplifier and keyer tube. The circuit is shown in Fig. 10-33. The 6DT6A is similar to the 6AU6, but has high grid-No. 3-to-plate transconductance. It has the advantage of complete plate-current cutoff with only a few volts negative on grid No. 3. Also, because grid No. 3 is a high-impedance element, the key-click filter need be nothing more than a simple RC network. A 0.02-μf. capacitor and a 220,000-ohm resistor are shown in Fig. 10-33. For "harder" keying the capacitor value should be reduced, and for "softer" keying it should be increased. The 3900-ohm plate-load resistor shown provides an output of about 15 volts. If more output is required, the value of the plate-load resistor can be increased to approximately 5600 ohms. The highest output is obtained by use of a 2.5-mh. radio-frequency choke in place of the plate-load resistor.

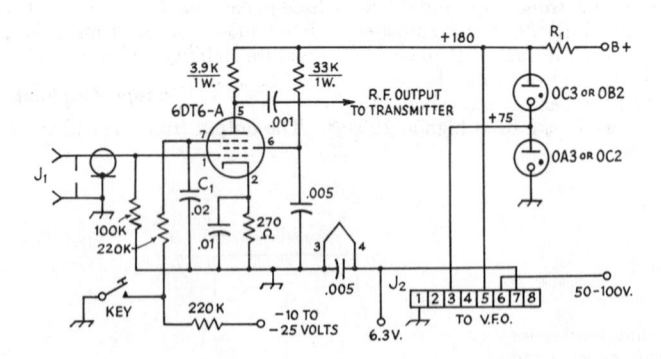

FIG. 10-33—Suggested circuit for keyed amplifier and voltage regulator. Capacitors may be ceramic or mica as desired.

C_1—Approximately 0.02 μf., paper. Increase capacitance for softer keying, decrease for harder keying.

J_1—Coaxial connector, chassis-mounting type.

J_2—Octal socket.

R_1—25-watt slider type; value dependent on voltage of supply source. Adjust for 30 ma. through VR tubes with key open.

A Practical V.F.O.

The additional amplifier stage can be incorporated in the transmitter with which the v.f.o. is to be used, if you can find room for it. An alternative would be to mount the v.f.o. assembly on a chassis enough larger than the v.f.o. box so that the 6DT6A and the regulator tubes could be accommodated. A 7 by 9 inch chassis would provide ample space.

Power Supply

Plate voltage for the v.f.o. should be obtained from a supply giving about 250 volts, using a pair of VR tubes as shown in Fig. 10-33. Voltages used are 75 volts for the oscillator and 180 volts for the cathode follower and amplifier. Lead No. 6 of the power cable may be connected to any convenient source of 50 to 100 volts so that the v.f.o. may be spotted on frequency without turning on the transmitter. On 80 meters the signal from the v.f.o. without cathode follower will be just enough to be used for spotting purposes; on the higher-frequency bands it will be necessary to energize the cathode-follower section and the 6DT6A amplifier.

Chapter 11

Transmitting Accessories

Getting a signal out into the air may take something more than a transmitter and an antenna. Obviously a telegraph key will be needed for code work, and a microphone for phone. But there are other items that will add immensely to the ease with which you can operate, and which sometimes will be essential to success.

Among these accessories are circuits for ensuring that the transmitter will deliver its maximum power to the antenna, for suppressing out-of-band radiation, for aiding the tuning-up process, and for easy changeover from transmitting to receiving. They are not complicated in construction.

Coupling the Transmitter to the Antenna

With very few exceptions, transmitters have final-amplifier tank circuits designed around loads of 50 to 75 ohms. The load is not restricted to *exactly* these figures, in most cases. A few manufactured sets come with no loading adjustment built in, and the load has to be just right for these sets to work properly. But whether or not this is the case, it is good practice to make the load—your antenna or transmission line—look like about 50 or 75 ohms as the transmitter sees it.

Some antenna systems do it for you automatically. With these (some of them are described in Chapter 6) you simply connect the transmission line to the transmitter's output terminals and you're in business. However, there are other systems that can't be handled so simply. One, for example, is the "random-length" antenna that many amateurs are forced to use, for lack of facilities for putting up anything better.

The random-length antenna is simply a wire of whatever length can be used. One end comes in to your operating position and the other is fastened to some support at as great a distance

FIG. 11-1—A simple transmatch for use with random-length antennas, using the circuit of Fig. 11-2. The coaxial fitting for the connection to the transmitter can be mounted on any convenient spot on the chassis wall. The chassis in the photograph is 3 by 5 by 10 inches. The knob on the front of the chassis is the control for C₁. The clip lead, which is 9 inches long, is connected to the input end of the coil. The clip is an E. F. Johnson type LC8. Feed-through insulators are used to hold the coil in place. A clip on the antenna lead can be used for connecting the antenna to the output end of the coil.

The Monimatch

and as great a height as can be managed.

If the station end of such a wire is connected to the "hot" output terminal of the transmitter the chances are good that no adjustment of the loading and tuning controls will load the final-amplifier to normal input. This is because the fed end of the antenna looks like something quite different from 50 or 75 ohms. The problem here is to change whatever load the antenna represents into a load that the transmitter can handle. This job can be done by a matching circuit—a **transmatch** (Chapter 6).

A transmatch suitable for random-length antennas is shown in Fig. 11-1. The circuit, Fig. 11-2, uses only a tapped coil and a variable capacitor. Note that one side of the circuit is connected to an earth ground. This can be a connection to the cold-water system of your house or to a driven ground stake (see Chapter 6). The earth is necessarily a part of an end-fed antenna system, so it pays to use as good a ground as you can get. If the connection is omitted the r.f. nevertheless will look for a way to get to earth, and the only path is through the transmitter and the a.c. line. The r.f. characteristics of such a path are usually unknown, and not what could be considered desirable in any case. So use a separate ground connection and keep it as short as you can.

The transmatch is connected to the transmitter through a length of coaxial cable. RG-58/U or RG-59/U will be satisfactory for moderate power. There are two general methods of adjusting the transmatch, one without and one with the aid of a standing-wave-ratio indicator in this coax link between the transmitter and transmatch. If you don't have an s.w.r. indicator such as the "Monimatch," first set the transmitter's loading control at minimum (maximum capacitance in a pi-network tank) and adjust the tank tuning capacitor to the dip in plate current that indicates resonance. Set the tap on L_1 in the transmatch at one end of the coil, and then turn C_1 through its complete range. If some setting of C_1 causes the plate current to change, leave it there and readjust the transmitter's tank tuning capacitor for the plate current dip. Then try increasing the loading by means of the transmitter's output control, always readjusting the tank tuning for the dip after each change in the loading control. If you can't reach the desired plate current at resonance, move the tap on L_1 a turn or two and go through the process again. Eventually you will reach a pair of settings for L_1 and C_1 that enable you to make the transmitter load up.

This is a cut-and-try process that will result in adequate loading, but it is not necessarily the best adjustment of the transmatch. Also, it will hold good only for the particular length of coax used between the transmitter and transmatch. If the length of this line has to be changed for some reason, the whole process will have to be gone through again. The optimum adjustment can be reached more surely by using a Moni-

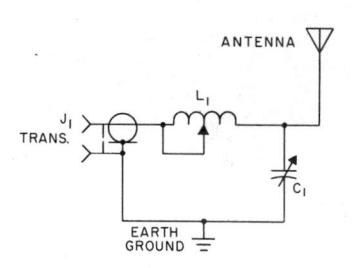

FIG. 11-2—Circuit diagram of the transmatch for random-length antennas.

C_1—140-$\mu\mu$f. variable (Hammarlund MC-140-S, E. F. Johnson 140R12, or equivalent).

J_1—Coax chassis receptacle, SO-239.

L_1—24 turns No. 12, 6 turns per inch, 3-inch diameter (Air Dux 2406).

match in the coax link. The next section describes how to build one. The method of using it is basically the same for all transmitters and transmatches, so it will suffice here to say that the transmatch of Fig. 11-2 is adjusted simply by trying different taps on L_1, along with different settings of C_1, until the Monimatch shows the null indicating that the antenna load is matched to the coax link.

The transmatch doesn't have to be installed close to the transmitter, although it is convenient to have it within reach if you change bands frequently. For r.f. reasons, it is better to have it close to the spot where the end of the antenna comes into the station, since this avoids running any appreciable part of the antenna around the room and may also make a shorter ground lead possible. The coax link can be any convenient length.

The Monimatch

The **Monimatch** is a combination of r.f. power output monitor and impedance-matching indicator. The cost of the device itself is nominal, but in addition to the circuit shown in Fig. 11-4 a d.c. microammeter (0–100 or 0–500 scale) is required. The Monimatch is a form of **reflectometer**, a type of instrument that is capable of giving indications proportional to the incident and reflected energy, respectively, on a transmission line. When the load at the output end of the line matches the line's characteristic impedance, no power is reflected from the load back toward the transmitter. Thus if the load impedance is adjusted so the reflected-power reading is zero, the line is matched.

Although the construction shown in Fig. 11-3 is quite simple, the action of the device is rather complicated in principle. It is a form of **bridge** circuit combining inductive and capacitive coupling, and is actually a short section of transmission line. The one shown is built in a 2¼ × 2¼ × 5-inch aluminum box (Bud CU-3004A), with coaxial sockets (SO-239) mounted on opposite ends. A 4⅝-inch length of ¼-inch copper tubing, the inner conductor of the Monimatch

FIG. 11-3—The Monimatch is built in a small aluminum box that can be installed in any convenient spot—even inside the chassis of a homemade transmitter or transmatch. This view shows one resistor at the right-hand end of its pick-up wire. The connection to the crystal diode is visible at the other end. The d.c. output leads from the diodes to the "forward" and "reflected" terminals are made with shielded wire. Pin jacks are used as terminals, mounted on the right-hand end.

line, runs directly from the inner-conductor pin of one socket to the other. The socket pins fit inside the tubing and are soldered in place. This can be done by first mounting one socket and soldering the tubing to its pin, after which the second socket can be mounted and soldered. Before mounting the sockets and tubing, make two spacers out of polystyrene or bakelite about 1/8 inch thick as shown in Fig. 11-5, and slip them over the tubing as shown in the photograph.

The outer conductor of the line is two strips of thin copper or brass, 5/8 inch wide and 4 7/8 inches long, mounted above and below the copper tubing as shown. These strips are soldered at the ends to lugs held by the mounting screws for the coax connectors. They rest on the insulating spacers, which maintain the proper separation.

The "pick-up" wires, shown by the heavy lines in Fig. 11-4, are 4-inch lengths of No. 14 tinned wire mounted in the slots in the spacers. They should be inserted the full depth of the slot. Two are used, one on each side of the copper tubing. The resistors, R_1 and R_2, are soldered to the opposite ends of the wires (Fig. 11-4), with the shortest possible leads both at the wires and at the chassis side. The crystal diodes are connected 3 3/8 inch from each resistor as shown. Keep these leads short, too, but don't overheat the diodes when soldering. A single-lug insulated tie point is used for the junction of each diode and its associated 0.001-μf. capacitor. Ground the capacitor directly to the coax socket.

The resistors must be the composition type, *not* wire-wound. Wire-wound resistors are available in the 1/2-watt size, so be sure you don't get the wrong kind.

If the dimensions given are followed carefully the Monimatch should work satisfactorily without adjustment. Its operation can be checked if a 50-ohm dummy load capable of dissipating the transmitter's power output is available.

Using the Monimatch for Transmatch Adjustment

The indicator part of the Monimatch is not built into the instrument itself, as it is often con-

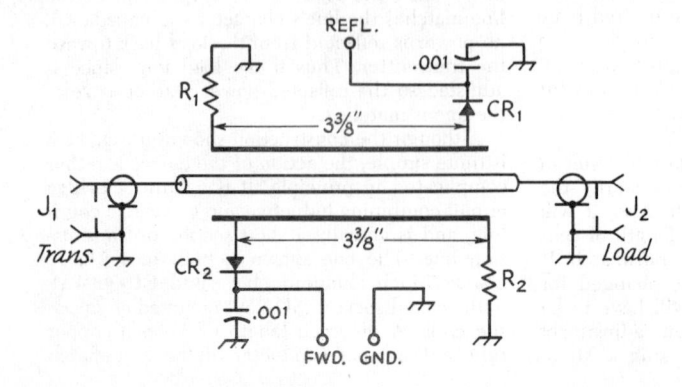

FIG. 11-4—The Monimatch circuit, with the strips forming the outer conductor omitted. Pick-up wires are shown by heavy lines. The 0.001-μf. capacitors are disk ceramic.

CR_1, CR_2—1N34A or similar.

J_1, J_2—Coaxial chassis receptacles (SO-239).

R_1, R_2—1/2-watt composition. For 50-ohm line, use 150 ohms. For 75-ohm line, use 100 ohms.

The Transmatch

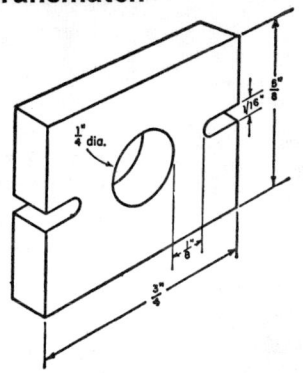

FIG. 11-5—Insulating spacers for maintaining proper separation between inner and outer conductors and pick-up wires. These may be made of polystyrene or bakelite 1/8 to 1/4 inch thick. Two are needed.

venient to make it up separately so it can be placed on the operating table where it can be seen easily. It consists of the meter, sensitivity control resistor and s.p.d.t switch shown in Fig. 11-6. Use the type of line (50- or 75-ohm) for which you have built the Monimatch, in making the coax connections shown in Fig. 11-6.

To adjust the transmatch, turn on the transmitter and set its tank tuning to resonance. (Throughout the procedure that follows, always keep the tank tuned to resonance.) Set S_1 in the "forward" position and adjust the sensitivity control, R_1, for a reading near maximum on M_1, or as much of a reading as you can obtain by adjustment of the transmitter controls. Then set S_1 in the "reflected" position and try various adjustments of the transmatch controls with the object of making the meter pointer drop to zero. When a zero or near-zero reading is obtained, put S_1 in the "forward" position and readjust the transmitter for maximum reading. Then switch back to "reflected," and check the adjustments again. After one or two trials the meter should read zero reflected power no matter how high it reads in the "forward" position. This represents the matched condition. Finally, adjust the transmitter controls for maximum output as shown by the highest reading forward. The rated plate current of the transmitter's final amplifier should not be exceeded, of course, in adjusting for maximum output.

S_1 can be left in the "forward" position to serve as a continuous indicator of output.

The sensitivity of the Monimatch increases with frequency. The following table is typical of the current readings to be expected with transmitter power outputs of 10 and 50 watts:

Band	10 Watts R.F.	50 Watts R.F.
3.5 Mc.	70 μa.	250 μa.
7 Mc.	200 μa.	1 ma.
14 Mc.	750 μa.	Over 1 ma.
21-28 Mc.	Over 1 ma.	Over 1 ma.

These are with R_1 set for maximum sensitivity (zero resistance). For most low-power transmitters a 0–500 microammeter will work very well, and even 0–1 ma. is usable on the 3.5-Mc. band if the transmitter power is 50 watts or more. The current can always be kept within the range of the meter by increasing the resistance of R_1.

Universal Transmatch

The transmatch circuit shown in Fig. 11-8 can be used for matching between the transmitter and any type of line used for feeding the antenna. It can also be used for random-length and coax-fed antennas as shown in Fig. 11-9.

The parallel circuit formed by L_2 and C_2 is resonant at the operating frequency and the taps on L_2 are used for adjusting the impedance ratio between the line and the coax input terminal, J_1. With correct adjustment the impedance at J_1 matches the coax line used as a link between the transmatch and the transmitter. Thus the proper load is offered to the transmitter. C_2 should be insulated from the chassis, and should have an insulated coupling between its shaft and the dial. C_1 gives a smooth adjustment of the effective coupling between L_1 and L_2 as a further aid in the matching adjustment.

Fig. 11-7 shows a suitable layout for the transmatch. The capacitors used have a plate spacing adequate for transmitter powers up to 150 watts or so. The chassis is of aluminum, $2 \times 7 \times 9$ inches, with a 6×7 inch front panel. Dials with at least a 10-division scale should be used so settings for different bands can be recorded and returned to without readjustment. The input connector, J_1, is mounted on the rear wall of the chassis, and is connected to L_1 by a lead running through a grommet underneath the coil socket.

The L_1 coil is mounted inside L_2, in each case. It is centered in L_2 and held in place with Duco cement. The leads from L_1 go between the turns of L_2 to the plugs on the mounting bar. Use spaghetti tubing over these leads to prevent accidental short circuits. The L_2 coils need no special mounting other than their leads (which go into the end prongs on the bar) if the leads are bent so the coil almost rests on the bar. Soldering the coil leads in the prongs will be a little easier if the nickel plating is filed off the ends of the prongs first. After soldering the leads in place be sure to clean off all traces of rosin

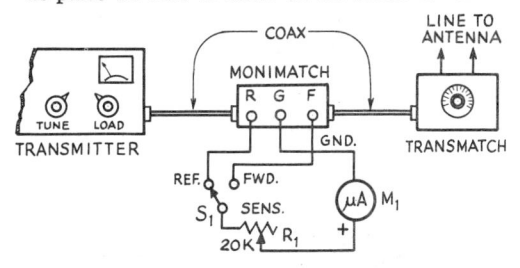

FIG. 11-6—Using the Monimatch in adjusting a transmatch. Adjustment procedure depends on the type of transmatch used, but is always aimed at getting the meter to read zero when S_1 is in the "reflected" position, while giving the highest possible reading in the "forward" position.

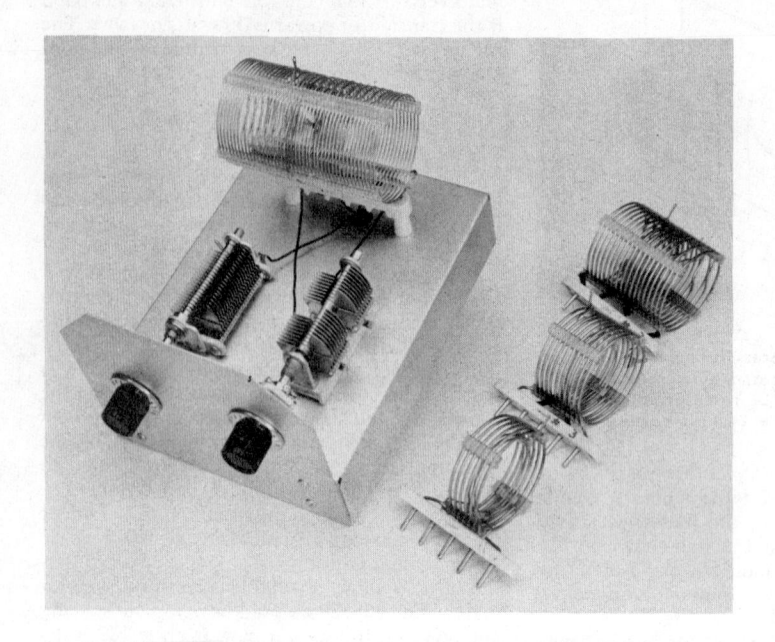

FIG. 11-7—Universal transmatch. It can be used with any antenna and transmission-line system. The jack bar (Millen 41305) for the plug-in coils is mounted on small ceramic stand-off insulators to allow the contacts to clear the chassis. Four coils cover the range 3.5 to 30 Mc.; a single coil suffices for both 21 and 28 Mc. All are mounted on Millen 40305 plug bars.

flux, since a coating of rosin will interfere with good contact between the prong and its jack.

Type 235-860 (E. F. Johnson) clips can be used for making the taps on L_2. Once the proper tap positions have been determined for a coil, solder a lug at each point so it projects outwards. The clip can be put on the lug sidewise and will fit snugly if its adjusting screw is tightened. This method makes coil changing easy as compared with using the clips directly on the coil turns.

No terminals are shown in Fig. 11-7 for connecting the transmission line to L_2 as these will depend on the type of line used. Binding posts or a screw-type terminal strip can be used for parallel-conductor line, and also for the random-length antenna. A coax line should have a separate chassis connector, which can be mounted on the rear wall.

The best way to adjust the transmatch is shown in the setup of Fig. 11-6, using a Monimatch as an indicator. The procedure is much the same as described earlier in this chapter. Keep the amplifier tank tuning at resonance throughout the tune-up process, and adjust the loading for a high reading on the Monimatch meter with the switch in the "forward" position. With open-wire line, set the taps on L_2 at equal distances from the center of the coil, using a few turns each side at the beginning. Switch the Monimatch to the "reflected" side, and try to bring the reading down to zero by varying both C_1 and C_2. Switch over to "forward" occasionally to make sure that power is going to the transmatch. If the reading won't go to zero, move the taps closer together and go through the process once more. Continue until there is a good zero or null reading with a high forward reading.

Usually it will be possible to get a good null over a range of tap positions on L_2. If so, always put the taps as far apart as you can. This will broaden the tuning of the transmatch. The frequency range, within a band, that can be covered without readjustment will depend a good deal on the type of antenna and transmission line you use. It is a good idea, when first putting the transmatch into operation, to see how much range you can cover before readjustment be-

FIG. 11-8—Circuit of the universal transmatch.

C_1—325-$\mu\mu$f. variable (Hammarlund MC-325-M).

C_2—140 $\mu\mu$f. per section dual variable (Hammarlund MCD-140-M).

J_1—Coax receptacle, chassis-mounting type SO-239.

L_1—10 turns per inch, 2-inch diameter, No. 16 wire (B & W 3907-1 or Illumitronic 1610T).
   3.5 Mc.: 10 turns
   7 Mc.: 6 turns
   14 Mc.: 3 turns
   21/28 Mc.: 2 turns

L_2—3.5 Mc.: 44 turns No. 16, 2½-inch diameter, 10 turns per inch (Illumitronic 2010T).
   Coils for 7 through 28 Mc. are 2½-inch diameter, No. 12 wire, 6 turns per inch (B & W 3905-1 or Illumitronic 2006T).
   7 Mc.: 18 turns
   14 Mc.: 10 turns
   21/28 Mc.: 6 turns

Harmonic Suppression

FIG. 11-9—Circuits for use with a random-length antenna or single-wire feed (A) and for a coax-fed antenna (B). A coax fitting can be installed on the chassis for the latter type of line.

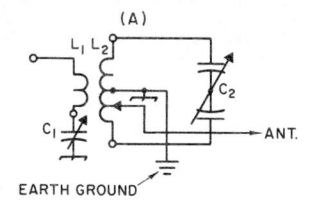

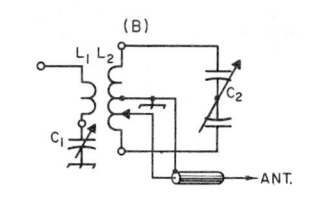

comes necessary. The "reflected" reading can be allowed to go up to 20 per cent or so of the "forward" reading without retuning. Almost any band can be covered with two or three settings of the capacitors. Log these band-segment settings and you won't have to rematch when changing frequency.

Fig. 11-8 shows optional connections from chassis to the center of L_2, or to the rotor of C_2. There will be no difference in the matching either way, or even if neither is used. This optional ground is provided so that harmonic radiation can be reduced. Depending on the feed-line length, better harmonic suppression may result when C_2 is grounded, and vice versa. The best way to determine this is to have a fellow amateur tune his receiver to the frequency of your second harmonic. A quick test or two will show which connection results in the least harmonic output.

For random-length antennas (Fig. 11-9A) or coaxial line (Fig. 11-9B) the center of L_2 should be grounded to the chassis. Only one tap is used with either of these systems, since they are not balanced to ground. The adjustment procedure is exactly the same as for the parallel-conductor line except that only one tap need be moved.

If you don't have a Monimatch, follow the same general adjustment procedure, but with the object of loading the transmitter to normal input. The use of the Monimatch is strongly recommended, though, because its use leads to the optimum adjustment quickly and easily.

Harmonic Suppression

Your license gives you the privilege of working in certain frequency bands; any other frequencies are out of bounds. Nevertheless, it is the nature of transmitters to put out power on more than one frequency at a time, particularly on harmonics of the intended frequency. You have to suppress all such out-of-bounds frequencies, or you may find yourself in the embarrassing position of having to make explanations to the FCC.

Usually, a transmatch such as was described in the preceding section will prove to be more

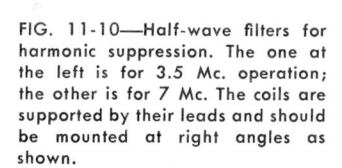

FIG. 11-10—Half-wave filters for harmonic suppression. The one at the left is for 3.5 Mc. operation; the other is for 7 Mc. The coils are supported by their leads and should be mounted at right angles as shown.

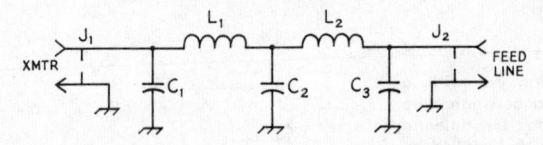

FIG. 11-11—Half-wave filter circuit. Constants for three Novice bands are given below.

C₁, C₃—3.5 Mc.: 820-µµf. mica, 500 volts.
       7 Mc.: 470-µµf. mica, 500 volts.
     21 Mc.: 100-µµf. mica, 500 volts.
C₂—3.5 Mc.:1500-µµf. (0.0015-µf.) mica, 500 volts.
    7 Mc.: 1000-µµf. (0.001-µf.) mica, 500 volts.
  21 Mc.: 200-µµf. (0.0002-µf.) mica, 500 volts.
J₁, J₂—Phono jacks.

L₁, L₂—3.5 Mc.: 11 turns No. 20, 16 turns per inch, 1-inch diam. (B&W Miniductor 3015).
    7 Mc.: 8 turns No. 18, 8 turns per inch, 1-inch diam. (B&W Miniductor 3014).
 21 Mc.: 7 turns No. 18, 4 turns per inch, ½-inch diam. (B&W Miniductor 3001).

than adequate for suppressing harmonics. However, if you don't use one, and do have a coax-fed antenna system—particularly an antenna of the multiband type—it is almost certain to be just a matter of time before you get one of those unwanted tickets from an FCC monitoring station. It's much better to be safe first than sorry afterward.

With some antenna systems the transmitter can be loaded up properly without a transmatch, so none is needed for matching purposes. A relatively simple filter, not requiring adjustment, will suffice for harmonic suppression in such cases. Fig. 11-10 shows a pair of such filters, one for the 3.5-Mc. band and the second for the 7-Mc. band; these two will suffice for most Novice operation. They cost very little. The container is a coffee can, which is just about the right size and provides suitable shielding.

The same circuit, Fig. 11-11, is used in both filters; only the values change for each band. The circuit is that of a **half-wave filter**, meaning that it is the equivalent of a half-wave section of transmission line. As you saw in Chapter 6, a half-wave line reproduces, at its input end, whatever value of impedance is connected to its output end. This means that the filter can be inserted in a line without causing any change in impedance on the input side, so no adjustment is needed. The transmitter can be loaded just as well after the filter was inserted in the line as before.

The circuit values given in Fig. 11-11 are optimum for use in coaxial lines, and the components will handle transmitters operating at the Novice power limit providing the standing-wave ratio on the line is no higher than 3 to 1. Any coax-fed antenna system of reasonable design should operate with an s.w.r. within these limits.

FIG. 11-12—Low-pass filter for reducing TV harmonics. The coils are mounted at right angles to minimize coupling between them. This filter can be left in the coax line on all bands below (and including) the 21-Mc. band.

TVI   215

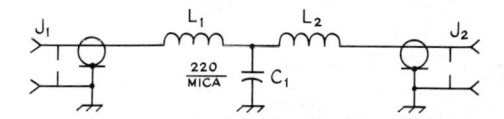

FIG. 11-13—Circuit of the low-pass filter. L_1 and L_2 are wound with No. 16 solid enameled wire. Each coil is 7 turns ½-inch diameter and ½-inch long. A ½-inch diameter drill shank or dowel rod can be used as a winding form. Leave an inch or so of lead length at the coil ends for connecting to J_1 and J_2 (phono jacks), and scrape the enamel from the ends of the leads before soldering. A 220-$\mu\mu$f. mica capacitor, ± 5 percent tolerance, should be used for C_1. The filter should be used with RG-58/U, RG-59/U, RG-8/U or RG-11/U coax.

The half-wave filter is basically a resonant circuit, so can be used only over a small range of frequencies such as the width of an amateur band. That is why two filters are needed for handling 3.5- and 7-Mc. operation. The filter shown uses phono jacks as connectors because they are cheap and it is easy to shift the line from one filter to the other. The coax cables must have mating phono plugs, of course.

The two filters are shielded from each other by a piece of tin (cut from another can) shaped as shown in Fig. 11-10 and soldered in place. In cutting the coils, allow about 1½ inch leads for making connections, and mount the coils in each filter at right angles, as shown. After assembly, replace the lid on the can to complete the shielding.

Television Interference

If the operation of a transmitter disturbs reception in a nearby television receiver, the reason can be tracked down to one (or both) of two distinct causes. One is that the transmitter may actually be radiating a signal in a television channel. Such a **spurious** signal may be a harmonic of the fundamental frequency, or perhaps a parasitic oscillation. Either way, it has no right to be there. Interference from this cause is something that the operator of the transmitter has to eliminate.

The second is no fault of the transmitter. The TV receiver just "folds up" when its antenna input circuits are hit by the extremely strong signal they will get from a nearby transmitter, even a low-power one. The receiver doesn't have enough selectivity to reject a powerful signal outside the TV channels. Different makes of receivers vary a great deal in their ability to reject non-TV signals.

Unfortunately, the effect on the TV screen is much the same whether the fault lies with the transmitter or receiver. However, experience has shown that when the transmitter is operating on the 3.5- or 7-Mc. bands the probable cause of TVI (television interference) is receiver overloading rather than harmonics from the transmitter. On 14 Mc. and above, transmitter harmonics, especially those that fall in Channels 2 to 6 (54 to 88 Mc.) become important. Receiver overloading, too, becomes worse, because the transmitting frequency is closer to the TV channels.

It is your responsibility, as the operator of a transmitter, to make sure that no TVI results from harmonics or parasitic oscillations. The transmitter always generates harmonics, so the problem is one of preventing them from reaching the antenna to be radiated. Good shielding around the transmitter circuits is the first essential, because without it harmonics may leak out in spite of all you may do in the way of adding selective circuits between the transmitter and the antenna system. Chapter 10 shows examples of transmitter construction with adequate shielding.

Once the shielding is good enough to prevent "spraying" of harmonics directly from the transmitter, circuits may be added externally to keep harmonics from reaching the antenna. The transmatch is an example of a selective circuit that will go far toward eliminating TVI. The half-wave filters described earlier will also help. However, if you work on 21 Mc. and the lower Novice bands, the filter described in the next section is recommended.

A Low-Pass Filter

A low-pass filter is one that allows everything

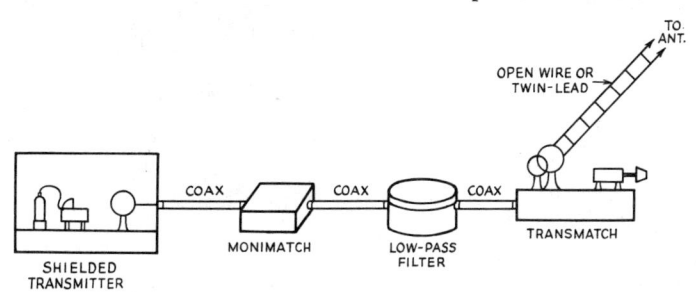

FIG. 11-14—How the low-pass filter is connected between the transmitter and transmatch. The Monimatch is needed for adjusting the transmatch to match the characteristic impedance of the coax line, since the voltage rating of the mica capacitor in the filter may be exceeded if the s.w.r. is high. The transmatch can be connected to a coax line feeding the antenna as shown in Fig. 11-9B.

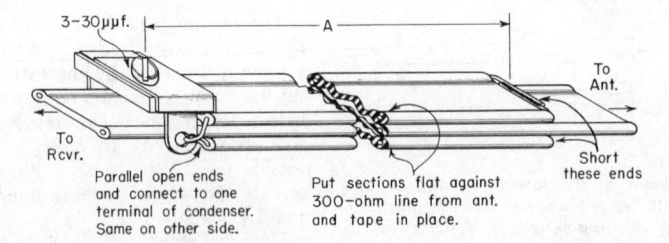

FIG. 11-15—Tuned trap for absorbing r.f. energy to prevent TVI. The trap uses two pieces of 300-ohm line, one on either side of the line to the TV receiver, tuned to the frequency of the amateur transmitter. Length A should be 40 inches for trapping out a 50-Mc. signal and 11 inches for 144 Mc. Adjust the capacitor for minimum interference, using a non-metallic screwdriver. Try different trap positions along the 300-ohm line to find the optimum spot, then fasten in place with tape.

below its **cutoff frequency** to pass through, but which suppresses all frequencies *above* the cutoff frequency. This type of filter can be left in the feed line all the time, on all bands below its cutoff frequency.

Fig. 11-12 shows an easy-to-make filter of this type. Its cutoff frequency is somewhat above 21 Mc., so it can be left in the line on all bands from 21 Mc. down in frequency. Like the half-wave filters, described earlier, it uses a coffee can as a shield. As a back-up for the transmatch, it should take care of TVI harmonic problems in any case where the TV signal is of sufficient strength to give a picture free from "snow."

The filter has to be used with coax input and output. The coax line can be the link between the transmatch and the transmitter, as shown in Fig. 11-14. The circuit data are given in Fig. 11-13. Fig. 11-14 shows how the Monimatch is connected in the line along with the filter. The Monimatch is practically essential when a filter is used, because the s.w.r. in the coax link must be low in order to protect the filter capacitor from breakdown.

TV Receiver Overloading

There isn't anything you can do to your transmitter to prevent receiver overloading. The remedy here is to keep your fundamental-frequency signal from getting into the TV receiver's front end. A **high-pass filter** is the answer to this. There are low-cost filters on the market for this purpose; a popular and effective one is the Drake TV-300-HP. This filter is made for use with 300-ohm twin line such as is used for TV lead-ins. It allows the TV signals to reach the receiver without attenuation, but discriminates against signals below 54 Mc. It should be installed in the TV receiver as close as possible to the antenna terminals so that the very minimum of exposed lead will be used between the filter and the antenna posts.

Checking TVI

The best way to resolve the question of whether the reason for TVI lies in a neighbor's receiver or your transmitter is to clean up any interference that may show on your own TV receiver. (You'll probably have to use a high-pass filter on it.) If you have no TVI at home, it is highly unlikely that transmitter harmonics or spurious radiations can be the cause of any interference that may occur in the vicinity. Thus you can be sure that your transmitter is "clean," and if there is still TVI the receiver needs a high-pass filter.

You may be able to demonstrate the effectiveness of a high-pass filter to a neighbor, but there is no reason why you should furnish it to him. Receiver shortcomings aren't your fault. The set owner should be able to get a filter through the dealer from whom he purchased the receiver, or from any local TV service organization. If there is a TVI Committee in your locality—check this

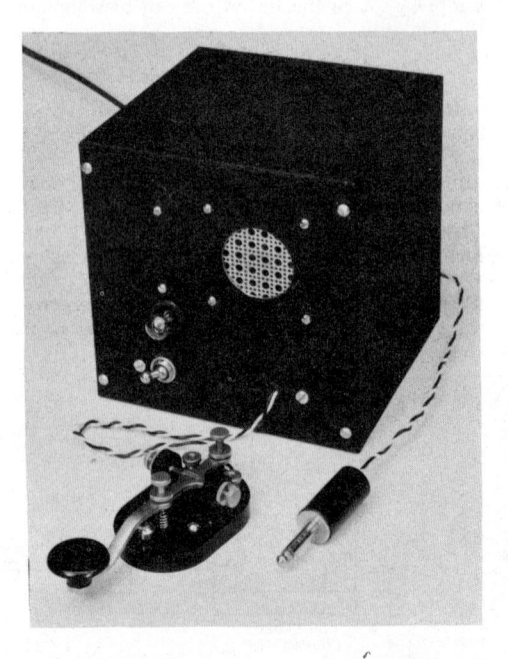

FIG. 11-16—Keying monitor and code-practice oscillator. The circuit, Fig. 11-17, uses a neon bulb audio oscillator, 6AQ5 amplifier, and a loudspeaker. The switch at the lower left-hand corner is S_1. V_1 is immediately above it.

TVI

TVI from V.H.F. Transmitters

It is unlikely that transmissions in the 50- and 144-Mc. bands will cause harmonic-type interference. A fourth harmonic of 50-54 Mc. can fall in Channels 11, 12, or 13, depending on the operating frequency. Since only one of these channels would be in use in a given locality, it is easy to avoid harmonic interference by the simple expedient of choosing a frequency that can have a harmonic only in a channel not in use. Harmonics will fall in Channel 11 when the transmitter is on 50–51 Mc., in Channel 12 when the output is on 51–52.5 Mc., and in Channel 13 from 52.5–54 Mc.

TVI from 50-Mc. operation is principally caused by TV receiver overloading and from spurious responses such as images. A high-pass filter will help. A high-pass filter is of no value when TVI results from 144-Mc. transmissions, since the transmitting frequency is already higher than the low group of v.h.f. TV channels. Spurious responses on the part of the TV receiver are the reason for TVI in this case. It is usually necessary to install simple traps such as are shown in Fig. 11-15 to alleviate v.h.f. TVI.

One type of TVI from v.h.f. transmitters *can* be the fault of the transmitter. Doubler and tripler stages may have output in a TV channel, either directly or on unused multiples of the fundamental crystal frequency. These can be radiated from the transmitter itself, if it is not properly shielded, or may leak through the final stage to the antenna. Check for them by disconnecting the antenna from the transmitter and tuning the TV receiver through Channel 2 to Channel 6. If you can see any interference in a picture your transmitter probably isn't shielded well enough.

Monitoring Your Sending

Most operators can send better code if they can hear what is being sent. If the station with whom you are working is on the same frequency you are, you should be able to hear your own signal whenever you press the key. However, the receiver may be overloaded by the transmitted signal, resulting in no beat note or perhaps a chirpy one. This can often be overcome by turning down the r.f. gain control while transmitting. Even so, the signal may be too loud for comfort. Also, if the other station is *not* on your frequency it becomes necessary to retune the receiver each time you stand by.

What is needed, then, is a monitoring method independent of either the transmitter or receiver tuning or volume. An audio-frequency oscillator or tone generator, keyed along with the transmitter, is commonly used. Fig. 11-16 shows how one can be constructed. The circuit, Fig. 11-17, is self-powered and so can be used independently of the transmitter for code practice. It has a built-in audio amplifier and loudspeaker, and has enough volume to be used for code classes.

The audio oscillator uses a neon bulb, V_1, along with resistors R_1, R_2, and capacitor C_2. When the bottom terminal of the bulb is con-

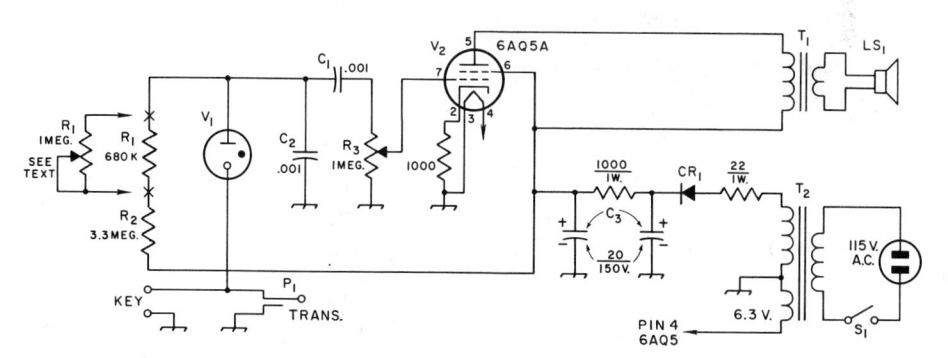

FIG. 11-17—Circuit diagram of the code-practice oscillator-monitor. Capacitances are in $\mu f.$, resistances in ohms; unless specified otherwise, resistors are ½ watt.

C_1, C_2—0.001-$\mu f.$ disk ceramic.
C_3—Dual 20-$\mu f.$ electrolytic, 150 volts.
CR_1—Selenium rectifier, 50 ma. (Sarkes-Tarzian type 50).
LS_1—2-inch p.m. speaker.
R_1—0.68 megohm, ½ watt.
R_2—3.3 megohms, ½ watt.

R_3—1-megohm control, audio taper.
P_1—Plug, to fit transmitter key jack.
T_1—Power transformer, single tube to voice coil, 5000 to 3.2 ohms (Knight 62G064).
T_2—Power transformer, 125 volts, 15 ma.; 6.3 volts, 0.6 amp. (Stancor PS8415, Knight 61G410).
V_1—Neon bulb, NE-21.
V_2—6AQ5A.

218   UNDERSTANDING

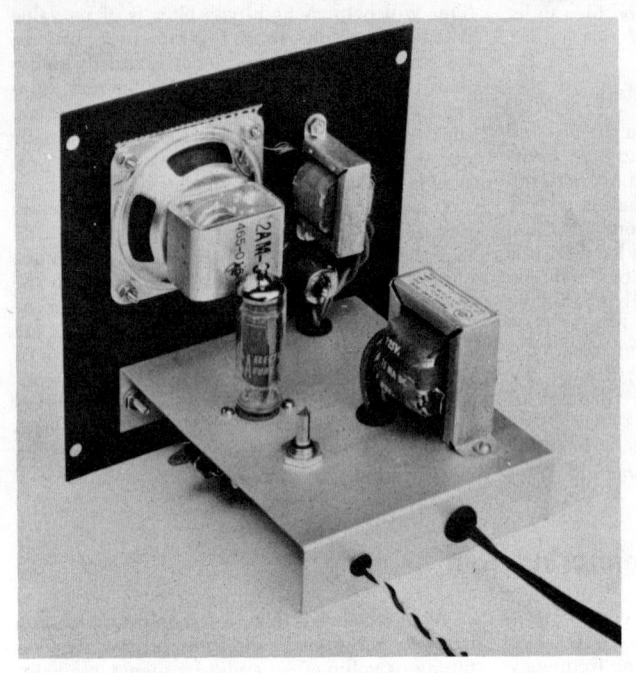

FIG. 11-18—Rear view of the code monitor. T_1, the output transformer, is on the panel to the right of the speaker. The transformer on the chassis is T_2.

nected to chassis ground through the key, V_1 conducts because C_2 is charged to the full voltage of the power supply, about 130 volts. However, the charge is rapidly used up in the bulb and the voltage drops below the value that will keep neon gas ionized. When ionization stops V_1 no longer conducts, whereupon C_2 again acquires a charge from the power supply through R_1 and R_2. The charge builds up relatively slowly to the point where the bulb again ionizes, after which the same thing happens all over again. The time constant of the circuit (see Chapter 2), together with the ionizing and extinction voltages of the neon tube, determine the rate at which the alternate charge and discharge occurs. With the circuit values given this rate will be in the audio-frequency range. The varying voltage across C_2 is coupled to the audio

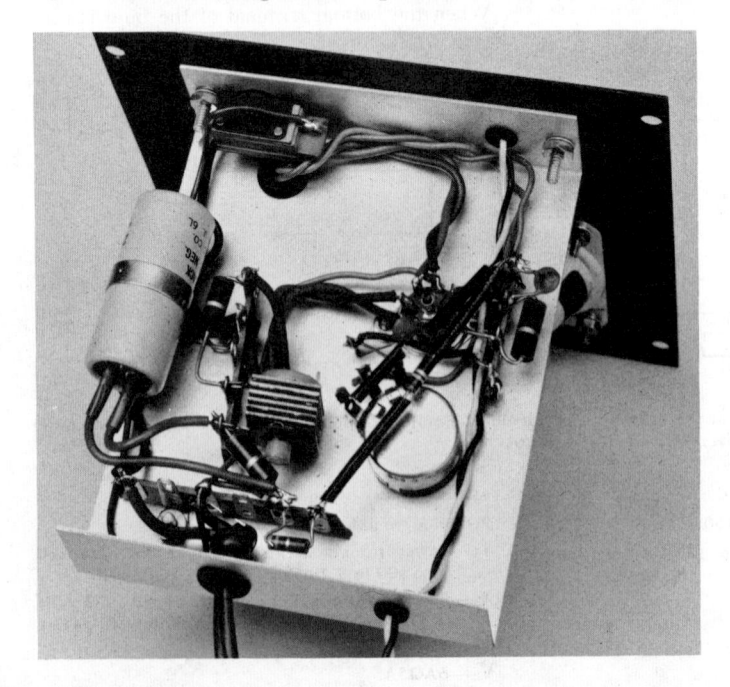

FIG. 11-19—Below chassis in the keying monitor. C_3 is at the left, with CR_1 to its right. The 6AQ5A socket and the gain control, R_3, are at the right-hand side.

Transmitter Monitoring

amplifier, V_2, through C_1 and is amplified before reaching the speaker through the matching transformer, T_1. The speaker volume can be set at a desired level by means of R_3.

The power supply uses a small transformer, with a half-wave semiconductor rectifier and a resistance-capacitance filter.

The construction of the code monitor is shown by Figs. 11-18 and 11-19. Most of the components are mounted on a $1 \times 5\frac{1}{8} \times 4$-inch aluminum chassis (Bud CB-1619). The speaker, output transformer, and neon bulb are mounted on the panel of a $6 \times 6 \times 6$-inch cabinet (Bud CU-1098). The power-supply components are mounted along one side of the chassis with the oscillator along the other side. The volume control is mounted on the chassis alongside the 6AQ5. If desired, this control can be mounted on the front panel.

The neon bulb is supported on the front panel by a ½-inch-diameter rubber grommet. Having the bulb on the panel shows you when the oscillator is working, since the bulb blinks on and off with keying.

The pitch of the tone can be changed by using different values for R_1. If you want to adjust the tone at will, substitute a 1-megohm control for the fixed resistor, as suggested in Fig. 11-17. This control could be mounted on the panel, as the lead lengths and layout are not critical.

To monitor your transmissions, plug P_1 into the key jack on the transmitter and connect the key as shown in Fig. 11-17. The circuit is for use with transmitters having cathode keying (most do) with one side of the key jack connected to the transmitter chassis. With other types of keying, especially in cases where the frame of the key jack is not grounded, it may be necessary to do the keying with a d.p.s.t. relay. The keying relay should be connected as shown in Fig. 11-20.

The "Matchtone" Keying Monitor

A very simple tone-oscillator keying-monitor circuit, devised by W4GF, is given in Fig. 11-21. It uses a transistor oscillator, and all the power it needs is obtained from the rectified current in the meter circuit of a Monimatch. A small audio transformer is used in the oscillator circuit, with one side of its push-pull winding furnishing the output coupling. The tone taken from this winding through C_2 should be fed to the grid of the first audio stage in the receiver. To avoid hum, a shielded lead should be used from the Matchtone to the audio tube's grid, with the shield grounded to the receiver chassis.

The pitch of the audio tone is controlled by the transformer characteristics, the capacitance of C_1, and the rectified current obtained from the Monimatch. A current of about 100 microamperes is needed to ensure oscillation. The pitch can be adjusted by means of R_2. The circuit values shown should result in a tone in the 500-

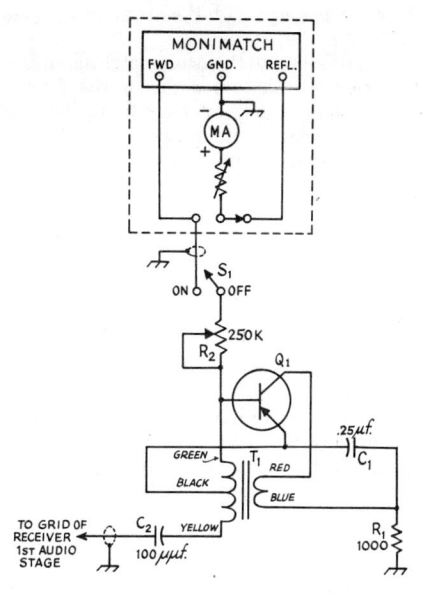

FIG. 11-21—"Matchtone" keying monitor. Section enclosed in dashed line is the Monimatch and its indicating circuit. Braid of shielded lead to audio grid should connect to receiver chassis.

C_1—Paper.
C_2—Mica or ceramic.
Q_1—2N109, CK722 or similar.
R_1—1000 ohms, ½ watt.
R_2—0.25-megohm volume control.
S_1—S.p.s.t. toggle.
T_1—Push-pull interstage audio transformer, 2:1 or 3:1 total grid to plate.

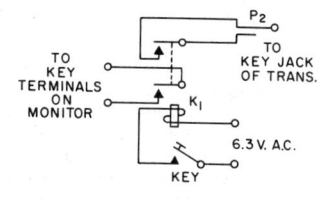

FIG. 11-20—When a keying relay is needed it should be connected as shown here.

K_1—Keying relay, double-pole, single-throw, 6-volt a.c. coil (Advance GHA/2C/6VA or equivalent).

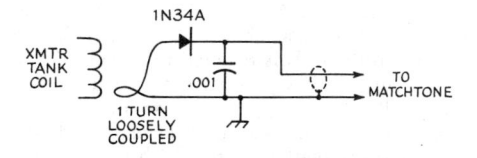

FIG. 11-22—Alternative power supply circuit for the Matchtone. Adjust the coupling between the 1-turn loop and the final tank coil for reliable operation of the monitor, then fasten the loop so it will not move. Very little coupling should be needed.

to 1000-cycle range with the average small audio transformer.

The connection to the Monimatch should go to the "forward" side, since this is the circuit in which the largest direct current flows. The meter in the Monimatch should be put in the "reflected" position if the forward current is low. This will permit the available current to be used for operating the Matchtone.

The circuit can be built in any convenient fashion.

If you don't have a Monimatch (although you should!) the power for operating the Matchtone can be obtained by coupling a turn of wire loosely to the transmitter tank coil with a crystal rectifier in series, as shown in Fig. 11-22. The d.c. output should be taken through a shielded lead, with the shield grounded to the transmitter chassis. This shielding is necessary to avoid the possibility of TVI caused by the rectifier, since a rectifier generates harmonics.

Other Useful Accessories

Simple R.F. Indicators

A handy thing to have around is some device that will show when r.f. is present in a transmitting circuit and, in a relative way, how much. Inexpensive lamps have long been favorites for this purpose.

A neon lamp will glow when touched to a "hot" r.f. point. It indicates in a rough way the amount of r.f. voltage at that point, thus is useful as a guide to optimum tuning when touched, for example, to the coil in a transmatch, or touched on open-wire feeders. The fingers should not touch the metal part of the bulb; hold the glass part only. And don't attempt to use a neon bulb on circuits inside the transmitter—the danger of accidental shock is too great.

ing up low-power stages in a transmitter. The loop can be made small enough to fit around the coil in the tuned circuit being checked. It is usually no problem to make it stick in place while the testing is going on. This can be done with the power off, so your hands don't need to be inside the transmitter when the power is on. If the loop is coupled to a small-diameter form, such as those having tuning slugs, it may be necessary to use a couple of turns to get sufficient coupling.

Often there is no provision for measuring plate current in low-power stages, so the lamp-loop is used instead of a plate meter as a resonance indicator. In one way its indications are more positive than the dip in plate current, because when the lamp lights there is no doubt that there is r.f. in the circuit.

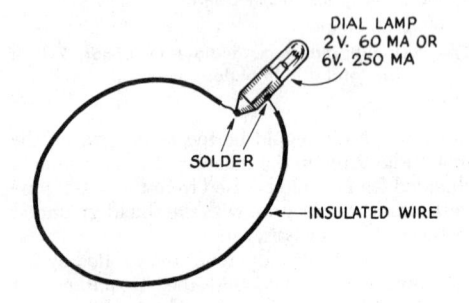

FIG. 11-23—The "Tune-Up" Loop. Wire of any variety can be used as long as it is insulated and rigid enough to hold the loop shape. Loop diameter is not critical and can be that of the coils to be checked.

Flashlight lamps and dial lamps make useful current indicators. A lamp connected to a one-turn loop of wire, as shown in Fig. 11-23, will light when the loop is coupled to a transmitting tank circuit. If a low-power lamp is used, such as the 2-volt, 60-ma. type (Nos. 48 or 49, pink bead), the filament will glow even though the power level in the circuit to which it is coupled is only a watt or two. These lamps burn out very readily when coupled to circuits with higher power, so a 250-ma. lamp (or one of larger rating still) should be used around such circuits.

The lamp-loop is particularly helpful in tun-

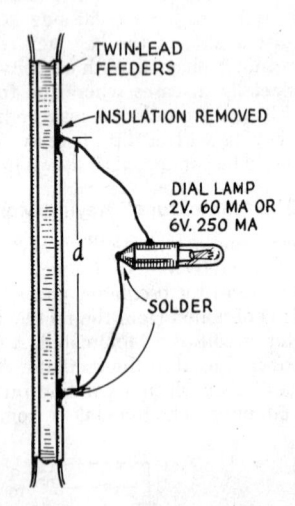

FIG. 11-24—Using a dial lamp as a feeder-current indicator. The distance d will depend upon the available power and the amateur band. Low power and low-frequency operation will require a greater length d than will high power and a high frequency. Where a tuned line is used (high s.w.r.), the position on the line (high-voltage or high-current point) will also influence the proper d. A greater length is required at a high-voltage point than at a high-current point. Use a distance d of about 1 foot for the first attempt, and increase the transmitter loading slowly to avoid burning out the bulb.

R.F. Indicators

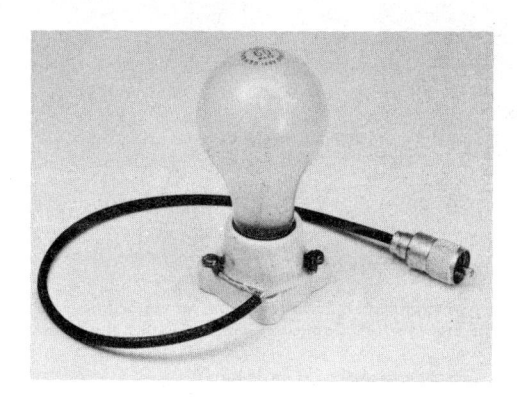

FIG. 11-25—An incandescent lamp as a dummy antenna. Lamp rating should be selected to be approximately equal to the expected power output of the transmitter.

Small lamps also can be used as substitutes for r.f. ammeters in parallel-wire feeder systems. One way of using a lamp for this purpose is shown in Fig. 11-24. Since the brightness of the lamp is proportional to the actual r.f. current in the feeder, any tuning condition that results in maximum brightness is the one giving the greatest power output. The plate-meter reading is not always wholly accurate in this respect.

Dummy Antennas

A **dummy antenna** is a device for absorbing the power output of a transmitter. Using one lets you test the transmitter without putting a signal on the air, as you would if you used a regular antenna for this purpose. (Using a real antenna for transmitter testing is not permitted by the regulations, except very briefly, and is inconsiderate of others besides.) The basic ingredient of a dummy antenna is resistance, since resistance absorbs power.

An ideal dummy would be a pure resistance matching the characteristic impedance of coaxial cable, 50 or 75 ohms, approximately. Small composition resistors are fine at frequencies up to 150 Mc. or even higher, but they can't handle much power. Two watts is the maximum for the largest commonly-available resistor of this type. Several resistors can be combined in series, parallel, or a combination of both series and parallel, for handling more power. However, it takes a great many of them to dissipate, say, 100 watts safely. Also, when a large number of them are wired together some of the good characteristics are lost—the wiring adds inductance and capacitance and the resistance is no longer pure.

Actually, most transmitter testing doesn't require a dummy of known resistance. Anything that will let you load up the final amplifier to normal input will do. The cheapest and most satisfactory resistor, in many respects, is an ordinary incandescent lamp. It will light up on r.f.

just as well as on 60-cycle a.c., at frequencies below 30 Mc. Its brightness will give you a fair indication of your power output. Simply choose a lamp that will light up to about normal brightness on the output of your transmitter, and compare it with one of the same rating in a regular 115-volt socket. Beware of estimating power output when the lamp isn't close to being normally bright, though. Just a couple of watts will make a 60- or 100-watt lamp show color!

The lamp can be put in a dime-store socket, for convenience in changing sizes. See Fig. 11-25. Use a foot or so of coax, with a fitting on one end, to connect the socket to your transmitter. The coax braid should go to the shell connection and the inner conductor should connect to the center stud of the lamp.

One disadvantage of the lamp is that its resistance changes with the temperature of the filament. This represents a changing load as you tune up, so you have to be careful in adjusting your loading and tuning controls. One thing you can be sure of—when the lamp is as bright as you can get it with any possible tuning adjustment, your transmitter is delivering all the power you can expect to get from it.

Electronic Transmit-Receive Switching

While it is always good practice to use the same antenna for both receiving and transmitting, to do so you have to shift the antenna or feeder connections back and forth. This can be done by manual switching or with an antenna relay. However, neither of these can be operated rapidly enough to follow keying, so in either case break-in operation becomes impracticable. But if the switching is done electronically it can be practically instantaneous. A device which does this is called an **electronic t.r. (transmit-receive) switch.**

Fig. 11-27 is the circuit of a simple switch of this type. The coax line from the antenna or from a transmatch is permanently connected to the

FIG. 11-26—Electronic transmit-receive switch, with built-in power supply. The components on top of the chassis are the power transformer and the switch tube.

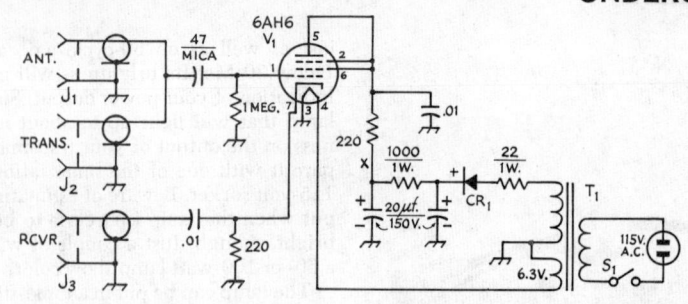

FIG. 11-27—Circuit diagram of the t.r. switch. Unless otherwise indicated, decimal values of capacitance are in µf., others are in µµf., resistances are in ohms, resistors are ½ watt. Capacitors marked with polarity are electrolytic.

CR₁—Selenium rectifier, 50 ma. (Sarkes-Tarzian Model 50).
J₁, J₂, J₃—Coax chassis receptacles, or phono jacks.
S₁—S.p.s.t. toggle.
T₁—Power transformer, 125 volts, 15 ma., 6.3 volts, 0.6 amp. (Stancor PS-8415, Knight 61G-410).

transmitter through J₁ and J₂. The inner conductor of the coax is connected to the grid of the t.r. switch tube, a 6AH6, through a small blocking capacitor. When the transmitter is idle (key up) incoming signals from the antenna reach the t.r. tube's grid, and are coupled to the receiver from the tube's cathode through J₃, so the receiver operates normally.

When the transmitter's key is closed the r.f. voltage on the transmission line is very much larger than the grid bias on the t.r. tube. (This bias is normally only a volt or two, obtained from the voltage drop across the 220-ohm cathode resistor.) This causes grid current to flow through the 1-megohm grid resistor, developing a negative grid voltage practically equal to the peak value of the r.f.—far more than enough to reduce the tube's plate current to practically zero. Thus there cannot be enough r.f. output from the cathode circuit to cause any damage to the receiver. But immediately on opening the key the grid-leak bias disappears and the receiving system is back in operation.

Coax cable should be used between the t.r. switch and the receiver, with the braid connected to the receiver chassis. This helps prevent stray pick-up of r.f. from the transmitter. If your receiver has doublet antenna terminals, ground one of them to the chassis and connect the inner conductor of the line to the other.

Figs. 11-26 and 11-28 show how the t.r. switch can be constructed. The principal point to observe is that the antenna and transmitter connectors should be close together, with the tube mounted alongside them so the shortest possible grid lead can be used. This helps prevent stray coupling between the transmitter and receiver.

It is convenient to incorporate the power-supply circuit shown in Fig. 11-27, but if your transmitter has an accessory socket from which the

FIG. 11-28—Underneath the chassis of the electronic t.r. switch. The tube and coax connectors are mounted close together to keep the exposed r.f. leads as short as possible. The rectangular component near the upper edge of the chassis is the power-supply rectifier, CR₁. The edge of the dual filter capacitor is just visible above it.

T.R. Switches

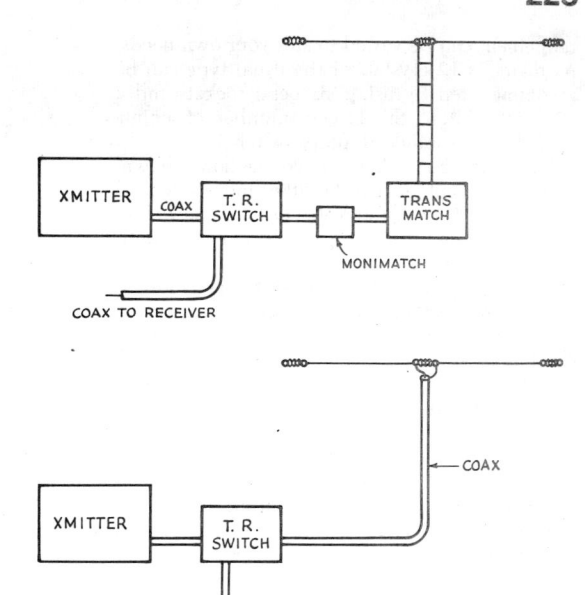

FIG. 11-29—The t.r. switch should be mounted closest to the transmitter when other equipment, such as a Monimatch, transmatch, or low-pass filter (not shown) is used.

power for the 6AH6 can be taken the built-in power supply isn't actually necessary. The tube's heater requires 0.45 ampere at 6.3 volts, and the plate supply should be about 100 volts. The current is about 10 ma. Plate voltage should be introduced at point X in the circuit if a separate supply is used.

Fig. 11-29 shows how the t.r. switch is connected to the rest of the equipment, in one case using a coax-fed antenna and in the other using a transmatch. Either way, the coax line into which the switch is inserted should operate at a standing-wave ratio of less than 2 to 1, for output power of the order of 100 watts. With a higher s.w.r. the voltage on the line may be too high for the safety of the switch tube.

Finally, the grid circuit of the tube is a recti-fier when the transmitter is on, and like all rectifiers it will generate harmonics of the output frequency. It can be the cause of TVI on that account. If so, a low-pass filter should be used on the antenna side of the t.r. switch.

A Handy Crystal Switch Assembly

Many Novice transmitters have only one crystal socket, or at most offer switching for no more than two or three crystals. Plugging and unplugging crystals and storing them so the right one can be found quickly is a chore. Fig. 11-30 shows an easily-made gadget for selecting the desired crystal from an assortment merely by turning a switch. As made up, it has provision for four crystals of the FT-243 holder type, plus a socket for the larger crystal holders. Obviously this ar-

FIG. 11-30—Crystal-switching assembly. It can be used with any transmitter having a crystal-controlled oscillator, simply by inserting the 300-ohm line plug in the transmitter's crystal socket. However, the lead to the plug should not be more than a few inches long, in order to ensure reliable crystal operation.

rangement can be varied to suit your own needs. As many as 12 crystals of the usual type can be accommodated by using six octal sockets and a 12-point switch, the largest number of points available on a standard rotary switch.

By using the socket pin connections shown, two FT-243 holders can be fitted into each octal socket. The 5-prong socket will take one holder having ¾-inch pin spacing. To add more sockets, simply continue the same plan of wiring. A 300-ohm line plug is used to connect the assembly to the crystal socket on the transmitter. The leads—300-ohm line can be used—should be kept as short as possible.

Before making the chassis ground connection shown in the diagram, check your transmitter's oscillator circuit. In some transmitters neither side of the crystal is grounded; in such case, omit the chassis connection. If one crystal socket pin *is* grounded in the transmitter, make sure that the chassis connection in this unit is led to the grounded side of the crystal socket.

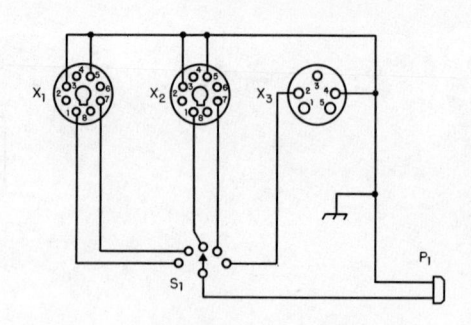

FIG. 11-31—Circuit diagram of the crystal-switching assembly. Other sockets can be added in parallel if a switch with more contacts is used.

P_1—Plug for crystal socket (Millen 37412, Mosley type 301).
S_1—Single-pole rotary, as many positions as needed.
X_1, X_2—Octal socket.
X_3—5-pin socket.

Chapter 12

The Power Supply

Except for a few measuring instruments, practically all radio equipment needs some form of power supply before it can function. The power source is generally the 115-volt a.c. line, but 115-volt 60-cycle a.c. is hardly ever usable without modification. This modification is the job that the power supply handles. You've seen many examples in the preceding chapters.

In vacuum-tube equipment one necessity is power for heating the cathodes. This is easy, because most tube heaters will work from "raw" a.c. A simple transformer can step the voltage down to the appropriate value—generally 6.3 volts for receiving and small transmitting tubes, and 5.0 volts for receiving-type power rectifiers.

But heaters, while certainly essential, usually take no direct part in circuit operation. The tube electrodes that do have such a part require d.c. power. Furthermore, it must be *good* d.c.—that is, unvarying in amplitude. Changing the 115-volt a.c. to this kind of d.c. calls for a rectifier and filter in addition to a transformer. The transformer changes the 115 volts to a more usable value; the rectifier changes the a.c. to d.c.; and the filter takes the rather rough d.c. from the rectifier and smooths out the unwanted fluctuations.

This territory should be familiar to you, having been explored in some detail in *How To Become A Radio Amateur*. The difference between **half-wave** and **full-wave** rectification was explained there, and the types of **filters** were described. Here we'll simply repeat the circuits, Fig. 12-1, of the rectifiers and give you a few figures to go with them. A d.c. voltmeter connected across the load resistance, R, will read the average values shown, provided R is large enough to hold the current down to a small value. With low current, the voltage drops in the transformer and the diode rectifiers are small enough to be neglected. However, these voltage drops can't be overlooked when the system is delivering all the power it is designed for. The voltage readings will be smaller by perhaps 10 to 20 per cent at the rated current.

Bridge Rectifier

A third type of rectifier circuit, frequently used in amateur transmitters, is the **full-wave bridge** shown in Fig. 12-2. The output current or voltage waveshape is the same as for the full-wave center-tap circuit. The *amplitude* of the output voltage, as compared with the voltage at the transformer secondary, is different. In the center-tap circuit *each side* of the center-tapped secondary must develop enough voltage to supply the d.c. output voltage desired. In the bridge circuit no center tap is needed. Dispensing with the center tap has a price, though—two additional rectifiers must be used.

In the bridge circuit rectifiers A and B are connected to the transformer secondary in the same way as the two rectifiers in the center-tap circuit. The second pair, C and D, is also connected in series between the ends of the transformer winding, but in reverse. When the upper end of the transformer winding is positive the current path is through A, the load R, and D back to the lower end of the winding. When the lower end is positive the path is through B, R, and C.

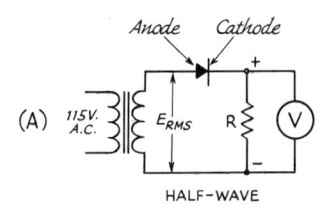

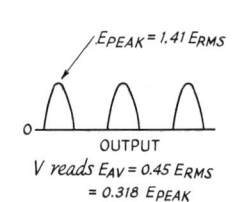

FIG. 12-1—Half-wave and full-wave center-tap rectifier circuits. The general symbol for a diode is shown; in practice a tube rectifier is often used. If R is a high resistance—perhaps 100,000 ohms—the voltmeter, V, will read the values shown at the right.

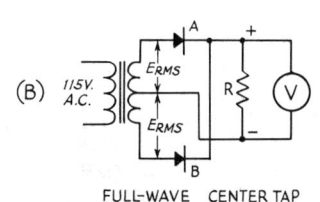

(A) HALF-WAVE
$E_{PEAK} = 1.41 E_{RMS}$
OUTPUT
V reads $E_{AV} = 0.45 E_{RMS}$
$= 0.318 E_{PEAK}$

(B) FULL-WAVE CENTER TAP
$E_{PEAK} = 1.41 E_{RMS}$
OUTPUT
V reads $E_{AV} = 0.9 E_{RMS}$
$= 0.636 E_{PEAK}$

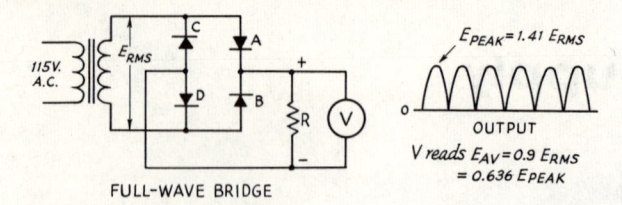

FIG. 12-2—Bridge rectifier circuit. Remarks under Fig. 12-1 apply also to this drawing.

Filters

Although the output of a rectifier is direct current, its amplitude fluctuates all the way from zero to the peak output voltage from the transformer. In the full-wave circuit the rate at which this happens is twice the supply frequency. That is, the **ripple** in the output voltage has a frequency of 120 cycles when the supply frequency is 60 cycles. The ripple frequency in a half-wave rectifier is equal to the supply frequency.*

If an attempt is made to use the rectifier output as is with, for example, a receiver, there will be nothing but a roaring hum from the loudspeaker. The fluctuations must be smoothed out so the ripple is substantially eliminated. This is where the filter comes in.

Capacitor and Choke Input

Filters having a capacitor next to the rectifier are called **capacitor-input** filters: Those having an inductance between the rectifier and capacitor are called **choke-input** filters. *All* power-supply filters use capacitors, although not all use chokes.

Both types of filters, with half-wave and full-wave rectifiers, are shown in Fig. 12-3. The drawings at the right give an idea of the form of the ripple on the d.c. output. The approximate output voltages also are indicated. In the case of the choke-input filter, C, the output voltage will have the value shown only when a large inductance is used. "Large" here means an inductance, in henrys, greater than the load resistance, R, divided by 1000. For example, if the load is 100 ma. at 300 volts, the value of R is 300/0.1 amp., or 3000 ohms. Thus L should be at least 3 henrys.

Ripple and Filter Sections

Notice that there are twice as many "bumps" in the d.c. output voltage per cycle of line frequency (the length of a cycle is indicated by the short vertical lines along the horizontal axis of the output drawing) when a full-wave rectifier is used. This is just another way of saying that the ripple frequency is doubled with full-wave rectification. Values of L and C in the filter need be only half as large with full-wave rectification as with half-wave rectification, for the same smoothing of the d.c. output.

---

*More accurately, these are the *fundamental* frequencies of the ripple. Actually, the ripple waveform is highly distorted, and there are many harmonic frequencies in it along with the fundamental.

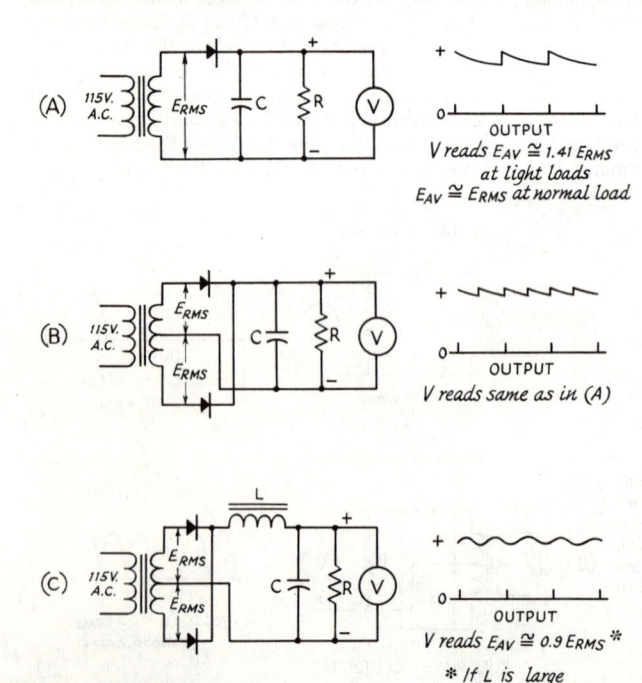

FIG. 12-3—Capacitor- and choke-input filters. The approximate shape of the ripple voltage remaining after filtering is shown by the drawings at the right. Divisions along the horizontal axis represent one cycle of the line frequency. Output voltage shown at C neglect drops in the transformer, rectifiers and filter choke.

Power Supplies

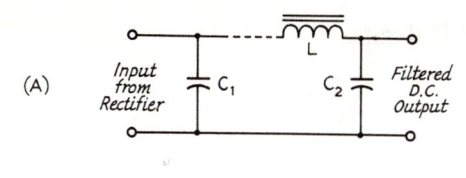

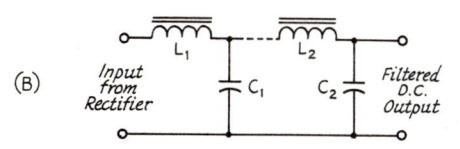

(A) Input from Rectifier — C_1 — L — C_2 — Filtered D.C. Output

(B) Input from Rectifier — L_1 — C_1 — L_2 — C_2 — Filtered D.C. Output

FIG. 12-4—Filter sections may be added as necessary to secure a desired amount of smoothing.

The inductance and capacitance can smooth out the d.c. because they store energy while the voltage from the rectifier is rising, and release it while the rectified voltage is falling. Less energy has to be stored, for the same over-all effect, if the charge-discharge periods are quick instead of drawn out over a longer period of time.

An *LC* combination such as is shown in Fig. 12-3 at C is called a **filter section**. Sections are added to the filter as shown in Fig. 12-4. Sections can be added either to the simple input capacitor (A) or to a choke-input section (B). As many sections as may be necessary to get the desired smoothing can be used.

How much inductance and capacitance are needed for good filtering? There is no definite answer. Your standard for "good" may differ considerably from the next fellow's. The intended use of the power supply has a lot to do with it. If you wear headphones for listening to weak signals, a hum that you find highly objectionable may not even be audible when a small speaker is used. As a general rule, receivers require more elaborate filters than transmitters.

Practical power supply circuits you'll find later in this chapter are representative. So are those given in other chapters in association with receiving and transmitting equipment. It does no harm to use the largest economical value of capacitance. The difference in cost between 40-μf. and 20-μf. electrolytic capacitors, for example, is rather small; it pays to use the larger value. This is less true of chokes, which tend to be expensive in the higher values of inductance. But the effectiveness of a filter section is proportional to *L* multiplied by *C*, so if you have to use a small inductance a correspondingly larger capacitance will compensate for it.

RC Filters

Another type of filter is shown in Fig. 12-5. This is called a **resistance-capacitance** or **RC** filter. The resistor serves the same purpose as the decoupling resistors discussed in earlier chapters. There is an a.c. (ripple) voltage drop in the resistor, so less ripple reaches the capacitor. This makes the capacitor's smoothing job easier. However, there is also a d.c. voltage drop in the resistor. In contrast, the d.c. voltage drop in a choke would be quite small.

Resistance-capacitance filters are used principally in circuits where the current is small enough to make the d.c. voltage drop in the resistor unimportant. An example of such a circuit is a resistance-coupled audio voltage amplifier.

The Voltage Doubler

A circuit that is frequently used in transmitter power supplies is the **voltage-doubler** shown in Fig. 12-6. This works by charging the capacitors through separate half-wave rectifiers on alternate halves of the supply cycle, and then discharging them in series through the load, *R*.

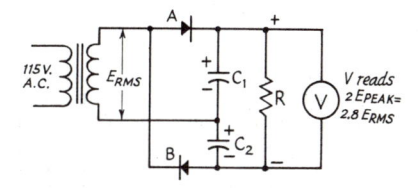

FIG. 12-6—The voltage doubler circuit. Two half-wave rectifiers, working alternately, charge capacitors connected in series.

[115 V. A.C. — E_{RMS} — C_1, C_2 — R — V reads $2 E_{PEAK} = 2.8 E_{RMS}$]

When voltage at the upper end of the transformer secondary is positive, C_1 is charged to the peak value of the a.c. voltage through rectifier A. When the lower end of the winding is positive, C_2 is similarly charged through B. The stored voltages on the two capacitors add together to give a total output voltage twice what one would expect to get from the half-wave rectifier of Fig. 12-3A. Both halves of the supply cycle are used, so this is a full-wave rectifier circuit. However, *each* capacitor has to be as large as the single capacitor used in the half-wave circuit, for the same smoothing. It is a particularly useful circuit when a moderately-high d.c. voltage is wanted, because the transformer voltage need be only half as great as in the half-wave circuit, and only one-fourth as large as the total secondary voltage required by the full-wave center-tap circuit. The voltage doubler is used in a number of manufactured low-power transmitters and kits.

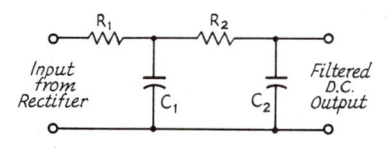

Input from Rectifier — R_1 — R_2 — C_1 — C_2 — Filtered D.C. Output

FIG. 12-5—Resistance-capacitance filter. It is principally useful where the load current is small, since the d.c. voltage drop in R precludes using a large resistance value for heavy currents. The RC filter makes a good "back-up" for an LC section in many circuits.

Rectifiers and Ratings

You're already familiar with the way a diode rectifies. A power-supply rectifier is just a diode big enough to handle the current and voltage needed in a power supply. Many sizes are available.

In choosing and using power-supply rectifiers there are some important factors that must never be overlooked. One is the **peak inverse voltage**, often abbreviated **p.i.v.** This is the maximum voltage applied to the rectifier in the *nonconducting* direction. For example, in Fig. 12-1B when rectifier A is conducting and B is not, the full transformer voltage is applied in the reverse direction to B, because there is very little voltage drop in A when it is conducting. Since the peak value of the voltage is 1.41 times the r.m.s. value, you can see readily that in Fig. 12-1B the p.i.v. is 1.41 times the *total* secondary voltage. Values for the various rectifier circuits are given below:

Circuit	P.I.V.
12-1A	$1.41 E_{RMS}$
12-1B, 12-3B, 12-3C	$2.82 E_{RMS}$
12-2	$1.41 E_{RMS}$
12-3A, 12-6	$2.82 E_{RMS}$ (max.)

In Figs. 12-3A and 12-6 the p.i.v. is the sum of the peak transformer voltage and the d.c. voltage stored in one filter capacitor. The maximum value shown is for light loads—that is, with little or no output current. For safety, the rectifier rating always should be chosen on the basis of no-load p.i.v.

The current rating of the rectifier also must be considered. Diodes usually are rated for two current values, the **average d.c. output current** and the **maximum peak current**. The former is the direct current flowing to the load, so if your load needs a current of 100 ma., you would choose a rectifier having at least that output-current rating. In full-wave rectifier circuits each diode carries half the current, so each can be rated for half the desired output current. When there are two diodes in one assembly, though, as in the case of many of the full-wave rectifier tubes used in receivers, the rating is based on what the *pair* can do.

The peak-current rating is the largest current that can be allowed to flow through the rectifier at any time during the rectified cycle. It is important with capacitor-input filters because most of the charging current into the first filter capacitor flows in a short burst, reaching a peak value several times the value of the d.c. output current. In a choke-input filter the peak current is very little larger than the d.c. output current, so with this type of filter the peak-current rating can be neglected.

High-Vacuum Rectifiers

The diodes used for rectification of the a.c. voltage may be vacuum tubes, gas-filled tubes, or semiconductors. You're already familiar with vacuum-tube diodes. Many power supplies use the types found in broadcast and television receivers. The 5Y3G, 5U4G, 6X4, and several others are typical. These contain two rectifiers in one envelope for full-wave rectification. There is also considerable use of single diodes such as the 6DE4.

The full-wave rectifiers usually have directly-heated cathodes rated at 5 volts and 2 or 3 amperes, depending on the tube type. Single diodes generally have indirectly-heated cathodes rated for 6.3 volts at a current depending on the size of the rectifier.

Mercury-Vapor Rectifiers

Mercury-vapor rectifiers are tubes into which a little mercury has been introduced after the air has been pumped out. When the cathode is heated some of the mercury vaporizes, forming a gas which ionizes when a voltage is applied between the plate and cathode.

Conduction in an ionized gas has a rather peculiar characteristic. The amount of current is practically independent of the voltage between the electrodes—in this case the plate and cathode. Or, stated another way, any value of current (within the physical limitations of the tube) can flow with little or no change in the voltage between the plate and cathode.* This voltage, called the **tube drop**, is a characteristic of the gas used, and in the case of mercury vapor is about 15 volts. In contrast, the tube drop in a vacuum diode may average 50 volts or more at its rated load.

Because of the low tube drop the power dissipated in the rectifier plate is less in a mercury-vapor rectifier than in a high-vacuum rectifier, for the same current. So the rectifier doesn't have to be as large. This is advantageous in high-power, high-voltage supplies. However, mercury-vapor rectifiers aren't needed in sup-

---

*The current has to be limited by resistance external to the tube; otherwise the tube would be destroyed.

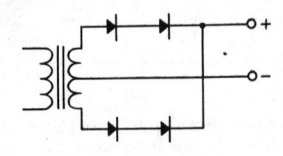

FIG. 12-7—Using two semiconductor diodes in series doubles the peak-inverse voltage rating.

Rectifiers

plies for transmitters such as are described in this book.

Semiconductor Rectifiers

Two types of semiconductor diodes are used in power supplies—selenium and silicon. The voltage drop in either type, even at currents of a half ampere or more, is only a few volts. There is no heater to consume power and raise the overall temperature, and the rectifier works the instant voltage is applied to it. The principal disadvantage is that the peak inverse voltage ratings are rather low for most types—of the order of 400 volts. (Some silicon types are rated for higher voltages—up to about 1000 volts.) A 400-volt p.i.v. rectifier is only good for about 130 volts, r.m.s., of *applied* voltage.

The low p.i.v. rating on semiconductor rectifiers limits the use of a single unit to rather low output voltage. However, two or more can be placed in series, as shown in Fig. 12-7. This divides the p.i.v. across the diodes, so the safe p.i.v. increases in proportion to the number of rectifiers. For example, if each of the rectifiers in Fig. 12-7 has a rating of 400 volts p.i.v. the total rating is 800 volts. From what was said earlier (Fig. 12-1B) the total transformer voltage that could be applied safely to this combination is 800/1.41, or approximately 560 volts—280 volts each side of the center tap.

Because the voltage drop in a semiconductor rectifier is so small, the diode itself will not limit the peak current. It is therefore good practice to connect a small resistance in series, as shown in Fig. 12-8, to limit the peak charging current in a capacitor-input filter. A value of 10 ohms (1-watt size) is sufficient with rectifiers rated at 100 ma. or more.

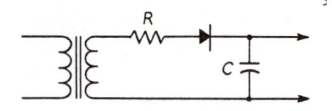

FIG. 12-8—Resistor R limits the peak rectifier current when the filter capacitor, C, is being charged. A value of 10 ohms is typical.

The cathode (straight line in the current symbol for the rectifier) is usually marked "+" or "Cath" on small silicon rectifiers. Selenium rectifiers may have a red dot instead of the plus sign to indicate the cathode. Be sure to identify the cathode correctly before wiring the rectifier into a circuit.

Caution!

Here's one caution that you should never overlook in using semiconductor rectifiers: *Never* exceed the p.i.v. rating. Even a *small* excess voltage is not safe. With vacuum tubes you can sometimes get away with running rectifiers at more than the p.i.v. rating (although we certainly don't recommend the practice) but not with selenium or silicon diodes.

Filter Components

Filter capacitors in amateur transmitting equipment in the power range we're considering in this book are nearly always the electrolytic type. They are far less expensive than paper-dielectric capacitors, but are not made to stand voltages higher than 450 or, in a few cases, 500 and 600 volts.

However, two or more units can be connected in series to raise the safe voltage. It is seldom necessary to use more than two in series in transmitters up to 150 watts or so. Of course, the total capacitance of the combination is less than that of either unit alone, as you learned earlier. Fig. 12-9 is a typical example. When connecting such capacitors in series it is advisable to make them identical. If the capacitances are different the voltages do not divide equally, and one capacitor may operate considerably over its rating while the other is well under. To make certain that the voltages do divide equally, identical resistors often are connected across each capacitor in a series string, as shown in the figure. These need not take much current, so values of the order of 100,000 ohms are common for these **equalizing resistors**.

Polarization

Electrolytic capacitors are **polarized**. This means that they must be connected in a d.c. circuit in only one way. The terminals are always marked to show which one goes to the positive side of the circuit and which to the negative. When the container is a metal can the can is usually the negative terminal. However, this should be checked before using an electrolytic. If the capacitor is connected the wrong way it will be ruined in short order.

All electrolytic capacitors have a small **leakage current**—a few milliamperes flowing through the capacitor itself. This is not serious in a power-

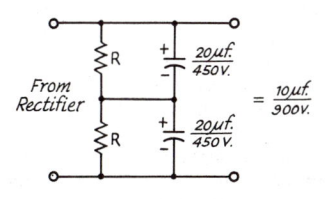

FIG. 12-9—Electrolytic capacitors connected in series to increase the voltage rating. Equalizing resistors, R, help divide the d.c. voltage properly.

supply filter, but prohibits using the capacitor in many other circuits. Also, the capacitor must always be used in a circuit in which the d.c. voltage is larger than the peak value of any a.c. voltage that may be superimposed. Electrolytics cannot be used on "raw" a.c.

Paper Capacitors

Paper-dielectric capacitors do not have these limitations. The leakage current is practically unmeasurable if the unit is not defective. Also, since these capacitors are not polarized they will work on either d.c. or a.c. These two factors make them well suited to such uses as interstage coupling in audio-frequency amplifiers. However, the larger values of capacitance, at the higher voltage ratings, are very much more costly than equivalent capacitance in electrolytics.

Filter Chokes

Filter chokes resemble transformers in appearance and general construction (see Chapter 2). However, there is only one winding. Also, the core laminations are not interleaved to provide the best possible magnetic path, as is the practice in transformer construction. In fact, a small **air gap** is made in the core, as shown in Fig. 12-10. (The gap in the drawing is much exaggerated; actually it would be only a small fraction of an inch, almost invisible, in an ordinary choke such as you would use in your equipment.) The reason for this is that the direct current flowing through the choke would tend to **saturate** the core to the

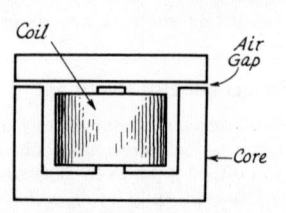

FIG. 12-10—Power-supply choke construction.

point where a change in current would no longer cause a corresponding change in the magnetic field in the core. Saturation reduces the effective inductance of the choke for alternating currents, so the choke no longer can smooth out the ripple in the rectified a.c.

Even with the air gap there is always some tendency toward saturation, with the result that the inductance of a choke steadily decreases as the direct current through it is increased. It is customary to rate chokes in so many henrys of inductance at such-and-such direct current. The inductance will be larger at smaller currents, and smaller at higher currents. Just how much larger and smaller is a matter of the choke design. Chokes in which these changes are small are rather expensive, for a given inductance rating. However, in power-supply filter design the principal thing is to have enough of everything that contributes to smoothing. So if the components have adequate ratings at the output current you want, there's no necessity for worrying about what may happen at smaller currents. The filtering just gets better when you reduce the load.

Voltage Regulation

The amount of direct current that a receiver or transmitter demands from the power supply is not always constant. A c.w. transmitter may take very little current with the key open, but with the key down it may load the supply up to the maximum current it is designed to give. The total current taken by a receiver circuit will vary depending on the setting of the r.f. gain control. These changes in current affect the supply's output voltage.

The change in output voltage with current is called the **voltage regulation** of the supply. It is usually expressed as a percentage of the output voltage at the rated load current. For example, the supply may be designed to deliver a current of 200 ma. at 600 volts. If the voltage rises to 800 when the output current is zero, the change is 200 volts and the regulation is 200/600. This is ⅓, or 33⅓ per cent.

Several things contribute to the voltage drop in the supply. One is the resistance and reactance of the transformer windings. Another is the voltage drop between the anode and cathode of the rectifier. A third is the resistance of the filter choke and any resistors that may be used for

filtering. The voltage drops in all of these increase with current. Finally, in a capacitor-input filter the output voltage depends on the amount of energy stored in the input capacitor and the rate at which the energy is released to the load. This rate increases with an increase in load current, causing the voltage to decrease accordingly. With no output current the voltage in a capacitor-input filter builds up to the peak value of the rectified voltage, or 1.41 times the r.m.s. voltage applied to the rectifier (Fig. 12-1).

A choke-input filter with a large value of inductance will prevent this voltage build-up at small output currents and will tend to hold the output voltage at the *average* amplitude of the rectified wave form. (Fig. 12-3C). The voltage drops mentioned above still cause the output voltage to drop off with increased current, of course.

The comparison between choke- and capacitor-input filters in their effect on voltage regulation is shown in Fig. 12-11.

Bleeders

In discussing choke-input filters we said that

Voltage Regulation

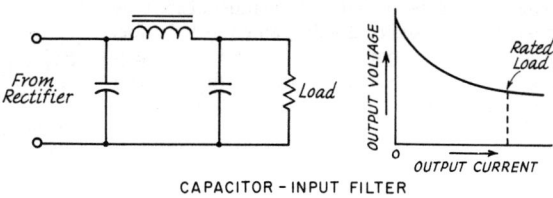

FIG. 12-11—The curves are typical of the way the output voltage varies with load current for the two types of filters shown.

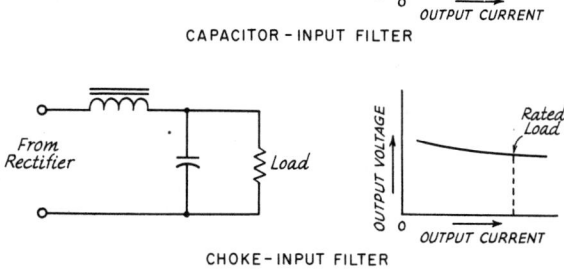

CAPACITOR-INPUT FILTER

CHOKE-INPUT FILTER

the input choke must have an inductance equal to or larger than $R/1000$ when a full-wave rectifier is used. If there is no R—i.e., no load on the supply and no output current—the output voltage is not held down to $0.9E_{RMS}$ but rises to $1.41E_{RMS}$. The regulation would be just as poor in this case as with a capacitor-input filter. The remedy is to **bleed** some current from the supply all the time, even when there is no current to the actual load.

This can be done by connecting a resistor across the output terminals of the supply. Obviously we don't want to waste any more power in this resistance than is absolutely necessary. The proper value of **bleeder resistance** is equal to the inductance in henrys multiplied by 1000. Thus the larger the choke inductance the higher the permissible resistance, therefore the less wasted power.

The bleeder serves another useful purpose: it discharges the filter capacitors when the supply is turned off. A bleeder should always be used, even with capacitor-input filters. It will prevent many an accidental shock—and the shock you can get from a filter capacitor is nothing to trifle with. Resistance values of 25,000 to 50,000 ohms are common. However, anything that will discharge the capacitors will do, and larger values are often used.

Bleeder resistors have to dissipate a fair amount of power, as a rule. Having decided on the resistance you want to use, calculate the power by Ohm's Law: $P = E^2R$. For instance, if the bleeder is 25,000 ohms and the voltage is to be 400, the power the resistor must handle is $(400)^2/25,000$, or a little over 6 watts. A 10-watt resistor should be used, to give a safety factor.

Voltage Regulators

Circuits or devices which hold the output voltage at a constant value regardless of changes in current are called **voltage regulators**. There are several varieties, but the only one that need concern us here is the gas-tube regulator, or **VR** tube. Earlier in this chapter it was pointed out that the voltage across an ionized gas tends to be independent of the current flowing through the gas. This property can be used to advantage in maintaining constant output voltage from a power supply.

How the tube operates can best be illustrated by seeing how the circuit, Fig. 12-12, is designed. Suppose the load wants 150 volts, regulated, at a current of 25 ma., and that the unregulated d.c. input voltage is 250. Allowing 5 ma. (the minimum current for stable operation) for the current through the VR tube, the total current through R is $5 + 25 = 30$ ma. The resistor R thus must drop 250 volts to 150 volts; that is, the drop in R must be 100 volts. By Ohm's Law, its resistance is $100/0.03 = 3300$ ohms.

Now if the input voltage rises to, say, 300 volts, the drop in R must increase to $300 - 150 = 150$ volts, to keep the output voltage constant at 150. The current through R therefore must be $150/3300 = 45$ ma. The load is still taking 25 ma., since the voltage across it has not changed, so the additional current, $45 - 25$ or 20 ma., must flow through the VR tube. In other words, the tube regulates the voltage by taking more current when the input voltage tends to rise, and by taking less when it tends to decrease.

A little thought will show that a design such as this should be based on the *lowest* input voltage likely to exist, if a minimum VR current of 5 ma. is used. Increasing the minimum current will permit regulation both ways, but remember that the *maximum* current through the VR tube should not be allowed to exceed the tube rating.

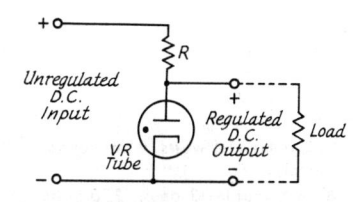

FIG. 12-12—Voltage-regulator tube circuit for maintaining constant output voltage.

The unregulated input voltage must always be higher than the **striking voltage** of the VR tube, which is usually about 25 per cent higher than the working voltage. The tubes commonly used in amateur equipment are listed in the table at the right.

Type	Working Voltage	Striking Voltage
0A3/VR75, 0C2*	75	105
0C3/VR105, 0B2*	105	135
0D3/VR150, 0A2*	150	185

*Miniature type, maximum current 30 ma. Others are octal based; maximum current 40 ma.

Practical Circuits

The various rectifier and filter circuits can be combined in just about any way you please, in order to arrive at some desired set of operating characteristics—output voltage, output current, filtering, and voltage regulation—provided the principles discussed are observed. However, experience has evolved several more-or-less standard arrangements.

Fig. 12-13 is typical of these. It uses a power transformer, T_1, having a center-tapped high-voltage winding, a 5-volt winding for heating the rectifier tube, and a 6.3-volt winding for the heaters of tubes in the associated equipment. There is quite an assortment of these transformers in the parts catalogs. They are generally known as "replacement" transformers because they were designed to be substituted for burned-out transformers in broadcast receivers.

A filter consisting of C_1, C_2 and L_1 is usually sufficient when this circuit is used with low-power transmitters. More filter may be needed for receivers, in which case L_2C_3 can be added as shown. If the circuits can operate at less than the full output voltage, a resistor can be used to replace L_2. Its value would be based on the desired output voltage and current, using Ohm's Law. For example, the supply may be capable of delivering 300 volts at 80 ma. at point X, but the equipment with which it is to be used needs only 250 volts at 80 ma. The difference therefore can be "lost" in a resistor. The resistance would be 50 (the difference in voltage) divided by the current, 0.08 ampere, or 625 ohms.

Fig. 12-14 is a circuit useful for small-current applications, such as "gadgets" using one or two tubes. T_1 is a "booster" transformer—so called because this type was introduced in the TV preamplifiers or "boosters" popular some years ago —having a low-current (10 to 25 ma.) secondary giving 115–125 volts, plus a 6.3-volt secondary for heating one or two tubes. Here, too, the amount of filtering needed depends on the application. R_2 and C_3 can be omitted if the filtering requirements are not severe.

The same type of transformer is useful for grid-bias supplies for receivers and amplifiers where the tubes to be biased do not take grid current. A Class A or AB_1 amplifier is an example. The

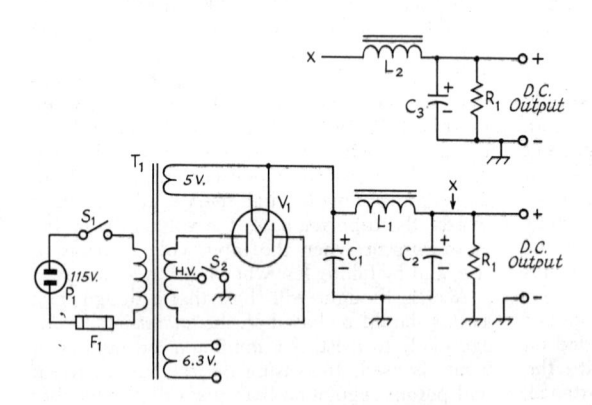

FIG. 12-13—Full-wave center-tap rectifier with capacitor-input filter. This circuit is suitable for receivers and low-power transmitters not requiring more than approximately 350 volts d.c. at 120 ma. Extra filter section, inserted at point X, will improve filtering for receivers but should not be needed if RC decoupling circuits are used in low-level audio stages.

C_1, C_2, C_3—40-μf. electrolytic, 450 volts d.c. working.

F_1—Fuse, 2 to 3 amperes, depending on power (see text).

L_1, L_2—Filter choke; value not critical; 8 to 16 henrys recommended, rated for current to be taken from supply.

P_1—Line cord and plug.

R_1—50,000 ohms, 10 watts, wire-wound (value not critical).

S_1, S_2—S.p.s.t. toggle, 3 amp., 250 volts. (S_2 is for turning plate power on and off while filaments remain heated; may be omitted if not needed).

T_1—Power transformer. High-voltage winding: for continuous operation, 750 volts c.t. max., at d.c. output current rating required; for intermittent load such as a keyed transmitter, 630 volts c.t., max., at required current. Filament voltages: 5 volts at 2 or 3 amp., depending on type of rectifier tube; 6.3 volts with minimum current rating equal to current required by equipment.

V_1—5Y3G for d.c. output currents up to 100 ma. 5U4G for currents over 100 ma.

Power Supplies

FIG. 12-14—Low-current, low-voltage supply for accessory equipment having light power demand. Number of tubes that can be handled depends on which of the two power transformers listed below is chosen. Alternative filter circuit for bias supplies is shown; the smaller power transformer will suffice for this purpose. The bias supply is suitable for amplifier circuits in which no grid current flows.

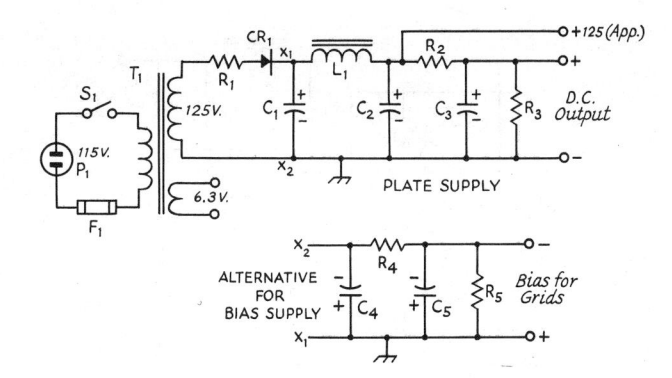

C_1–C_5 inc.—20- to 40-μf. electrolytic, 150 volts.

CR_1—Selenium rectifier, 50 ma., 130 volts r.m.s.

F_1—0.5-ampere fuse.

L_1—20 henrys at 15 ma. or 16 henrys at 50 ma., depending on power transformer used.

P_1—Line cord and plug.

R_1—10 ohms, 1 watt.

R_2—Select to drop voltage to a desired lower voltage, when required. A choke similar to L_1 may be substituted if largest possible output voltage is wanted. R_2 and C_3 needed only for audio circuits where hum must be very low.

R_3—0.1 megohm, 1 watt.

R_4, R_5—Total resistance app. 25,000 ohms. Select R_5 for desired bias voltage, making up the difference in R_4. Use 1-watt resistors.

S_1—S.p.s.t. toggle or slide switch.

T_1—For bias supply and equipment using not more than two 0.3-ampere-heater tubes: 125 volts at 15 ma.; 6.3 volts at 0.6 amp. For heavier demand: 125 volts at 50 ma., 6.3 volts at 2 amp.

resistances of R_4 and R_5 can be proportioned to give the desired bias voltage. Bear in mind that the d.c. voltage between the points X_1X_2 will be about 10 per cent higher than the r.m.s. secondary voltage of the transformer when its rated d.c. current flows through R_4 and R_5. R_5 can be an adjustable resistor or control if smooth adjustment of bias is necessary.

"Economy" Power Supply

The replacement transformer is built for continuous operation over long periods of time. Amateur transmitters are operated intermittently, and the plate power supply actually is delivering full output for only a relatively small fraction of the total time during an operating period. This means that the supply transformer runs at a temperature considerably below what it is designed to stand. You can take advantage of this low **duty cycle** to get more power from the transformer than its normal rating.

Fig. 12-15 is the circuit of an "economy" plate supply operating on this principle. The circuit is used in a number of transmitters in this book. The replacement transformer, T_1, is made to deliver double its normal d.c. output voltage by using a bridge rectifier. Intermittently, the rated current can be taken even though the rating was based on center-tap rectification (where each half of the winding delivers half the total current). Two of the rectifiers in the bridge must have their cathodes connected to opposite ends of the high-voltage winding. Tubes with indirectly-heated cathodes are used here so the heaters can be run from the 6.3-volt winding on the transformer. The remaining rectifier is the usual two-diode type operating from the 5-volt winding.

A second positive voltage—approximately half the high-voltage output—can be taken from the center-tap on the high-voltage winding in this circuit. This can be used for the plates of low-power stages in the transmitter. If such an output voltage is not needed, L_1, L_2, C_1 and C_2 can be omitted.

Choke-input filters are used in both outputs. This not only improves the voltage regulation but also reduces transformer heating because the rectifier peak currents are low.

Resistors and Voltage Dividers

In many cases you may need a lower voltage than the supply develops. One instance is the screen voltage for a tetrode or pentode. Except for the audio power stage (the pentodes used for this usually operate with the same voltage on plate and screen) screen-grid tubes operate with less than half as much voltage on the screen as on the plate. The customary way of reducing the voltage is by a series dropping resistor; the way to determine the right value was explained in earlier chapters.

Whatever the reason for needing a lower voltage, the dropping resistor method can be used *provided* either that the current is constant, or that it is permissible for the voltage to vary if the current is subject to change. Suppose you have a 250-volt supply and need 100 volts at a load of 15 ma. By Ohm's Law the resistance that will do it is $R = E/I$, or 150/0.015 (to drop 150 volts at 15 ma.). This is 10,000 ohms. But suppose that the current can be expected to vary

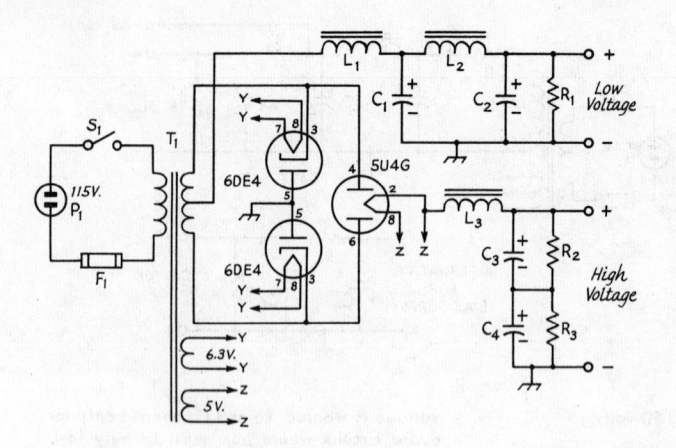

FIG. 12-15—"Economy" power supply for transmitters. Delivers approximately 600 volts, depending on power transformer voltage, at intermittent loads up to 120 ma. Low voltage is approximately 300 at 30 to 40 ma.

C₁, C₂—20-to 40-µf., electrolytic, 450 volts working.
C₃, C₄—40-µf. electrolytic, 500 volts working.
F₁—3-ampere fuse.
L₁, L₂—8 or 16 henrys, 50 ma.
L₃—10.5 henrys, 110 ma.
P₁—Line cord and plug.
S₁—S.p.s.t. toggle, 3 amp., 250 volts.
T₁—Power transformer, high-voltage winding 700–750 volts, c.t., 120 to 150 ma.; 5 volts at 3 amp.; 6.3 volts at 3.5 amp. or more. (Separate filament transformer must be used for equipment if transformer is rated at 3.5 amp. since two 6DE4s require 3.2 amp.)

between 5 and 20 ma. At the lower current the voltage drop will be only 10,000 × 0.005, or 50 volts; at the higher current it will be 10,000 × 0.02 or 200 volts. Instead of 100 volts, you have 250 − 50 = 200 volts with the low current and 250 − 200 = 50 volts with the high current. Such a shift in voltage with current may be far outside tolerable limits.

This situation can be improved by using a **voltage divider** instead of a series resistor, as shown in Fig. 12-16. The idea here is to make the current through R_2 as large as possible compared with the current to the actual load connected to the low-voltage terminals. Suppose it is made three times as large in the example above. Then the current through R_2 will be 45 ma., and since the voltage is to be 100, R_2 will be 100/0.045 = 2200 ohms. The current through R_1 is 45 + 15 ma., or 60 ma., and since a 150-volt drop is required, R_1 becomes 150/0.06 = 2500 ohms. The voltage change with a change in current through the load will be smaller, although it still won't be zero by a considerable margin. (It isn't possible to calculate it exactly without knowing exactly how the load behaves when the voltage applied to it is changed.)

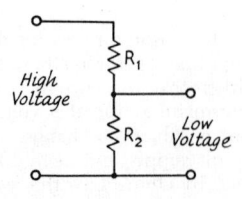

FIG. 12-16—Voltage divider, for obtaining a low voltage from a higher-voltage supply.

This improvement carries with it a very appreciable waste of current and power. As you can see, with the dropping resistor the power taken from the supply was 250 volts times 0.015 ampere, or 3.75 watts; with the voltage divider this rises to 250 × 0.06, or 15 watts. The divider resistors have to dissipate the extra power, and more heat is generated in the equipment.

Generally speaking, it is best to avoid the use of voltage dividers if the current is more than a couple of milliamperes. They are wasteful, and much better voltage regulation can be secured by using VR tubes whenever the voltage and current are within VR capabilities.

Fusing

If a component breaks down, either in the power supply itself or in the equipment it is serving, the result may be a short-circuit, or near short, on the high-voltage supply. This puts a heavy load on the power transformer and rectifier. Both may be permanently damaged if the overload persists for any length of time.

A fuse in the transformer primary circuit will protect the equipment from serious damage when this happens. Fuses are shown in the practical circuits of Figs. 12-13, -14 and -15. The fuse may be installed in the power supply itself—mountings of various types are available—or in the line plug ("fuse-in-plug" type). The fuse rating should be related to the total power the supply is supposed to deliver. For example, a single transformer might supply the following, under normal conditions, to a transmitter:

Power Supply Construction

Plate power—600 volts at 200 ma. = 120 watts
Rectifier cathode—5 volts at 3 amp. = 15 watts
Transmitter
   heaters—6.3 volts at 2 amp. = 12.6 watts
            Total   147.6 watts

Allowing about 20 per cent for transformer and other losses, the total primary power then would be, in round figures, 180 watts. At 115 volts, the primary current will be approximately 180/115, or somewhat over 1.5 amperes. You shouldn't try to fuse more closely than about twice the normal current, so a 3-ampere fuse would be reasonable in such a case. A lower-current fuse would tend to blow on slight overloads such as might occur with temporary mistuning, or on transients set up by switching.

In choosing a fuse for a supply incorporating two or more transformers, as when separate filament transformers are used, remember to sum up *all* the estimated power if you use just one fuse for the entire supply.

Bypassing

Receiver power-supply transformer primaries are often bypassed to the chassis by capacitors of about 0.005 or 0.01 µf., as shown in Fig. 12-17. This is done to help prevent r.f. noise on the power line from getting into the receiver circuits. Paper capacitors rated at 400 volts or more are suitable; when using them, make sure that the end marked "outside foil" is connected to the chassis. Sometimes a narrow band is marked around the capacitor at the end to be grounded, instead of the legend.

If you use ceramic capacitors, get the type intended for use on a.c. The ordinary variety, while rated for 500 volts or more on d.c., does not always stand up well when used on the a.c. line. But whatever the type used, put the fuse and on-off switch on the *line* side of the capacitors, as shown in the diagram.

Layout and Construction

In building a power supply you don't have the short-lead problems that go with an r.f. layout. Neither do you have to worry about stray couplings within the supply itself. The frequency is too low for such things to bother. There are two principal concerns—insulation, and heat.

Use wire with heavy-enough insulation to eliminate any possibility of breakdown. The leads on the power transformer will give you a good clue to what is adequate. The hook-up wire you buy also has a voltage rating, usually specified in the catalogs or on the package. Almost all such wire will stand at least 1000 volts, so the wire itself seldom need worry you. More important is your way of *using* the wire. Use insulated tie-points or standoff insulators wherever you join two wires together, in order to support the connection, and make sure that no bare conductor —wire or solder—can short-circuit to the chassis or another conductor. Ordinary bakelite tie-points—with no solder overflow—will take voltages up to about 1000 without difficulty.

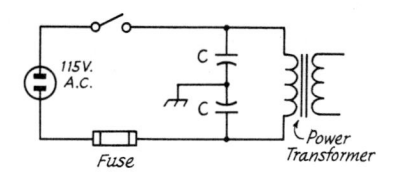

FIG. 12-17—Bypassing the transformer primary to reduce line noise.

Try to give the high-temperature components, such as rectifier tubes and power resistors, a spot on the chassis where the air can circulate around them. A location near the chassis edge is best. Don't mount electrolytic capacitors right alongside such components. Electrolytics deteriorate at high temperatures, so mount them in a place where they won't get direct heat from the hotter elements of the supply.

If the supply is part of a complete piece of equipment, as is the case with many of the transmitters described in this book, mount the supply components near one edge of the chassis and as far as possible from r.f. circuits, such as oscillators, whose performance is affected by temperature. You'll find more on this subject in Chapters 8 and 10.

Safety

One of the most important "don'ts" in power supply construction—and in the construction of any equipment, for that matter—is this: Don't have any exposed leads or terminals! Make sure that *nothing* with voltage on it is outside the set where it can be touched. Use covered terminals, or better still, sockets and plugs for external connections. And *don't* ever make the mistake of wiring a male connector so it has "live" prongs. Such a connector should always be wired to be "dead" when disconnected. Use the female type, such as a socket, so your fingers can't possibly touch a hot lead if the power happens to be on when the connector is pulled out.

Chapter 13

Modulators and Speech Amplifiers

A good many beginners have trouble grasping the ideas that lie at the bottom of modulation. And why not? There isn't any easy way to picture the process by which radio-frequency sidebands develop out of audio frequencies.

But this confusion needn't extend to the modulator itself. A modulator is just an audio-frequency amplifier. The simpler ones are quite like the ones you find in common every-day broadcast receivers. And even the bigger ones, such as might be used for plate modulation of transmitters of the power levels we're considering in this book, are the same in principle as those you find in hi-fi gear. The fact is that the hi-fi amplifier usually has a much more complicated circuit than is needed for modulating an amateur transmitter.

The object of modulator circuit design is to obtain enough audio power output for modulating the r.f. amplifier, with as little distortion as circumstances will permit. The audio power developed by a microphone is far too small to be useful directly. It has to be amplified up to the necessary level. The audio stage that acts directly on the r.f. amplifier is called the **modulator**, and the amplifier stages between the microphone and modulator make up the **speech amplifier**.

There are innumerable audio-frequency amplifier circuits, as well as combinations of tubes and component values. For our purposes they can be boiled down to a few "standard" circuits that will do everything needful.

The Speech Amplifier

The audio output voltage that you can get from amateur-type crystal and ceramic microphones—the types generally used—is small. Something like 30 millivolts (0.03 volt) is typical, for average voice intensity and a speaking distance of a couple of inches. What the modulator

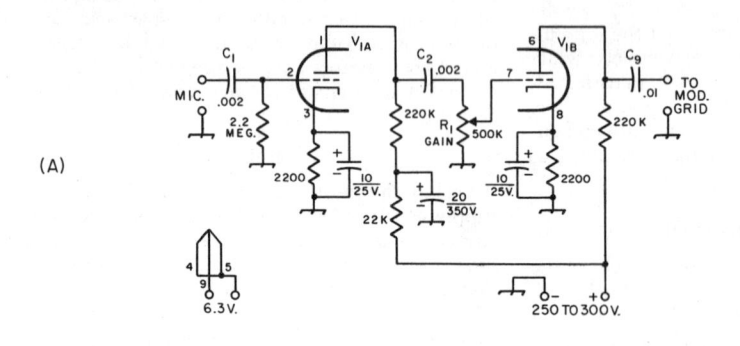

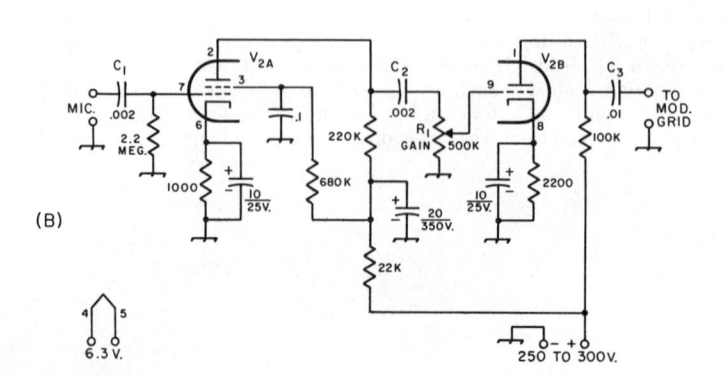

FIG. 13-1—Two commonly-used speech amplifier circuits. Either one has enough voltage gain to drive the grid of a Class A single-tube modulator, using resistance coupling to the modulator grid. Capacitances are in $\mu f.$, resistances are in ohms; fixed resistors are ½-watt rating. Capacitors with polarity marked are electrolytic; others may be paper, mica, or ceramic, rated at 400 volts or more. R_1, the gain control, should have an audio taper.

V_1—12AX7 or 7025.

V_2—7199 (6AN8 may be used; see Fig. 13-5 for pin connections).

236

Speech Amplifiers

needs in the way of audio driving voltage on its grid (or grids, when two tubes are used in push-pull for plate modulation) depends on the modulator tubes and the modulation system. It may range from as little as 5 or 6 volts (peak) to as much as 100 volts. In the latter case the speech amplifier is called upon to amplify the audio voltage in the ratio of 100 divided by 0.03, or somewhat over 3000 times.

The larger values of voltage are needed only with push-pull modulators. These customarily have their grids driven through a center-tapped transformer that steps up the voltage by a factor of at least 2 to 1. Thus in nearly all cases the tube gain in the speech amplifier needs to be only about 1500.

Practical Circuits

The two speech-amplifier circuits in Fig. 13-1 have voltage gains of well over 3000, or more than twice as much as is needed. It is always advisable to have something in reserve, but it doesn't pay to design too much gain into an amplifier. Excess gain can bring on troubles like hum, feedback and "howling," microphonics, and so on—most of which can be avoided by using a minimum number of stages and providing just a moderate safety factor in gain.

Only one dual tube is needed in either circuit of Fig. 13-1. Which circuit to choose is a matter of personal preference; in actual performance, there isn't much to distinguish one from the other. In the form shown in Fig. 13-1 the output tube is resistance-coupled to the modulator. This is satisfactory when the speech amplifier is driving a single-tube modulator such as would be used for screen modulation of the r.f. stage. When a push-pull stage is to be driven, transformer coupling should be used, as in the circuits given a little later.

In both circuits the gain control, R_1, is between the two amplifiers. This is the optimum place, because it tends to give the best ratio of voice signal to hum and other noise that may originate within the amplifier, as compared with putting the gain control right at the microphone input. The gain control shouldn't be put in a later stage, such as the modulator grid, because there is some danger that the second stage in the speech amplifier may be over-driven if the amplitude of the signal at its grid can't be controlled.

The 22,000-ohm resistor in the plate circuit of the first stage is a decoupling resistor, bypassed by the 20-μf. capacitor. These R and C values don't have to be followed exactly; they can be varied over a 2 to 1 range, at least, without affecting the operation of the amplifier. Also, the 10-μf. value for the cathode bypass capacitors is not sacred. Preferably, it shouldn't be *smaller* than 10 μf., but any larger value will be OK.

Values of the grid and plate resistors can be varied, too, but here whatever you do will affect the amplifier gain. It's best to stick to those given.

Screen Modulator

The circuit given in Fig. 13-2 is recommended for modulating the screen of an r.f. amplifier. It is a perfectly straightforward method of applying audio power to the screen of the r.f. stage. It isn't tricky to adjust, as the controlled-carrier circuits are if you want minimum distortion and a high percentage of modulation. With this circuit, as we said in Chapter 5, the carrier output is a mite less than what it is *possible* to obtain with controlled-carrier, but the difference is negligible. And since you can be sure of 100 per cent modulation, your talk power often will be greater with this circuit than it will be with some forms of carrier control.

The circuit includes provision for switching between c.w. and phone. For phone you have to lower the screen voltage as described in Chapter 5. R_1 lets you do this. The complete adjustment method has been described in that chapter. All we need to say here is that you first adjust the r.f. amplifier with the switch in the c.w. position, as described in Chapter 5. Then switch over to phone and adjust the slider on R_1 to cut your plate current to one-half its c.w. value. *Don't* touch R_1 with the power on! Shut off power, make a trial adjustment of the slider, and then switch on again to see whether you're near the right plate-current value.

Combining Modulator and Speech Amplifier

Either of the speech amplifier circuits in Fig. 13-1 can be used. Choose the one you want and simply tie the output connections to the corresponding input connections in Fig. 13-2.

FIG. 13-2—Screen modulator circuit. Capacitors are electrolytic.

R_1—Slider-type wire-wound resistor.

S_1—2-pole, 2-position rotary, or double-pole, double-throw toggle.

T_1—Driver transformer, 1 to 1 turns ratio, total primary to total secondary, primary 40 ma. (Stancor A-4752 or equivalent). Use total secondary.

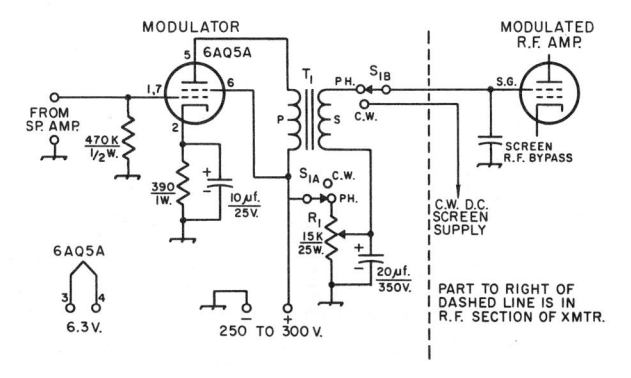

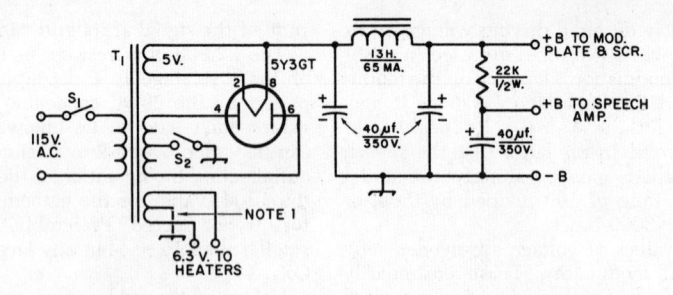

FIG. 13-3—Power-supply circuit for screen modulator and speech amplifier. Capacitors are electrolytic. Any filter choke having an inductance of 10 henrys or more and capable of carrying 60 ma. d.c. can be used instead of the value specified.
Note 1: See text discussion on grounding of heater circuits for minimum hum.

S_1, S_2—Single-pole single-throw toggle. T_1—Power transformer; 500 volts c.t., 70 ma.; 5 volts at 2 amp.; 6.3 volts at 2.5 amp.

The same power supply can be used for both the speech amplifier and modulator. The supply should deliver about 80 ma. at 250 to 300 volts d.c., plus 6.3-volt heater power for whatever tubes are used. If you can get the power you need from your transmitter's power supply, fine. If not, Fig. 13-3 is a typical circuit.

Bear in mind the check for 100 per cent modulation with this circuit: the plate current of the modulated r.f. amplifier should just show a small flicker on your voice peaks. Adjust the speech amplifier's gain control so the plate current flickers only on *peaks*. If the plate current really jumps when you talk, you're overmodulating. An r.f. output indicator using a diode rectifier and d.c. milliammeter won't show any change in reading, if the meter is linear. But if you use a lamp as a dummy antenna, you'll see the brightness come up with every word. The lamp responds to the *power* in your signal, but a linear voltmeter reads *average voltage*. As you will remember from Chapter 5, the average amplitude doesn't change with modulation, although the power actually increases.

Building Speech-Amplifier Modulators

There are fewer sensitive points in audio than in r.f. equipment, so you can use just about any chassis layout you please, within reason. Your principal antagonist is hum, which usually can be traced to the tube heater circuits. It isn't a serious problem in speech equipment, though, because we don't need to amplify the very low frequencies that would be needed for good reproduction of music. In fact, it is good practice to discriminate against these lower frequencies in the amplification, both to avoid hum and to give your voice more "punch" on the air. This can easily be done by using rather small values of coupling capacitance between resistance-coupled stages. In the circuits of Fig. 13-1, C_1 and C_2, the audio coupling capacitors, are only 0.002 μf.—quite small, for audio work.

Some care must be used in installing the heater wiring. Keep the heater leads away from the other amplifier wiring as much as you can. It is a good idea to run heater wires along a fold in the chassis, and have them leave the fold only where it is necessary to bring them out to tube sockets. Keep the heater wiring close to the chassis, and use a steel chassis in preference to aluminum. The steel tends to reduce the magnetic field around the wiring.

You can connect one side of the heater winding of the supply transformer to the chassis and similarly ground one heater pin on each tube socket, and in most cases you'll have no trouble from hum when this is done. An alternative arrangement which sometimes is better, especially when the amplifier gain is high, is to ground the center-tap on the transformer's heater winding (if it has such a center-tap) and *not* ground the heaters of the tubes. The leads from the transformer winding to the tube heaters should be twisted together so the magnetic fields around the wires cancel each other as much as possible.

It is also good practice to use a shielded lead from the microphone connector to the grid of the first speech-amplifier stage. Just a tiny amount of hum coupled to the first tube's grid will show up only too plainly in the modulator's output, because the voltage gain from the first grid to the output is high. The shield around the grid lead must be grounded to the chassis. Ordinary shielded wire can be used; you don't have to make any special shields.

Modulators

Preventing R.F. Feedback

If a little r.f. voltage gets into the speech amplifier or modulator it can raise hob with the audio system. The microphone input circuit is the most vulnerable spot, because the microphone cable acts as a small antenna for picking up stray r.f. A shielded microphone cable is an absolute necessity, and so is a shielded connector on the mike cable. A shielded mating connector must also be used on the speech amplifier chassis. (All these things must be done to prevent stray hum pick-up, too.) Inexpensive phono connectors are satisfactory, and so are the regular microphone connectors you'll find in the catalogs.

Don't have any exposed wiring in the audio circuits. Keep all the wiring inside the metal chassis. It's a good idea, too, to use a shield on at least the input tube in the speech amplifier, and a simple r.f. filter as shown in Fig. 13-4 also helps.

Even when these points are carefully observed you may have trouble from **r.f. feedback**. It's easily recognized because a squeal is set up whenever you try to open up the audio gain control. (You may not hear it yourself, unless you're monitoring your transmissions, but the fellow at the other end certainly will!) If the audio system has been carefully shielded, feedback is almost always caused by a "hot" chassis — the transmitter case is not at ground potential for r.f. Sometimes you can actually draw r.f. sparks from the chassis.

It's easy to establish whether a "hot" chassis is the cause of r.f. feedback. Disconnect the antenna from the transmitter and substitute a dummy antenna. Monitor your signal. If your audio shielding is adequate, turning up the gain won't cause any howl. (If the shielding *isn't* adequate, you may get one, in which case you'll have to "cool down" the amplifier before you can put the setup on the air.) If there's no howl, the set itself is satisfactory. The blame lies in your antenna system.

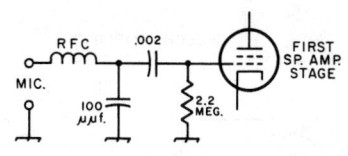

FIG. 13-4—Filter for preventing r.f. from reaching the grid of the first speech amplifier tube. The filter consists of the r.f. choke and 100-$\mu\mu$f. capacitor. The *L* and *C* values are not critical, but the choke should be one that is effective in the frequency range of the transmitter. An inductance of 1 to 2.5 mh. is satisfactory for frequencies below 30 Mc. At v.h.f. a 50-μh. choke can be used.

Most transmitters are installed at some distance from actual ground. The path to real earth is too long to be ignored; it isn't negligibly small compared with the transmitting wavelength. A ground connection can't bring the transmitter chassis actually to ground potential for r.f. That being so, the ground lead will try to become a part of your antenna system if it's given a chance. To overcome this, and prevent r.f. feedback, you have to isolate the transmitter chassis from the antenna system.

Most "hot-chassis" r.f. feedback troubles can be charged to hooking the transmission line directly to the r.f. output terminals of the transmitter. Sometimes it helps to take the ground connection off the transmitter, and sometimes doing this makes matters worse. But *don't* omit grounding the chassis; this is an important safety precaution, beside which feedback troubles are insignificant. The way to fix it is to use a transmatch. The transmatch, properly installed and adjusted, will cure the feedback no matter how far your transmitter is from actual earth. Chapter 11 tells you how to make one and how to use it.

Plate Modulators

We really should say "plate-and-screen" modulators, because final r.f. amplifiers are nearly always screen-grid tubes, and the screen must be modulated along with the plate. If it isn't, the modulation won't be linear. However, it's easier to call it "plate" modulation; just remember that "and screen" is understood whenever the shortened term is used.

A circuit that will suffice for any transmitter running at r.f.-stage inputs up to at least 120 watts is shown in Fig. 13-5. The thing that determines the audio power output from the circuit is your choice of modulator tubes, V_2V_3, and the voltages at which you run them. The same speech amplifier will serve in all cases. Table 13-I lists combinations of tubes and voltages for various output powers. In all cases the modulator tubes are operated as **Class AB**$_1$ amplifiers.*

The speech amplifier is identical with the pentode-triode circuit of Fig. 13-1, with one exception: the second stage is not resistance-coupled to the modulator but is coupled through an audio transformer having a center-tapped secondary. This is necessary because the modulator tubes must be operated in push-pull. The dual-triode circuit of Fig. 13-1 is not suitable in this case. The reason is that the tubes are high-μ triodes, which do not work well with transformer

---

*As mentioned in Chapter 2, a Class AB amplifier is one operated between the Class A and Class B conditions. The subscript 1 means that the amplifier is not driven into grid current, and so takes no power from the preceding amplifier stage.

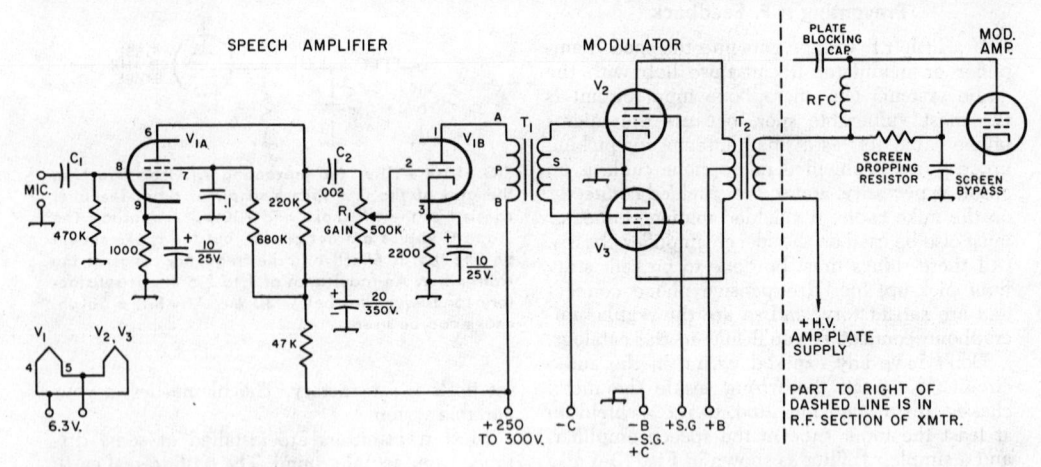

FIG. 13-5—Plate modulator circuit using a push-pull Class AB$_1$ modulator. Capacitances are in µf.; capacitors with polarities marked are electrolytic; others may be paper, mica, or ceramic, 400-volt or higher rating. Resistors are ½ watt.

R$_1$—Audio-taper control.

T$_1$—Interstage audio transformer, single plate to push-pull grids; turns ratio 1 to 2 or 1 to 3, primary to total secondary (Stancor A-52-C, A-53-C, or equivalents).

T$_2$—Multi-match modulation transformer, power rating according to modulator tubes selected.

V$_1$—6AN8 (7199 may be used; see Fig. 13-1 for pin connections).

V$_2$, V$_3$—See Table 13-1.

coupling. In the pentode-triode combination V$_{1B}$ is a medium-µ triode; this type does quite nicely with transformer coupling.

Fixed bias is shown on the modulators in Fig. 13-5. In general, it will be possible to get more power output if you use fixed bias instead of cathode bias. However, in cases where enough power can be obtained without a bias-voltage supply, it is easy to substitute a cathode-bias resistor as shown in Fig. 13-6.

Output Power

The output figures given in Table 13-I are for the tubes alone. There will always be some loss of power in the modulation transformer, since it isn't perfect, and there may also be some small reduction in power output because you can't always get exactly the right turns ratio in the modulation transformer, even with the "multi-match" type. The modulator should have a power-output capability equal to one-half the total d.c. power input to the modulated stage (its d.c. plate voltage multiplied by the sum of the plate and screen currents). Thus if this input is 60 watts the modulator should have an output of 30 watts. To allow for the losses mentioned above the modulator tubes should be able to deliver 10 to 20 per cent more than this to the transformer primary—perhaps 35 watts in this example.

If you don't find exactly the power-output figure you want in Table 13-I, take the next larger one. A modulator can always be operated at *less* than its maximum capability; just keep your speech amplifier gain down to the point where you get the output you need and no more. Operating a modulator below its maximum ratings will result in a cleaner signal because the distortion nearly always decreases at lower output levels.

Voice Waves *vs.* Sine Waves

If you measure the average current in a half cycle of a pure tone (a sine wave) the value will be found to be approximately 0.64 times the maximum or peak value of the wave, as you saw in Chapter 2*. But voice waves have some odd-looking shapes, and they don't follow the same

FIG. 13-6—Cathode-resistor bias for the Class AB$_1$ modulator. See Table 13-I for data. A and B in this circuit connect at A and B, respectively, in Fig. 13-5.

*A half cycle is specified because the average value of current or voltage in a whole cycle of any a.c. wave is zero. There is just as much current flowing in one direction during the positive half cycle as flows in the other direction during the negative half cycle. Taken separately, each half cycle has to have the same average amplitude as any other, whether positive or negative.

Modulators

Table 13-1
Push-Pull Class AB₁ Modulator Tube Data

Mod. Tubes (2)	Plate Volts	Screen Volts	Grid Bias Volts‡	Cath. Res. Ohms	P-to-P Load Ohms	No-Signal Cathode Current Plate ma.	No-Signal Cathode Current Screen ma.	Maximum-signal Cathode Current* Plate ma.	Maximum-signal Cathode Current* Screen ma.	Power Output Watts
6AQ5A	250	250	−15	——	10,000	70	5	79	13	10
6BQ5	250	250	——	130	8000	62	7	75	15	11
	300	300	——	130	8000	72	8	92	22	17
7189	400	300	−15	——	8000	15	1.6	105	25	24
7868	300	300	−12.5	——	6600	74	10	116	28	24
	350	350	−15.5	——	6600	72	9.5	130	32	30
	450	350	−16.5	——	6600	60	7.2	142	26	38
	450	400	−21	——	6600	40	5	145	30	44
	450	400	——	170	10,000	86	10	94	20	28
6L6GC	360	270	−22.5	——	6600	88	5	132	15	26.5
	450	400	−37	——	5600	116	5.6	210	22	55
7027A	450	350	−30	——	6000	95	3.4	194	19.2	50
	540	400	−38	——	6500	100	5	220	21.4	76
	380	380	——	180	4500	138	5.6	170	20	36
	425	425	——	200	3800	150	8	196	20	44
807 or 1625	400	300	−30	——	6800	56	2	143	16	36
	500	300	−32	——	8200	44	1	141	15	46
	600	300	−34	——	10,000	36	0.6	139	15	56
	750	300	−35	——	12,000	30	0.5	139	16	72

*Current with sine-wave tone signal; see text.
‡Adjust bias for specified no-signal plate current.
Bases: 6AQ5A, 7-pin miniature; 6BQ5, 7189 and 7868, 9-pin miniature; 6L6GC and 7027A, octal; 807, 5-pin medium.

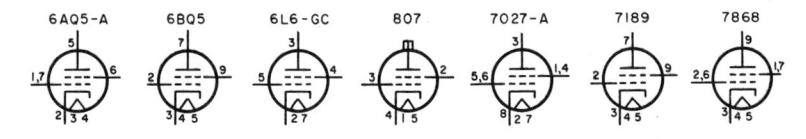

rule. A voice wave might look like the lower drawing in Fig. 13-7. Here the average current is only 0.3 times the maximum value, although the maximum itself is exactly the same as in the sine wave in the upper drawing. If the peak is 100 ma., for example, the average current with the sine wave would be 64 ma. and the current for the voice wave would be 30 ma.

There is another point, too. Voice intensity always varies throughout a word. The peak value we have in mind here is the *largest* peak that is reached at *any* time. This largest peak is what determines the maximum percentage of modulation. Therefore in the example above a meter capable of reading the average current would register 30 ma. only now and then—on what are termed **voice peaks**. The rest of the time the current would be appreciably lower.

Voice waves such as these are what the modulator develops and delivers to the r.f. amplifier through the modulation transformer. As we have said, the average current or voltage over a *whole* cycle is zero, so there is no average change in the d.c. current or voltage in the modulated amplifier. The current and voltage decrease just as much on the modulation down-swing as they increase on the up-swing, so a d.c. meter needle doesn't move when you modulate; it can't follow the rapid variations. But conditions are different in the modulator. Here a d.c. plate-current meter would be connected in the plate-supply lead to both tubes. Remember that in a push-pull amplifier the plate current of one tube swings up on one half cycle and the plate current of the other swings up on the other half cycle. From where the meter sits, this looks like full-wave rectification, since the plates of the tubes are in parallel for d.c. The current is always flowing in the same direction through the meter, thus the meter reads the sum of the average currents in both tubes.

Modulator Operating Currents

If the tubes are biased to cutoff (pure Class B

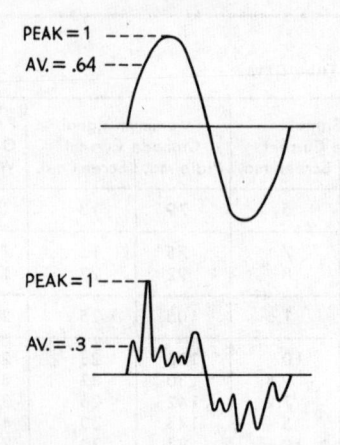

FIG. 13-7—Sine-wave and voice-wave currents or voltages that reach the same peak amplitude have widely-different average values.

operation) there will be no plate current when there is no signal on the grids. When a signal is applied, the plate meter will read in exactly the way we have described above. That is, it will read about half as much with a voice signal as with a pure-tone signal, when both have the same peak value. If the modulator bias is not enough to cut off the plate current with no signal there will be a **resting current** which shows on the meter when you're not talking.

Modulators generally do operate with some value of resting current, since this results in less distortion than operation right at cutoff. However, in all Class AB_1 amplifiers the plate current, when the modulator is delivering full output, is larger than the resting current. Table 13-I gives both the resting current and current at full output for various selected sets of operating conditions. The full-output figure is for a *tone* signal, not voice. With voice, the current won't rise that high. If the resting current is large, as in the third set of operating conditions given for 7027-As, there will be little observable change from the resting current of 138 ma. even when you hit a voice peak. On the other hand with the 807 modulators, there is a large difference between resting and full-output currents, and the current on voice peaks will rise momentarily to around half the tone full-output value—to 60 or 70 ma. in this case.

The Modulation Transformer

With a wide selection of modulator tubes and r.f. amplifier tubes available, together with an assortment of plate voltages and plate currents, no single design will serve as an all-purpose modulation transformer. Chapter 5 has showed you how to calculate the modulation-transformer impedance ratio (or turns ratio). Generally, you can select the proper ratio by using the chart that accompanies a multi-match transformer. These charts are based on impedances, and are usually arranged so you can find the proper primary connections for a desired plate-to-plate load impedance for the modulator, and then go along a row or column to find the secondary taps to use for the modulating impedance your r.f. amplifier will represent. Don't forget to include the screen current in calculating the modulating impedance of the r.f. tube.

The values rarely will come out exactly as you want them. You may have decided, for example, to use a pair of 7868 tubes for modulating a 6146. According to Table 13-I, the 7868s should have a plate-to-plate load of 6600 ohms to deliver 38 watts output. The closest figure on the chart might be 6000 ohms. The 6146 might be running at maximum phone ratings of 600 volts on the plate, a plate current of 112 ma., and a screen current of 8 ma.—a total of 120 ma., and a total power input of 72 watts. The modulating impedance is 600/0.12, or 5000 ohms. Now since the impedance *ratio* required is 6600/5000, or 1.32 to 1, the 6000-ohm plate-to-plate figure should be divided by 1.32 to find the *apparent* secondary load that will give the best match. This is 4550 ohms, so if the chart has a figure in the vicinity of 4500 ohms for the secondary load, those taps are the ones to use.

Note that it is the impedance *ratio* that is most important, not the impedances in ohms as given by the transformer chart. However, you should always pick out the primary taps that will be as close as possible to the desired plate-to-plate figure. This will ensure that the transformer will work properly. Then, after making the calculations as described above, if you find that you can't get the exact impedance ratio you want, take the nearest one. If there is a choice between two ratios that are about equally above and below the right figure, take the smaller one. This will reflect a lower plate-to-plate load on the modulator tubes* and will give you somewhat more power output than Table 13-I shows. The penalty for this is slightly more distortion. If you use a higher ratio, raising the plate-to-plate load, the power ouput will be reduced to some extent. But don't carry this to the point of selecting a ratio that will make the plate-to-plate load resistance differ from the right figure by more than 20 per cent if you can avoid it.

Average and Peak Power

A plate-modulation system for voice transmission should be designed to supply peaks of audio power that will be sufficient to modulate the transmitter 100 per cent. The average power outputs in Table 13-I contain the peak power necessary for modulating twice that much d.c. power input to the r.f. amplifier. They are based on a pure-tone signal because that type of signal is the simplest and easiest to use as a reference.

Actual voice power, on the average, is less

---

*Remember that the actual impedance looking into the primary of a transformer is determined by the turns ratio and the secondary load resistance -- see Chapter 2.

Modulator Power Supplies

than half that in a pure tone having the same peak power, because of the waveform differences discussed above. The result is that the modulation transformer really doesn't have to handle as much audio power as it is rated for. This lets you get away with using a smaller transformer than you might need if you were going to transmit pure tones. The principal limitation is the fact that the secondary has to carry the direct current that flows to the modulated stage. Some transformers are rated for this direct current, but in some cases the ratings are not given.

In practice, transformers rated at about half the power developed by a sine-wave signal have been used without harm. That is, a transformer rated at, say, 25 watts can be used for voice work with a modulator capable of a pure-tone output of 50 watts.

Modulator Power Supply

Translated into plate-supply requirements for the modulator, the behavior of plate (or cathode) current on voice signals can help make things easy. To estimate the current the supply will have to deliver, divide the maximum-signal current by 2 and compare the result with the resting current. The supply should be capable of delivering the larger of these two currents. If the screen current is to come from the same supply, add it to the plate current.

If, in the operating conditions you select, the maximum-signal current (as given in Table 13-I) divided by 2 is appreciably larger than the resting current, you can get away with a still lighter supply, since this figure represents the voice-peak demand. The average current over a longer period—while you say several words, for example—is still less. Because of this, the power-supply transformer won't heat appreciably more than it does when delivering the resting current alone.

What you *do* need, when the modulator plate current fluctuates widely with a voice signal, is a large capacitance on the output side of the power-supply filter. The current peaks in this type of operation have to come from the final filter capacitor, because the current can't change rapidly enough through the filter choke. This is so regardless of the nominal d.c. rating of the power supply. At least 20 µf. in the output filter capacitor is called for, and more is desirable.

Any ordinary power-supply circuit, such as those described in Chapter 12, can be used for the modulator. If the steady current is not large (low resting current) you can often take the modulator plate power from the supply that feeds the r.f. part of the transmitter, if that supply is reasonably conservative in design. This means, of course, that the modulator then has to work at the same plate voltage as the modulated r.f. stage, so you have to choose your modulator tubes on that basis. If you don't need all the audio power the tubes are capable of giving at that voltage, just don't drive them quite so hard. An amplifier can always be operated at less than its maximum output. Then you won't need as much plate current for the modulator as would be required for maximum output—which is all to the good when a single supply is used for both r.f. and modulator.

Grid Bias for the Modulator

Although it's a bit of a nuisance to have to furnish fixed grid bias—and you do have to use it if you want to get the most power from the modulator tubes, as Table 13-I shows—it's relatively easy with a Class AB$_1$ amplifier. This type of amplifier doesn't take grid current, so the resistance of the bias supply can be quite high without doing any harm to the operation of the amplifier. The required bias can be developed by a d.c. current in the neighborhood of one milliampere from the bias supply. Since the power

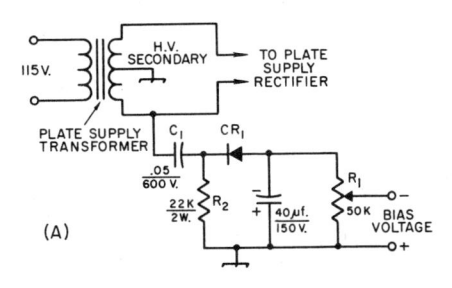

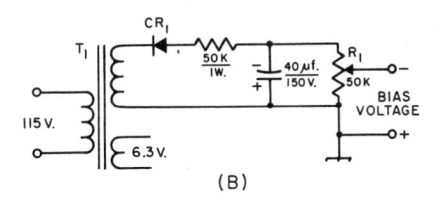

FIG. 13-8—Bias-supply circuits for Class AB$_1$ modulators. A—Bias voltage taken from one side of the plate winding of the power transformer. B—Separate supply using a small transformer with half-wave rectifier. Either circuit will deliver sufficient bias voltage for the modulator tubes listed in Table 13-I.

C$_1$—0.05-µf. paper, 600 volts.

CR$_1$—Selenium rectifier, 20 ma. or more, 400 volts p.i.v.

R$_1$—Linear-taper control.

R$_2$—22,000 ohms, 2 watts.

T$_1$—Power transformer, 125 volts, 15 ma. (6.3-volt winding furnished on this type of transformer not needed for bias supply).

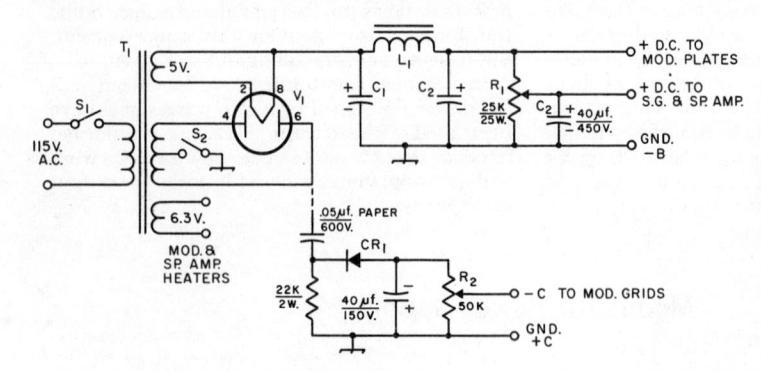

FIG. 13-9—Modulator and speech amplifier power supply. This circuit can be used for modulator tubes requiring up to 400–450 volts. See Table 13-II for data on C_1, C_2, L_1, T_1 and V_1. Bias supply may be omitted if not needed.

CR_1—Selenium rectifier, 20 ma., 400 volts p.i.v.
R_1—Bleeder/voltage-divider resistor, slider type.
R_2—Linear-taper control.
S_1, S_2—S.p.s.t. toggle.

is so small, it can be taken from some existing source with no danger of overloading it.

A simple scheme making use of the plate-supply transformer is shown in Fig. 13-8A. A capacitor, C_1, is connected in series with R_2 across one side of the high-voltage secondary. The values of these two components are selected to result in a d.c. output voltage, across R_1, that is somewhat higher than the bias required. The a.c. voltage is rectified by CR_1 and filtered by an electrolytic capacitor. This simple filter is ample because the d.c. output current flowing in R_1 is so small.

The method of Fig. 13-8A is useful for power-transformer secondary voltages up to 400 volts r.m.s. each side of the center tap. This includes practically all of the "replacement"-type power transformers. When the power supply has a regular transmitting-type plate transformer (for higher-voltage output) it is better to use an entirely separate bias supply. The one in Fig. 13-8B uses an inexpensive small transformer with a half-wave rectifier and a simple RC filter. The 6.3-volt winding on the transformer can be used for anything you like, within its current rating. The primary of the transformer should be connected in parallel with the primary of the filament transformer for the modulator tubes; this ensures that the tubes will be properly biased before plate current can flow.

Screen Supply

The screen voltage does not have to be taken from the supply that furnishes the power for the modulator plates, but this is generally the most convenient source for it. Extra components can be saved when the plates and screens operate at the same voltage. This is possible, as you can see from Table 13-I, for audio power outputs up to more than 30 watts. Higher-power output requires raising the plate voltage above the maximum rated screen voltage for the tubes, with most modulator tubes.

When the screen voltage has to be lower than the plate voltage a voltage divider can be used. It is desirable to have constant voltage on the screen at all signal levels, and especially important that the voltage at *maximum* signal be right up to the value given in the table. If the screen voltage is low you can't get the output power that the table lists. (Of course, this is also true of the plate voltage. At those times when the modulator is delivering full output the plate voltage, too, must have the value specified.)

If you use a voltage divider to get the screen voltage, adjust the tap on the divider so the voltage is right when you're modulating 100 per

Table 13-II
Power-Supply Data (Fig. 13-9)

Power Transformer, T_1		Filter Choke, L_1		Capacitors C_1 and C_2		Rect. Tube V_1	D.C. App. Output Volts at	
Sec. V. C.T.	D.C. Ma.	L @	Ma.	C in μf. @	Working Volts		50 ma.	Max. T_1 Rating
800	200	4.5 h.	200	30	500	5U4GB	500	400
750	150	7 h.	150	40	450	5U4GB	470	380
720	120	7 h.	150	40	450	5U4GB	450	400
600	90	10.5 h.	110	40	450	5Y3GT	315	275

Note: D.c. output voltages given are for guidance only; actual voltages will differ somewhat, depending on transformer and filter choke resistance, and will vary with line voltage. Figures are based on line voltage of 115.

Modulator Power Supplies

cent. The voltage will rise somewhat when there is no audio signal, but if the voltage-divider resistance is reasonably low the rise will not be serious enough to cause any operating difficulties.

Complete Supply for Low-Power Modulators

Fig. 13-9 is a power-supply circuit that combines the features just discussed. This type of circuit can be used with modulator/speech-amplifier systems that do not require more than approximately 400 volts at currents up to 200 ma. This actually would include most of the modulator combinations listed in Table 13-II up to voltages as high as 450, since the actual maximum load current, with voice, will be between 100 and 150 ma. at most. At this current the voltage from the combination given in the first line of Table 13-II should be in the vicinity of 450 volts on voice peaks.

The bias supply section of the circuit can be left out if you don't need fixed bias for your modulator tubes.

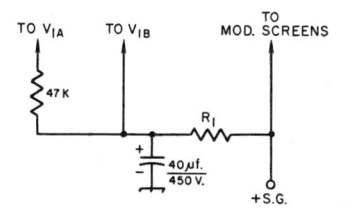

FIG. 13-10—Addition to circuit of Fig. 13-5 when a common power source is to be used for the modulator and speech amplifier and the modulator screen voltage is over 300 volts. R_1 should drop the screen voltage to 250 to 300 volts for the speech-amplifier tubes. Allow 10 to 12 ma. for the speech-amplifier current and calculate the required resistance by Ohm's Law. A typical value for dropping 400 volts to approximately 300 would be 10,000 ohms, 1 watt.

R_1 should be used whether or not the required screen voltage is lower than the plate voltage. It bleeds the charge from the filter capacitors when the set is turned off. If the modulator plates and screens work at the same voltage, the tap on R_1 can be used for the speech amplifier, and should be adjusted to give 250 to 300 volts.

By combining Figs. 13-5 and 13-9 you have a complete modulation system. When the modulator screen-grid voltage is approximately 300, the +B terminal in Fig. 13-5 can be connected directly to the "+S.G." terminal. If the screen voltage is appreciably over 300, use a series resistor and another capacitor as shown in Fig. 13-10.

More About Screen Voltage

If your modulator plate voltage is 500 volts or more, you can take the screen voltage from another power supply. Quite frequently a supply giving 300 volts or so is used for the r.f. exciter stages in a transmitter. Since the modulator screen current is not large, the screens can be connected to such an existing supply.

However, there is one precaution that must be observed when doing this. Voltage should *never* be applied to the screens unless the grid-bias voltage is applied simultaneously or has been applied beforehand. With screen voltage alone the screen current will be excessive and may burn out the screen. Also, if the plate and screen voltages go on without control-grid bias the plate current will be sky-high, again possibly wrecking the tubes. The grid bias must go on *no later* than the screen voltage. Taking the proper precautions here is a matter of how you handle the switching of your power supplies. It's easily worked out, usually.

You can leave the plate voltage on whether or not you have grid bias, so long as there is no screen voltage. The plate current will be too small to do any damage when the screen isn't operating.

Regulated Screen Voltage

When the screen voltage is to be taken from a supply delivering 500 volts or more, it is a good idea to use voltage-regulator tubes. If the voltage has to be dropped 200 volts or more from the plate voltage, the regulation isn't likely to be good enough when the voltage is obtained from a simple resistance divider.

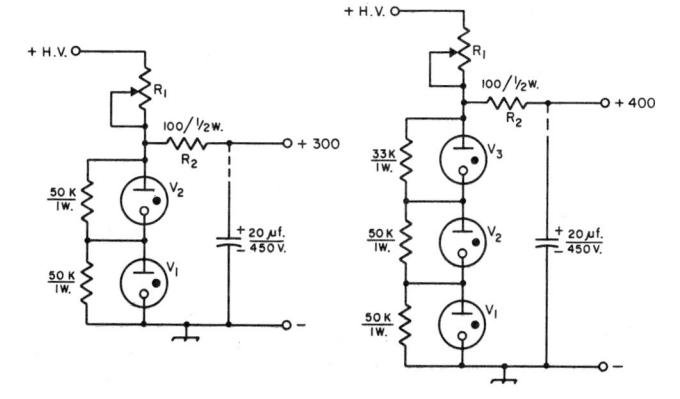

FIG. 13-11—VR tube regulator for screen voltage taken from a high-voltage plate supply. See text for discussion of use of capacitors. Resistors across VR tubes are voltage equalizers to insure proper firing.

R_1—Dropping resistor; see Chapter 12 for information on selection and adjustment.

R_2—For suppressing oscillation.

V_1, V_2—OA2 or OD3, depending on current to be regulated.

V_3—OB3.

The method of using and adjusting VR tubes has been described in Chapter 12. The miniature types can be used if the maximum-signal screen current as given by Table 13-I isn't over 15 or 20 ma. The larger tubes should be used if the maximum screen current is over 20 ma. Considering the current ratings on the two types of VR tubes —30 and 40 ma. respectively—this may seem ultraconservative. Actually, it is not. The screen currents given in the table are *average* currents—what a d.c. meter would read when the screen current is varying at audio frequency. The *peak* of this varying screen current is always higher than the average current, by a factor of 2 or more.

In fact, if the average screen current is over 20 ma. it is advisable to connect a large capacitance across the VR circuit as shown in Fig. 13-11. The capacitor will store enough energy to handle current peaks in excess of those which the VR tubes can regulate by themselves.

Applying Plate Modulation to Manufactured Transmitters

The kind of plate-modulation equipment we have discussed in this chapter readily can be used with low-power manufactured or kit transmitters made for c.w. only or having a screen modulator. The effective carrier output power is seldom more than 10 or 12 watts (and sometimes less) with controlled-carrier screen modulation (see Chapter 5). This can be increased to 40 or 50 watts by using plate modulation, if the transmitter has one of the popular beam tetrodes—6146, 6DQ5, 6DQ6, etc.—in the final stage.

The simplest method is to build a separate modulator, complete with its own speech amplifier and power supply. The circuits of Figs. 13-5 and 13-9 may be combined for this purpose. With an entirely separate modulator you don't need to decipher the transmitter's switching system. It is necessary only to open the d.c. leads to the r.f. amplifier's plate and screen, as shown in Fig. 13-12, and bring out four leads. When S_1 in this diagram is in the "c.w." position the transmitter will operate exactly as it did originally. And it can easily be restored to its original circuit if you come to the trading-in point.

The wires can be brought out in regular four-wire cable which, for convenience, can be terminated in a female four-prong cable connector. A male four-prong connector on the modulator chassis will let you disconnect the two whenever necessary.

The part of the circuit to the left of the dashed line should be incorporated in the modulator. The function switch on the transmitter should be set to the c.w. position. Then S_1 lets you use either plate-modulated phone or c.w. as you wish.

S_2 is the phone send-receive switch. It opens the modulator cathode circuit so the tubes won't take current while you're listening. The second pole is connected to a plug which goes to the transmitter's key jack, to turn the r.f. end on and off.

R_1 is the screen dropping resistor for plate-and-screen modulation. Its value should be selected to put the normal d.c. screen voltage, based on c.w. operation, on the r.f. amplifier. This is usually between 150 and 200 volts, but the resistance value will depend on the plate voltage used in your transmitter. Measure the voltage on the screen with the transmitter oper-

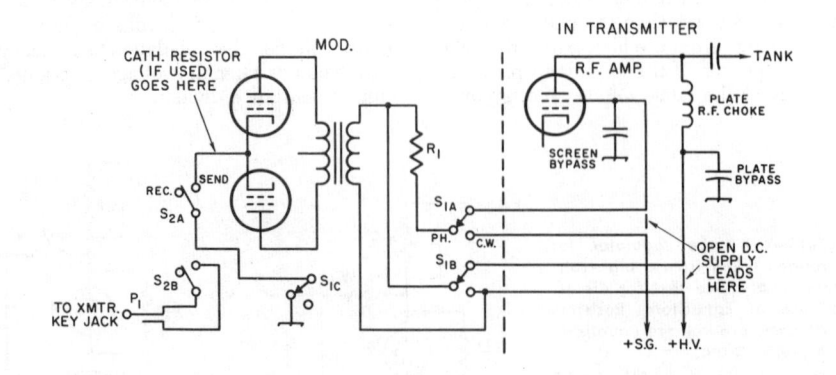

FIG. 13-12—Combining the plate modulator with a manufactured or kit transmitter. This method requires a minimum of change in the transmitter.

P_1—Plug to fit key jack on transmitter.

R_1—Screen dropping resistor for modulated r.f. amplifier; see text.

S_1—3-pole, 2-position rotary switch.

S_2—D.p.d.t. toggle.

Modulator Metering

ating on c.w., using the set just as it comes. Then figure out the voltage drop needed in R_1 with an assumed screen current of about 10 ma. Using this trial value at R_1, put S_1 in the phone position, and measure the screen voltage again. If it is lower, reduce the resistance of R_1; if higher, raise the resistance.

Before putting the transmitter function switch in any other than the c.w. position, always throw S_1 in Fig. 13-12 to c.w., and when you're actually using c.w., turn off the a.c. power to the modulator. S_1 has a third pole, in series with S_2, to make sure that when the switch is in the c. w. position the modulator cathode circuit will be open. This protects the modulation transformer from high-voltage surges that will occur when the modulator is fully operative with no load on the transformer secondary. If much sound reaches the microphone under such conditions the audio voltage developed in the modulation-transformer primary sometimes becomes large enough to break down the insulation.

In using plate modulation, first tune up the transmitter as you would for c.w. Then switch in the modulator and adjust the voice level as described later in the section on modulation checking.

Using Meters with the Modulator

At least in the initial testing stages, you should measure the plate or cathode current of a Class AB_1 modulator. It doesn't make much difference which you measure, but whichever one it is should be checked against the values given in Table 13-I for the no-signal condition. Add the plate and screen currents together if you're checking the total cathode current.

Fig. 13-13 shows how the meters should be connected, M_2 for cathode current, M_3 for plate current alone. The milliammeter should be one having a full-scale range somewhat above the values given for the maximum-signal current in Table 13-I.

When fixed bias is used on the modulator grids, measuring the cathode or plate current lets you set the grid bias to the right value. The bias figures in Table 13-I are approximately correct, but minor variations in tube characteristics make it advisable to adjust the bias to give the specified plate or cathode current. The bias should *not* be set by measuring it with a voltmeter, if supplies such as are shown in Fig. 13-8 are used. The voltage will decrease when the voltmeter is connected, and rise when it is disconnected.

A meter in the plate or cathode circuit also can be useful as a modulation indicator. Once you find out how high the needle kicks when your voice is modulating the transmitter 100 per cent, you can watch the meter to make sure that the reading never goes above that point on the scale.

The microammeter, M_1, shown in the grid-return circuit in Fig. 13-13 is not an operating

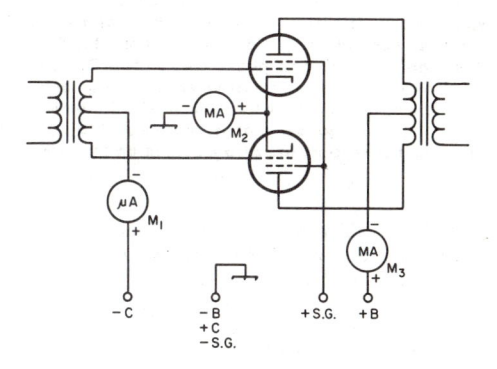

FIG. 13-13—Connecting meters in the Class AB_1 modulator circuit. M_1 is used for indicating overdriving of the grids. M_2 is for measuring total cathode current, M_3 for measuring plate current (either may be measured, but measuring both is not necessary).

necessity, but is helpful as a check on the operation of the modulator. A Class AB_1 amplifier should never be driven into grid current, so this meter can only show whether or not you're overdriving the modulator. The pointer should stay on zero all the time. If it flicks upward, you're hitting the modulator grids too hard. The meter should be a sensitive one—500 microamperes or less full scale—since a larger range won't be sensitive enough to show small amounts of grid current. Your audio waveform will be "clipped" with the start of grid current, causing distortion.

Plate Modulator for the TV/Surplus Amplifier

The 150-watt amplifier shown in Chapter 10 can be operated at 120 watts input on phone with plate modulation. A modulator for it can be built rather economically, because the plate power supply that goes with the amplifier has enough excess capacity to take care of the modulator along with the amplifier. Fig. 13-14 is a view of a modulator which was built at very low cost. The most expensive item in a modulator is always the modulation transformer; this one came from a military-surplus MD-7 modulator, a member of the "ARC-5" group. These can generally be picked up for a few dollars, complete with tubes, although sometimes they are a little hard

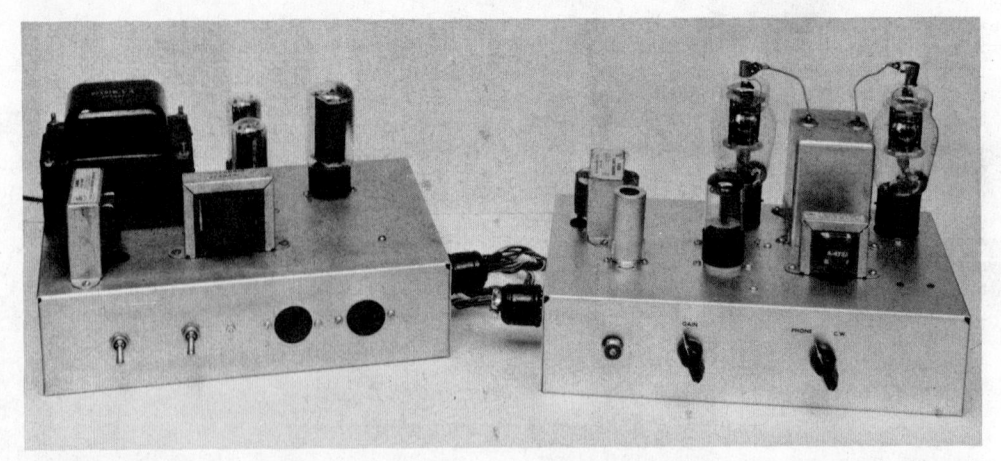

FIG. 13-14—Inexpensive plate modulator using a military surplus modulation transformer and other salvaged parts (right). The power supply at the left is a modified version of the supply for the 150-watt amplifier (Chapter 10). Components along the front of the modulator chassis are the microphone connector, audio gain control, and phone-c.w. switch. On top of the chassis, the modulation transformer is between the two 1625s. The driver transformer is in front of it, with the 12J5 speech amplifier to the left. The filament transformer is at the rear left. The shielded tube is the dual-triode speech amplifier.

On the power-supply chassis, the on-off and send-receive switches are the toggles on the front wall. Two of the output connectors are also mounted on this wall; the third one is at the rear.

to find. It's worth looking around for one, though, if you want a low-cost phone transmitter.

The circuit, Fig. 13-15, is intended for use with a crystal or ceramic microphone, and is similar to the plate-modulator circuit discussed earlier in this chapter. The differences result principally from making use of as many parts as possible from the MD-7. There is an extra stage in the speech amplifier, and the excess gain is cut down by using a step-down transformer, T_1, to the modulator grids. Also, the cathode resistor of the second stage, V_{1B}, is unbypassed, giving some negative feedback which reduces the gain of this stage.

Grid bias for the first stage is obtained by a somewhat different method than was shown earlier. There is no cathode bias; instead, a high-resistance grid leak is used, and enough electrons flying off the cathode hit the grid to cause a small grid current to flow. This develops about 1 volt of negative bias in the grid resistor. The advantage of this method is that it saves a cathode resistor and bypass capacitor. The disadvantage is that the effective grid-to-cathode resistance is relatively low, so the microphone is working into a considerably lower resistance than the 2.2-megohm resistor. This reduces the output voltage of the microphone, but it is not a serious consideration in this case because the speech-amplifier gain is more than adequate to make up the difference.

Another way in which this circuit differs from the one described earlier is in the method of coupling the audio power to the r.f. amplifier. The MD-7 transformer has a special secondary for introducing audio power into the r.f. screen circuit. With such a winding it is unnecessary to modulate the screens through a dropping resistor. This avoids wasting some of the audio power in the resistor. R_2 in this circuit drops only the d.c. screen voltage; it is bypassed for audio by C_4 when the phone-c.w. switch, S_1, is in the phone position. (The modulation transformer also has a "side-tone" winding, terminals 4 and 5, which is left unused.)

Power Supply

A few changes should be made in the power supply (Chapter 10) that goes with the r.f. amplifier. The revised circuit is shown in Fig. 13-17. A low-voltage output (approximately 300 volts d.c.) is taken from the center tap of the power transformer's high-voltage secondary. This is filtered by L_2 and the two 40-μf. capacitors, and is used to supply the speech amplifier tubes and the modulator screens. It may also be used for the r.f. exciter, if the exciter is the low-power transmitter described in Chapter 10 (Fig. 10-1). Two extra octal sockets, J_2 and J_3, are added. The former handles all the power for the modulator. J_3 carries the plate voltage for the exciter, along with heater power for the exciter tubes. Thus the one power supply takes care of the entire transmitter.

If the transmitter of Fig. 10-1 is used to drive the amplifier, only one tube should be used in the final stage. It will have plenty of output for exciting the 1625s. Removing one tube lightens the drain on the power supply.

The modulator, Fig. 13-15, has its own filament transformer, T_4. This is used because some salvaged TV power transformers may not have enough current capacity to handle all four 1625s, with the two 6.3-volt windings in series.

Modulator Construction

The power-supply circuit, Fig. 13-17, includes a relay, K_1, for send-receive switching. A double-pole relay is necessary because of the dual voltage outputs. S_3 controls the relay, and a pair of terminals in parallel with S_3 allows you to use an external switch for send-receive, in case the power-supply chassis has to be installed in a spot where it isn't easily accessible.

Modulator Construction

While it might be possible to use the MD7 chassis for the circuit, a much neater job can be made by mounting the parts on a new chassis. The one shown in Figs. 13-14 and 13-16 is aluminum, $3 \times 7 \times 12$ inches. The layout needn't be followed exactly, although it is a good idea to follow the same general scheme. Use shielded wire to connect J_4 to Pin 2 of V_{1A}, and ground the shield at both ends. This will minimize hum pick-up by this lead.

C_2 is a metal-cased dual electrolytic capacitor from the MD7. Connections to the 20-μf. section should be made to the terminal marked "20-μfd." (positive) and to the case (negative). The center terminal is the positive side of the 5-μf. capacitor, and the remaining one is the negative side. C_4 is a 1.2-μf. capacitor in a metal case; the single terminal connects to S_1 in this circuit and the case is grounded to the chassis.

There are six terminals on the bottom of the modulation transformer, with an identifying number marked on the case alongside each terminal. When making the six holes in the chassis for the terminals be sure to mark the terminal number alongside the hole on the underside of the chassis. The No. 4 terminal is the case of the transformer and should be connected to the chassis.

R_2 is the screen dropping resistor for the r.f. amplifier. Moving it from the amplifier to the modulator chassis simplifies the cabling between the two. The only change required in the r.f. amplifier itself, aside from moving R_2, is to add a lead to Pin 5 of the amplifier power plug as shown in Fig. 13-18.

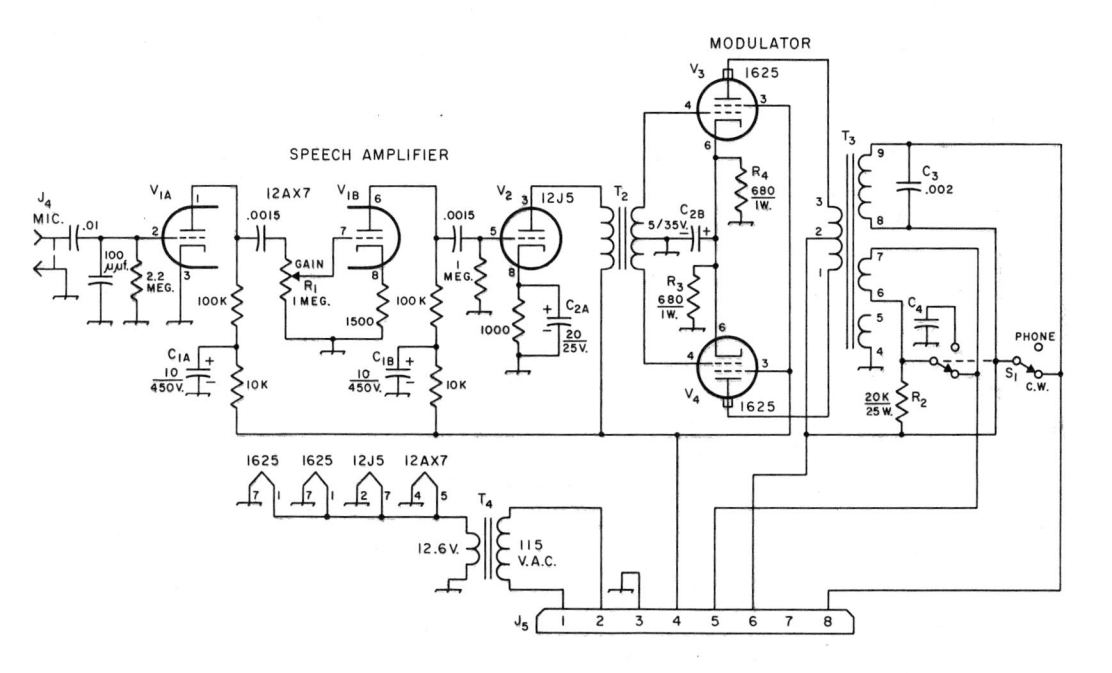

FIG. 13-15—Circuit diagram of the plate modulator. Unless otherwise indicated, capacitances are in μf., resistances are in ohms, resistors are ½ watt. Capacitors not listed below can be paper, mica, or disk ceramic.

C_1—Dual 10-μf. 450-volt electrolytic.
C_2—Dual electrolytic from MD7; see text.
C_3—0.002-μf. paper or disk, 1000 v.
C_4—1.2-μf. electrolytic from MD7.
J_4—Microphone connector (Amphenol type 75-PC1M).
J_5—Octal plug, male chassis-mounting type (Amphenol type 86-CP8).
R_1—1-megohm control, audio taper.
R_2—20,000 ohms, 25 watts; see text.
R_3, R_4—680 ohms, 1 watt, from MD7.
S_1—Ceramic rotary, 1 section, 2 poles, 6 positions, 2 positions used (Centralab PA-2003).
T_2—Driver transformer, single plate to push-pull grids. Ratio 3:1 primary to ½ secondary (Stancor A-4723).
T_3—Modulation transformer from MD7; see text.
T_4—12.6 v., 2.0 amp. (Knight 61G420, Triad F-26X, Stancor P-8130).

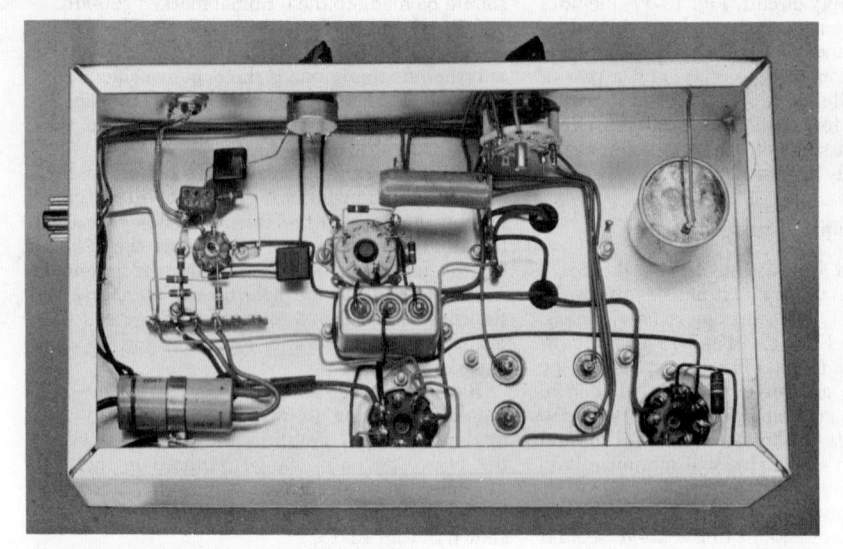

FIG. 13-16—C_2, taken from the MD7, is in the center of the chassis in this bottom view of the modulator. Just above C_2 is R_2, the screen dropping resistor. The capacitor in the upper right-hand corner is C_1, also taken from the MD7.

Operating the Combination

To test the system, use a lamp dummy antenna on the r.f. amplifier and connect the exciter to the amplifier, using a short length of coax cable. Plug the power cables into the appropriate sockets, put the c.w.-phone switch, S_1, in the c.w. position, and turn on S_2. After the tube heaters have warmed up, close S_3 and adjust the exciter's loading and tuning controls to give approximately 8 ma. final-amplifier grid current.

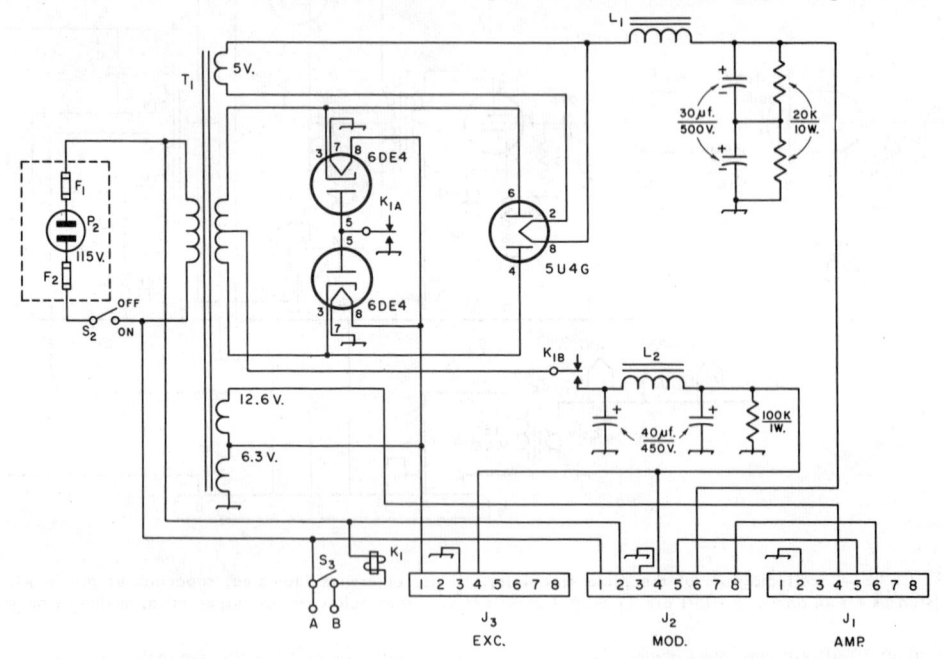

FIG. 13-17—Power-supply circuit. Capacitances are in μf., capacitors electrolytic. Resistances are in ohms.

J_1, J_2, J_3—Octal sockets. J_3 not needed if low-power transmitter mentioned in text is not used as the r.f. exciter.

K_1—D.p.d.t. relay, 115-volts a.c. (Potter & Brumfield KA11AY).

L_1—App. 2 henrys, 300 ma. (taken from TV receiver).

L_2—15 henrys, 75 ma. (Stancor C-1002 or equivalent).

P_2—Line plug, fuse-in-plug type. Fuses F_1 and F_2, each 5 amp.

S_2, S_3—S.p.s.t. toggle.

T_1—Power transformer salvaged from old TV receiver.

Modulation Checking

Adjust the loading on the 1625 amplifier for a plate current that represents approximately 120 watts input. However, do not exceed 200 ma., since this is the maximum rating for the two tubes. The adjustment procedure so far is just the same as for ordinary c.w. operation.

Next, set S_1 to the phone position (turn off the plate voltage while you do this) and talk into the microphone at your normal voice level. Turn up the gain control, R_1, until the lamp brightens up on voice peaks. This shows that the amplifier is being modulated. To check on the modulation, watch the plate meter of the r.f. amplifier. Increase the gain until you see the needle give a small kick on voice peaks, then back off on R_1 until this happens only occasionally. A pronounced change in amplifier plate current, or continual wobbling of the pointer while you talk, is a certain indication of overmodulation. Keep the gain down and avoid splattering.

An additional check on the modulation can be made by measuring the modulator tubes' plate current. This can be done by temporarily taking the plus-B lead off terminal 2 of the modulation transformer and inserting a d.c. milliammeter having a 0-100 range. The resting plate current should be about 50 ma. On voice peaks that modulate the transmitter 100 per cent the plate current will kick up to about 60 ma. at the most.

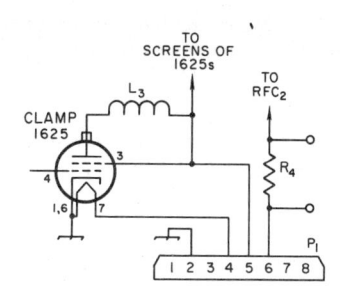

FIG. 13-18—R.f. amplifier screen-circuit modification, after moving screen dropping resistor to modulator chassis.

Checking Your Modulation

There is only one certain way to check modulation, and that is by using an oscilloscope. However, a scope is a relatively expensive piece of equipment, and unless you're pretty familiar with how it works and what to look for, it can be misleading, too. We aren't recommending that you start out with one, therefore. There are some pretty good ways to tell whether or not your phone signal will pass inspection without the scope. The indicators are right in equipment you already have. All you have to do is to make the right use of them.

Modulated-Amplifier Plate Current

The statement has been made a number of times, both in this chapter and in Chapter 5, that when a constant carrier is properly modulated the d.c. plate current of the modulated amplifier does not change. Modulation varies the current up and down, but the average value, which is what the d.c. plate meter reads, stays the same. The statement applies to either screen or plate modulation. (By "plate" modulation we mean, of course, plate-and-screen modulation when the modulated r.f. tube is a tetrode or pentode.)

Thus the modulated-amplifier's plate meter is a modulation indicator, in a limited sort of way. That is, you can be reasonably certain that you are modulating properly if the plate current remains steady while you talk. It doesn't tell you anything about the *percentage* of modulation, though, until you overshoot the mark. It's only when the modulation becomes nonlinear that the meter will begin to fluctuate.

Carrier Shift

Nonlinearity in a modulated amplifier means that the average d.c. input, and with it the average carrier level, has changed. This is why the plate meter shows a change in current. A change of this sort is called **carrier shift**. If the modulated amplifier is properly loaded and has the right drive and grid bias, and if the modulator has enough audio output to modulate it more than 100 per cent, carrier shift will occur when you overmodulate. The plate current will jump up above the unmodulated level. An upward shift in plate current can be caused by other kinds of nonlinearity, but overmodulation is the principal one.

If the plate current shifts downward when you hit voice peaks, there are at least three possible reasons. One is that the modulation characteristic of the r.f. amplifier isn't linear up to 100 per cent modulation (see Chapter 5). This means that you can't reach the peak-envelope power that you should. The principal cause of this is insufficient r.f. drive, not enough grid bias, and loading the amplifier too heavily. Stick to the tube ratings for grid current, grid bias, plate and screen voltage, and input. If you do you shouldn't have this trouble.

A second cause—and a common one, although often overlooked—is that the d.c. plate voltage drops a bit when you modulate, because of voltage regulation. This is prone to happen when you run the r.f. and audio all from the same power supply. You can check it by measuring the voltage while you talk. If the voltage decreases in the same proportion as the plate current, there's the reason for the downward carrier shift.

The third cause is found in the shape of the audio waveform from the modulator. Voice waveforms often are not symmetrical in amplitude. That is, the peaks on the "negative" part of the cycle may not have the same amplitude as peaks on the "positive" part. If the "negative" peak is larger and is the one that swings the r.f. amplifier plate voltage downward, you can have downward overmodulation even though you aren't modulating 100 per cent upward. In this situation the average plate current to the r.f. stage decreases. Reversing the audio polarity by reversing the connections to the secondary of the modulation transformer will change a downward carrier shift into an upward carrier shift, if this lack of symmetry was responsible. In either case, the audio gain should be kept below the point where the carrier shift occurs, because the shift indicates overmodulation.

Modulator Plate Current

Since the plate current of a Class AB modulator always is greater at maximum signal than when there is no signal at its grids, the modulator plate meter can be used as a modulation level indicator. It isn't ideal, because it can't tell you anything about peaks. These depend on the speaker's voice. No two voices are alike, and two that would give exactly the same reading on the meter may reach quite different peak amplitudes; one might be overmodulating while the other is well under 100 per cent modulation.

However, if you check the modulator platemeter reading against the r.f. amplifier's plate current, you can find the modulator reading that just causes carrier shift with *your* voice. Then if you keep the modulator current below that reading you should be modulating correctly. As we have pointed out earlier, this meter reading will be well below the figure it would reach with pure-tone modulation, so you can't go by the latter figure.

R. F. Voltmeter

Some transmitters have r.f. output indicators, usually just a semiconductor diode rectifier and a d.c. meter coupled in some way to the output circuit. These can be used like the plate meter in the r.f. stage *provided* the readings are reasonably linear. They will *not* be linear if the resistance in the d.c. circuit to the meter is less than about 5000 ohms, with a 0-1 milliammeter. A series resistance of 10,000 ohms is better still.

The point about such indicators is that the readings with and without modulation *should be the same*. The average r.f. voltage and current, like the d.c. plate current, is not changed by linear modulation. A change in the meter reading indicates nonlinearity. That is why you have to be careful about the linearity of the r.f. indicator itself. If the meter isn't linear, its indications will shift even though the transmitter may be perfect.

Again, this is a limited type of indication; if there is no change, both the transmitter and the meter are OK. But if there *is* a change you don't know which to blame, unless you have some way of independently checking either or both. Comparing the r.f. indicator with the plate meter can give you a clue. If both behave the same way, the finger of suspicion points to the transmitter itself as the culprit. If they don't, you have the problem of finding out which is right.

Controlled Carrier

You can disregard what we've said above about meter readings if your transmitter has controlled-carrier screen modulation. With this system the meter readings tell you little or nothing —except that if the reading goes up to something approximating what it reads on c.w. you've gone far afield into the nonlinear region. This may *look* good on the meter, but there isn't much else that can be said in its favor.

In a general way, the meter readings with controlled carrier are comparable with the readings of a Class AB_1 modulator plate meter. The r.f. amplifier plate current or the relative r.f. output will kick up to perhaps a third or half of the c.w. reading when the transmitter is being modulated to its maximum linear capability. Only an oscilloscope can show when the actual limit is reached, and it requires some experience to interpret the scope patterns correctly.

The best way to check such a transmitter, in the absence of the scope and experience in looking at scope patterns, is to use your receiver as described in the next section. You'll be able to tell the level at which splattering begins, and that is the most important thing you have to watch out for. Make a note of the meter reading that corresponds to the splatter point, and don't let the pointer go any higher than that when you talk.

Using Your Receiver

The ultimate test of a phone signal is what it's like on the air. This doesn't mean that you have to depend on reports from other stations. In fact,

Modulation Checking

you shouldn't. If you have a reasonably good receiver you can hear for yourself what the signal is like.

To do this you have to make your own signal small enough, in the receiver circuits, so that no stage in the receiver will be overloaded. First, put the transmitter on a dummy antenna, with the modulator turned off. Disconnect the receiver from the antenna. At this point, if your receiver is reasonably well shielded, you should be able to tune in the transmitter's signal, with the receiver's a.g.c. operating, at a level that is no greater than the stronger signals you normally hear. If the receiver has an S meter, the meter reading should be in the middle range. If the signal is too strong, try short-circuiting the antenna terminals, and throw the antenna trimmer off tune. Possibly you may have to pull out the r.f. amplifier tube—anything that will reduce the input enough to give you a signal of no more than moderate strength.

Checking for Hum

Turn off the modulator completely, including the tube heaters. Listen carefully to the unmodulated carrier, tuned in "on the nose." Does it have a noticeable hum, as compared with the hum you hear on other stations' carriers? If so, the r.f. plate supply isn't filtered well enough. But if it's far-enough down to be satisfactory, turn on the modulator, with the speech-amplifier gain control turned off. Did the hum come up? If it did, your modulator plate supply no doubt needs more filter.

If the transmitter passes these two tests, wrap the microphone in a blanket, or otherwise insulate it from stray sound, and slowly advance the audio gain control. As you get up toward maximum gain you'll probably hear some hum. However, there should be some setting at which the hum isn't serious, and if this or a lower setting is what you find you use for normal speaking intensity, you can give yourself a passing grade on hum. If not, there are some suggestions earlier in this chapter and in Chapter 7 on reducing hum. Usually it's the heater circuit of the first speech amplifier that is responsible, although it isn't impossible that the plate filtering for the speech amplifier is too light.

Modulation Quality

Once the hum question is settled you'll have to use headphones for listening. The loudspeaker is practically certain to start a howl because of acoustic feedback to the microphone.

At this stage you need a helper, because you can't simultaneously talk and listen to your modulation. Have your assistant talk into the microphone at about the same loudness, and from the same distance, as you yourself would. Start with the speech-amplifier gain at a low setting, and slowly advance the control while you listen critically to the quality. Watch the r.f. amplifier's plate meter—or other indicator—as described earlier. The voice quality should not change as the gain is increased (although the signal will become louder) up to the point where the plate meter begins to flicker on peaks. Keep the audio gain in the receiver at a comfortable headphone level while making this test.

When the modulation indicator does show one of the signs we have identified earlier with overmodulation, you still may not find any really pronounced change in quality, although if you listen carefully you'll be able to detect a little roughness that wasn't there before. However, if roughness is there *before* the meter shows any indication of overmodulation, the modulator

USE A DUMMY ANTENNA

itself is generating distortion. The probable reason here is inadequate audio power output, which may have a number of causes. One is that the modulator is just not big enough for its job; you haven't picked the right tube combination. Or the voltages may not be up to what is required, even though the tubes would be capable of giving the needed output at the specified voltages. Or you may not have picked the right multimatch tap combination to give the proper turns ratio in the modulation transformer. Not forgetting, of course, that a mistake may have been made somewhere in wiring, or a component was used that is either defective or does not have the value you thought.

Splatter

So far, if everything has checked out, you have the makings of a good phone signal. The final test is for **splatter**, those "burps" that occur *outside* the channel a phone transmission needs. This channel, for an a.m. signal, is approximately 6 kc. wide, extending 3 kc. on both sides of the carrier frequency. Anything you hear outside that range should be very weak compared with the proper audio output from your signal.

Here you have to turn off the a.g.c. Let the receiver's audio gain control stay at the setting you've been using, and set the manual r.f. gain control at the point that makes your signal have the same loudness that it had with the a.g.c. on, at the transmitting audio level that gave what you estimated to be 100 per cent modulation. Slowly tune the receiver off to one side. If the receiver has reasonably good phone selectivity, the signal will drop off rapidly when you tune a few kc. away. It should sound clean until it disappears, and there should be no "burps" at any time—and especially not in the region where you can't hear the signal itself. If you do hear such "burps," cut back on the speech intensity until they disappear. Check this against whatever meter you may be using for a modulation indicator, so you'll know how far you can go when you put the transmitter on the air.

In making the splatter test, use the highest selectivity the receiver offers. The sharper the better, because you can't really measure signal bandwidth with a broad-band receiver. It might pay to refresh your memory on this by looking at Chapter 3 again.

Using the B.F.O.

An even better splatter and bandwidth test can be made with the help of the receiver's b.f.o. Set the receiver's controls as you would for c.w. reception, using high audio gain and adjusting the manual r.f. gain control for a beat of moderate strength when the carrier is unmodulated. Use the highest selectivity the receiver can give. Again get your helper to modulate the transmitter at what you estimate to be the 100-per cent level. Tune the receiver off to one side, listening carefully to the beat tones as you go. The beats will sound clean if the signal is properly modulated. You may hear some weak ones when you get outside the normal 3-kc. channel, but these are normal sidebands if they still sound clean.

Outside the channel, any strong beats will be overmodulation spatter. These beats *don't* sound clean. They're harsh-sounding—real "burps," usually. If you have trouble distinguishing between the two kinds, run up the speech-amplifier gain until you're really overdriving the modulator (remember that these tests are made on a dummy antenna!) and you'll quickly catch on to the difference. With a bad case of overmodulation, intermittent beats, corresponding to voice peaks, will be heard over a wide frequency range. Having identified splatter, turn down the speech-amplifier gain to the point where there no longer is any. Then you can put your transmitter on the air with the assurance that it will meet FCC requirements and won't cause unnecessary interference.

Carrier Frequency Shift

You can check for one other thing by using the b.f.o. as described above: a shift in carrier frequency when modulation is applied. Crystal-controlled transmitters rarely show this effect, but it is not uncommon when a v.f.o. is used. The usual reason for it is power-supply voltage regulation, when the whole transmitter is run from the same supply. The voltage decreases because the supply is more heavily loaded during modulation peaks, and the change in voltage causes a slight change in v.f.o. frequency.

Adjust the receiver for a low-frequency beat tone on the carrier, with no modulation. Then have your helper modulate the transmitter normally while you listen carefully to the carrier beat. The tone will shift on modulation peaks if the v.f.o. is being affected by the modulation. It requires a little concentration to detect the shift, because the modulation itself sets up other beat tones which may be confusing. For this reason the test should be made with the maximum selectivity available.

Using voltage regulation—a VR tube, for example—on the oscillator plate and screen voltage should cure this type of frequency shift. If the v.f.o. supply is already regulated, the shift may be caused by r.f. getting back from the amplifier into the v.f.o. circuits. Better shielding and more isolation is the answer to this. Any v.f.o. phone transmitter needs at least one good buffer stage along with regulated voltages on the oscillator.

Mechanical Helpers

Checking a phone transmitter takes time, especially when it is first being tested and you find that some things need correcting. Nonamateur helpers to talk into the microphone may be willing at first, but their enthusiasm usually isn't lasting. Besides, you can get more uniform modulation by mechanical means, some of which are available in many households.

One is the record player. The output of a crystal or ceramic pickup, usually between 0.5 and 3 volts, will be too high to be put directly into the microphone input of a speech amplifier. It can be stepped down through a resistance voltage divider. Connect a 0.22 megohm resistor directly across the microphone input terminals and a 2-megohm resistor in series with the "hot" lead from the pickup, as shown in Fig. 13-19.

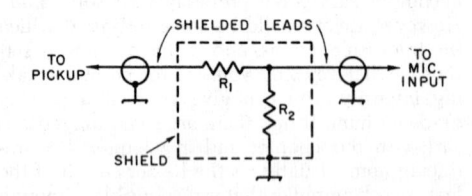

FIG. 13-19—Input attenuator for connecting a crystal or ceramic phonograph pickup to the microphone input terminals of the speech amplifier. R_1—2 megohms; R_2—0.22 megohm. The same circuit may be used when audio voltage is taken from the voice coil of a radio speaker, in which case R_1 can be 100 ohms and R_2, 10 ohms. Shielding is probably not needed in the latter case.

The resistors and the connecting "hot" leads must be shielded to prevent hum pickup. If the pickup is magnetic the output is too low, so you have to take off the signal from a suitable point in the phonograph's preamplifier. If it is taken after the gain control it is easy to set the level properly. Leads should also be well shielded in this case.

Modulation Checking

Another tireless helper is a tape recorder. Record a voice—preferably your own, although a radio program will do. Many recorders have a preamplifier take-off which can be used as is. The connecting cable and plugs must be shielded.

Even a radio receiver will serve, if you can clip on to the speaker voice-coil terminals. At normal volume you can expect something of the order of 0.25 volt at this point, so a 10-to-1 voltage divider is about right. Since it is a very low-impedance circuit, the divider can consist of a 10-ohm resistor across the microphone input terminals and a 100-ohm resistor in series with the "hot" lead to the voice coil. Shielding isn't so important here.

In all three systems you need a ground connection—which you should have anyhow, for safety (see Chapter 1). The record player and tape recorder have the advantage that you can use the loud-speaker on your receiver. If you can muffle the radio, you may be able to use the speaker with it, too, but this isn't always possible.

Chapter 14

Making Measurements

Whole books could be (and have been) written on electrical measurements. The subject has a fascination all its own. However, in this chapter we'll not attempt to do more than touch on a few measurements aimed at helping you adjust your equipment for optimum results.

What Accuracy Means

At the outset, it may be useful to give you a little perspective. There is no such thing as an "absolute" measurement. No matter how good the measuring equipment and the care with which it is used and read, the best that one can do is to say that the true value of a measurement lies between a pair of limits. Inside those limits the value is uncertain. An accurate measurement is one in which the limits are close together. The farther apart the limits are, the more "fuzzy" the measured value.

The limits, for a piece of measuring equipment, usually are specified in terms of **percentage accuracy,** but sometimes other methods are used. One example should suffice. You have a 0-100 d.c. milliammeter. If it is a good-quality instrument it will probably have the scale divided into 50 parts, each representing a 2-milliampere change in current. Suppose in measuring the plate current of a tube the pointer comes to rest between the divisions representing 54 and 56 ma.

What is the Current?

If the pointer seems to be about half-way between the two divisions you would naturally say that the current is 55 ma. But on looking up the specifications on the instrument you find that it is rated to be accurate to "within plus or minus 2 per cent of full scale." The full-scale reading is 100 ma., and 2 per cent of that is 2 ma. So you have a built-in uncertainty of plus or minus 2 ma.—one whole division on the scale. The best you can say is that the *actual* current should be between 55 − 2 and 55 + 2—that is, between 53 and 57 ma.

That isn't bad accuracy, for many measurements. An error of 10 per cent or even 25 per cent or more won't make any practical difference in many cases. For example, the capacitance used in a power-supply filter can have any value within a wide range, just so long as it is large enough. But there are other cases where the accuracy of measurement has to be exceedingly good. The frequency limits of the band in which your transmitter operates is such a case. If you're going to work near the edge, you have to *know* just where that edge is. Here an error of just a few cycles in several million per second can get you into trouble with the FCC.

The instruments described in this chapter will be accurate enough for their intended purposes. This does not mean that you can expect your measurements to agree absolutely with those made by highly-specialized and often highly-expensive equipment. It *does* mean, though, that the results you get, in terms of adjustment of the equipment with which these instruments are used, will be just as good as they would be with measuring gear of higher accuracy.

D. C. Measurements

The same basic instrument is used for measuring direct current and voltage. No doubt you already own a **milliammeter** or **voltmeter.** They come in various shapes and sizes. All of them use electromagnetism to make the pointer move across the scale. Some—the inexpensive kind—simply have a coil through which the current

A "multimeter," or combination volt-ohm-milliammeter, is practically indispensable if you do any building or trouble-shooting. This is an inexpensive commercial version, available in kit form, using a 1000-ohms-per-volt instrument. It will also measure resistance.

Measurements

flows, exerting a magnetic pull on a soft-iron vane to which the pointer or "needle" is attached. The larger the current the greater the magnetic attraction and the greater the pointer movement. The scales on these instruments are **nonlinear**. That is, the divisions are spread out more at one end of the scale than the other.

Better-quality (and higher priced) instruments have the coil pivoted between the faces of a permanent magnet. This is called a **D'Arsonval movement**. It is capable of higher accuracy—although this depends on the care in construction—and the scale is linear.

Milliammeters should be selected with an eye to the range of currents to be measured. The accuracy is greatest in the upper half of the scale. Thus if you want to measure currents in the neighborhood of 150 ma. an instrument having a full-scale range of 200 or 250 ma. would be optimum. But measuring a current of, say, 10 ma. with such a meter would be considerably less accurate—remember the example above!—and a 25-ma. meter would be much better.

Multirange Instruments

Very often you do need to measure currents of widely-different values. The plate current of a tube may be around 50 times as great as the grid current; nevertheless, it is advantageous to measure both with the same instrument, rather than having to purchase two separate meters with appropriate ranges. This can easily be done, because a low-range meter can always be made to measure currents *greater* than its own full-scale reading. The range can be increased by adding **shunts** across the meter terminals.

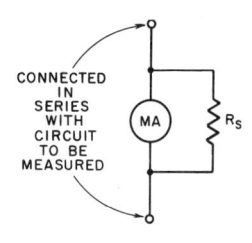

FIG. 14-1—How a shunt, R_S, is connected for extending the full-scale range of a milliammeter.

A shunt is simply a resistance placed in parallel with the meter so the current has two paths to follow. By diverting some of the current through the shunt there is less to go through the meter. If you know how much more goes through the shunt than through the meter, you simply read what the meter says and then multiply by the shunting factor to find the actual current. For example, in Fig. 14-1 if four times as much current goes through the shunt, R_S, as through the meter, the meter is getting only one-fifth of the total current, so the current in the circuit is five times the meter reading.

Values for Shunts

To know what resistance to use in the shunt for a desired scale multiplication you first have to know the resistance of the meter itself. This is probably given in the manufacturer's data on the meter. It is often of the order of 50 ohms for a 0-1 milliammeter, decreasing as the full-scale current range increases. The shunt resistance, R_S, is then equal to the meter resistance divided by the desired current multiplication minus 1. For instance, if you have a 10-ma. meter having a resistance of 5 ohms and want to increase its range to 100 ma. ($10\times$ multiplication) you divide 5 by $(10 - 1)$ or 9, so the shunt resistance required is 0.55 ohm.

The resistance of a shunt is usually rather low, as this example shows. You can easily make your own from small-gauge copper wire. No. 30 is suitable for most shunts. It has a resistance of 0.105 ohms per foot, so to make a 0.55-ohm shunt you would need $0.55/0.105 = 5.3$ feet of wire. It can be scramble-wound on a piece of matchstick to make a small coil.

If you don't know the resistance of the meter, start out with a *short* piece of wire as a shunt. Measure a current near full scale without the shunt, then put on the shunt and observe the new reading. If you want a $10\times$ shunt, change the wire length until the reading drops to one-tenth its value without the shunt. The same method can be used for other multipliers. However, it is best not to go more than 10 times in one jump. The first $10\times$ multiplier can be used to find the right shunt for the next, and so on.

Voltmeters

A voltmeter is simply a milliammeter with a high resistance in series, Fig. 14-2. The current through the resistor is measured and then, by Ohm's Law, the voltage drop across it can be found. The combination is connected across the circuit in which the voltage is to be measured.

In most cases we don't want the voltmeter to load the circuit it is measuring. That is, we want the current the meter takes to be very small compared with the current flowing in the circuit. For this to be so the meter itself usually has to be quite sensitive. A full-scale range of 1 milliampere is about the largest that is commonly used. Meters that give a full-scale deflection with as little as 50 microamperes are used in many test instruments.

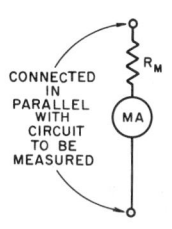

FIG. 14-2—A voltmeter is a milliammeter (or microammeter) with a high resistance, R_M, in series.

The resistance needed for the voltmeter (resistor R_M in Fig. 14-2 is called the **multiplier**) is easily calculated. It is equal to the full-scale voltage to be read divided by the full-scale meter current in amperes. One volt divided by 1 milliampere (0.001 ampere) is 1000 ohms, so a voltmeter using a 0–1 milliammeter is called a "1000-ohms-per-volt" meter. With such a meter you simply multiply the desired full-scale voltage by 1000, to find R_M. Thus a 500-volt voltmeter would require $500 \times 1000 = 500{,}000$ ohms, or 0.5 megohm, for R_M.

In making a voltmeter multiplier keep in mind that the power that has to be dissipated in the resistor may not be negligible. The power is equal to the voltage multiplied by the current, and in the example above would be ½ watt. However, small resistors sometimes have a maximum voltage rating as well as a maximum power rating, so look up both ratings before selecting a resistor. Actually, it costs only a few cents to be conservative and use two resistors in series where you might get by with only one.

Measuring Resistance

Ohm's Law points the way to measuring resistance with a milliammeter. All we need to do is apply a known voltage to the resistor and milliammeter in series, measure the current, and the resistance is equal to the voltage divided by the current. The known voltage can be taken from one or two small dry cells. These give 1.5 volts each.

A 0–1 milliammeter can be used this way for measuring resistances ranging from about 100 to 50,000 ohms, with useful accuracy. This covers a great many of the resistance values you need to use in transmitting and receiving equipment. For higher values of resistance it is necessary to use a meter with more sensitivity—50 or 100 microamperes full scale—and for lower values a more complicated circuit arrangement is required.

The circuit used for resistance measurement is given in Fig. 14-3. If the unknown is a very low resistance, connecting it across the terminals AB practically amounts to short-circuiting them. The safe meter current would be greatly exceeded in that case if it were not for the series resistor R_1, which is selected so that the meter just reads full scale when terminals AB are shorted. R_1 can be an adjustable resistor, if you wish, so the current can be adjusted to make the meter read *exactly* full scale with AB shorted. This adjustment helps compensate for the change in battery voltage as the dry cells run down with age and use.

Combination Instruments—the Volt-Ohm-Milliammeter

You can't be in ham radio very long without feeling a need for a test instrument—completely independent of your transmitter, receiver, and other communication equipment—that will let you measure voltage (both d.c. and a.c.) resistance and, with lesser urgency, current. Although

More expensive, but more versatile, is the 20,000-ohms-per-volt v.o.m., such as the one shown above. These instruments also are available in kit form, at an appreciable saving in cost as compared with an assembled instrument.

you can get separate instruments for each of these measurements, it is far more convenient to have everything in one compact multirange unit.

The same basic meter movement can be used, its range being extended by supplementary circuits such as the shunts and multipliers described earlier. This suggests that the assembly can be built up from scratch by buying the necessary parts. However, we don't recommend it, unless you get one of the special kits made just for the purpose. Much of the value of an instrument is in its calibration and the convenience with which its scales can be read. In this field, you're badly

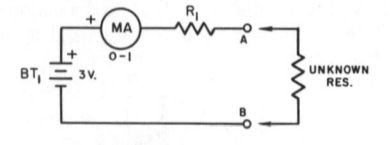

FIG. 14-3—Simple ohmmeter circuit. The resistance of R_1 should be equal to the battery voltage divided by the full-scale reading of the milliammeter. For a 3-volt battery and 0–1 milliammeter, R_1 is 3/0.001, or 3000 ohms. If a variable resistor is used it should have somewhat larger resistance (5000 ohms would be suitable in this case) to allow leeway for adjustment. A composition control can be used but an equivalent wire-wound control will be more stable.

Measurements

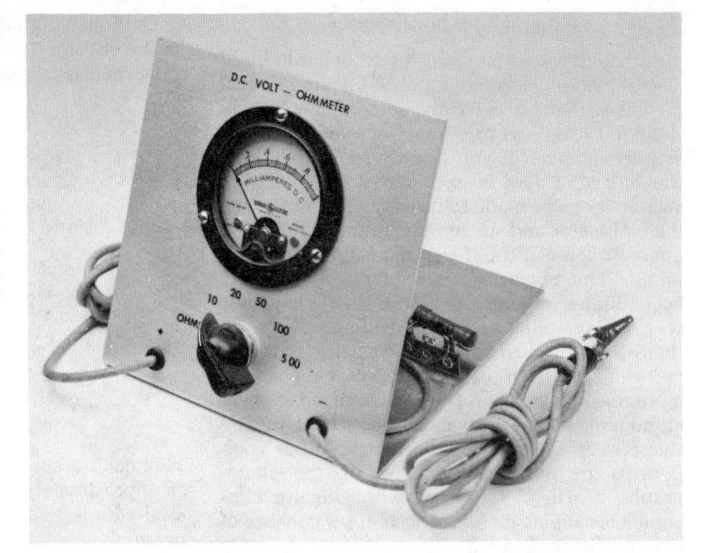

FIG. 14-4—Multirange voltmeter made from a 0–1 milliammeter, switch, and series resistors. There is also an ohms scale useful from 100 to 47,000 ohms.

handicapped by not having the facilities available to manufacturers. Furthermore, if you buy the parts separately and assemble them, you wind up by spending more than you would for a kit—or even, in many cases, more than for a completely-assembled instrument.

The **volt-ohm-milliammeter**—or **v.o.m.**, as it is usually called—is available in a variety of sizes, sensitivities, and ranges. The simplest generally has several d.c. voltage ranges up to 500 or 1000 volts, with a resistance of 1000 ohms per volt. The useful "ohms" range will be of the order of 50,000 ohms in these instruments. Direct-current ranges, if included (they may not be) may go up to 250 ma. or so. There are usually a.c. voltage scales corresponding to the d.c. voltage scales. As you go up the price ladder the number of ranges increases and so does the sensitivity; the better instruments of this type have 20,000-ohms-per-volt d.c. voltage scales, higher-voltage a.c. scales, and more elaborate ohmmeter circuits that not only can measure low values of resistance but also extend the high range up to a megohm or more. In the end, it pays to buy the most versatile instrument you can afford. However, a low-cost "handy tester" will take care of a lot of jobs for you.

FIG. 14-5—Circuit diagram of the d.c. volt-ohmmeter. Resistors are ½ watt.

Voltage ranges are 10, 20, 50, 100 and 500 volts with the meter and resistance values specified. The resistors are connected in series to add up to the required value at each position of the range switch.

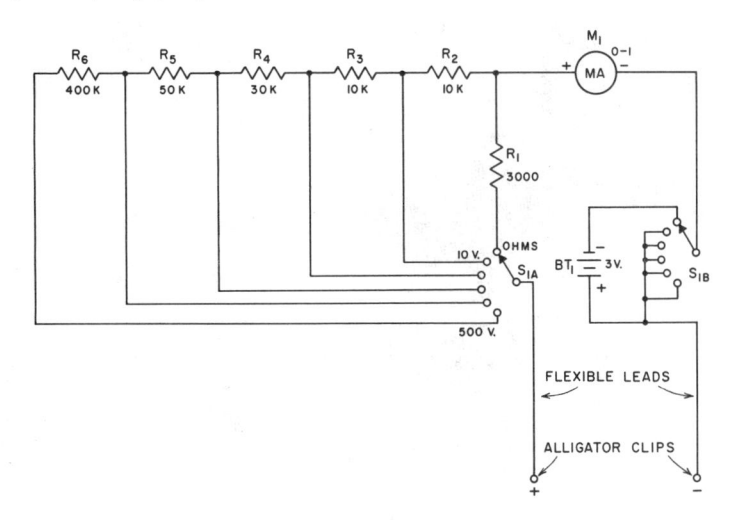

BT_1—3 volts; two 1½-volt penlite cells in series.
M_1—0–1 d.c. milliammeter.
R_1—3000 ohms, 5 per cent.
R_2—10,000 ohms, 5 per cent.
R_3—10,000 ohms, 5 per cent.
R_4—30,000 ohms, 5 per cent.
R_5—50,000 ohms (51,000 ohms, 5 per cent).
R_6—0.4 megohm (0.39 megohm, 5 per cent; 1-watt size suggested for more conservative operation).
S_1—2-pole, 6-position rotary (Mallory 3226J).

Homemade Volt-ohmmeter

Although we advise against setting out to build a combination instrument with a full complement of scales, there is certainly no reason why you shouldn't make use of a low-range milliammeter or microammeter if you already have one. With the aid of a few inexpensive resistors and a switch, it can be made to do duty as a multirange d.c. voltmeter and as an ohmmeter. Fig. 14-5 is a circuit you can use if you have a 0–1 milliammeter. The operating principles have already been discussed. Figs. 14-4 and 14-6 show one way of assembling such an instrument, but since there is no need to follow any particular layout you can build it up in any way you choose. The mounting shown is a 5 by 10 inch piece of sheet aluminum, about 1/16 inch thick, bent in a V shape so it can support the meter. Other components are mounted as shown in the photographs. A pair of test leads with alligator clips completes the assembly. The battery consists of two penlite cells with soldered connections to the top and bottom terminals. (Don't hold the soldering iron on the cell any longer than is necessary to make the connection. Also, make sure that the negative connection is actually to the zinc container of the cell; some cells have a foil covering which must be removed to get at the case.)

Resistor values having 5 per cent tolerance, as specified in Fig. 14-5, will make the voltmeter accurate enough for most testing. However, if the milliammeter is rated at 2 per cent you may feel it worth-while to use 1-per-cent tolerance **precision resistors**. They cost quite a bit more than the ordinary type, though.

The following table shows the value of an unknown resistor connected to the terminals when the switch is in the ohmmeter position:

Meter Reading (Ma.)	Ohms
1.0	0
.97	100
.86	500
.75	1000
.60	2000
.50	3000
.375	5000
.30	7000
.23	10,000
.20	12,000
.14	18,000
.10	27,000
.06	47,000

This calibration applies when any 1-ma. meter of approximately 50 ohms internal resistance is used with a 3-volt source and a 3000-ohm series resistor.

A. C. Measurements

Meters for measuring 60-cycle a.c. voltage and current are similar in principle to those used for d.c. There is rarely a need for measuring current at the supply-line frequency, so you will seldom if ever see an a.c. ammeter or milliammeter in an amateur station. A.c. voltmeters frequently are used in transmitters having high-power tubes, because good tube life requires that the filament voltage be kept at the rated value. Instruments of this type usually are accurate only at the frequency for which they are designed—60 cycles, generally.

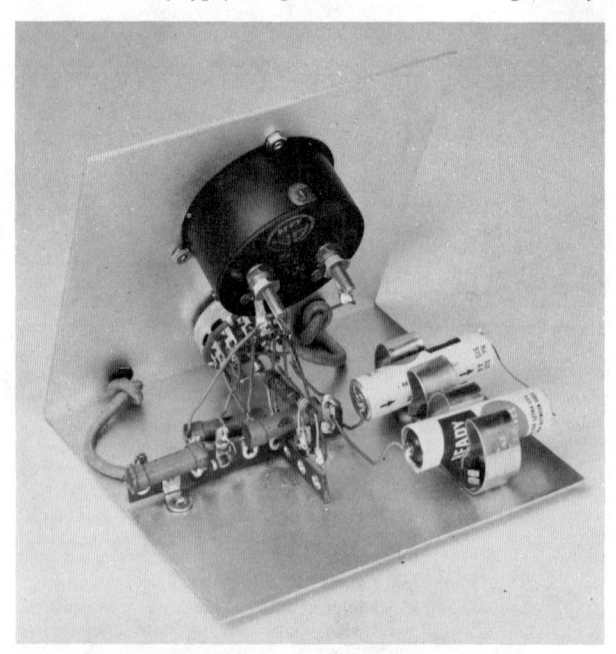

FIG. 14-6—Rear view of the volt-ohmmeter. Resistors are supported on tie-point strips and the switch contacts. The spring clips for the flashlight cells are holders for broom handles, tools, etc., available at any hardware store. Make sure that the cell cases are insulated from the clips.

Measurements

A different type of voltmeter is incorporated in the multipurpose test instruments mentioned earlier. The basic meter in these instruments must measure d.c., and to make the meter useful on a.c. an a.c. voltage must first be converted to d.c. A semiconductor (copper oxide) full-wave bridge rectifier is often used for this (Fig. 14-7). The copper-oxide rectifier a.c. voltmeter will work well over the audio-frequency range from 50 to 5000 cycles.

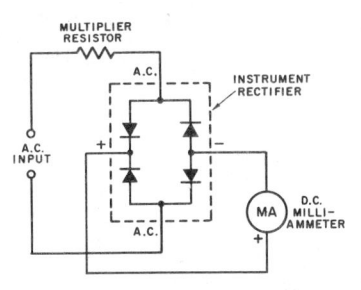

FIG. 14-7—One type of a.c. voltmeter using a d.c. meter movement with a rectifier.

The readings of all types of a.c. meters must be taken with reservations. The scales on such instruments are always calibrated in terms of the r.m.s. value of a sine wave. But rectifier-type instruments do not actually respond to the r.m.s. value, so the readings are within the rated accuracy of the instrument only when the voltage is sinusoidal. Generally, it is safe to assume that the voltage will be sinusoidal if it is coming from the a.c. line directly or through a transformer. Thus you should get a reasonably accurate reading on, say, the 6.3-volt heater winding of a transformer. Measurements on audio circuits, particularly voice waveforms, are useful only in a relative sense—i.e., stronger or weaker; in general, the scale reading means little without supplementary information about the waveform.

R. F. Voltage

The rectifier scheme is a particularly good one for measuring radio-frequency voltages. Ordinary instruments will not do for frequencies in the r.f. range. R.f. circuits, particularly tuned ones having relatively high impedance—as is the case in most receiving and transmitting equipment—are acutely sensitive about having things connected to them. Test leads have too much inductance and capacitance; they detune the circuit so much that measurements are worthless.

On the other hand, a germanium diode (the 1N34A is typical) has a capacitance of only about one micromicrofarad and physically is so small that it can be installed right in the r.f. circuit where the voltage is to be measured. Then only d.c. leads need be brought out to go to an ordinary milliammeter. Fig. 14-8 is a typical circuit. The part to the right of the dashed line is the voltmeter; the part to the left is merely one type of connection that could be made. The leads from the r.f. circuit through CR_1 and C_1 would be made as short as possible by installing these components right at the circuit being measured—usually installing them permanently. Sometimes the voltmeter components are made up in the form of an **r.f. probe** which can be touched directly on the circuit to be measured, with negligible lead length in the r.f. part. The d.c. leads can be any length desired.

In amateur equipment the r.f. voltmeter is used principally for *relative* voltage indications. An example is the r.f. output indicator used as a tuning aid in many transmitters. For this purpose it isn't necessary to know the actual voltage, since the practical purpose of the instrument is fully realized when you make the meter pointer go as far toward maximum as possible.

In some r.f. measurements—e.g., in measuring s.w.r.—it is necessary that the meter indications be linear, if the measurements are to mean anything. Even here, though, you don't have to know the *actual* value of the voltage—it is sufficient to know that the voltage will be directly proportional to the scale reading. The principal purpose of R_1 in Fig. 14-8 is to make the meter linear. With little or no added resistance in the circuit the meter indications tend to be proportional to the *square* of the r.f. voltage applied to the rectifier. It takes about 10,000 ohms at R_1 to make a reasonably linear voltmeter when M_1 is a 0–1 milliammeter.

R. F. Current

The instrument used for measuring r.f. current is a **thermocouple r.f. ammeter**. A thermocouple is a junction of two dissimilar metals which has the property of generating a small direct current when heated. In the r.f. ammeter the radio-frequency current flows through a short length

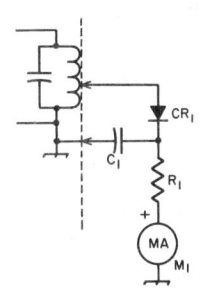

FIG. 14-8—An r.f. voltmeter circuit. CR_1 usually is a 1N34A or a similar type diode. The value of C_1 is usually 0.001 to 0.01 µf.; a ceramic capacitor is recommended. M_1 can be a 0-1 milliammeter. R_1 (see text) can be variable if the sensitivity of the system is to be adjusted (the larger the resistance value the lower the current); a 10,000-ohm composition control is typical.

In the circuit above, the coil in the r.f. circuit provides a d.c. return to ground for the rectifier. If the circuit does not offer a d.c. return, an r.f. choke should be connected from the upper end of the rectifier to ground.

of resistance wire which heats up in proportion to the r.f. power it dissipates. This heat is applied to the thermocouple, and the current generated by the junction flows through a d.c. milliammeter. The face of the instrument is calibrated in r.f. amperes.

An r.f. ammeter is expensive when purchased new, but a variety of these instruments have been available at low prices through military

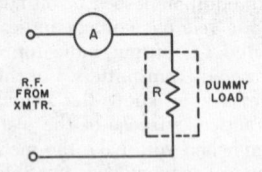

FIG. 14-9—Using an r.f. ammeter to measure current in a dummy antenna of known resistance, for power measurement. The power is equal to I^2R. A current of 1.5 amp. flowing in a 50-ohm resistance is therefore $2.25 \times 50 = 112$ watts.

surplus. It is rarely necessary to use one in an amateur transmitter; the principal value of an r.f. ammeter in amateur work is in measurement of r.f. power (Fig. 14-9). For this purpose it is necessary to have a load of known resistance, and to know that it is a "pure" resistance. By Ohm's Law, the r.f. power in such a load is equal to the square of the current multiplied by the resistance of the load.

Aside from such measurements, it usually suffices to know the *relative* change in current when you're adjusting a transmatch or the transmitter tuning. A flashlight or dial lamp is a cheap, but satisfactory, indicator for this. The brightness of the lamp shows whether an adjustment has resulted in more or less r.f. output. Ways of using the lamps for this purpose are described in Chapter 11. The circuit given in Fig. 14-10 is typical for checking feeder current.

Instrument Effects

You can have an instrument of excellent quality, capable of a high degree of accuracy, and still make measurements that are wide of the mark.

Why? Because every instrument becomes a part of the circuit in which the measurements are being made. To the extent that the presence of the instrument *changes* the circuit conditions,

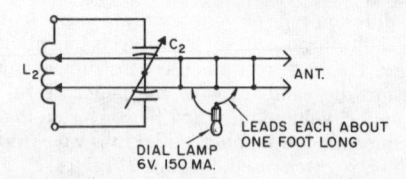

FIG. 14-10—Using a dial lamp as a substitute for an r.f. ammeter, to show relative r.f. current. In measurements of this type the actual value of current need not be known; in fact, the current in amperes will be different at different frequencies in tuned feeder systems, for the same power. The important thing is to get as much current as possible into the feeder at the frequency in use.

as compared with those existing when it isn't present, the indications will be in error. Of course, if you leave the instrument in the circuit permanently this doesn't matter. But it *does* matter when a test instrument is used temporarily for checking.

Errors of this nature are more likely to be appreciable in measuring voltage than in measuring current. Fig. 14-11 is an illustration. By Ohm's Law, the voltage across each 100,000-ohm resistor in A must be 50 volts when 100 volts is applied to the two in series. If we try to measure the voltage across the lower resistor, R_2, with a 1000-ohms-per-volt meter, its resistance, also 100,000 ohms for a 100-volt scale, is in parallel with R_2. The resistance of this combination is only 50,000 ohms, so now there is only 33 volts across R_2, as shown in B. This is what a perfectly accurate meter would read.

Now if a 20,000-ohms-per-volt meter is used instead, the voltmeter resistance (still assuming a 100-volt scale) is 20 times as large as R_2, so the effect on the total resistance is small. The meter now reads 48.7 volts, again quite accurately, but the indicated voltage still isn't quite what the *actual* voltage would be with the meter disconnected. A 10-megohm voltmeter reduces the error still more, as shown at D.

Observe, however, that if R_2 had been connected directly to a 50-volt source, such as a battery, having very low internal resistance, the added current taken by the voltmeter would have only a negligible effect on the actual voltage. In that case all three instruments would give the same reading, 50 volts. It is the presence of a large *series* resistance, R_1, that is responsible for

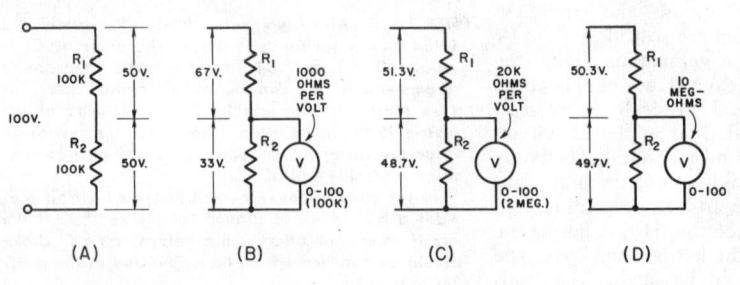

FIG. 14-11—Effect of voltmeter resistance on indicated voltage in a typical case. Although all three meters may be equally accurate, each indicates a different voltage because the presence of the *meter itself* changes the voltage across R_2.

Measurements

the error. Nevertheless, if the voltmeter's resistance is very large compared with the resistance across which it is measuring voltage, the error will not be large even if the series resistance is high. "Very large" here means at least ten times (which may cause an error as large as 10 per cent in the reading) and preferably at least 20 times (which cannot cause more than about 5 per cent error in the worst case).

A comparable error in current measurement results when the resistance of a milliammeter is large compared with the series resistance of the circuit in which it is inserted. A 50-ohm instrument introduced into a 25-ohm circuit obviously would reduce the current by a very considerable amount. But this isn't so likely to happen in practice with ordinary instruments. A 50-ohm instrument would be a low-range one, and therefore would be used in low-current circuits—which are usually ones having fairly high resistance, at least as compared with the meter resistance. But "not likely" doesn't mean "never"; so don't take it for granted that current measurements will always be free from instrument effect.

The Vacuum-Tube Voltmeter

A type of instrument that is relatively free from the effects discussed above is the **vacuum-tube voltmeter**. It is beyond our scope here to go into details, so it must suffice to say that by and large the "v.t.v.m.," as it is usually called, is essentially an amplifier for d.c. voltages, combined with an indicator. Its principal advantage is that the vacuum tube requires no power from the source of voltage being amplified (this was discussed in Chapter 2) and therefore a voltmeter using a tube doesn't load the circuit being measured. Practical v.t.v.m.'s don't quite meet this ideal, since it is necessary to have a d.c. return in the grid circuit. Nevertheless, the voltmeter resistance for d.c. measurements is very high; 11 megohms is a more-or-less standard resistance for these instruments.

The most economical way to get a good v.t.v.m. is to buy a commercially-produced kit. There are several of these on the market, all in the same price range. It would cost you a good deal more to buy the parts individually, and you would not have the calibrated scales that go with the commercial jobs. Most instruments of this type have five or six d.c. voltage ranges, from a low range of 1.5 or 3 volts to a high range of 1000 or 1500 volts. They also have a comparable number of ohmmeter ranges, capable of measuring a few ohms to many megohms. The extreme

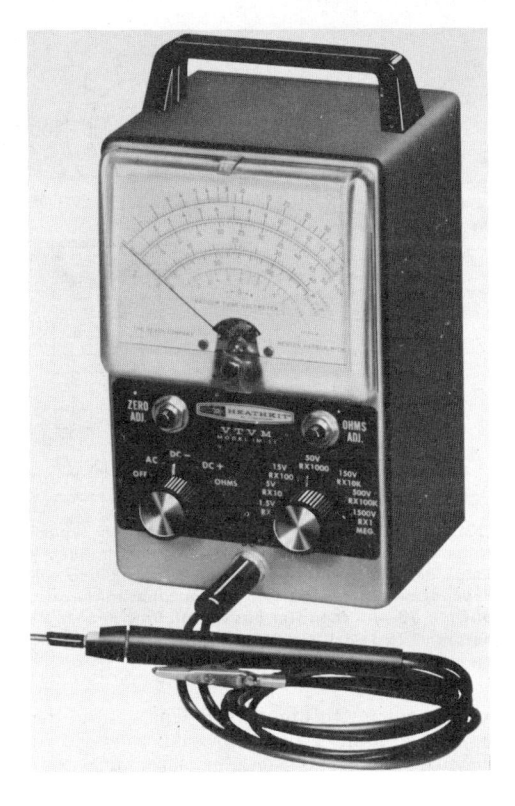

For many measurement purposes the vacuum-tube voltmeter has advantages not matched by conventional meters. Many amateurs consider it the indispensable servicing tool. The kit shown above, like most instruments of this type, has a large meter face for easy readability.

ranges cannot be duplicated by the v.o.m. described earlier. Also, there is usually a rectifier for converting a.c. to d.c. for the purpose of a.c. voltage measurements. The v.t.v.m. generally reads the *peak* value of the a.c. wave, although the meter is calibrated in r.m.s. based on a sine wave.

If you do much testing and experimenting you'll find a v.t.v.m. an invaluable part of your measuring equipment. Its disadvantage is that it needs a.c. power (although there are battery operated models available, at higher cost) and is usually suitable only for measurements where one terminal of the instrument can be grounded. These are not serious handicaps for most work.

Frequency Measurement

Literally, you can comply with the FCC regulations governing your transmitter's frequency without actually measuring it. The regulations don't require you to know the exact frequency on which you are sending. They *do* require that it be within a *band* of frequencies, and you must have some way of telling whether or not you are operating inside an authorized band. The system

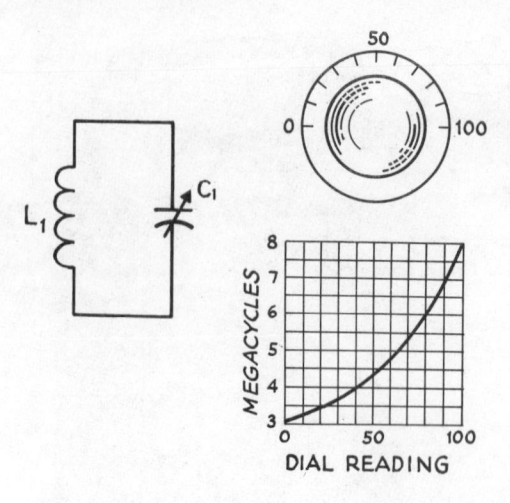

FIG. 14-12—The wavemeter circuit and a typical calibration. L_1C_1 can be selected for any frequency range desired. The range covered with a single coil depends on the maximum/minimum capacitance ratio of C_1; a 50-μμf. capacitor has a useful range of a little over 2 to 1; a 140-μμf. capacitor will tune over a bit more than 3 to 1.

for determining this must be independent of the transmitter itself. That is, you cannot depend on the frequency marked on a crystal used in your transmitting oscillator. It may be quite accurate, but in itself it isn't enough to satisfy the FCC.

It takes only simple equipment to locate the limits of these frequency bands with very high accuracy. Such equipment is based on using harmonics of a high-stability oscillator. The oscillator can be adjusted to exact frequency by using the government's standard-frequency transmissions over the National Bureau of Standards station WWV. It is something of a paradox, though, that while such a simple **frequency standard** will locate a band edge for you to within a few cycles of the exact frequency, it gives you no assurance whatsoever that your transmitter is inside the right limits. You might actually be transmitting on the 80-meter band, for example, when you *think* your transmitter is "putting out" in the 40-meter band. The frequency standard takes care of the cycles, but you need something else to find the right *mega*cycles.

Adequate frequency checking, then, has to be done in two steps: First, it must be determined that the transmitter's output is in the right band. Second, it must be established that the frequency is within the right limits *inside* that band. It is possible to do both without any auxiliary equipment at all, after you've gained a little experience. But it's a help to have simple frequency-measuring gear on tap.

The Wavemeter

The very simplest device for measuring frequency is a resonant circuit—a coil and capacitor. You learned in Chapter 2 that the current in such a circuit will be largest when the circuit is tuned to the frequency of the r.f. energy introduced into it. If you hook a coil and variable capacitor in parallel, Fig. 14-12, the circuit can be tuned over a range of frequencies by varying the capacitance from maximum to minimum. If the capacitor is provided with a numbered dial, this **wavemeter** can be calibrated so you know the frequency corresponding to each dial setting. The calibration can be drawn on graph paper as suggested in the figure.

Parenthetically, the name wavemeter dates back to the time when radio was known as "wire-

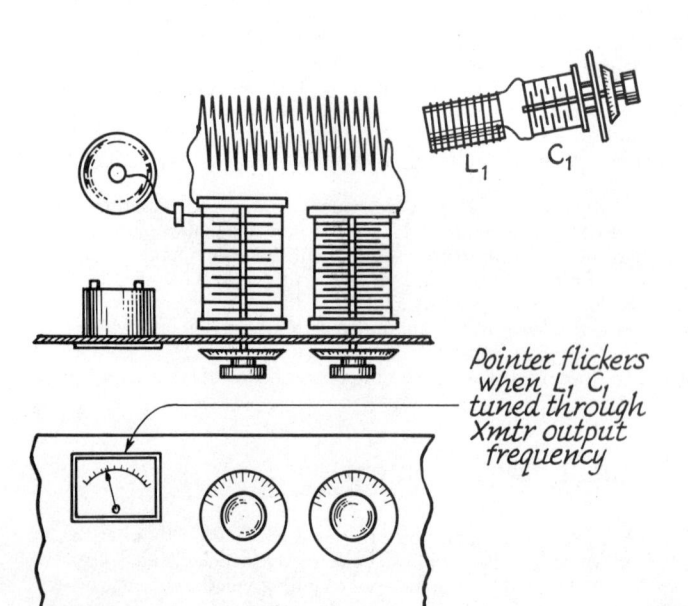

FIG. 14-13—How the absorption wavemeter is used for checking transmitter frequency. CAUTION: Don't allow your hands or any part of the wavemeter to come in contact with "live" circuits.

Measurements

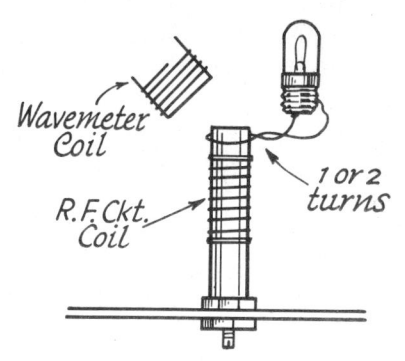

FIG. 14-14—Using a dial lamp and loop as a resonance indicator in low-power circuits.

less" and it was the universal custom to think in terms of wavelength instead of frequency. The remnants of this practice are still with us—we speak of the "80-meter" and "40-meter" bands, for instance. But no one seriously thinks of measuring wavelength. Frequency can be specified and measured much more accurately.

The wavemeter is used by bringing its coil near a circuit carrying r.f. energy. Checking the output frequency of a transmitter would be done by the method shown in Fig. 14-13, where the wavemeter coil is loosely coupled to the final tank coil. In cases like this it isn't necessary to have any form of current or voltage indicator connected in the wavemeter circuit itself. When the wavemeter is in resonance it takes a little energy out of the amplifier circuit. This disturbs the circuit operation and the amplifier's plate milliammeter will give a small "kick" as you tune the wavemeter through resonance. The same thing will happen if the meter on the transmitter is one that indicates r.f. output instead of plate current. A wavemeter used in this way is called an **absorption wavemeter**.

Some circuits, especially the low-power intermediate stages in transmitters, may have no plate- or grid-current indicators. In such a case the method shown in Fig. 14-14 can be used, if the power in the tank circuit is a watt or so. A 60-ma. dial lamp will light on quite low power. When the wavemeter is tuned through resonance the lamp will get dimmer because some of the energy is absorbed by the wavemeter. This method is particularly useful for low-power oscillators and amplifiers during construction.

When the tuning of an intermediate stage of a complete transmitter is being checked the grid-current meter of the final stage can be used. The grid current will "dip" when the wavemeter is in resonance with the frequency of the intermediate stage to which it is coupled.

The "Band-Finder"

Figs. 14-14 and 14-15 are photographs of a simple absorption wavemeter covering the range from 3 to 270 Mc. It takes six coils to cover this range without any gaps. Continuous frequency coverage is desirable because occasionally a transmitter stage may be tuned up on a frequency considerably removed from the intended band; without continuous coverage it would be difficult to determine just where the mistake was made.

For convenience in handling, the Band-Finder is mounted in the smallest size Minibox—2¼ × 2⅛ × 1⅝ inches (Bud CU-2100A). The four-prong socket for the plug-in coils is centered on one end. The tuning capacitor (the circuit is simply that of Fig. 14-12) is mounted on the adjacent side of the box, positioned so that the rotor lug makes contact with the No. 1 socket prong and one stator rod is right alongside the No. 4 prong. Not more than about 1/16 inch of lead should be needed for making the connection between the stator rod and the socket.

If this "no-lead" construction is duplicated and the coils are made exactly as shown in Fig. 14-17,

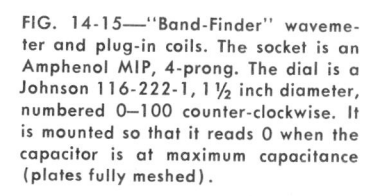

FIG. 14-15—"Band-Finder" wavemeter and plug-in coils. The socket is an Amphenol MIP, 4-prong. The dial is a Johnson 116-222-1, 1½ inch diameter, numbered 0–100 counter-clockwise. It is mounted so that it reads 0 when the capacitor is at maximum capacitance (plates fully meshed).

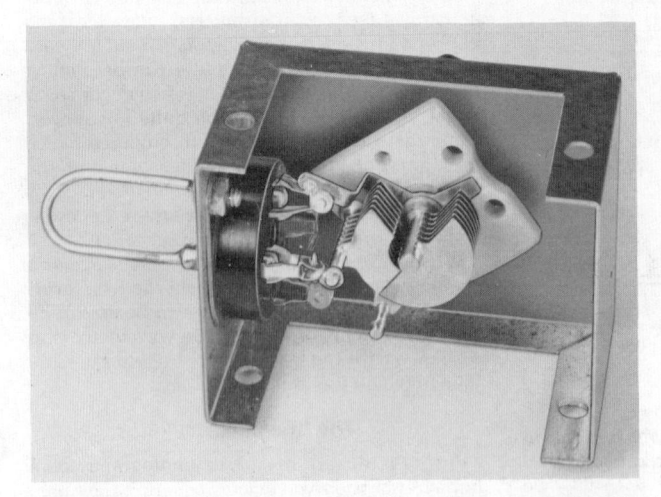

FIG. 14-16—Inside the Band-Finder. The coil socket is mounted with the edges of the metal plate parallel to the top and bottom edges of the end of the box. The No. 1 prong is on the inside—nearest the fold in the metal. If the capacitor shaft is mounted in the exact center of the top, positioning the capacitor (Hammarlund HF-50) as shown will reduce the lead length practically to zero. This is necessary if the calibration data in the table are to be useful.

the calibration data given in the table will be accurate enough so a direct calibration won't be needed. For this to be so, however, the components and construction specified *must* be followed *exactly*. The highest-frequency coil, No. 6, is a "hairpin" made from a piece of No. 12 wire as shown in Fig. 14-16, using the large prongs broken out of an old tube base or 4-prong plug or form. When using this coil, be sure that the rims of the prongs are seated directly on the socket.

With a tuning capacitor having semicircular plates, such as this one, the low-capacitance half of the scale tends to be crowded. For this reason calibration points are given for each 5 divisions in this half; in the high-capacitance half points 10 divisions apart suffice. The points should be plotted on graph paper and a smooth curve matched to them. Sufficient accuracy (and all that the dial size warrants) will be achieved if the graph scale is 50 divisions on paper having 20 divisions to the inch. This will give one division for each of the 50 divisions on the dial itself. The frequency scale in each case also should be spread over 50 divisions, since a square graph gives optimum readability consistent with accuracy.

FIG. 14-17—Band-Finder coil construction. Except for the highest-frequency (No. 6) coil shown at the left, all coils are wound on 1-inch diameter 4-prong bakelite forms (Millen 45004). Holes for ends of coils are drilled directly above large prongs, 90 degrees apart around the form circumference. All coils are wound with No. 24 enameled wire, with uniform spacing between turns to fill out space between holes for leads. Pull leads tight in pins before soldering.

Coil No.	Range, Mc.	No. of Turns	A (in.)	B (in.)
1	3–7.5	60*	1/8	1 7/16
2	6–15	29 1/4	1/8	1 3/8
3	11.5–30	13 1/4	3/16	1 3/16
4	27–73	4 1/4	1/4	3/4
5	60–170	3/4	1/4	1/4
6	87–270	See drawing above.		

* Close-wound; start at A and wind in direction of small arrow. Coils with spaced turns may be uniformly spaced by winding wire or string of appropriate thickness between turns; then tighten coil by pulling leads through pins, remove spacing wire or string and lacquer coil to hold turns in place.

"Band-Finder" Wavemeter Calibration in Megacycles

Dial Setting	Coils					
	1	2	3	4	5	6
0	2.9	5.8	11.3	27	59	87
10	3.0	6.1	11.7	28	62	90
20	3.2	6.3	12.4	29	65	95
30	3.4	6.7	13.2	31	68	100
40	3.6	7.2	14.0	33	73	107
50	3.9	7.8	15.0	36	80	117
60	4.3	8.5	16.5	40	88	131
65	4.5	9.0	17.5	42	93	139
70	4.8	9.5	18.5	45	99	148
75	5.1	10.2	19.7	47	106	160
80	5.5	10.9	21.3	51	116	175
85	6.0	12.4	24.3	56	129	195
90	6.6	13.4	26.4	64	146	225
95	7.5	15.1	29.5	72	167	260
100	7.7	15.5	30.1	73	172	270

Measurements

FIG. 14-18—Simple band checker for use with coaxial line. It is built in a 2⅛ × 3 × 5¼ inch Minibox. The band-switch knob is on the left, capacitor knob on the right. The dial-lamp indicator fits into a ⅜-inch rubber gromet. Phono jacks are used as inexpensive connectors for the coax cables; regular coax chassis connectors can be substituted.

Your transmitter can be used to check the calibration. If its output is in the Novice 80-meter band, for example, a check by the method of Fig. 14-13 should show resonance slightly below 50 on the wavemeter scale, using the No. 1 coil. Similarly, with the transmitter in the 40-meter Novice band there should be a kick in the transmitter's meter reading when the wavemeter dial is at about 95. On the No. 2 coil the meter should resonate with the 7-Mc. output of the transmitter at approximately 40 on the dial. This type of check can be carried through on all bands the transmitter can cover.

If you have or can borrow a calibrated grid-dip meter you can make an actual calibration for each coil. Set the wavemeter dial successively at the numbers listed in the table and find the frequency with the grid-dip meter. The use of a grid-dip meter is discussed a little later in this chapter.

The Band Checker

It's handy to have a "gimmick" permanently connected to the transmitter to tell you for sure which band you're on. A wavemeter with a lamp indicator can easily be used for this if the transmitter is working into coax line. Figs. 14-18 and 14-20 show how one can be built. The circuit is given in Fig. 14-19. Three values of inductance are available for covering 3.5 to 30 Mc. with a variable capacitor. L_1 is coupled to the inner conductor of the coax line by a one-turn link. This transfers enough energy to make the lamp glow when the circuit is tuned to the operating frequency.

If the specifications given in Fig. 14-19 are followed exactly, no special calibration of the circuit is needed. The tuning ranges will be approximately 2 to 1. In the first position of the switch the circuit covers the 3.5-Mc. band at the high-capacitance end of the tuning scale and the 7-Mc. band at the low-capacitance end. The second position of the switch similarly gives coverage of 7 and 14 Mc., while the third covers 14 to 30 Mc. The 21-Mc. band is in between on this last range. The lettering on the panel in Fig. 14-18 indicates the approximate position of each band on each scale. The high-capacitance end is reached with the knob pointing horizontally to the left.

The band checker should be connected between the transmitter and the coax feed line to the antenna. Before using it, it is a good idea to put it through its paces with a dummy antenna. A 115-volt lamp can be used for this; connect the shell of the lamp base to the chassis and the center stud to the inner connection on J_2. The transmitter's output goes to J_1. Using the proper

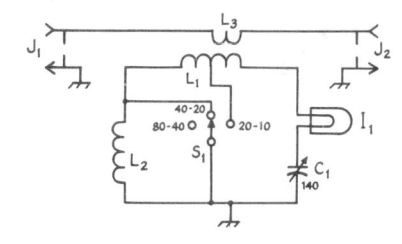

FIG. 14-19—Circuit diagram of the band checker for coax lines.

C_1—140-$\mu\mu$f. variable (Hammarlund HF-140).

l_1—Dial lamp; up to 25 watts input, No. 48 or 49 (60 ma.).
    25-50 watts input, No. 47 (150 ma.).
    50-75 watts input, No. 46 (250 ma.).

J_1, J_2—Phono connector.

L_1—15 turns No. 20, 16 turns per inch, 1-inch diam. (B & W Miniductor 3015).

L_2—22 turns No. 24, 32 turns per inch, 1-inch diam. (B & W Miniductor 3016).
    Tap for 7-14-Mc. range is at junction of L_1L_2; tap for 14-28-Mc. range is 5 turns from l_1 end of L_1.

L_3—Coupling loop; see Fig. 14-14.

S_1—Single-pole three-position switch (Centralab type 1461).

FIG. 14-20—Inside the coax band checker. C_1 is at the left, S_1 at the right (a two-pole switch is shown, but only one pole is used.) L_1 is between C_1 and S_1. L_2 is mounted behind S_1. The coupling turn L_3, is the loop of insulated wire on the near end of L_1. Leads to the lamp are soldered to the shell and base; no socket is needed although one could be used if desired.

crystals in each case, put the transmitter successively on all five bands and find the position of C_1 at which the lamp lights brightest on each. Mark the panel at each such spot so you know where to set the knob on C_1 when preparing to tune up for actual operation on each band.

Get the habit of first setting the band checker to the right band when you plan to shift your transmitter. If the lamp doesn't light after tuning up, the transmitter controls may not be set properly for the band you want. If the lamp does light you know you're OK.

Frequency Markers

Now we come to the second step in frequency checking—determining your transmitter's whereabouts *inside* an amateur band. For this some accurately located guideposts or **markers** are needed. These are signals whose frequencies are known very accurately, and which can be tuned in by your receiver.

Some receivers come equipped with a 100-kc. crystal-controlled oscillator just for this purpose. The **crystal calibrator**, as it is called, generates a harmonic at every multiple of 100 kc. The harmonics get weaker as we go higher in frequency, but they are strong enough to be used as marker signals on any amateur band through at least 30 Mc.

Harmonics are always *exact* whole-number multiples of the oscillator frequency. Therefore if the oscillator is set precisely on a known frequency, the harmonic frequencies also are known with the same percentage accuracy. When the oscillator is on exactly 100 kc. its 35th harmonic is 3500 kc., the 36th is 3600 kc. and so on. The edges of most amateur bands also are multiples of 100 kc., which is one reason why an oscillator on 100 kc. is useful. However, some sub-bands, including a few used by Novices, end with a 50-kc. figure. To check the edges of these bands it is desirable to have multiples of 50 kc. This requires a 50-kc. oscillator.

A 50-kc. Frequency Standard

To be most reliable, the oscillator in a frequency standard or **marker generator** should be crystal-controlled. Crystals for 100 kc. are readily available at relatively low cost, but 50-kc. crystals are rather rare. To get around this, the standard shown in Figs. 14-21 and 14-23 uses a 100-kc. oscillator followed by a frequency-dividing circuit which reduces the output frequency to 50 kc. The frequency divider used in Fig. 14-22 is called a **multivibrator**. It is a two-tube resistance-coupled circuit with the output of each tube driving the grid of the other. This feedback results in self-oscillation at a frequency determined, in general, by the time constants of the *RC* circuits.

A multivibrator in itself is a very unstable oscillator. Its frequency tends to wander continually when the circuit is **free running**. The feature that makes it useful is this: If a signal from a stable oscillator is introduced into the circuit, the multivibrator will **lock in** at some exact submultiple (divisor) of the controlling frequency. Thus we can use a signal from a crystal-controlled 100-kc. oscillator to lock a multivibrator on 50 kc., and the frequency stability will be equal to that of the crystal oscillator. The divisor in this case is 2, but 3 or 4 or even higher numbers could be used. In some frequency standards multivibrators divide by 10, with a 100-kc. oscillator controlling 10-kc. oscillation in the multivibrator. However, markers closer than 50-kc. apart aren't needed for our purposes.

The 100-kc. crystal oscillator in Fig. 14-22

FIG. 14-21—A frequency standard that generates marker signals for frequency checking. Marker intervals of either 100 or 50 kc. are available. The chassis is aluminum, 5 × 4 × 2 inches. The power transformer is at the left on top. The two tubes, from the front, are V_2 and V_1, Fig. 14-22. The 100-kc. crystal is at the rear.

can be used either by itself or to control the 50-kc. multivibrator. The oscillator circuit is a more-or-less standard one, and you can build it separately if you want just 100-kc. harmonics. The frequency can be adjusted over a small range by means of C_1. This adjustment compensates for minor circuit variations which affect the frequency slightly, and enables setting the oscillator on exactly 100 kc. The method of doing this is discussed a little later. S_{2A} connects the oscillator either to the output terminal, when just 100-kc. markers are wanted, or to the multivibrator when 50-kc. markers are required.

The frequency standard shown here includes a small power supply, making it independent of other equipment. However, power for it can be be taken from the accessory socket on your receiver. The heaters of the two tubes take 0.45 amp. at 6.3 volts, and the plate requirements are 15 ma. at about 130 volts. The power-supply circuit shown uses a semiconductor rectifier and an RC filter. The power switch, S_1, has three positions; off, heaters on and plates off, and everything on. The "plates off" position is quite desirable since the markers may interfere with your regular reception if the oscillator cannot be shut off. At the same time, it is desirable to keep the tube heaters up to temperature so you can make a quick check when necessary. Separate s.p.s.t. toggle switches could be used to do the same switching job.

The output coupling capacitor, C_6, from the multivibrator can be chosen to give a satisfactory marker-signal strength in your receiver. A few $\mu\mu$f. should suffice. The capacitor in this standard is simply a turn of insulated wire wrapped around the lead from pin 6 of V_{2B} to the 1500-ohm resistor. There is no metallic connection at this point; the wire itself connects to the output terminal. The output terminal should be connected to the antenna post on your receiver, and the chassis of the receiver and standard should be connected together.

Using the Frequency Standard

The frequency standard gives you a series of signals of known frequency, and checking a transmitter frequency is simply a process of comparing the transmitter's signal with the markers. Your receiver is an essential part of this process.

A marker signal by itself gives no indication of its frequency. All you actually know is that each such signal is an exact multiple of 50 kc. Thus the first step is to *identify* a particular marker signal so you can be sure, for instance, that it is on 7150 kc. and not on 7300 or some other 50-kc. multiple.

This is done by comparing the markers with signals of known frequency. Government or commercial stations near the amateur bands, as well as amateur signals themselves, can be used for this. Examples of government signals that are useful are NSS on 4005 kc. and CHU on 7335 kc. Suppose you want to use CHU to find the 7-Mc. band. Tune in the time signal that this station transmits, turn on the frequency standard with S_2 set for 100-kc. output, and slowly tune *lower* in frequency until you run into the first marker. This must be 7300 kc. Note the receiver dial reading and continue tuning lower until you hear the next marker, which will be 7200 kc. Note this dial setting and then go on in the same way until you finally reach 7000 kc. At this point you have the dial settings for each 100-kc. point throughout the band.

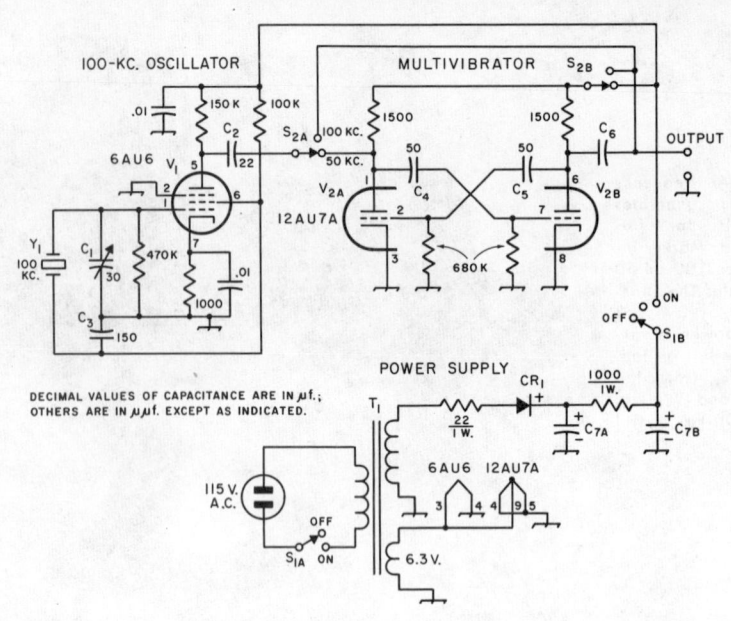

FIG. 14-22—Circuit diagram of the frequency standard. Resistances are in ohms, resistors are ½ watt unless specified otherwise. Fixed capacitors with decimal values are disk ceramic, others are mica, with the exception of C_6 (see text) and C_7 which is electrolytic.

C_1—3–30 μμf. mica compression trimmer.
C_2—22-μμf. mica.
C_3—150-μμf. mica.
C_4, C_5—50- or 47-μμf. mica.
C_6—See text.
C_{7A}, C_{7B}—20-μf.-per-section dual electrolytic, 150 volts.
CR_1—Silicon rectifier, 400 volts p.i.v. (International Rectifier SD94, Sarkes-Tarzian 1N1084, RCA 1N540, GE 1N1695).
S_1—Single-pole, three-position wafer switch with a.c. switch mounted on rear (Centralab 1465).
S_2—Two-pole, three-position wafer switch, two positions used (Centralab 1472).
T_1—Power transformer, 125 volts, 15 ma., 6.3 volts, 0.6 amp. (Stancor PS-8415).
Y_1—100-kc. crystal.

Now turn S_2 to 50-kc. output, with the receiver still tuned to 7000 kc. Tune slowly upward in frequency and the first marker you come to will be 7050 kc. The dial setting should be about midway between those you found for 7000 and 7100. Continue and find the other odd 50-kc. frequencies until you have your receiver calibrated at each 50-kc. point throughout the band. (This calibration will be more accurate than the one on the receiver's dial; the latter is useful principally as a guide.) Now you can easily spot the exact limits of the Novice 7150-7200-kc. band, for example. Any signal that appears on the dial between those two points must be in the Novice band—and conversely, a signal either higher or lower in frequency than those two markers must be *outside* the Novice band.

The same method is used on any band you need to calibrate. In many cases you can tell where the extreme ends of an amateur band should be because you won't find amateur signals outside those limits. The marker at that point will give you the exact band-edge frequency.

In calibrating a receiver in this way it is desirable to have a reasonably-strong marker signal, but not one so strong that it tends to overload the receiver. (This can happen on the lower-frequency bands if the coupling between the standard and the receiver is too great.) Tinker with C_6 until you get a marker of a bit greater strength than the average amateur signals you hear on that band.

V.F.O. Calibration

The application of this and similar methods to calibrating other equipment, such as a v.f.o., should be obvious. In the case of a v.f.o. you may be able to get calibration points at smaller intervals than 50 kc. For example, if your v.f.o. covers 3500–4000 kc. you may be able to hear its fourth harmonic by tuning your receiver over 14,000–16,000 kc. The frequency standard will give you signals at 14,000, 14,050, 14,100 and so on. If you tune in the *harmonic* of the v.f.o. in each case the calibration points at the v.f.o.'s *fundamental* frequency will be one-fourth of 50 kc., or 12.5 kc., apart.

By using higher-order harmonics of the v.f.o. the calibration points will be still closer together. Using the fifth harmonic you can get points exactly 10 kc. apart on the fundamental by this method.

Measurements

FIG. 14-23—Under-chassis view of the frequency standard. The power supply components are grouped at the left hand side of the chassis. At the right are the components for the 100-kc. oscillator and multivibrator circuits. The capacitor at the lower right corner is C_1. A two-terminal strip (on the chassis back) is used for the output connections.

Checking Transmitter Frequency

To check your own transmitter's frequency, tune it in on your receiver and see which marker signals are on either side of it. The frequency will be somewhere between those two marker frequencies. You don't need to know anything more than that it is between 7150 and 7200 kc., say, if you're a Novice; the FCC only requires you to be safely inside the band.

Making a frequency comparison of this sort is simple in principle, but in actual practice it can be confusing if you don't go about it properly. Receivers will give all kinds of spurious responses when hit by extremely strong signals such as the one from your transmitter. One of the easiest things to do is to tune in one of these spurious signals—which *won't* be on the right spot on the dial—and make the mistaken assumption that you're listening on your actual frequency. If you've used an absorption wavemeter to check the transmitter you'll know you're in the right *region*. To get the right *frequency*, disconnect the antenna from the receiver, turn down the r.f. gain, and then tune around where you know the signal should be. Under these conditions it will be the strongest signal you pick up, and if the receiver is well shielded this signal will be the *only* one you'll hear in that region. Keep the receiver's r.f. gain down to the level where your own signal is about the same in strength as ordinary signals with the antenna connected. Then compare the frequency with the markers.

How to Use WWV

Thanks to standard-frequency transmissions by the Bureau of Standards, your crystal frequency standard can be checked at any time. This is done by comparing a harmonic of the standard with any of the WWV or WWVH transmissions that may be receivable at the time. WWV transmits on 2.5, 5, 10, 15, 20 and 25 Mc.; WWVH is on 5, 10, and 15 Mc. Except for a silent period of a few minutes each hour, these transmissions are continuous. Time ticks and various types of modulation, including standard audio frequencies of 440 and 600 cycles, are used.* Only the carrier frequencies are of interest to us in checking radio frequencies, and in making such checks the modulations are confusing rather than helpful. Fortunately, there is a minute or so during each five-minute period when there is no modulation except for the one-second time ticks. These are not bothersome.

WWV is near Washington, D. C., and WWVH is in Hawaii. Any carrier frequency you can hear can be used. The best one will depend on your distance from one or the other of the stations and the time of day. In general, the lower frequencies are better at short ranges and during the evening hours. In daytime it may be necessary to listen for one of the higher frequencies if you are a few thousand miles from either station. If your receiver is one having continuous tuning throughout the high-frequency spectrum, finding the best signal at any time is simply a matter of trying them all. The

---

*Complete details can be found in the chapter on measurements in *The Radio Amateur's Handbook* published by the American Radio Relay League.

receiver calibration should put you in the right frequency region, and the signal itself is easily recognized because of the continuous tone modulation and the time ticks.

If yours is an amateur-bands-only receiver, the WWV converter described in Chapter 9 will bring in either the 5- or 10-Mc. transmissions. One or the other of these two will be found useful at most places in the United States, especially during the evening.

To adjust your crystal calibrator, first tune in WWV with the calibrator off, but with the oscillator tube heater on and the circuit thoroughly warmed up. Turn off the receiver's beat oscillator. Wait for the tone on WWV to stop, leaving nothing but the time ticks and the usual background noise that you hear on an unmodulated carrier. Then turn on your crystal standard. If it is not exactly on frequency, you will hear either a low-pitched audio tone or a "whoosh-whoosh" pulsation in the background noise. The tone or pulsation is the beat between WWV's carrier and the harmonic of your crystal standard. The presence of such a beat indicates that your standard is off-frequency by the number of cycles per second in the beat tone or pulsation. Now adjust C_1, Fig. 14-22, to bring the oscillator frequency into zero beat (no tone or pulsation) with WWV. A very slow pulsation—one or two per second—may be the best you can hold. This is plenty good enough. WWV's transmissions are accurate to about 1 cycle in 100 *megacycles*. If you can get within a cycle or two per second at, say, 5 megacycles, and hold it there, your standard is doing very well indeed. However, simple crystal oscillators do drift with temperature and other factors, so it is a good idea to check against WWV every now and then.

This same procedure can be used with any crystal calibrator, homemade or manufactured. Most of them will have some built-in provision for adjusting the frequency, corresponding to C_1 in Fig. 14-22.

The beat between your crystal oscillator and WWV will be most easily observed if its harmonic and the WWV signal are about the same strength. If one is very much stronger than the other the beat may be hard to detect. Use the WWV signal that comes closest to matching the strength of the oscillator harmonic, or adjust the coupling between the oscillator and the receiver to match the strength of the WWV signal.

The Grid-Dip Meter

When you construct any equipment with tuned-r.f. circuits in it, such as a transmitter, receiver, transmatch, or the like, you need to know whether these circuits are capable of being tuned over the intended range. Until you have the equipment actually working the absorption wavemeter doesn't help, because there has to be r.f. current in the circuit before the wavemeter can function. For checking a "dead" circuit the measuring device has to supply its own r.f., and must also have some sort of built-in indicator to tell when it is in resonance with the tuned circuit being checked.

The **grid-dip meter** or **grid-dip oscillator** (g.d.o.) is just such an instrument. It is a low-power oscillator having a sensitive d.c. meter in series with the oscillator tube's grid resistor for measuring its grid current. The g.d.o. operates like an absorption wavemeter in reverse. That is, the circuit *being checked* absorbs some r.f. power from the grid-dip meter when the latter's tuned circuit resonates with it. This causes the grid current to flick downward, or "dip."

Grid-dip meters usually have plug-in coils so a wide frequency range can be covered. You can build such a meter from a manufactured kit, several brands of which are available. The kits compare favorably in cost with the cost of equivalent new parts, and in addition have the calibration scales supplied. However, if you already have some suitable components you can easily construct your own, since the circuit is not at all complicated.

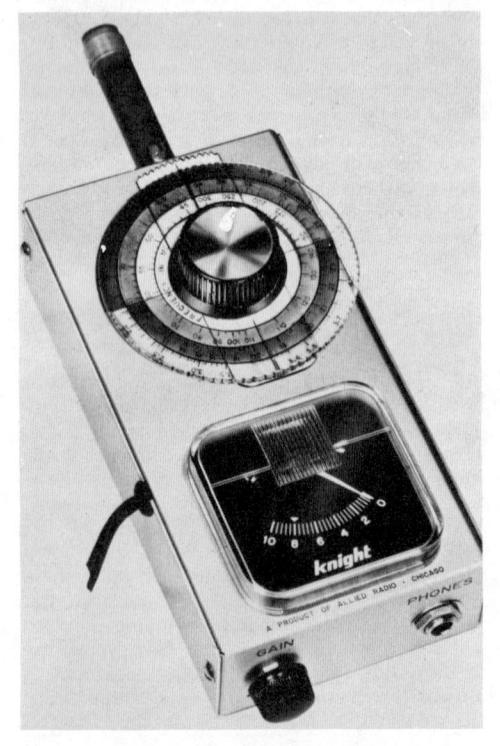

Grid-dip oscillator built from a kit. Commercial instruments generally have self-contained meter and power supply, as does this one.

Measurements

Figs. 14-24 and 14-26 are two views of a compact grid-dip meter covering 1.6 to 160 Mc. with six plug-in coils. The power supply and meter are in a separate unit, Fig. 14-27. Most commercial grid-dip meters combine the meter and power supply with the oscillator, but separating them in this way makes the "business" end of the assembly quite small and light, which is often an advantage when the instrument is to be used in spots where space is at a premium.

The construction shown in Fig. 14-26 keeps lead lengths as short as possible so the meter will work well through the high end of the range. The tuning capacitor, C_1 in Fig. 14-25, is mounted so its terminals are practically touching the coil socket prongs. The oscillator tube is mounted on a small aluminum bracket positioned so that the leads to the tuned circuit are short, but at an angle so that the tube can be removed from its socket. The connection between the cathode prong and the corresponding prong on the coil socket is a flat strip of copper, which has less inductance than a wire of the same length.

Coils for the two lowest ranges are hand-wound directly on the forms. All other coils are inside the forms, which are used principally as plug-in mountings.

The tuning dial is a disk of ⅛-inch Plexiglas, 2½ inches in diameter, with a hairline indicator scratched on it. The transparent disk allows putting the frequency calibrations on a card glued to the top of the box. This gives more room for lettering than would be the case if the calibration were put on the dial itself. Extending the disk over the edge lets you operate the tuning control with the thumb of the same hand that holds the box.

The power supply circuit, Fig. 14-28, uses a small transformer with a selenium rectifier and RC filter. A potentiometer, R_2, is used for adjusting the output voltage. Adjustable voltage is desirable because the grid current is different on different frequencies. R_2 lets you keep it in the middle part of the meter scale, where the dip is easily observed. The four-conductor cable connecting the oscillator and power supply can be any reasonable length; the only voltage drop of consequence will be in the heater leads in a long cable. The power unit can be placed wherever it is convenient for watching the meter while you check a circuit with the oscillator.

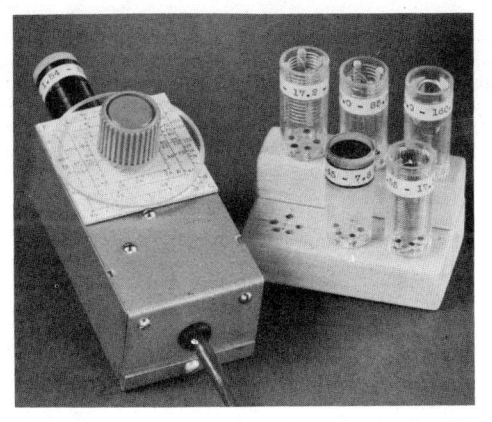

The instrument may be calibrated by listening to its output with a calibrated receiver. The calibration should be as accurate as possible, although "frequency-meter accuracy" is not required in the applications for which a grid-dip meter is useful.

In using the grid-dip meter for checking the resonant frequency of a circuit the coupling should be set to the point where the dip in grid current is just perceptible. This reduces interaction between the two circuits to a minimum and gives the highest accuracy. With too-close coupling the oscillator frequency may be "pulled" by the circuit being checked, in which case different readings will be obtained when

FIG. 14-24—Compact grid-dip meter and coils. Milliammeter and power supply are in a separate mounting, Fig. 14-27. The tuning knob can be fastened to the transparent disk by a machine screw or cement.

Coil Data, L_1

Freq. Range	Turns	Wire	Diameter	Turns/inch	Tap*
1.59- 3.5 Mc.	139	32 enam.	¾ in.	Close-wound	32
3.45- 7.8 Mc.	40	32 enam.	¾ in.	Close-wound	12
7.55- 17.5 Mc.	40	24 tinned	½ in.‡	32	14
17.2 - 40 Mc.	15	20 tinned	½ in.‡	16	5
37 - 85 Mc.	4	20 tinned	½ in.‡	16	1⅓
78 -160 Mc.	Hairpin of No. 14 wire, ⅜ in. spacing, 2 inches long including coil form pins. Tapped 1½ in. from ground end.

* Turns from ground end.
‡ B. & W. Miniductor or equivalent mounted inside coil form.
Coil forms are Amphenol or Allied Radio 24-5H, ¾-in. diameter.

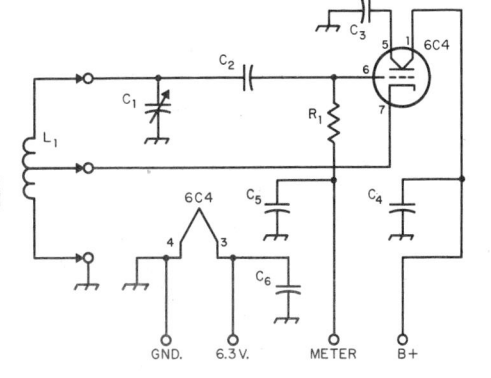

FIG. 14-25—Circuit diagram of the grid-dip meter.

C_1—50-μμf. midget variable (Hammarlund HF-50).
C_2—100-μμf. ceramic.
C_3, C_4, C_6—0.001-μf. disk ceramic.
C_5—0.01-μf. disk ceramic.
R_1—22,000 ohms, ½ watt.

274 UNDERSTANDING

FIG. 14-26—The grid-dip meter is built on one piece of a two-piece box measuring 4 × 2 1/8 × 1 5/8 inches. (Bud type CU-2102A is equivalent to the box shown except for flanges along the sides.) Wires in the cable connecting to the power supply are soldered to the four-terminal tie-point strip at the left. C_3, C_4, and C_6 are grounded to a soldering lug at the left of the socket in this view. Leads to these capacitors should be made as short as possible. C_5 is under the grid resistor. C_2, visible above the tuning capacitor, goes from the stator of C_1 to the tube socket. The leads are covered with spaghetti to avoid possible short circuits.

resonance is approached from the high side as compared with approaching from the low side.

Calibrating and Using the Grid-Dip Meter

The grid-dip oscillator generates an r.f. signal that can readily be picked up by your receiver. If you have a manufactured general-coverage receiver, calibration of the g.d.o. is easy. You simply set the receiver to an appropriate frequency, indicated by its dial, and adjust the grid-dip meter tuning until its signal comes in. This gives you one calibration point. By working upward continuously through the spectrum you can calibrate the meter up to the high-frequency limit of the receiver's tuning range, changing the g.d.o. coils whenever necessary. No special coupling to the receiver should be needed. Just setting the grid-dip meter alongside the receiver should give you a strong-enough signal.

If your receiver doesn't have general coverage, the Band-Finder wavemeter can be used. In this case you adjust the wavemeter tuning to various frequencies as shown by your calibration curves. Couple the g.d.o. coil to the wavemeter coil and adjust the g.d.o. tuning for the dip in grid current at each frequency. Use the loosest coupling that will give a definite dip. If the coupling is too tight, the tuning of one circuit will pull the tuning of the other and the calibration won't be as accurate as it could be.

In using the grid-dip meter for checking tuned circuits the procedure is the same as in calibrating from the absorption wavemeter; simply tune it for the dip in grid current, using the loosest possible coupling to the circuit being checked. If you get no dip on one coil, try another; the circuit being investigated may not tune to the frequency you think it does. Trans-

FIG. 14-27—The power supply and milliammeter are mounted in a meter case. The knob on top is for plate-voltage adjustment. Power supply parts can be arranged in any convenient way.

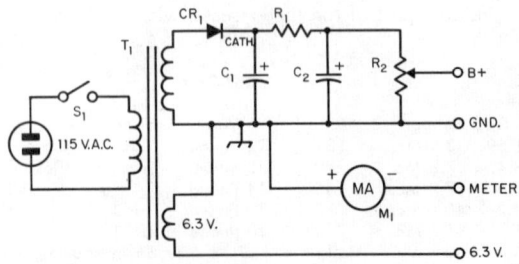

FIG. 14-28—Circuit diagram of the power supply for the grid-dip meter.

C_1, C_2—16-μf. electrolytic, 150 volts.
R_1—1000 ohms, 1/2 watt.
R_2—0.1-megohm potentiometer.
T_1—Power transformer, 6.3 volts and 125 to 150 volts, 15 ma. (Stancor PS-8415, Knight 61G410 or equivalent.)
CR_1—20-ma. selenium rectifier.
M_1—0-1 d.c. milliammeter.

Measurements

mitter or receiver r.f. circuits can be pretuned by this method, either during construction or after completion. Knowing that the circuits are tuned properly gives you a head start in getting a new piece of equipment working.

Because it generates an r.f. signal, the grid-dip meter also is useful as a **test oscillator**—an oscillator that provides a steady signal of known frequency for receiver circuit alignment and similar jobs. The output is great enough so that no special coupling to a receiver is needed. Whenever a small amount of r.f. is called for— and you will run into plenty of such cases— the g.d.o. is a convenient signal source.

The grid-dip meter also can be used as an absorption wavemeter if the plate voltage to the oscillator tube is turned off. The grid-cathode circuit of the tube acts as a rectifier, so the meter will read when the g.d.o. coil is coupled to a circuit carrying r.f. current—provided, of course, that the g.d.o. circuit is tuned to resonance. The diode loads the tuned circuit to some extent, so the circuit is not as selective as it is in the simple wavemeter. However, if the power supply is shut off completely, the circuit can be used as a plain absorption wavemeter; the method is the same as with the Band-Finder. Used in this way the power supply can be disconnected entirely from the meter of Fig. 14-24.

Other Measurements

The measurements described in this chapter are the most important ones for the beginner. The more experienced amateur will find many other types of measurements useful, but these are outside the scope of this volume. You can find more information in *The Radio Amateur's Handbook*, if you want to pursue the subject. It is worth while to do so, because measurement of the performance of circuits and equipment is a vital part of the technical side of radio. Examples of other measurements that are frequently needed in amateur work are finding values of inductance, capacitance, r.f. resistance, impedance, and similar circuit constants; audio-frequency voltage, power and frequency measurements; r.f. power and voltage; and the standing-wave ratio on transmission lines.

Measuring s.w.r. with good accuracy requires special care. The Monimatch circuit described in Chapter 11, although inherently capable of s.w.r. measurement, is seldom reliable for this purpose in practice. (But it is quite satisfactory for impedance-matching adjustments, as well as giving relative indications of r.f. output.) Fortunately, there is seldom a real need for accurate measurement of s.w.r. unless you are experimenting with antennas and transmission lines.

A most versatile instrument for many kinds of testing and measurement is the **cathode-ray oscilloscope**. Eventually you may want to own one, but whether you do or don't you should have at least a general idea of what the "scope" can do and how it works.

The Oscilloscope

The oscilloscope is built around a cathode-ray tube much like the one that gives you the picture in your TV receiver. The tube isn't as big— a 7-inch diameter display face is about the largest ordinarily used—and its internal construction is different.

You've probably read enough about television to know that the picture is drawn on the face of the tube by a rapidly-moving spot of light. The bright spot is formed when the fluorescent material on the glass face of the tube is struck by a narrow beam of electrons projected from a "gun" in the neck of the tube. An oscilloscope tube has the same kind of gun and a similar **fluorescent screen**. The fluorescent material belongs to the same family as that used in fluorescent lamp tubes that you see everywhere.

The electron beam can be made to change its course inside the tube if it is subjected to a magnetic field. In the TV tube the magnetic field

This oscilloscope kit uses a 3-inch cathode-ray tube and has the features—amplifiers and sweep circuits —needed for general oscilloscope measurements. More elaborate kits also are available.

HORIZONTAL

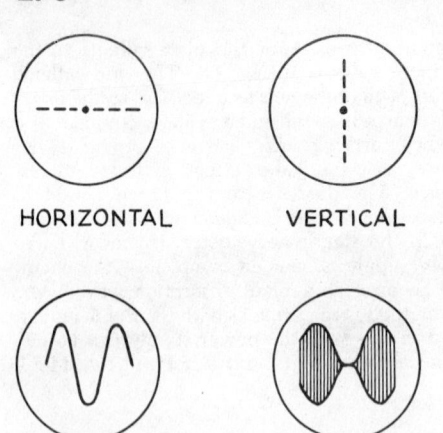

VERTICAL

A.C. VOLTAGE ON VERTICAL

MODULATED R.F. VOLTAGE ON VERTICAL

FIG. 14-29—With an alternating voltage applied only to one set of deflection plates of the oscilloscope the moving spot traces a straight line, as shown in the two upper drawings. With voltages applied to both sets of plates a pattern is formed because the position of the spot then depends upon the instantaneous values of the voltages on both sets. The two lower drawings show typical examples.

is set up by a current flowing through a "deflection coil" wound around the neck of the tube. Actually, two such coils are used, one for moving the beam sidewise and the other for moving it vertically.

A different method is used in the oscilloscope tube. The beam is deflected by subjecting it to an *electric* field. To do this a pair of **deflection plates** is placed inside the tube so the beam has to pass between them. When a voltage is applied to the plates, the beam is attracted to the plate with the positive charge and repelled by the one having the negative charge. With two pairs of plates at right angles, the beam can be moved both sidewise and up and down. This **electrostatic deflection** has the advantage, over the magnetic method, that very high frequencies—far up into the radio-frequency region—can be applied to the plates. Also, the power required for deflection is insignificant.

In the absence of deflection voltages the oscilloscope controls are adjusted so that a small bright spot appears in the center of the screen. Then, if the tube is properly oriented, an a.c. voltage applied to the "horizontal" plates will cause the spot to move from side to side, as in Fig. 14-29. If a similar voltage is applied to the "vertical" plates, with no deflection voltage on the horizontal plates, the spot will move up and down, also as shown in the figure. The speed at which the spot moves is exactly proportional to the rate at which the voltage is changing. In the course of one a.c. cycle the spot will move in one direction, say to the right, while the voltage is increasing from zero. When the positive peak of the cycle is reached and the voltage begins to decrease, the spot will reverse its direction and go to the left. It will continue in this direction, passing through the center of the screen at zero voltage, until the negative peak of the voltage is reached. Then it will reverse again. This continues as long as the voltage is applied to the deflection plates. The action is the same if the voltage is applied to the vertical plates, except that the spot moves up and down.

The oscilloscope becomes useful when deflection voltages are applied *both* horizontally and vertically. Suppose the spot is made to move at a uniform rate of speed horizontally, while an alternating voltage is applied to the vertical plates. Then the moving spot traces a pattern of the voltage just like those you've seen earlier in this book. This is shown at the lower left in Fig. 14-29. You can actually look at the waveform of the voltage and tell whether it is a sine wave or some other form.

One final example from among the innumerable things that the oscilloscope can show is found at the lower right in the figure. This is the kind of pattern that results when a modulated r.f. signal is applied to the vertical plates while the spot is moving horizontally at uniform speed. The outline of the pattern traces the modulation envelope of the signal. From it you can tell what the percentage of modulation is and whether the signal is distorted. Inside the modulation envelope the pattern appears to be filled in with light. This is because the beam actually is moving up and down with each r.f. cycle. As there are usually many hundreds—or thousands—of r.f. cycles in each audio-frequency cycle of the modulation envelope, their traces on the screen blend together, forming a solidly-lighted area.

The oscilloscope tube requires a number of auxiliary circuits—power supply, amplifiers for the voltages applied to the horizontal and vertical deflection plates, special circuits for generating deflection voltage, and so on—if its full capabilities are to be exploited. As a general rule, building such a scope from one of the several varieties of kits now available is far preferable to assembling parts and building from scratch. On the other hand, a scope used just for modulation checking takes little more than the power supply (which frequently can be taken from the transmitter itself) and a few resistors, in addition to the cathode-ray tube. You can easily put together an oscilloscope of this type. Suitable circuits, along with the method of using the finished scope, are given in *The Radio Amateur's Handbook*.

Chapter 15

Antennas and Masts

All the basic information you need for putting up a workable antenna system is contained in Chapter 6. It will pay you to digest the principles discussed there, because antennas and transmission lines, like transmitting and receiving circuits, will be found to act quite rationally when you know how they're supposed to work.

This chapter contains a few concrete suggestions for putting up simple antennas, including the construction of the antennas, the kinds of transmission lines (if any) to use with them, and the methods of supporting them.

The Center-Fed Dipole

The center-fed antenna is the "old stand-by" for amateurs who work the lower-frequency bands—3.5 and 7 Mc. especially. If there is room to run one in a straight line and it can be erected at a height of 30 feet or more, you can count on its doing an excellent job for you. It's just as capable a radiator at the higher frequencies, too, and many amateurs have done eminently good long-distance work with it on 14, 21 and 28 Mc., with power under 100 watts. On these DX bands you can't expect to jump in and compete for the "rare" DX with stations running high power and beam antennas. Nevertheless, you can work just as far, in actual distance, and get your share of good signal-strength reports.

We pointed out in Chapter 6 that you don't have to measure the length of a center-fed dipole to a fraction of an inch if you use an open-wire transmission line, plus a transmatch. It's the most flexible system there is. You can use the same antenna on practically any frequency, even frequencies for which it is considerably shorter than a half wavelength. It is shown in practical form in Fig. 15-1.

The antenna or "flat-top" part of the system should be about a half wavelength long at the lowest frequency you want to use, if that is possible. But if the length has to be shorter, just put as much in the air as you can.

Multiband Operation

For working on all bands from 3.5 to 28 Mc. there are even some advantages in using a length somewhat less than a half wavelength at 3.5 Mc. A good all-around length for this purpose is 100 feet, because this length will tend to be non-directional on 14 Mc. On this band it is a 1½-wavelength antenna having the directional pattern shown in Chapter 6. On 7 Mc. the antenna will be somewhat directional **broadside** —that is, at right angles to the direction of the wire. On 3.5 Mc. it will be almost non-directional. Directional effects on 21 and 28 Mc. will not be clearly evident with an antenna of this length, unless you have some way of comparing it with another antenna. For practical purposes, it is an "all-around" system on all bands.

It should hardly need saying that an antenna will have the best chance of working well if it is up in the air by itself. Nearby wires, metal gutter piping, and similar conductors won't give it much assistance in getting your signal out, and may do considerable harm to the antenna's radiating properties. Steer clear of them if you can. Nevertheless, a poorly-located antenna is better than none at all!

Location and Supports

The antenna can be supported at the ends by buildings, trees, or anything you may be able to use. Later in this chapter there is information on putting up poles and masts. A mast may be the real answer to holding up at least one end of the antenna. Many operators fasten one end of the antenna to the house, or to a short pole on the roof, and build a mast far enough away to permit using the desired antenna length. The spot for the mast can be chosen so the antenna will be in the clear.

You can slope the antenna wire if your two

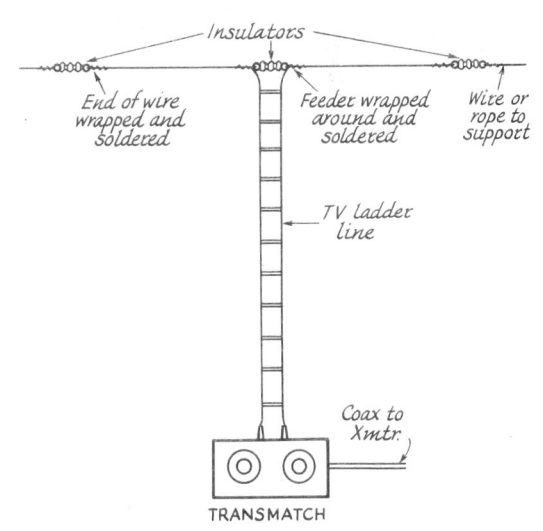

FIG. 15-1—Center-fed dipole with open-wire feeders. The total antenna length preferably should be in the neighborhood of one-half wavelength at the lowest frequency to be used, but if space doesn't permit such a length it can be reduced to as little as one-quarter wavelength. See text.

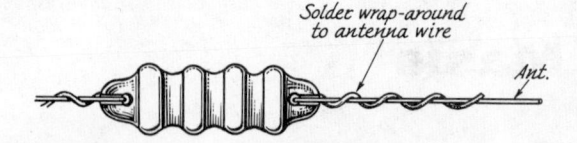

FIG. 15-2—Securing the antenna wire to an insulator. The support wire, at left, is generally stranded steel wire, obtainable at hardware stores, and does not need soldering.

supports aren't the same height. A sloping antenna usually will radiate best in the direction of the downslope. You can take advantage of this effect by deliberately sloping the wire to get a desired "best" direction. The directivity is not so marked that you'll lose out too badly in the opposite direction, except possibly on the highest-frequency bands.

Materials and Construction

Use No. 14 or No. 12 copper wire for the antenna. It can be either bare or enameled. It is worthwhile to use the hard-drawn variety, if available. Although it is harder to handle because it tends to be springy, it has greater strength.

For center feed cut two equal lengths, one for each of the two halves of the antenna. Scrape the ends clean for at least 12 inches. Feed the ends through the eyes of the insulators and wrap five or six inches of end around the wire. See Fig. 15-2. Solder this wrapped joint. Wrap the ends of the feeder wires around the inner antenna ends, on each side of the center insulator, and solder them, too. A heavy (100-watt or larger) iron or a small torch should be used for this job. Because the wire conducts the heat away rapidly, a small iron won't make a good joint. Do your soldering indoors where there won't be any breeze to cool the iron and joint. Also, in cold weather it's almost impossible to solder outdoors with an ordinary iron.

The insulators in the antenna should be the "strain" type, in which the wires go through holes ("eyes") in the ends. Don't use the "egg" type here; egg insulators are fine for use in guy wires but there is considerable capacitance between the wires fastened in them. This capacitance is undesirable in the antenna itself. Either glass or ceramic is suitable, and the size doesn't matter a great deal at the r.f. power levels we're concerned with in this book. At one time you could find antenna insulators in every hardware and "5 & 10" store, but most broadcast receivers today use built-in antennas, so insulators aren't so commonly available. You'll probably have to get them through an outlet that handles radio components.

The most convenient type of open-wire transmission line is the TV "ladder" line. Use the kind that has the widest spacing between wires—1 inch. Ladder line comes on a reel, when you buy a standard length of 100 feet. Don't unroll it until the antenna has been put together and is ready to be pulled up. If you're careful in unrolling it then, the line will have relatively little tendency to twist. It can't be prevented from twisting a little, though, unless it is pulled up really tight, with individual tension adjustment on each wire. This isn't possible when the line hangs from the center of the antenna. A gradual twist doesn't matter, as long as the wires maintain their spacing throughout the length of the feeder.

An open-wire line also can be made up using **spreaders**. Porcelain ones are available having lengths of 2, 4 and 6 inches. The 4-inch size is probably the easiest to use for runs of moderate length. The line wires can be No. 12 or No. 14 copper, and should be fastened to the spreaders as shown in Fig. 15-3. The tie wires don't have to be quite as thick as the line wires, although the same size can be used for both. Don't solder these joints. Just wrap the tie wires tightly. Then you can slide the spreader along the wire to adjust the spacing for a shipshape job.

Feeder Length

You remember the old saw about the right length for table legs—long enough to reach the floor. The question of feeder length has a similar answer—long enough to reach from the antenna to the transmitter or transmatch. Feeders can always be made to work, regardless of their length.

However, open-wire feeders will make a better load for a transmatch if the impedance they present to it is close to being a pure resistance. For this to be so the whole system—not just the antenna—must be resonant, or not very far from resonance. We've discussed this a little in Chapter 6. It's less confusing to consider the length to be that of *one* half of the antenna plus the length of *one* of the feeder wires. The sum of these two should be some whole-number multiple of a quarter wavelength.

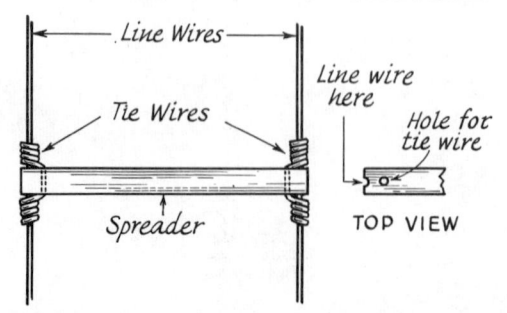

FIG. 15-3—Method of fastening feeder wires to line spreaders. It is best not to solder these joints because a slight amount of slippage will be helpful in aligning the feeder wires for uniform separation when the antenna is put up.

Antennas

Table 15-I
Quarter Wavelength in Feet and Inches at Various Amateur-Band Frequencies.

Frequency	Length
3500 kc.	66 feet, 11 inches
3600 "	65 " 0 "
3700 "	63 " 4 "
3800 "	61 " 7 "
3900 "	60 " 0 "
4000 "	58 " 6 "
7000 kc.	33 feet, 6 inches
7100 "	33 " 0 "
7200 "	32 " 6 "
7300 "	32 " 1 "
14.0 Mc.	16 feet, 8 inches
14.15 "	16 " 6 "
14.3 "	16 " 4 "
21.0 Mc.	11 feet, 2 "
21.2 "	11 " 0 "
21.4 "	10 " 11 "
28.0 Mc.	8 feet, 4 inches
28.4 "	8 " 3 "
28.8 "	8 " 2 "
29.2 "	8 " 0 "
29.6 "	7 " 11 "

Table 15-I gives the length of a quarter wavelength in feet and inches for various frequencies in the bands from 3.5 to 28 Mc. If you want to use the antenna on all these bands it is best to choose 3600 kc. as your "basic" frequency. Lengths based on this frequency will be reasonably close to resonance in some part of each of the five bands.

To see how the table works out, suppose that you've selected an antenna length of 100 feet. The total length of half the antenna plus one feeder wire will be some multiple of 65 feet (a quarter wavelength at 3600 kc.). Since half the antenna length is 50 feet, the shortest length of feeder you would use would be 65 − 50 = 15 feet. This feeder length probably would be too short to be practical, so you might take the next length, 2 × 65 feet or 130 feet. Then 130 − 50 = 80 feet is the feeder length.

As another example, suppose you're principally interested in c.w. work on 7 Mc. and up. Since this kind of operation is in the low ends of the bands, a reasonable choice for the basic frequency would be 7050 kc. Interpolating in Table 15-I, the length would be 33 feet, 3 inches for one side of the antenna (the total length would be 66 feet, 6 inches.) Since one side of the antenna is a quarter wavelength, the feeder length itself can be any multiple of one-quarter wavelength. Thus you could use a feeder 33 feet, 3 inches long, or one 66 feet, 6 inches long, or 99 feet, 9 inches, and so on.

In passing, we might point out that the 100-foot antenna mentioned above can be scaled down for use on 7 Mc. instead of 3.5 Mc. as the lowest frequency. Then it will have a total length of 50 feet—25 feet on a side. This length will be substantially nondirectional on 28 Mc. but will be somewhat directional broadside on 14 Mc., with a small gain over a half-wave dipole on that band. The various examples given here are shown in Fig. 15-4.

Physically, you may not need all the feeder length that Table 15-I shows to be optimum. If not, there are two things you can do. One is to make the feeder take a path from the antenna to the transmatch that does use up all its length. It should not be rolled up in a coil, but any path that doesn't bend it back sharply on itself will be satisfactory. Part of it can even be strung up around your operating room if some of the length has to be used up indoors.

The second thing you can do is to cut the

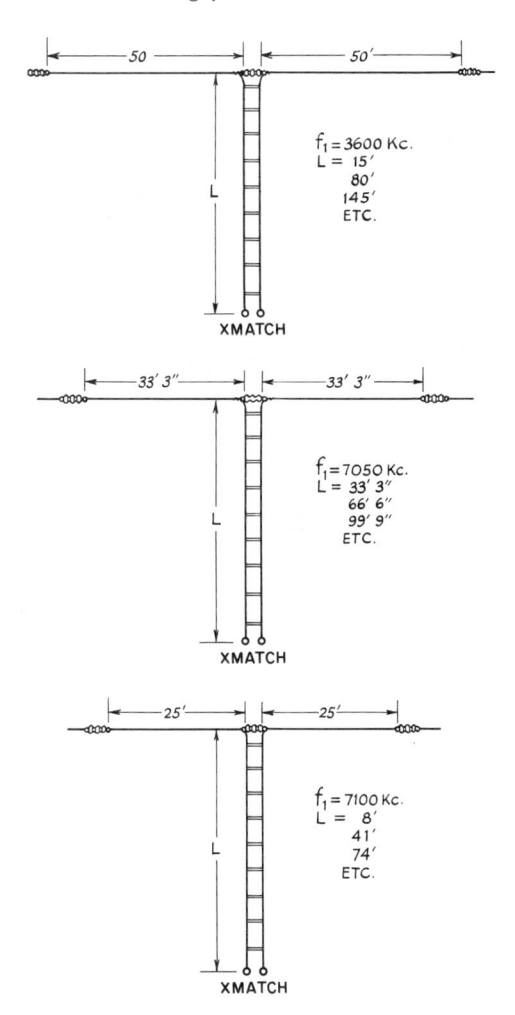

```
|←――― 50 ―――→|←――― 50' ―――→|
              ┬
              │                f₁ = 3600 Kc.
              │                L = 15'
              L                    80'
              │                   145'
              │                   ETC.
              ┴
            XMATCH

|←― 33' 3" ―→|←― 33' 3" ―→|
              ┬
              │                f₁ = 7050 Kc.
              │                L = 33' 3"
              L                   66' 6"
              │                   99' 9"
              │                   ETC.
              ┴
            XMATCH

|← 25' →|← 25' →|
          ┬
          │                    f₁ = 7100 Kc.
          │                    L = 8'
          L                       41'
          │                       74'
          │                       ETC.
          ┴
        XMATCH
```

FIG. 15-4—Some suggested dimensions for center-fed dipoles. These antennas can be used on all bands from 3.5 to 30 Mc., but the lower one is a little shorter than desirable for 3.5 Mc.

FIG. 15-5—Using a TV line stand-off insulator for supporting ladder line.

feeder to the length you need for coming into the station. As we said earlier, any feeder length can be made to work. Some lengths will give difficulty; it may be hard, or even impossible, to find taps on the transmatch that will give you proper loading if you happen to hit one of the "bad" lengths. These are in the region around lengths that represent a multiple of an eighth wavelength instead of the quarter-wavelength multiples we have recommended. Avoid lengths of this order and you should have little trouble.*

In determining feeder length, include the section inside the station—right up to where the line connects to the transmatch. This should be the same type of line that you used outside.

Feeder Installation

The feeder is in the strong electromagnetic field around the antenna, since it has to run right to the antenna in order to connect to it. This field will induce an r.f. voltage in the feeder wires, and in turn the voltage will cause a current to flow in the wires. This current is not re-

*Whatever the length, the line can always be loaded at its input end to become the electrical equivalent of a multiple of a quarter wavelength. Methods of doing this are described in *The Radio Amateur's Handbook* and *The A.R.R.L. Antenna Book*.

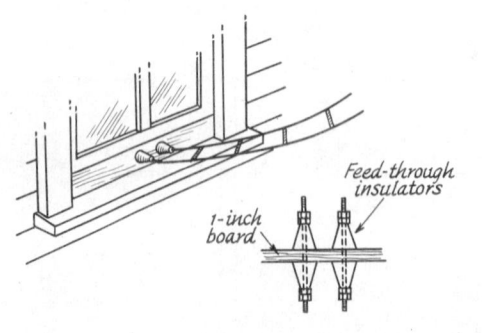

FIG. 15-6—Feeder connections can be brought into the station by means of feed-through insulators mounted in a 1-inch board that fits into the window frame. The board can go on top of the upper window sash instead of below the lower one as shown.

lated to the transmission-line current discussed earlier. In fact, the induced current flows in the *same* direction in both wires, because both wires are in the same field. For this current the wires simply act as though they were the same conductor, or two conductors in parallel. This **parallel current** will cause radiation just as current in the antenna itself causes radiation.

We want the antenna to do the radiating, not the feeders. To minimize current of this type the feeder must be placed where the voltage induced in it will be least. If the system is perfectly symmetrical, mechanically and electrically, the pickup will be zero when the feeder is exactly at right angles to the antenna. Such perfection is rare, if it ever exists at all, but certainly the induced voltage will be least when the feeder leaves the antenna at right angles to the antenna wire. If it is at all possible, the feeder should continue in a straight line for a quarter wavelength before any turns or bends are made in it. Do the best you can in this respect. In any case, don't have the feeder parallel to one side of the antenna and close to it; this is the worst condition.

Bringing the Feeder Into the Station

Open-wire feeders can be supported along the way from the antenna to the entrance to the station, if any points of support offer themselves. It is better to support them than to let them dangle. If they have to come near any metal, such as rainspouting, make them cross at right angles if the installation permits. If they must run parallel to conductors of this nature keep them at least a foot away, and make the parallel run as short as you can.

It does no particular harm to run feeders along nonmetallic structures. TV ladder line can be supported by the screw-type insulators made for TV work, as shown in Fig. 15-5. You may have to open the eye of the screw a little in order to force the spreader into the insulator. A pair of gas pliers will do this easily, and the same pliers can be used to force the eye down on the insulator to hold it tight after the assembly has been installed.

A time-honored method of bringing open-wire feeders into the station is shown in Fig. 15-6. The dimensions of the board should be such as to fill the width below the window, which rests on the top. A four-inch height is usually enough. This leaves a gap between the two sashes which must be filled somehow to keep out cold air in winter or insects in summer. Filling the space with glass wool or cotton batting will take care of this. Or a piece of sheet rubber tacked to the top of the lower sash and cut to fit the opening can be used; this will let you open the sash when necessary.

The feeders can be grounded for lightning protection on the outside of such an assembly. See Chapter 1 for grounding methods.

Antennas

Coupling to the Transmitter

The transmatch provides the means for matching the input impedance of the feeder to the coax line that goes to the transmitter. Transmatch adjustment has been described in Chapter 11. You should also read Chapter 6 to become familiar with the principles.

Other Dipole Systems

The center-fed dipole with open-wire feeders is the most versatile system you can use, since it lets you work on several amateur bands with negligible loss of r.f. power in the feeders. However, there are several other dipole arrangements, some of which may be better suited to the physical conditions existing at your station. Most of these are basically one-band systems, but there are ways in which they can be adapted to multiband operation.

The End-Fed Dipole

One which is not inherently a one-band antenna is the end-fed dipole. As you saw in Chapter 6, any dipole a half wavelength long at one frequency will resonate as a one-wavelength (full-wave) antenna at approximately twice that frequency, as a 1½-wavelength at three times the frequency, and so on. Thus an antenna resonating as a half-wave dipole at, say, 3600 kc. will resonate again at approximately 7200 kc., at 14,400 kc., and on up the scale. It doesn't matter whether such an antenna is fed at the center or the end, so far as resonance goes. The end-fed system is shown in Fig. 15-7.

However, you can't take liberties with the antenna length, as you can with center feed. If the antenna itself isn't resonant, the feeder currents will be unbalanced, and the line won't act wholly like a real transmission line. Unbalance will cause it to radiate, too, so it becomes part of the antenna. Also, only one line wire is connected to the antenna; the other simply stops between two insulators, as shown in the drawing. This unsymmetrical termination for the line is itself enough to cause a parallel current to exist on the line, even when you're operating on the exact resonant frequency of the antenna.

Aside from this, the system is much like the center-fed arrangement. Open-wire feeders should be used, because the standing-wave ratio is high and the power loss in any other type of line would be excessive. Although the feeder can be any length, multiples of one-quarter wavelength are preferred. Such lengths will make the line look like a pure resistance, or nearly so, to the transmatch circuit. The considerations here are the same as those outlined in the discussion on feeder length for the center-fed antenna. Suitable feeder lengths can be taken from Table 15-I. The figures in that table also can be used for the length of the antenna itself, except that for any given frequency the length given must be multiplied by 2. That is, an antenna for 3600 kc. (plus the multiples of that frequency) should be $2 \times 65 = 130$ feet long, one for 21.2 Mc. should be $2 \times 11 = 22$ feet long, and similarly for other frequencies.

With end feed, the antenna has somewhat different directional characteristics when operated on harmonics, as compared with the center-fed wire. However, the differences are not likely to be very evident in practice. Also, the direction toward the open end of the wire is favored slightly, as compared with the opposite direction.

Adjustment of the transmatch for coupling to the transmitter is the same as with center feed.

Matched Dipoles

The advantage of matching the antenna and transmission-line impedances is that the feeder length has no effect on the input impedance of the line. Strictly speaking, this is true only when the match is perfect. In practice, the match isn't perfect—except possibly at one single frequency. Therefore there is some change in line input impedance with length. However, the change is

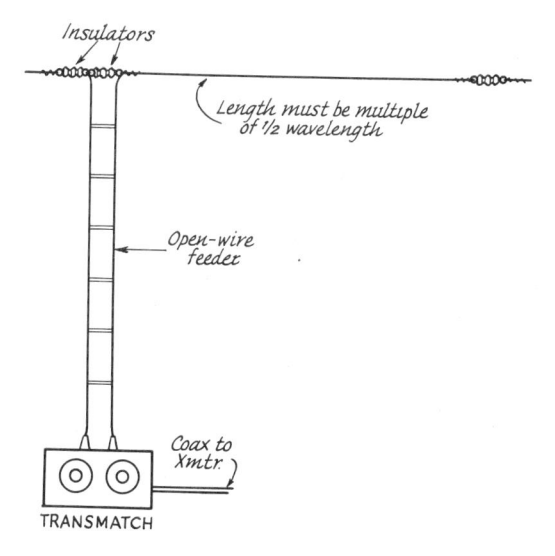

FIG. 15-7—End-fed antenna with open-wire feeders. This system requires that the antenna length be resonant at the operating frequency, either on the antenna's fundamental (half wavelength) or its harmonic resonances. For operation in the Novice bands the following lengths are recommended:

Band	Antenna	Feeder
3.5, 7 and 21 Mc.	130 ft.	A multiple of 65 ft.
7 and 21 Mc.	65 ft.	A multiple of 33 ft.
21 Mc.	22 ft.	A multiple of 11 ft.

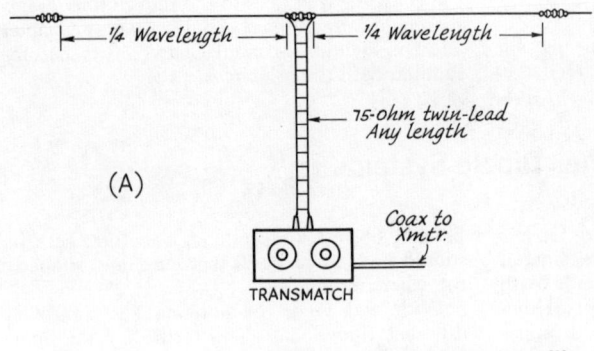

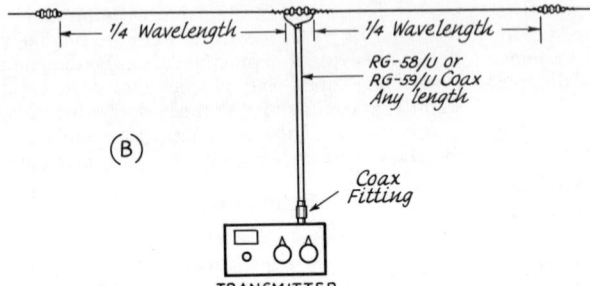

FIG. 15-8—Matched antennas. These are generally useful only on the band for which the antenna length is a total of one-half wavelength. The dipole must be resonant.

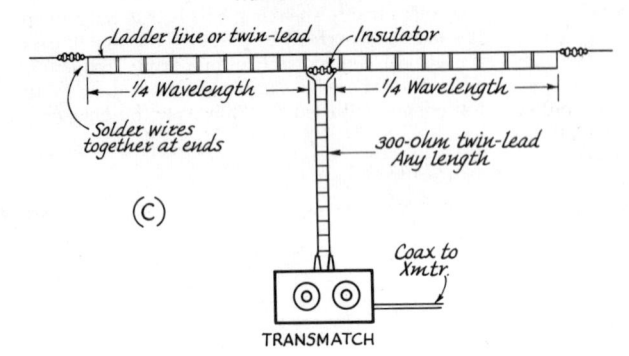

small enough so that the pi-network tank circuit in the transmitter can be adjusted for full loading on the final amplifier, over the range of frequencies for which the match is reasonably close, with any length of line you may use.

In return for this convenience your operation is restricted to a relatively small range of frequencies around the one for which the dipole is resonant. A range of about 200 kc. is representative on the 3.5-Mc. band; that is, you can work up to 100 kc. above or below the resonant frequency with no intolerable increase in standing-wave ratio. On 7, 14 and 21 Mc. the whole band can be covered (a separate dipole must be used for each band, of course). On these three bands it suffices to make the antenna resonant at the mid-frequency; there is no need to choose a particular part of the band unless you have no interest in working anywhere else in it. On 28 Mc. you can also cover the band with a single antenna cut for the center, but with a higher s.w.r. at the ends than on the three next lower bands.

The two common types of matched antennas are the single-wire dipole shown in Fig. 15-8 at A and B, and the folded dipole shown at C. In all three cases the length of one-half of the antenna should be one-quarter wavelength at the selected frequency. This length can be taken from Table 15-I.

Single-Wire Dipole

Since the center-fed dipole is a balanced antenna, it should be fed with balanced—that is, parallel-conductor—line if at all possible. There is a transmitting-type twin line available for this purpose (see Chapter 6). To keep the system balanced throughout, a transmatch should be used at the input end. It should be connected and adjusted as described in Chapter 11.

Actually, the transmatch could be omitted and the line could go directly to the output terminals on the transmitter. This wouldn't be any worse than the coax-cable system shown at

Folded Dipoles

FIG. 15-9 — Multiple-dipole antenna. This system uses a number of half-wave dipoles in parallel. The one that is resonant at the operating frequency takes practically all the power. The remaining ones affect the impedance match to some extent, but not seriously enough to cause the s.w.r. to be excessively high.

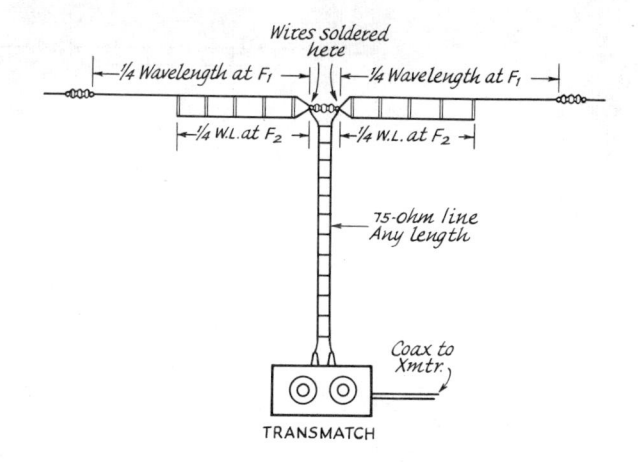

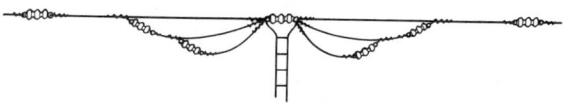

ANOTHER METHOD OF CONSTRUCTION

B, but would sacrifice the balance that can be maintained by using twin line.

The principal reasons for using coax (Fig. 15-8B) instead of twin line are three: coax is easier to find in radio stores, the operator doesn't want to use a transmatch, and he fondly believes that all the r.f. will be inside the cable. Unfortunately, the outside of the cable is inherently part of the antenna with this method. If you're lucky in your choice of cable length and in your grounding system this may give you no obvious difficulties. If you're unlucky, there may be r.f. feedback troubles (see Chapter 13) in addition to the other undesirable effects that go with a radiating feeder.

The Folded Dipole

The folded dipole, Fig. 15-8C, is a little more complicated to construct than the single-wire dipoles. Two wires are needed in the antenna itself. TV ladder line can be used for the dipole part. As an alternative, solid-dielectric twin line can be used. The heavy-duty twin line is best mechanically, especially if the antenna is fairly long as at 3.5 and 7 Mc. Twin line of the ordinary TV variety can be used for the feeder.

Like the 75-ohm parallel-conductor feed shown in Fig. 15-8A, the folded dipole is a balanced system and the power should be fed to the transmission line through a transmatch in order to keep it balanced. Chapter 11 describes the transmatch adjustment.

The folded dipole is a good single-band antenna. Its frequency coverage, for a given s.w.r. at the ends of the range, is a bit greater than that of the single-wire dipole. The reason for this is that the two dipole wires tend to act like a single thick conductor. A thick antenna is always broader-tuning than a thin one.

Regular TV stand-offs can be used for supporting the 300-ohm line along a building or at other points where an anchorage can be used. The line should not be allowed to get closer than an inch or so to other conductors that it has to pass, and should not be parallel to such conductors if it can be avoided. Also, the line should go away from the dipole at right angles, to avoid r.f. pickup from the antenna's field.

Multiple-Dipole Antenna

A number of schemes have been advanced for using one antenna system for operation in several bands, with acceptable impedance matching between the antenna and line on the frequencies for which the system is designed. Probably the simplest of these is the multiple-dipole antenna shown in Fig. 15-9. This system merely uses a number of dipoles, one cut for each band, connected in parallel at the feed point. Only two are shown in the drawing, but three or even more can be used.

A two-dipole antenna can be built quite easily from TV ladder line. The longer of the two wires takes all the mechanical strain. The shorter one is made by cutting the second wire at the proper lengths from the center. At the center insulator the two wires on each side are joined together and connected to the transmission line. The unused wires and insulators beyond the ends of the shorter dipole should be clipped off.

Separate single-wire dipoles can be used instead of the ladder line. The dipoles do not have to be parallel to each other. The shorter ones can be suspended from the longest one by letting them form loops as shown in the lower drawing. A separation of a few inches is ample; the principal thing is that the outer ends of each dipole

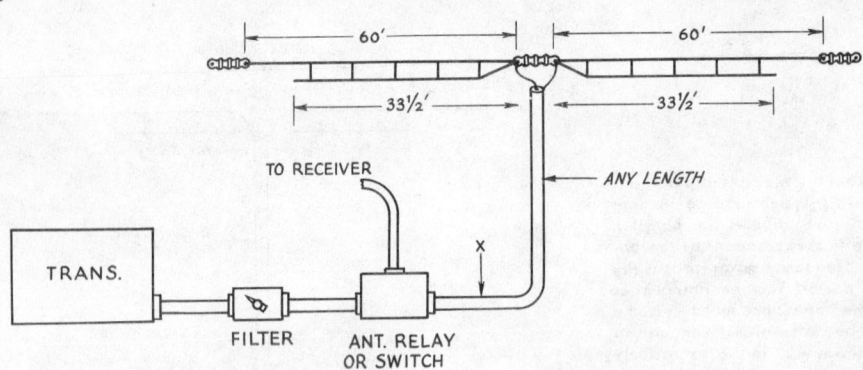

FIG. 15-10—Two-dipole antenna for three Novice bands. Although a coaxial feeder is shown, 75-ohm twin line can be used instead. Coax feeder may be either RG-58/U or RG-59/U. Complete system shown includes a half-wave filter assembly, for preventing harmonic radiation, and a t.r. switch. Low-pass filter for TVI reduction, if required, should be inserted in feeder at X.

must be insulated from the other wires. The wires can be pulled off separately to the supports, making enough of an angle with each other so they cannot touch. Separate supports, not necessarily in the same line, can be used if available.

Such an antenna does not match the line equally well on all the bands for which the dipoles are cut. In fact, it does not result in as good a match on *any* band as you could expect from a single dipole on that same band. However, it does well enough in this respect to let you use any convenient length of feeder, and the feeder losses will not be large at frequencies close to the resonant frequencies of the dipoles. The range of frequencies, on each band, over which the s.w.r. on the feeder will be low is smaller than is the case with the systems shown in Fig. 15-8.

The preferable type of feeder for this antenna is 75-ohm twin line, since the antenna itself is symmetrical about the center. The transmatch will help maintain line balance. However, the antenna will work with 75-ohm or 50-ohm coax line, with the same limitations as discussed earlier.

The advantage of such an antenna—that it will operate on a number of bands—is also a disadvantage. It will not reject harmonics of the transmitting frequency that happen to fall in the range of one of the dipoles. The transmatch will discriminate against these. If a transmatch is not used, a single-band half-wave filter (described in Chapter 11) should be connected between the transmitter and the feeder.

Three Novice Bands With Two Dipoles

The antenna shown in Fig. 15-10 combines the multiple-dipole idea with harmonic operation to give three-band operation with only two dipoles. We mentioned in Chapter 6 that a center-fed antenna working at the third harmonic (three times its fundamental frequency) has an impedance in the vicinity of 100 ohms. This is

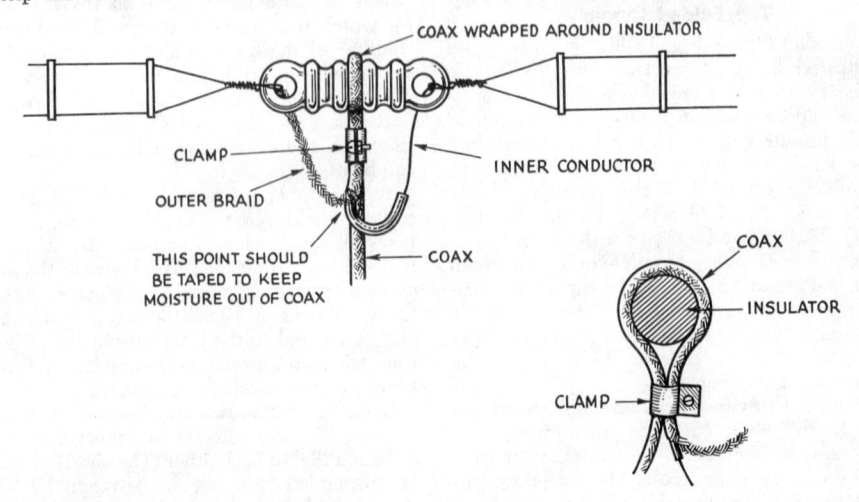

FIG. 15-11—Suggested method of anchoring coax at the center insulator. The same type of cable support can be used with any of the coax-fed antennas shown earlier.

Multiple Dipoles

not too bad a mismatch for coaxial cable or 75-ohm twin line. If the antenna length is adjusted so that the third harmonic is in the Novice 21-Mc. band the antenna also will work well in the 7-Mc. band. The feeder can be any length.

The system shown here uses a 7-Mc. antenna in this way, and has a 3.5-Mc. dipole in parallel with it. The lengths of the two dipoles have been experimentally adjusted for optimum results in the three low-frequency Novice bands.

You have your choice of three kinds of transmission line for the feeder—50-ohm coax (RG-58/U), 75-ohm coax (RG-59/U), or 75-ohm twin line. The advantages and disadvantages of coax and parallel-conductor line are the same as with other balanced systems described earlier.

The dipoles are made from TV ladder line having 1-inch spacing. Fig. 15-11 shows how the inner ends are connected to coax line, together with a method of anchoring the coax to the insulator so there is no mechanical strain on either the inner conductor or the braid.

A single 100-foot roll of ladder line is enough for the two dipoles. Cut the roll as shown in Fig. 15-12. The outside ends of the roll go to the center insulator of the system. After using six inches for bending back through the insulator and wrapping, one half-dipole is 33½ feet long. It needs no support at the open end. The remaining piece should be cut down so the length between the holes in the inner and outer insulators is 60 feet, as in Fig. 15-10.

Like other multiband systems designed to match the antenna to the feeder, this one will radiate any harmonic power that may get out of the transmitter. To be safe, you need to do something about filtering out all frequencies except the one you want. When the feeder is 75-ohm twin line a transmatch should be used for the sake of maintaining line balance, and the selectivity of the transmatch will be adequate for suppressing the undesired frequencies. A transmatch also can be used with coax (Chapter 11) but as an alternative you can use half-wave filters. This is the method shown in Fig. 15-10.

Only two filters are needed, one for 3.5 and the other for 7 Mc. These will take care of harmonics in the high-frequency spectrum—the harmonics that travel for long distances. Fig. 15-14 gives the circuits. As these are single-band filters the proper one must be switched in each time you change bands. S_1 does this, and also has a straight-through position for 21 Mc.

The two half-wave filters in Fig. 15-13 are built in a small aluminum chassis. The four coils can be cut from a single length of coil stock. Allow an extra couple of turns on each coil when cutting; when these are unwound they provide leads for wiring to S_1. Inexpensive phono jacks are used as connectors for the coax cable in the pictured assembly, but regular coax connectors can be substituted if you wish. The leads between the switch and the inner terminals of the connectors are coax line with the vinyl jacket removed. The braid should be grounded to the chassis at both ends of the lead. A similar piece of coax is used for the straight-through connection between the two sections of the switch. Alternate switch contacts are grounded, as shown in Fig. 15-14, to reduce capacitance between the active contacts.

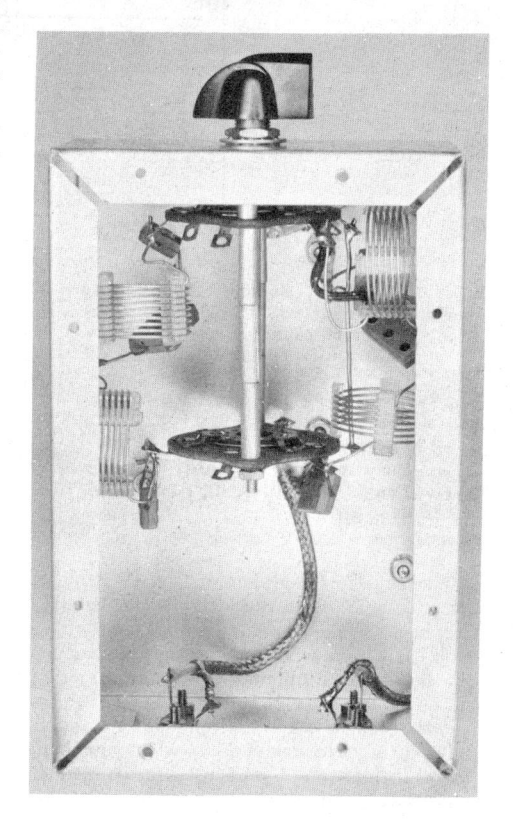

FIG. 15-13—Half-wave filters for the antenna system of Fig. 15-10. The 3.5-Mc. components are along the right-hand edge of the 3 × 4 × 6 aluminum chassis in this view. The 7-Mc. filter is at the left. Note that the coil axes are at right angles; this reduces stray coupling between them. A chassis bottom plate should be used to complete the shielding.

FIG. 15-12—How to cut a 100-foot length of ladder line to make the two dipoles of Fig. 15-10.

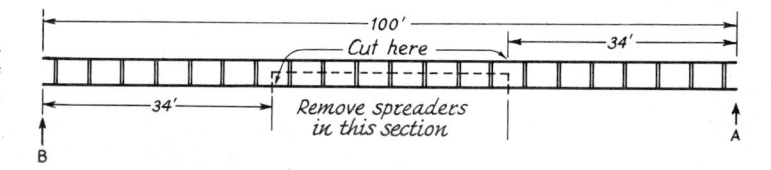

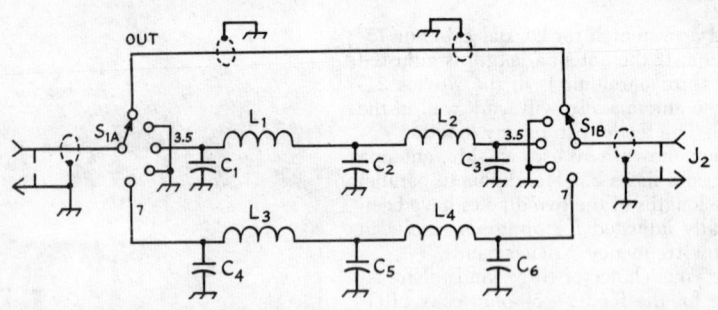

FIG. 15-14—Circuit diagram of the half-wave filters for 3.5 and 7 Mc. The mica capacitors listed below should be rated at 1000 volts or more for transmitters running more than about 75 watts input. For 75 watts or below, 500-volt capacitors are satisfactory.

C_1, C_3—750-$\mu\mu$f. mica.
C_2—1500-$\mu\mu$f. mica.
C_4, C_6—500-$\mu\mu$f. mica.
C_5—1000-$\mu\mu$f. mica.
J_1, J_2—Phono jacks or coax chassis fittings.
L_1, L_2—2 μh.; 8½ turns No. 20, 1-inch diam., 16 turns per inch.

L_3, L_4—1.2 μh., 6 turns No. 20, 1-inch diam., 16 turns per inch. (All four coils can be made from a single length of B & W Miniductor, type 3015.)

S_1—Rotary, 2 sections, 5 positions, 1 pole per section (Mallory Hamswitch type 151L).

In setting up the system of Fig. 15-10, install the filter near the transmitter so the length of coax between the two can be kept short. If you don't use a send-receive relay, the feeder can go directly from the antenna to the output side of the filter. The filter is reversible, so either connector can be used for the input side.

Don't forget to switch filters when you change bands! If the wrong filter is used the components may be damaged.

No filter is provided for 21-Mc. work because harmonics of this frequency don't get very far away from the station. They *can* cause TVI in your vicinity. To suppress harmonics in the TV channels a low-pass filter should be used. This type of filter can be left permanently in the feeder, since it passes all frequencies below its cutoff frequency.

When Space Is Limited

So far in this chapter we've taken it for granted that you'd have all the space you need for putting up a half-wave dipole at a reasonable height above ground. Only too often, though, what the city dweller sees from the station window isn't exactly in the "ideal location" class.

Don't let yourself be discouraged by such a situation. It has been faced—and licked—by plenty of other amateurs. Makeshift antennas often work surprisingly well. Here's something you can take comfort from: Almost any antenna you put up will be at least as efficient as the *best* antenna you could install on a car for mobile operation. And even at the 75-watt Novice limit you have quite a bit more power than the average mobile transmitter. But in spite of inefficient antennas and low power, thousands of hams have had lots of fun operating mobile.

The principle to follow, when you can't put up a half-wave dipole, is to get out as much wire as you can, as high as you can. Then add loading to the system to make it resonant at your operating frequency. That's about all there is to it. In this section we have a few suggestions that may be useful to you, but in the end they boil down to applications of the first two sentences of this paragraph.

The "Random-Length" Antenna

The term "random length" is used to describe any antenna that simply consists of a wire running out of the station to whatever length is feasible. No special feeder is used; the r.f. power is applied at the station end of the wire.

A ground connection is essential, so the system as a whole corresponds to the grounded antenna described in Chapter 6. As an antenna of this type has already been discussed at some length in that chapter, we don't need to go into it further except to repeat that it is a good idea to make the wire about a quarter wavelength long at the lowest frequency you want to use. This is in the neighborhood of 60 feet for the 3.5-Mc. band. Don't worry if you can't get the entire length in a straight line; bend it if you have to, but don't make the bends at acute angles if there is some way to avoid it. Also, the wire doesn't have to be horizontal; it can slope either up or down away from the point where it enters the station. The main thing is to get as much of it as high in the air as possible.

Loading and coupling circuits for use with antennas of this kind are discussed in Chapter 6, and you will find an example of one in Chapter 11. The construction of the antenna—wire, insu-

The Vertical Antenna

lators, lead-in, and so on—should be the same as recommended for dipoles earlier in this chapter. And do your best about that ground connection—see Chapter 6.

The Vertical Antenna

Although you may have very little ground space available you often can find room to go up. A vertical antenna is the answer in that case. Electrically, it is practically the same as the random-length antenna. You use the same method of coupling power into it, and use the same method of grounding.

Reams have been written about the radiation patterns and other characteristics of vertical antennas. These discussions are all based on an antenna sitting on perfectly-conducting ground, with nothing nearby to disturb the pattern. Your antenna won't be like that. Most likely it will have houses, wires, pipes, and probably trees right within arm's length, and the ground below it will act more like a resistor than a conductor. The chances are, too, that the bottom of the antenna won't even be close to the earth. So don't look for anything approaching the theoretical performance of a vertical antenna. It will be more accurate to think of it as a random-length antenna that goes up instead of sidewise.

Make it as tall as you can. Thirty feet is a good length, but requires guying. The antenna can be made out of aluminum tubing, about an inch in diameter, but a good substitute can be found in the TV mast sections that are readily available at radio supply houses. These are usually 1¼ inches in diameter and have one end swaged for fitting into another section as the mast is built up. Ten feet is a standard length, so two such sections will give you a 20-foot antenna and three sections will make up 30 feet.

If the construction of your house permits, the bottom of the antenna can be fastened to the wall just outside the window of your operating room. Fig. 15-15 shows how TV wall brackets can be used for supporting it. If the upper bracket can be high enough to hold the top section you will need no guys. Just fit the sections together and, if possible, drill them for self-tapping screws in the part that overlaps. This will add some mechanical strength and will help the electrical contact. However, if any joints come above the upper wall bracket, use rope guys at the top. It isn't safe to have a section supported only by the joint at its lower end. Plastic clothesline makes a good guy rope and needs no insulators. The guys should be anchored at some solid point that will let them make at least a 30-degree angle with the antenna.

When brackets are used as in Fig. 15-15 the antenna should be insulated from them. One simple way is to strip the wires out of a length of 300-ohm twin line, leaving a polyethylene tape that you can wind around the antenna where it is held by the U-bolt. Make the wrapping as thick as the width of the U-bolt will per-

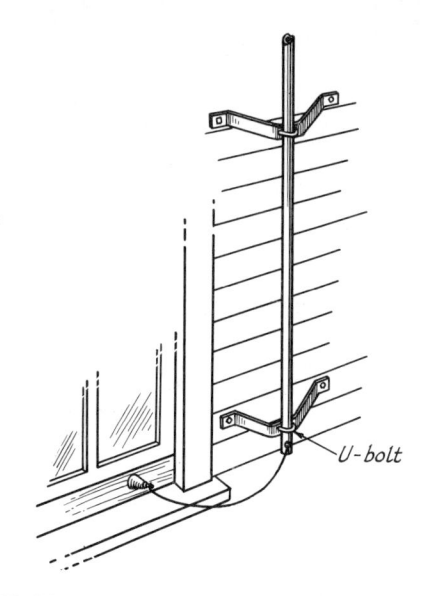

FIG. 15-15—Using TV wall brackets for supporting a vertical antenna made of metal tubing. The antenna should be insulated from the brackets as described in the text.

mit, to reduce the stray capacitance introduced by the wall bracket.

Since you're treating this vertical as a random-length antenna, you don't need to worry about the length of the wire that goes from the bottom of it to your loading and coupling apparatus. It's more important to get the antenna up in the air as much as you can.

Combining Vertical and Horizontal

The performance of a vertical antenna such as just described depends a great deal on the ground connection. You have no way of knowing whether or not you have a "good" ground, in the r.f. sense. If you can eliminate the ground connection as a part of the antenna system you're that much better off. Fig. 15-16 shows how it can be done. Instead of a ground, the system is completed by a wire—preferably, but not necessarily, horizontal—of the same length as the antenna. This makes a balanced system somewhat like the center-fed dipole.

It is desirable that the length of each conductor be of the order of 30 feet, as shown in the drawing, if the 3.5-Mc. band is to be used. At 7 Mc., this length doesn't really represent a compromise, since it is almost a half wavelength on that band. Because the shape of the antenna differs from that of a regular half-wave dipole, the radiation characteristics will be different, but the efficiency will be high on 7 Mc. and higher frequencies. Although the radiating part is only about a quarter wavelength at 3.5 Mc. the efficiency on this band, too, will be higher than it would be with a grounded system. If

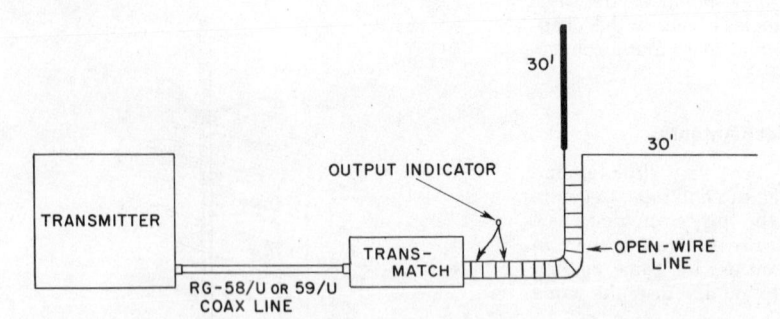

FIG. 15-16—Vertical and horizontal conductors combined. This system can be used on all bands from 3.5 to 28 Mc. with good results.

you're not interested in 3.5 Mc. and can't use the dimensions shown, the lengths can be reduced. Fifteen feet in both the vertical and horizontal conductors will not do too badly on 7 Mc. and will not be greatly handicapped, as compared with a half-wave dipole, on 14 Mc. and higher.

The vertical part can be mounted as already described. However, if you can put it on the roof of your house the extra height will be worthwhile. Fig. 15-17 suggests a simple base mounting using a soft-drink bottle as an insulator. Get one with a neck diameter that will fit into the tubing used for the vertical part of the antenna. To help prevent possible breakage, put a piece of some elastic material such as rubber sheet around the bottle where the tubing rests on it.

The wire conductor doesn't actually have to be horizontal. It can be at practically any angle that will let you pull it off in a straight line to a point where it can be secured. Use an insulator at this point, of course.

TV ladder line should be used for the feeder in this system. On most bands the standing-wave ratio will be high, and you will lose a good deal of power in the line if you try to use coax, or even 300-ohm twin line. The transmatch shown in Fig. 15-18 will work with practically any line length you find it necessary to use.

The transmatch circuit, Fig. 15-19, differs a bit from those shown in earlier chapters, although it does the same job (and other transmatch circuits could be substituted). It makes use of a type of tuned circuit known as a **multiband tuner**, and is capable of tuning all amateur bands from 3.5 to 30 Mc. Briefly, the lower frequencies, 3.5 and 7 Mc., are covered by L_3, C_2 and C_3. On these bands L_2 is too small to have much effect, so C_3 acts approximately as though it were in parallel with C_2. On the bands from 14 to 30 Mc. L_3 is too large to have much effect, so C_2 and C_3 act more or less as though they were in series across L_2. Thus on the two lower bands power is taken from the circuit through the coupling coil L_4, while on the three higher bands the output is taken from L_1. The load that is offered to the transmitter is controlled by C_1, in conjunction with the tuning of the multiband circuit.

A $2 \times 7 \times 9$ inch aluminum chassis is used for the transmatch shown in Fig. 15-18. C_1, at the left, must be insulated from the chassis, so it is mounted on ceramic stand-off insulators (Millen 31001). C_2 and C_3 are mounted directly on the chassis. Make the mounting holes large enough so you have a little play for aligning the two capacitors if you use a solid shaft coupler. The flexible type will give you a little more leeway in this respect. It isn't necessary

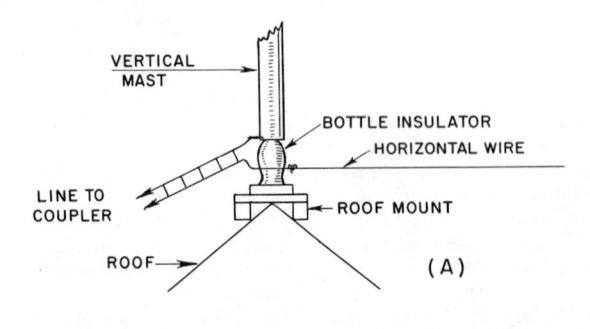

FIG. 15-17—One method of mounting the vertical section on a roof-top. The mounting base dimensions can be adjusted to fit the pitch of the roof. The 1×1 pieces should fit snugly around the bottom of the bottle to keep it from shifting position.

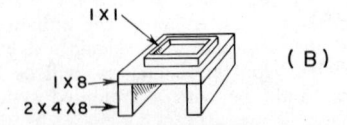

The Transmatch

FIG. 15-18—Five-band transmatch using multiband tuner. The two sets of coils are mounted at right angles to minimize coupling between them. Input and output connectors are at the rear end of the chassis.

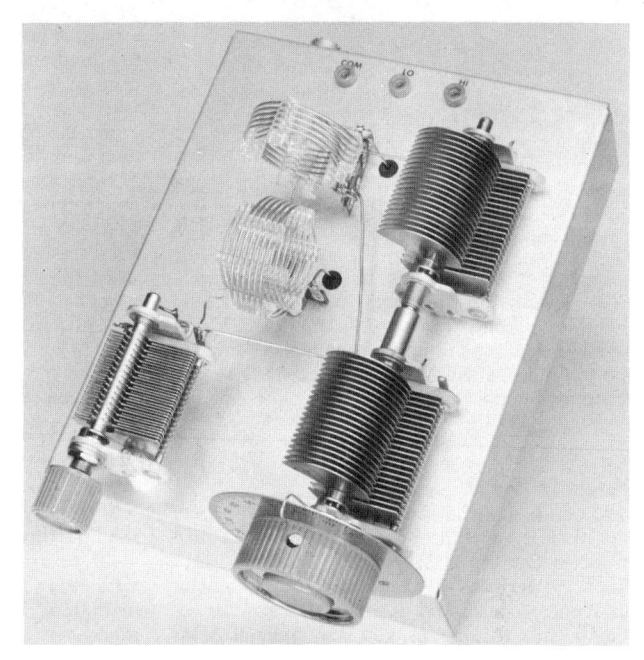

that an electrical connection be made through the shafts since the frames are both grounded to the chassis.

The coils are mounted on tie-point strips. Connections from L_2 and L_3 go directly to the tuning capacitors above the chassis. The connections from the output coils, L_1 and L_4, run through rubber grommets underneath the chassis to the tip jacks used as connectors for the transmission line. The connection to the rotor of C_1 similarly runs through a grommet to the inner terminal of the coaxial connector for the line to the transmitter. The ends of the feeder can be soldered into phone tips for plugging into the jacks, or regular pin plugs can be used instead.

This system can be tuned up by using a Monimatch in the coax line between the transmitter and transmatch, as described in Chapter 11. If you don't have a Monimatch, the r.f. output indicator shown in Fig. 15-20 should be used. Clip the lamp leads on one feeder wire, making sure that the clips make good contact. Put them about a foot apart.

FIG. 15-19—Circuit diagram of the multiband transmatch.

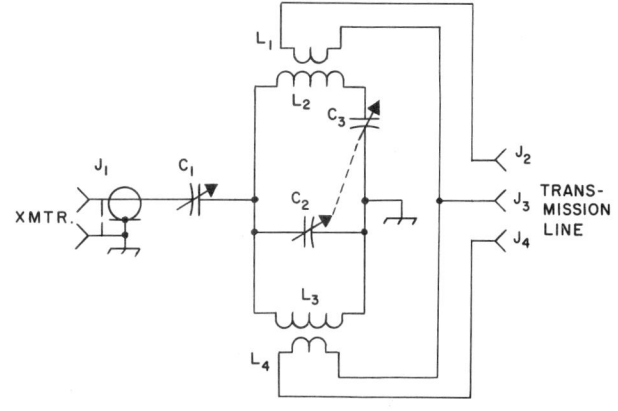

C_1—250-$\mu\mu$f. variable capacitor (Hammarlund MC-250-M).
C_2, C_3—325-$\mu\mu$f. variable capacitor (Hammarlund MC-325-M).
J_1—Coax chassis receptacle, SO-239.
J_2, J_3, J_4—Tip jacks.
L_1—5 turns No. 18, 1¼-inch diam., 10 turns per inch, mounted inside L_2 (Illumitronic Air Dux 1010).
L_2—5 turns No. 16, 1¾-inch diam., 10 turns per inch (Illumitronic Air Dux 1410).
L_3—10 turns No. 18, 1¼-inch diam., 10 turns per inch, mounted inside L_4 (Illumitronic Air Dux 1010).
L_4—6 turns No. 16, 1¾-inch diam., 10 turns per inch (Illumitronic Air Dux 1410).

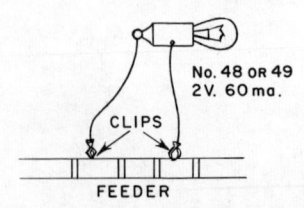

FIG. 15-20—R.f. output indicator using a dial lamp.

Plug the feeder wires into the appropriate pair of jacks for the band you wish to use— J_3J_4 for 3.5 and 7 Mc., J_2J_3 for 14, 21 and 28 Mc. Set C_1 at minimum capacitance, turn on the transmitter, and adjust the final-tank tuning capacitor for the resonance dip in plate current.

Then swing C_2C_3 through its range and find the setting where the plate current rises. Increase the loading with the loading control on your transmitter until its setting is approximately what you found to be needed with a lamp dummy antenna—remembering, of course, to keep the final tank in resonance.

Next, increase the capacitance of C_1 a little and retune C_2C_3, watching the lamp indicator. Let your transmitter's loading control alone, but keep the tank tuning in resonance. Continue this until the lamp lights brightest at the normal amplifier plate current. Record the dial settings that give you the most output with normal plate current on each band so you can return to them without having to go through the process again when you change bands.

V.H.F. Antennas

Antennas for the v.h.f. bands are the same, in principle, as those used on the lower frequencies. But on 50 Mc. a half wavelength isn't much greater than the floor-to-ceiling height in an average house, and at 144 Mc. a half wavelength isn't as long as you are yourself. This makes a big difference in what you can build.

Amateurs who do most of their operating on v.h.f. invariably use beam antennas, mostly of the Yagi type. These antennas give a worthwhile gain in power, both in transmitting and receiving. However, as you saw in Chapter 6, this gain is obtained in one principal direction at the expense of taking a loss in other directions. Unless you are interested in working only in one direction, this means that you have to rotate the antenna so it can be pointed at stations at various points around the horizon.

Because v.h.f. beam antennas are comparatively small and light, they can be handled without difficulty by the ordinary TV-antenna rotator. Also, the hardware and accessories for TV Yagis make assembling a 6- or 2-meter beam fairly easy. In fact, except for the element lengths amateur beams for these two bands are much like the TV beams you see everywhere.

In this book we won't go into beam construction, except for a very simple one described later for 144-Mc. work,* the antennas we'll discuss here are ones that you can practically throw together to get on the air. Although they aren't the last word for v.h.f. operation, they will let you make plenty of contacts in your local area.

Where to Put It

First, however, a few words about the antenna location. The most important thing is height. Height has more to do with your local range than anything else. A dipole well up in the air, above the buildings in its vicinity, will get out better than a beam that has to be installed at a low height (although the beam would be better than the dipole at the *same* height).

Of course, the location itself has a lot to do with it. If you're on a hill that looks down on everything for miles around, you can cover the area even with a low antenna. If your station is in a hollow you can't expect to do as well, but don't be discouraged on that account. Put the antenna on a pole on the roof, as high as you can get it.

Folded Dipoles

Among the simple antenna structures the folded dipole represents a good choice for v.h.f. It matches 300-ohm line fairly closely, and 300-ohm line has lower losses than coax. This difference in line loss can be a factor when you're trying to get the most you can out of a simple antenna. The open-wire type of line has least loss of any, so it is worthwhile to use it. There is a "300-ohm" variety as well as the wider-spaced line rated at 450 ohms. Either can be used; the difference in characteristic impedance is not enough to make an observable difference in losses.

When you install a v.h.f. dipole, folded or not, place it so that its length runs at right angles to the direction you want most to work. Although the dipole pattern is quite broad, there is little radiation off the ends—that is, along the direction of the antenna itself. If you can't turn the antenna at will, point these "nulls" where they will have the least effect on your coverage.

Folded Dipole for 50 Mc.

Fig. 15-21 shows how a folded dipole for 50 Mc. can be made from open-wire line using a support fashioned from 1×2 wood. An 8-foot piece of 1×2 (this is a standard length) is used as a crossarm for supporting the dipole. TV line stand-offs are screwed into the arm, as shown in the drawing, to hold the dipole. The 8-foot arm is a shade short of the length of the antenna, so the ends of the dipole extend out a few inches.

---

*You can find plenty of v.h.f. beam designs in *The Radio Amateur's Handbook* and *The A.R.R.L. Antenna Book.*

V.H.F. Antennas

FIG. 15-21—Construction of a folded-dipole antenna system for 50-Mc. work, using TV ladder line with a light wood support. Solid-dielectric 300-ohm twin line can be substituted for the ladder line.

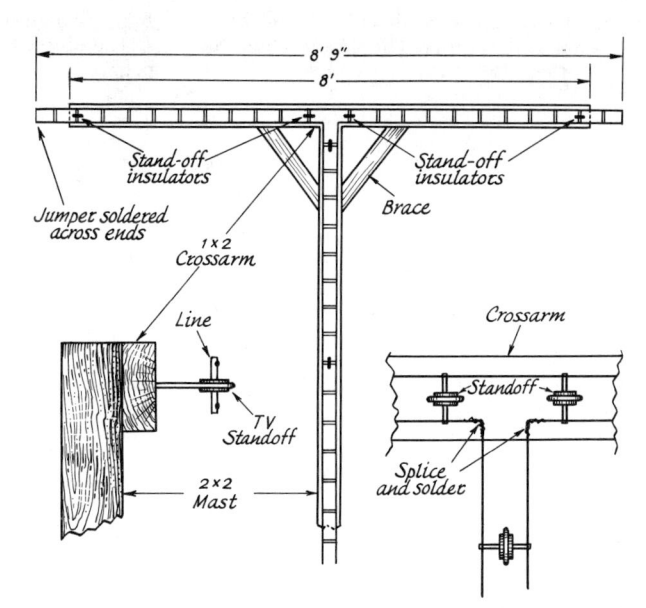

The line is stiff enough to be self-supporting for this short distance. The same type of line is used for the feeder, similarly supported from the mast holding the crossarm.

The arm can be fastened to a wooden pole with wood screws. Two pieces of 1 × 2 used as 45-degree braces will keep it from tending to rotate under strains such as might be set up by wind or snow. The pole itself can be a length of 2 × 3 lumber. It is feasible to use lengths up to 20 feet if you have the room to use three guys, fastened to the pole near the top and sloping down at an angle of 30 degrees or more with the mast. They should be anchored at approximately 120-degree intervals so the mast is equally well guyed for wind coming from any direction.

If the 2 × 3 can be solidly fastened to the wall of the house it can extend 10 feet or so above the uppermost support without guying.

The folded dipole doesn't have to be made from open-wire line; the web-type 300-ohm line can be used instead. The construction can be practically the same as that shown in Fig. 15-21. The losses in the feeder will be a bit higher, but this won't matter particularly if the line is not more than 50 feet long.

144-Mc. Folded Dipole

It shouldn't take more than a few minutes to put together the 144-Mc. folded dipole shown in Fig. 15-22. You need a piece of 1 × 2 wood to use as a mast. An 8-foot piece can be used with three guys, if the antenna is to stand on a roof. Or a 12-foot piece can be nailed to a wooden wall so that 8 feet or so stick up over the edge of the roof, without guys.

The antenna conductor is aluminum TV ground wire, which is about $5/32$ inch in diameter. Drill a $5/32$-inch hole in the 1 × 2 and push the wire through. Then bend the ends around as

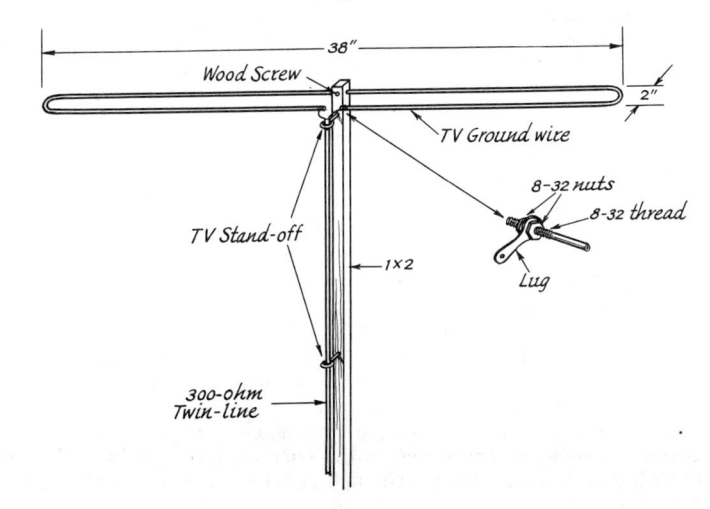

FIG. 15-22—An easily-built 144-Mc. folded dipole using wood support and TV ground wire. It is fed with 300-ohm twin line.

shown and cut them so there is about an inch separation at the center. The 300-ohm line connects to these two ends. A wood screw can be used to hold the center of the long conductor tightly, as shown. The feeder ends that are connected to the open ends will help support this side of the dipole.

The aluminum wire is about the right diameter for threading for 8-32 nuts, so the method of making connections shown in the small drawing can be used if you have an 8-32 die. Aluminum can't be soldered readily, so some form of mechanical connection has to be used. Another way is to make a small clamp of aluminum strip, as shown in Fig. 15-24. In either case, use a soldering lug and solder the feeder ends to the lug.

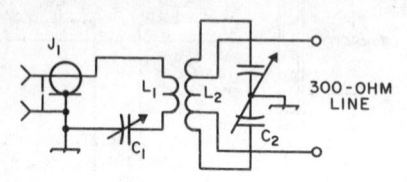

FIG. 15-23—V.h.f. transmatch circuit, for use with folded-dipole or other antennas having balanced feed and 300-ohm line. It should be built on a metal chassis, with ground connections made direct to the chassis.

C_1—50 Mc.: 100-$\mu\mu$f. variable (Hammarlund MC-100 or equivalent).
  144 Mc.: 50-$\mu\mu$f. variable (Hammarlund MC-50 or equivalent).
C_2—50 Mc.: 35 $\mu\mu$f. per section, 0.07-inch spacing (Hammarlund MCD-35SX or equivalent).
  144 Mc.: Same as for 50 Mc.; but for easier tuning, reduce to 2 rotor and 2 stator plates in each section.
L_1—50 Mc.: 4 turns No. 18, 1 inch diameter, 4 turns per inch (Air-Dux 808T or equivalent).
  144 Mc.: 2 turns No. 14 enam., 1 inch diameter, 1/8 inch spacing.
L_2—50 Mc.: 7 turns No. 14, 1/2 inch diameter; 4 turns per inch (Air-Dux 1204 or equivalent). Tap 1 1/2 turns from each end.
  144 Mc.: 5 turns No. 12, 1/2 inch diameter, 7/8 inch long. Tap 1 1/2 turns from each end.
Note: In the 50-Mc. coil assembly, L_1 is mounted inside L_2 at its center. In the 144-Mc. assembly, L_2 is centered inside L_1. Use short leads between L_2 and C_2.

TV ground wire is rather soft, and bends easily. A better wire, structurally, is aluminum garden wire sold in hardware and garden-supply stores. Although a little thinner—the diameter is 1/8 inch—the wire is considerably stiffer. Its diameter isn't large enough to be threaded for 8–32 nuts, but the wire will take a 6–32 thread.

V.H.F. Transmatch

The folded dipole uses balanced line, and the method of coupling it to the transmitter should preserve this balance. If the transmitter has a coaxial output connector, a simple transmatch circuit can be used. The system is the same as described for the lower frequencies in Chapter 6.

Fig. 15-23 gives a suitable circuit.

The object here is to transform the input impedance of the transmission line into 50 or 75 ohms, as seen by the coax line connected to J_1. The Monimatch described in Chapter 11 can be used on 50 and 144 Mc. provided it is carefully adjusted to give a proper null when a good dummy antenna is used for testing it. A 2-watt resistor— 50 ohms or 75 ohms, depending on the kind of line you will use— can be mounted inside a coax connector plug to make such a dummy. The values of the terminating resistors in the Monimatch may have to be varied somewhat to get the best null. As an alternative, on both frequencies you can find the proper coupling by using an r.f. output indicator such as the dial lamp described earlier in this chapter. Clip the lamp across a few inches of one side of the 300-ohm line and adjust C_1, C_2 and the transmitter loading controls for maximum lamp brightness at the normal loaded plate current for the final amplifier.

It is a good idea to do this by starting with C_1 at low capacitance, letting its setting alone while you vary C_2 for maximum brightness. Keep the amplifier plate circuit in resonance. Then let both C_1 and C_2 alone and vary the transmitter's loading and tuning controls for maximum loading. If the plate current can't be brought up to the normal value, increase the capacitance of C_1 a little and go through the process again. Use the smallest setting of C_1 that will give you maximum output (brightest lamp glow) at normal final-amplifier input.

Inexpensive 144-Mc. Beam

The three-element Yagi beam in Fig. 15-23, like the folded dipole just discussed, can use aluminum garden wire or TV ground wire for the elements. The boom for this antenna is also a piece of 1 × 2 wood. The reflector and director are straight pieces inserted in holes separated as shown, and held firmly by wood screws used as set screws.

The driven element, made from the same type of wire, is bent in the form of a hairpin on one side. This is like one side of a folded dipole, and the hairpin makes the driven element match 50-ohm coaxial line quite well. The line is connected as shown in the drawing. RG-8/U is the preferable type of line for this antenna. RG-58/U can be used, but its losses are higher. However, the difference between the two types won't be marked if the feeder is short—say less than 25 feet long—and the smaller line is a good deal easier to handle. With either type, be sure to weather-proof the open end at the antenna. Wrap the end completely with plastic electrical tape.

This antenna can be supported on a light wooden pole if it doesn't have to be rotated. However, it can also be mounted on a section of TV mast, using a U-bolt as shown. Hollow the wooden boom out slightly so it will seat firmly on the mast section.

Antenna Supports

FIG. 15-24—Three-element 144-Mc. beam made from aluminum wire and 1 × 2 wood boom. The open end of the driven element can be threaded for connection to the coax center conductor as in Fig. 15-22.

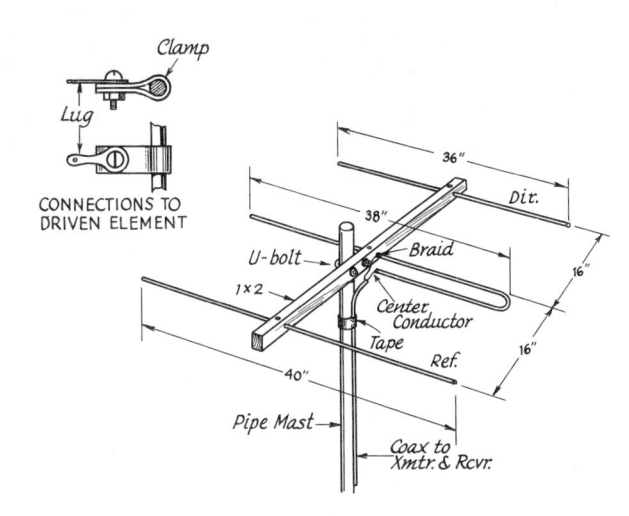

You can vary this construction to make it more solid mechanically. (The TV ground wire is rather soft and likely to bend under strain.) The elements can be made of aluminum or dural tubing, hard-drawn copper tubing, or other stiff conductor. The boom, too, can be metal tubing such as is used in TV antennas. TV hardware can be used to hold the elements to a metal boom, and the boom to the mast, when such materials are used. The elements do not have to be insulated from the boom, since they are supported at the center where the voltage is very low.

Supporting the Antenna

Height has a good deal to do with the performance of an antenna, as you must realize if you've read Chapter 6 and the remarks earlier in this chapter. Rather fortunately, height doesn't have the importance at 3.5 and 7 Mc. that it does on the higher frequencies. A low antenna will radiate mostly upward—at angles greater than 45 degrees with the ground. On the two lower bands a signal that goes up at a high angle will bounce back from the ionosphere to land on earth at distances up to 300 miles away. Then it bounces off the ground up to the ionosphere again, coming down at a still greater distance. This multihop way of traveling accounts for most of the work you'll do on these bands.

On 14 Mc. and higher frequencies such high-angle radiation is always wasted. On these bands the wave must "take off" at 30 degrees or less to come back at all. The higher the frequency the lower the angle you need. The angle gets lower as you increase the antenna height in *wavelengths*, not in feet. Here again Nature works in our favor, because an antenna that is, say, a half wavelength high at one frequency will be a full wavelength high at twice the frequency, and so on.

In practice, a height of 30 to 60 feet is in the optimum region for *any* band where the ionosphere plays a part.

You can get by with even less than 30 feet if you have to. However, nearly everyone can put an antenna at least 25 feet or so in the air. This height puts it in the class where, usually, at least one end can be supported by your house. The other end can be held up by another building (if you have permission to use it) far enough away, or by a tree, or if necessary by a pole you put up yourself. The pole doesn't have to be an elaborate structure, especially if it can be erected on top of something. A detached garage makes a good base. It should give you a start of 12 to 15 feet, and generally will have enough roof area to let you put guys on a simple wood pole that can be up to 20 feet tall.

Anchoring

Wire antennas are not heavy, and even when pulled up tight the strain can easily be supported by a husky screw eye or hook, if it can be sunk 1½ or 2 inches into good solid wood. Screw eyes can be used similarly for anchoring guy wires from light poles up to 20 feet long. Often other anchorages for guy wires offer themselves; use anything that seems solid enough and which gives you the opportunity for wrapping the wire around it a couple of times so it won't slip.

A length of 2 × 3 makes a good pole for heights up to 20 feet or so. It isn't heavy, but with three solidly-anchored guys it will hold any wire antenna you may use with it. Fig. 15-25 shows a simple type of construction, using fittings that you can buy at a hardware store. The pulley and halyard make it possible to raise and lower the antenna easily. With the antenna up, only two guys are really necessary, placed 90 to 120 degrees apart back of the antenna, as shown

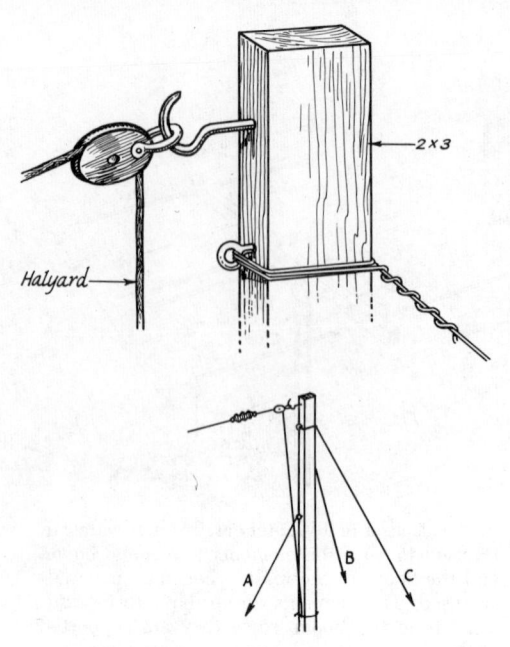

FIG. 15-25—Using a length of 2 × 3 as a support for a horizontal antenna. Lengths up to 20 feet can be used with a single set of guys arranged as shown.

The base is a length of 2 × 4 set solidly in the ground. The 2 × 4 should be buried to a depth of about three feet and should extend at least three feet above the ground level. Use a plumb line to make sure that it is vertical when setting it. Pack rocks around it before refilling the hole with dirt; this will make a solid foundation which is helpful when the mast is being raised. After it is up and the guys are in place the stress on the base is practically all straight down, so there is really no need for an elaborate footing.

The mast should be assembled on the ground (with the antenna halyard affixed; the method shown in Fig. 15-25 can be used for this) and fastened to the ground post with the lower carriage bolt. The guy wires, which should be made long enough to reach their anchorages while the mast is down, also should be attached. Push the mast straight up at the middle, letting it pivot on the bolt. When it gets up high enough, have an assistant slide a ladder under it to hold it. Keep working it up, moving the ladder down the mast as you go. Stop now and then to pull up on the guys. Keep the front guy short enough so the mast won't swing all the way over and come down on the other side, but leave it loose enough so it can't interfere as the mast goes up. When

at B and C. The third guy, A, can be part way down the pole so it will be out of the way of the pulley and halyard, and should pull off in about the same direction as the antenna. Its purpose is to hold the pole up when the antenna is down.

The guys can simply be wrapped around the pole near the top. One is shown in the drawing. To keep them from slipping down they can be run through a screw eye as shown. This avoids putting any strain on the screw eye. Stranded steel wire such as you can buy in a hardware store (a typical type has four strands of about the same diameter as No. 18 copper) has ample strength for guying a pole of this type. Guy wire also is available from radio supply houses.

Plastic clothesline can be used for the halyard. The plastic kind stands the weather better than ordinary cotton clothesline. If you do use cotton, get the type having a steel core; it has greater strength and lasting qualities. But whatever type you use, be sure to use enough. Remember that you need to be able to reach the free end of the halyard when the antenna is down!

Simple Wooden Mast

If you have to set a mast on the ground, the structure shown in Fig. 15-26 is easy to build and erect. Three 22-foot lengths of 2 × 3 are used to make a mast 40 feet tall. It makes a sturdy support for a horizontal-wire antenna when properly guyed. The mast is pivoted at the base for ease in raising (and lowering, if necessary), the pivot being the lower carriage bolt (A) shown in the drawing.

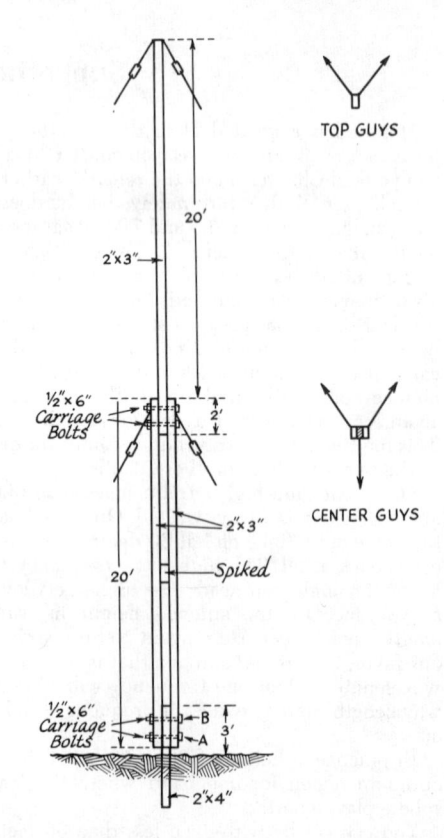

FIG. 15-26—This mast construction is suitable for heights up to 40 feet, for supporting a horizontal antenna. Pivoting the mast on the lower bolt, A, simplifies the job of raising and lowering.

Masts

the mast approaches the vertical, pulling on the top guys will bring it up the rest of the way. Slip in the upper carriage bolt (B) and the mast will stand by itself (if the base is solid) while you adjust the lower set of guys. Don't give the top guys their final tightening until the antenna is up, as they must work against the pull of the antenna.

The guy wires should be anchored about 20 feet from the base of the mast, for a height of 40 feet. If shorter 2 × 3s are used so that the completed mast is less tall, this distance can be correspondingly reduced.

Insulating Guy Wires

Guy wires are like antenna wires—if their length happens to be near resonance at your operating frequency they can pick up energy from the antenna's field and re-radiate it. This may do no harm, but then again it might distort the radiation pattern you hope the antenna will give.

To circumvent this, guy wires are often "broken up" by inserting insulators at strategic points. The idea is to avoid letting any section of guy wire have a length that is near resonance at any operating frequency or harmonic of it. Egg-type insulators are generally used as a safety measure. With these insulators the wires are looped around each other so that if the insulator breaks the guy will still hold.

General practice is to put an egg in each guy a foot or so from where it is attached to the pole or mast, and another a foot or so from the point where it is anchored. Whether any more are needed depends on the remaining length. In most cases it is sufficient to insert insulators in such a way that no section of the wire will have a length that can be divided evenly by either 16 or 22.

Guy Anchors

Trees or buildings at the right distance from the base of a mast can be used for anchoring guys. However, they don't always "just happen" to be present. In such case a guy anchor can be sunk into the ground.

A six-foot length of 1-inch iron pipe driven into the ground at about a 45-degree angle makes a good anchor. It should be driven toward the mast, with 12 to 18 inches protruding from the ground for fastening the wires. Fig. 15-27 shows the general idea, with a suggested way of fastening the guy so it can't slip off the anchor.

Weather Protection

A wooden mast or pole needs the same protection against weather as the exterior woodwork on a dwelling. New wood should be given three

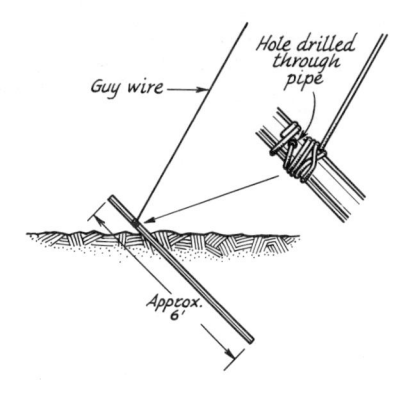

FIG. 15-27—Pipe guy anchor. One-inch or 1½-inch iron pipe is satisfactory for a mast such as is shown in Fig. 15-26.

coats of a good paint made for outdoor use, with the first coat thinned so it sinks in well. Follow the directions on the paint container.

Masts made from several pieces of lumber, such as the one shown in Fig. 15-26, should have their individual parts painted separately before assembly. The heads of nails or screws should be sunk into the surface far enough so they can be puttied over. These, too, should be painted—after assembly.

Hooks, screw eyes and similar hardware should likewise have a coat of paint to prevent rusting.

Other Types of Masts

Amateur ingenuity has conjured up lots of mast and tower designs over the years. It is possible, for example, to make a satisfactory pole, 40 feet or so high, out of lengths of metal downspouting. Wooden lattice construction has long been a favorite for masts and towers. A number of such designs will be found in *The A.R.R.L. Antenna Book,* and new ones keep coming along regularly in *QST*. Then there are TV masts, built up in sections of metal tubing and provided with guy rings and other hardware—readily available and not at all costly. These can be run up to heights as great as 50 feet, with proper guying.

In fact, anything you can put up, and which will *stay* up, will be satisfactory. Never forget that you can't afford to take any chances with safety. A falling mast can result in accidental injury, or even death, to anyone that happens to be in its path. The taller and heavier the mast the more necessary it becomes to build in plenty of safety factor—in size of guy wire, type of anchorage, and in the strength and cross section of the materials used in the mast or tower itself.

Chapter 16

Operating Your Station

Now that you have this amateur radio station of yours, what are you going to do with it? Rag chew, handle traffic, work DX, keep schedules with friends? One reason that ham radio is so fascinating to so many people is that it has many different facets, and one or more of these will surely strike your fancy.

Ham radio is just like any other sport or hobby—the more skilled you are in the various techniques, the more fun you can have and the more satisfaction you can get out of it. As an example, consider the automobile driver. A good driver gets where he is going safely and without waste of time, he does not interfere with others on the road, and he enjoys a certain amount of self-satisfaction through his knowledge that he *is* a skillful driver. The inept driver, on the other hand, does not in the least enjoy his drive. He gets all balled up in traffic jams, and frequently endangers his own life and the lives of others on the road. And then there is the discourteous driver who has no regard for the rights of others and so often turns a pleasure drive into a disagreeable experience. No doubt you can recall examples of each of these drivers in your own experience.

Now let's try to talk about some of the rules of the road in ham radio so that your on-the-air operation will be an enjoyable experience for yourself and for the others who share the amateur frequencies with you.

Although there are several other modes of communication, phone and c.w. are most used by hams. A few amateurs have radioteletype equipment, and an even smaller group of experiments use amateur television and facsimile. But if we learn about some of the better techniques for phone and c.w. operating, we'll be ready to tackle the other modes if and when.

Basic Phone Procedure

Let's talk about phone operation first. Even though the equipment required is a bit more complicated, talking is the way we conduct most of our every-day communication with our families and associates, so it seems logical that not much would have to be said about the technique of operating on phone. But there are, in fact, a few pitfalls to be avoided.

First of all, you must obey the basic FCC regulations regarding identification of your station and the station or stations with whom you are communicating. This is all clearly spelled out in the rules and regulations for amateur stations. (We assume that everyone has a copy of *The Radio Amateur's License Manual;* but if not, you can order one for 50¢ postpaid from ARRL, Newington, Conn. It contains complete information on the amateur rules and regulations.)

You are required to give the call of the station you are calling or working at the beginning and end of each transmission, or at the beginning and end of a series of transmissions if each of these transmissions is of less than 3 minutes duration. In addition, if you are the long-winded type and are carrying on a monologue which lasts for more than ten minutes, you must identify yourself and the station you are working each ten minutes.

The FCC regulation specifies that you first give the call of the station you are working, and then your own call.

For example, if your call is W1AAA and you are calling or working W6AAA, your identification would be thus:

"W6AAA, this is W1AAA," or
"W6AAA from W1AAA."

Most of the time amateur operators use some sort of phonetics—some of these are "cute", some have no business being used on the air, and some serve the purpose of aiding in the identification of your call letters. While it's all well and good to come up with a set of words that fit your call letters (such as W1LVQ identifying himself as W1 Lovely Virgin Queen) a major use of phonetics is to aid in the identification of your call during times when interference may be heavy. During such times, there is a definite advantage in using a standard phonetic alphabet which is known and recognized by everyone. The ARRL has adopted such a phonetic alphabet, as follows:

ADAM	JOHN	SUSAN
BAKER	KING	THOMAS
CHARLIE	LEWIS	UNION
DAVID	MARY	VICTOR
EDWARD	NANCY	WILLIAM
FRANK	OTTO	X-RAY
GEORGE	PETER	YOUNG
HENRY	QUEEN	ZEBRA
IDA	ROBERT	

Example: W1LVQ . . . William 1 Lewis Victor Queen. (Copies of this, postcard size for use in your radio room, are available free from ARRL, Newington, Conn. Ask for Operating Aid #1.)

296

C.W. Operation

Phonetic alphabets using geographical names are not as good, because they can cause confusion when a station is giving not only its call letters but also its location.

OPERATING PROCEDURES FOR PHONE AND C.W. ARE NOT ALWAYS IDENTICAL

You will hear many amateur phone operators using c.w. operating signals and abbreviations during their conversations. Although there may not really be anything wrong with this procedure, it *is* more natural to use the same sort of vocabulary on phone that you would use in your face-to-face conversations. And, of course, the charm of phone operation is the ability to talk naturally without any artificial aids, either mechanical or verbal. As an example, when you stand by and instruct the other station to go ahead and transmit, instead of using the telegraphic signal "K", why not simply say "go ahead"?

There is a greater trend toward voice-controlled operation, particularly with those amateurs who have single-sideband equipment. This does away with the cumbersome system of one operator talking for a few minutes, and then the other operator taking a turn. With voice control, the transmitter is turned on as the operator starts the first syllable (with proper adjustment no intelligibility is lost) and the transmitter turns off whenever the operator pauses for half a second or so. This enables the two stations to carry on a rapid-fire conversation, and is the most natural form of voice communication.

C.W. Operation

Although many amateurs start out on c.w. and later change over to phone, c.w. operation actually requires some greater skills than voice. First, to be a good c.w. operator, you must be able to copy the code well. Not necessarily very fast, but you must be able to copy well enough so that it is not a chore. After you have acquired sufficient skill, you will not copy the *code,* you will copy the *message.* The beginner often finds this hard to believe as he struggles with his code practice, but code is just like another language.

C.w. operators, in order to conserve time, are great users of abbreviations and, like phonetic alphabets, there are certain standard abbreviations which most people stick with. In addition, there are Q signals, which are internationally-recognized operating signals providing for the exchange of a great deal of information between operators who do not understand each other's native language. Amateurs use Q signals mainly as a way of expressing whole ideas in a group of three letters, thus cutting down on transmission time. (See the list inside the back cover.)

One trait that identifies many a beginning c.w. operator is that of using phone lingo on c.w. For example, you will sometimes hear a c.w. station spelling out each word just like this, "So what do you say? Now over to you, W6AAA DE W1AAA K."

The more experienced operator, however, would send, "Wat say W6AAA DE W1AAA K."

Or he might send, "Hw? W6AAA DE W1AAA K."

The "Hw?" is a c.w. man's abbreviation meaning "how", and so it translates into "how now" or "how about you" or "what do you say?"

By using these common c.w. abbreviations, you can easily double your effective speed of communication. Thus, even though your code speed might be only 15 words per minute, you could really communicate at a much higher rate by the proper use of these abbreviations. We say "proper use", because if you use non-standard abbreviations, the other guy may have a heck of a time figuring out what you are saying unless he's at least as smart as you are! (See page 305.)

So, study these c.w. abbreviations and use them on the air, and you'll find more pleasure in your c.w. operating.

Besides the Q signals and standard c.w. abbreviations, there are a number of operating signals which have been adopted internationally in order to speed up c.w. operation and to clarify operating procedures. They are, in a way, additional abbreviations.

K means "go ahead and transmit"
AR means "end of transmission"
SK means "end of communication"
KN means "go ahead and transmit (a particular station only) . . . others keep out"
CL means "this station is closing down"

These signals are used as follows:

K   is an invitation to the station you are working to go ahead and transmit.

AR indicates the end of a message. It would come at the end of a formal message, as we will describe later, or it might come at the end of an informal exchange, just before you sign your call letters and stand by for the other station. For example: ". . . Hw is wx? AR W6AAA DE W1AAA K"

SK indicates the end of your communication with a particular station, and that you are expecting no further transmissions from him. *Example:* ". . . TNX QSO es 73 SK W6AAA DE W1AAA"

CL would be sent by you at the very end to indicate that you were closing down your station for the time being and would not answer calls from anyone else.

KN is used to show that you are standing by for a particular station to transmit to you, and that you do not want anyone else to call.

There is no magic wand that you can wave to transform yourself into a good c.w. operator. As with any sport, to acquire the skills requires practice, and lots of it. You can get plenty of practice simply by getting on the air and operating. But try not to mimic the *bad* habits that you may hear. If you want to hear what good code sounds like, for example, tune in the ARRL Hq. station, W1AW, which conducts code practice transmissions every night at various speeds from 5 to 35 words per minute. (The schedule of these transmissions is carried every month in *QST*, or you may write to ARRL Hq.)

You may also get free of charge from ARRL a little booklet entitled *Operating an Amateur Radio Station*, which is full of helpful hints on the right way to operate, both phone and c.w.

What to do With Your Operating Time

There are many ways you can spend your time on the air. In your first months of amateur operation, as you improve your skills and understanding of amateur radio, you may be satisfied to ragchew with other hams. You will find excitement and pleasure as you contact others throughout the country. Sooner or later you will be bitten by other species of the ham radio bug, and you will find other interests in ham radio. You may take a fling at working DX (distance), or handling traffic, or you may take part in some of the many contests that occur each year.

Ragchewing

It really shouldn't be necessary to tell anyone how to ragchew, but it does seem as though many an amateur becomes tongue-tied when he gets on the air. This may be why you hear so many stereotyped QSOs as you listen across the ham bands. On either phone or c.w. you will hear stations limiting their conversational excursions to a recital of the weather, their locations, how much power their transmitters have, signal reports, and then fervent hopes that each will see the other again.

You can become a first-class ragchewer by using a bit of ingenuity in your QSOs. After you have covered the customary preliminaries of signal reports, locations, power, names, and the like, branch out. Mention some of your other activities, the interesting thing that happened to you on the way to the coliseum last night, what kind of work you do, or a choice bit of technical info that you happened across. If you give the other fellow a few such inviting leads, you may find that you have much in common. He may know how to lick that ignition noise that has had you stumped, or you may both be math teachers in high school and he knows what to do about the parent that doesn't believe in the new methods of teaching the subject. Then again, maybe gardening happens to be the opener.

GARDENING HAPPENS TO BE THE OPENER

If you're going to be a ragchewer, be a good one! Polish up your phone or c.w. operating techniques and it will be a pleasure to *converse* with other hams.

Working DX

The thrill of working DX is perhaps one of the greatest lures of amateur radio. Every amateur, every pioneer in radio, has wanted to contact someone else just a little farther away.

Just what *you* mean by DX depends on your experience and the band you operate. When you first start off in ham radio, a hundred miles may be DX, but later you will become accustomed to working every corner of the globe. Then, DX comes to mean not merely some place that is far away, but a place that is rare, where there are not many amateurs. The first time you work a station in, say, England, it will be DX! to you. But later on, you will find that there are hundreds of English stations, and you can work them by the dozen. It is no longer DX!

But whatever your definition of DX, the skills required are pretty much the same, except that when you start to hunt the rarer DX, you have to sharpen up those skills. They are the skills that you must acquire by practice, and by keen observation of what goes on.

This business of *listening* to what is going on before you operate your transmitter is most fundamental, and yet is a principle most often

LISTEN— AND THEN LISTEN SOME MORE

DX Operating

violated by a beginner. If you were a fisherman, before you began casting you would carefully analyze a stream, the weather, what had happened the last time you fished this area, and so on. That is, you would if you really wanted to catch fish. If you were satisfied merely to sit by the bank and relax, then you wouldn't be chasing DX.

So listen across the band. What are other fellows working? Does some particular section of the world seem to be coming in better right now? Do you hear some certain station that you would like to work? If so, study his operating habits. Sure, this will take a little extra listening, but your investment of time will be more profitable in the long run. Does this fellow answer calls on his own frequency, or does he prefer stations to call him somewhat higher or lower? Perhaps he varies his technique, first answering higher and then lower. Discover his operating habits, and then you will have a better chance of working him. Does he like long calls, or short snappy calls?

One thing you should remember is that as a U. S. ham you are one of some 250,000—there are many more U. S. hams than in all the rest of the world. Thus, except under unusual circumstances, it is not worthwhile for you to call "CQ DX" if you want to work rare DX. You must be the hunter, rather than the hunted. If a rare station in Africa, say, signs his call letters once or twice, half the world will be calling him. *You* can call "CQ DX" half the night and never get an answer from that sort of DX. However, you *can* have many enjoyable QSOs with less rare DX.

Besides listening on the band to determine what is going on, skilled DX operators pay close attention to the gossip of DX. They hear other stations telling each other what DX is on the air, and what frequencies are used. They read the DX columns in the various ham magazines, and perhaps subscribe to one of the several excellent DX club bulletins. They keep little notes of the times and frequencies that the various rare stations operate, and monitor these spots regularly.

DXing is a very competitive sport. You will find this out the minute you hear some rare bit of DX and begin calling it. You will find that a thousand other hams heard the DX station at the same instant and are also calling. Most of these hams are outstanding operators and have the very finest of stations. They have high power, excellent antennas, and a vast amount of experience. Some of them, however, are just like yourself—they are not running kilowatt transmitters and they are not the world's best operators. You can compete on an even basis with this latter group, and you can work up to competing with the top group, by using your head. *Listen* to the band and find out what is going on. *Analyze* a DX station's operating habits so that your on-the-air time will be most productive. *Be patient.* Because there are so many other stations on the air, your call may not be heard first, or second, or for four hours—despite your best efforts, *when* you are heard by the DX station through all the interference may be entirely a matter of chance.

Handling Traffic

Aside from working "DX", one of the first things that ham radio operators started doing was the relaying of messages or traffic. This, in fact, is how the American Radio Relay League came to be formed and how it got its name. In the early days, "DX" was 25 miles, and if there was a friend 50 miles away to whom you wished to send a message, you had to find some other ham in between who would relay the message.

There are many amateurs today who get almost their entire pleasure from ham radio through the handling of traffic, the relaying of messages, not only for other hams but also for the general public. In order to increase the reliability of this amateur traffic handling, a vast volunteer system of nets has been built up under ARRL supervision, covering the whole United States and Canada. By writing to ARRL you can obtain a mimeographed bulletin listing all the phone and c.w. nets known to be meeting regularly. From this you can find out when the net for your section meets, and you can listen in and hear messages being handled. The operators of most of these nets work at modest speeds, and they will welcome you to the net. Report in!

In order to channel this traffic nationwide, we have a National Traffic System. The NTS provides for each section net to send a representative into its designated "region" net (usually covering a call area), and region nets in turn send representatives into their "area" net (roughly covering a time zone). In this way, if you are in, say, Florida and want to originate a message for California, you would send it to some station in your section net, whence it would be passed to the region and area nets in turn. Area nets are tied together transcontinentally by a group of skilled c.w. operators called the Transcontinental Corps (TCC), who would put the message into the Pacific Area Net, whence it would go down through the region and section

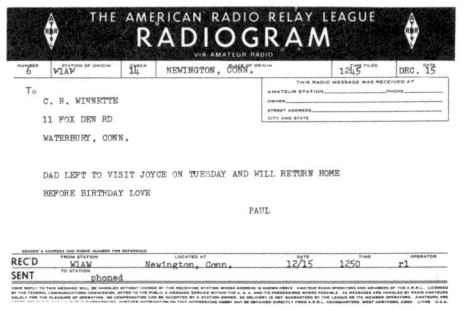

Here is an example of a plain-language message in correct ARRL form. The preamble is always sent as shown: number, station of origin, check, place of origin, time filed, date.

nets of destination to ultimate delivery. More details on this system from ARRL if you are interested.

In traffic handling, as in every other phase of ham operating, there are certain customs and procedures. We can touch on those briefly here, and you can become more familiar with them by listening to or participating in some actual net sessions.

Each net is controlled by a Net Control Station—he is in charge, calls the roll to obtain lists of traffic to be handled, designates which stations shall handle which traffic, and keeps records of net operations which he forwards to ARRL. One of the first things you will notice is that special Q signals are used on these nets. They supplement the international Q signals, and a list of them can be obtained by writing ARRL and asking for Operating Aid #9.

The essence of good net operation is brevity. The net control station will use the Q signals to convey the maximum communication in the minimum time, and will not permit net members to rag chew until all traffic has been cleared. (Even then, informal chats should be removed to some other frequency, because another net may meet on the same frequency after the first net has "secured".)

In order that traffic can be handled rapidly and accurately throughout the country, standardization is important. Amateur radiograms have a definite form that should be followed, and one easy way to insure that the necessary information is included in each message in the proper order is to use ARRL message blanks. The sample on the previous page shows the proper form, and each item has a reason for being included. The message shown would be transmitted on c.w. as follows: "NR 6 W1AW CK 14 Newington Conn 1245 Dec 15 1962 BT C R Winnette AA 11 Fox Den Rd AA Waterbury Conn BT Dad left to visit Joyce on Tuesday and will return home before birthday love BT Paul." (The AA is a separator between parts of the address.)

The servicing information on the message blank shows when W1AAA received this message and that he delivered it by telephone.

The first line of the message is known as the preamble, and this information identifies the originating station, the time and date of origination, and the file number. If someone along the networks experiences difficulty in handling the message, the information in the preamble allows a service message to be headed back to the originator so that the difficulty may be explained or cleared up.

There's more to the handling of messages than we have room to cover here. If you'd like more info on this phase of ham radio, drop a line to ARRL headquarters.

Awards

Everyone likes some sort of tangible evidence of his accomplishments, trophies of one sort or another. So it is in ham radio—there are dozens and dozens of awards that you can qualify for. Some are of considerable stature and require

THERE ARE MANY AWARDS

much skill, while others are on a much lower level of accomplishment. Some are issued by ARRL, some by various local radio clubs, by foreign ham radio societies, by various ham radio publications, and so on. We can't possibly list all of the awards, and so we'll restrict ourselves to some of those issued by ARRL.

For the ragchewer, there is membership in the Ragchewer's Club (RCC). To qualify you must chew the rag for 30 minutes with someone who is already an RCC member. Then you *both* send postcard reports of the QSO to ARRL and a certificate will be mailed to you.

If you can work all 50 states of the U. S. A. and receive the confirming QSL cards, you can apply to ARRL for the WAS (worked all states) award. Send the cards, a list of the states and stations worked, and return postage to ARRL Hq. Or write for a copy of Operating Aid #8, which gives a list of the states and all the rules.

For the DXer there is membership in the DX Century Club. The DXCC certificate is one of the most sought-after awards in ham radio, and requires many hours of patient hunting of DX, for you must submit QSL confirmation of contact with 100 different countries. For the complete rules and application procedure, and for an up-to-date copy of the ARRL Countries List, write to ARRL Hq.

The International Amateur Radio Union (of which ARRL is the headquarters society) issues the WAC (Worked All Continents) award, for submitting proof of contact with the six continental areas of the world (North and South America, Europe, Africa, Asia, and Oceania). There is an endorsement for all-phone operation.

Traffic handlers can receive awards for handling certain minimum numbers of messages each month, and continuing high performance in this traffic work can earn you a handsome medallion. These are not easy awards to earn, but well worth while.

There are a number of other awards issued by ARRL, in addition to the dozens and dozens

Contests

issued by other organizations and clubs. The active ham could literally paper the walls of his shack with the certificates that are available for various ham radio operating achievements. Some are awarded only for outstanding achievements, some are for much lesser feats, but all are fun to get.

Contests

Contests are a big thing in ham radio, even though not every amateur may take part in them. But the enthusiasm of those that do is pretty contagious, and you'll want to check into these someday.

Most of the contests provide the same basic goal—work as many different stations in as many different places as possible in a given period of time. The annual ARRL Sweepstakes is a competition between amateurs in the various operating sections of ARRL (the U. S. and possessions and Canada), and is a great contest for those who are looking for additional states toward their Worked All States award. It takes place each November.

In the annual ARRL DX Competition on two separate weekends each February and March, U. S. and Canadian amateurs endeavor to work as many foreign countries as possible, and during the contest period the ham bands are loaded with DX stations from all corners of the globe.

There are a number of v.h.f. contests each year, in which contest activity is limited to the bands above 50 Mc. These contests were originally started in order to encourage activity on the higher amateur frequencies, and now rival some of the other contests in popularity.

The annual ARRL Field Day is one of the year's most popular activities—more individuals take part in this than in any other ham radio event. Field Day is a week-end affair in June designed to test portable and emergency gear. Individuals and clubs take their gear out into the field and set up camp, and over a 24-hour period from Saturday night to Sunday night contact as many other similar groups and home stations as possible. There are special score multipliers for low-power transmitters, and for having all a.c. power supplied by other than the commercial mains. The fun of testing your equipment and operating skill under these field conditions is pretty hard to beat.

There are many other contests, some sponsored by the League and some by various other clubs and organizations. In fact, if you become a contest fanatic, it would be possible to take part in some sort of contest nearly every weekend of the year. They are all tests of your operating skill and your station equipment.

In all contests, someone has to win and someone has to lose. As in any sport, there are some who are so skillful that they can win almost any race they run. Yet this need not detract from your enjoyment of the contests. Let each one be a test of *you*, in which you try to improve your previous performance, and your operating skills will gradually improve. You will have the chance to work states and countries which are not on the air in such profusion as during contests. You will find ways in which you can improve the operating convenience of your station.

Amateur radio justifies its occupancy of valuable radio frequencies because of the public service it renders, and contests are another way in which amateurs can improve their operating skills and thus become better able to serve in both routine matters and emergencies.

ARRL Operating Appointments

For those who like to "belong" and take part in organized activities, there are a number of ARRL leadership and station appointments. There are Section Communications Managers who are elected by ARRL members in the various geographical areas into which the U. S. and Canada have been divided. Each SCM appoints interested amateurs to posts of responsibility and operation. There are Official Relay Stations, Official Phone Stations, Official Bulletin Stations, Official Observers, Official Experimental Stations, Route Managers, Emergency Coordinators, and so on. Each of these appointments appeals to a specific type of amateur operation, and for complete information you should ask ARRL to send you a copy of *Operating an Amateur Station*. The address of *your* Section Communications Manager will be found in *QST* each month, always on page 6.

Emergency Operating

Each amateur, regardless of his interest in certain special phases of operating, will find in emergency communications preparedness a challenge to develop his operating skills to the utmost. Testing of emergency equipment in on-the-air operating during a June Field Day has been mentioned: this is competitive and fun. A Simulated Emergency Test is also sponsored by ARRL each fall. This is often a subject for good local publicity for amateur radio, and for all those who take part. You will find it stimulating to employ amateur radio and your know-how in such exercises.

We can bring our general operating skills to a

very high point by engaging in the various aspects of hamming that we have already mentioned. But unless we get into emergency groups and help develop those local plans, we can hardly give the best account of ourselves if there is suddenly a situation involving a communications emergency. *During* the emergency is no time to practice!

Fortunately, *every* amateur can register his personal and equipment availability in the Amateur Radio Emergency Corps in advance. Here again, contact your Sections Communications Manager for full information, and write to ARRL for the booklets *Operating an Amateur Radio Station* and *Emergency Communications Manual.*

A lot of the excitement and thrill of operating with an AREC group can come when an Emergency Coordinator makes some of his communications exercises tie in with community public service projects. There are many opportunities for this sort of activity. Some of the examples which have been written up in *QST* include parade monitoring, road and Hallowe'en patrols, health-fund drives, weather warning, civil defense, and special tests for the Red Cross or other agencies.

No understanding of amateur radio can be complete without understanding that emergency communications is a very important public service rendered by amateur radio operators.

Keeping a Log

A log is required of you by the FCC. You must log all calls, listing the time, date, frequency band, operator's signature (that's you unless you have a visitor), the power input, whether phone or c.w., and the time of ending of any contact. A convenient form to use is the ARRL log, which has provision for recording all the information required by FCC.

However, you should keep a log not just because it is FCC-required but because of pride in the operation of your station. It's an excellent place to keep a record of adjustments to and experiments with your receiver and transmitter. By keeping records in your log of

WE MUST KEEP A LOG OF OUR STATION'S OPERATIONS

your antenna changes, you can get an idea of whether your over-all results are better or not.

QSL Cards

For almost all of the awards, it is necessary that you submit evidence to prove your eligibility, and so you hope that each fellow that you work sends you a card and that the card has the essential information contained on it. Likewise, those fellows that you work certainly hope that *you* QSL and that your card is adequate. Most hams take quite a bit of pride in their QSLs, and there are a great number of printers throughout the country who specialize in the production of attractive QSLs. Many of them advertise each month in the classified ad section in *QST*.

However, even the classiest QSL will be a failure if it does not contain the minimum of information required for the various awards. Here are some of the things that should be included:

Date and time of QSO.
Frequency band.
Mode of emission.
Call sign of station worked.
Call and location of your own station.
Signal report.

These items will satisfy most award hunters, although occasionally a fellow will be plugging for something that requires you to list what radio club you belong to, or what county you are in, and so on.

DATE TIME	STATION CALLED	CALLED BY	HIS FREQ. OR DIAL	HIS SIGNALS RST	MY SIGNALS RST	FREQ. MC.	EMIS- SION TYPE	POWER INPUT WATTS	TIME OF ENDING QSO	OTHER DATA		QSLs	
											NAME	S	R
11-16-62													
1815	W4IA	X	3.65	589	569X	3.5	A1	250	1843	Tfc-rec'd 6, sent 10			
1920	CQ	X				7	"	"					
1921	X	W4TWI	7.16	369	579		"	"	1932	Vy heavy QRM on me		X	X
2125	W8UKS	X	3.83	59	47	3.9	A3	100	2205	Sam			
11-18-62													
0705	VK4EL	X	14.03			14	A1	250		Answered a W6			
0709	ZL2ACV	X	14.07	339	559X		"	"	0720			X	
0721	X	KA2KW	14.07	469X	349		"	"	0733	First KA		X	
0736	CQ	X					"	"					
0737	X	W6TT	14.01	589	589C		"	"	0812				

Time

Make every effort to fill out the cards accurately and legibly. Many awards will not be issued if a submitted card has been altered, for there is no way for the administrator of the award to know whether the alteration has been made by the sender or receiver of the card. If you make a mistake in filling out the call on a card, or some other piece of information, make out a whole new card. This may save much embarrassment for the recipient.

If you didn't get the address of the fellow over the air, you can get it from the *Radio Amateur Call Book Magazine*, a quarterly publication that lists all U. S. licensees. Another edition lists Canadian and overseas stations. Cards for U. S. stations you must mail directly to the individual, but cards for overseas stations may be handled either direct or via one of the foreign radio society QSL bureaus. These bureaus are listed in *QST* each June and December. Also, a monthly column in *QST* devoted to news notes about DX stations often lists supplementary addresses for some of the newer and rarer stations. If you become a DX hound, you'll follow these news notes very carefully each month.

ARRL also maintains a QSL bureau system for *incoming* DX cards. Foreign stations and foreign QSL bureaus bundle up their cards (this saves them postage) and ship them to the individual QSL bureau managers in each U. S. and Canadian call area. These call area QSL bureau managers are listed every other month in *QST*, and what you should do is send your district QSL bureau manager a stamped, self-addressed envelope of the proper size (about 4½ by 9 inches). You place your own call sign prominently in the upper left-hand corner of the envelope, and whenever the QSL bureau manager finds that he has a bunch of QSLs for you, he'll drop them in the mail. Whereupon you promptly send him another envelope.

Time

We've mentioned "time" now and again in this book. It's a pretty important item, because you are required to log the times of QSOs, you want to record the QSO times on your QSL cards, you want to know when to make schedules with other stations, and so on. Yet, despite the importance of time, it's an item that has created a great deal of confusion, and this has come about because of the profusion of time zones throughout the world and because of "standard" time and "daylight saving time."

ARRL advocates the use of Greenwich Mean Time, and uses GMT for most of its contest and Operating News reports. If *everyone* used Greenwich time, one source of confusion in ham radio would be eliminated. If you made a schedule with an Australian station for 1430, you would not have to wonder whether he had in mind 1430 Australian standard time, or Eastern Daylight Saving time, or whatever. And when you got a QSL from the station in Hawaii, you would not have as much difficulty in locating him in your log if you *knew* that the time recorded on his QSL was GMT, corresponding to the GMT in which you kept your log.

Greenwich time is based on the prime meridian at Greenwich, England, and you can convert your local time to Greenwich time by knowing what time zone you are in. ARRL has a conversion chart which is yours for the asking—request Operating Aid #10.

In Conclusion

As we close this final chapter of *Understanding Amateur Radio,* we are going to emphasize some of the things that we said at the beginning of the chapter.

Ham radio is a constantly growing hobby, with some 250,000 U. S. citizens now licensed. This number of stations in our crowded bands poses a serious threat to our enjoyment of ham radio if we do not all operate courteously and intelligently. Yes, this is a free country, but our freedom does not give us the privilege of ignoring the rights of others.

Before you transmit, listen—to see whether you will be disrupting some communication already in progress on the frequency that you have selected. There *might* be a net there, handling traffic. There *might* be an emergency drill or an actual emergency operation. The courteous thing would be to graciously move your own frequency and let these fellows carry on.

Your good manners in ham radio would include insuring that the signal you emit is clean and free of interference-causing spurious radiations. The amateur bands are so crowded that there is no room for rough notes, key clicks, chirpy signals, splattering voice signals. In the same vein, you are doing the other fellow and ham radio a favor if you give honest reports on his signal. If he has some fault with his signal, call him—diplomatically, of course. And if someone tells you that something is amiss with *your* signal, don't get angry about it. See if you can't locate the difficulty.

In the long run, your enjoyment of amateur radio, and the enjoyment of the other thousands of amateurs who share the bands with you, will be considerably enhanced if you and they will adhere to The Amateur's Code, which we reproduce inside the front cover.

Have Fun!

There are many different facets to ham radio. You may get most of your enjoyment from this hobby by working DX, experimenting on the very high frequencies, handling traffic, taking part in contests, or by just plain ragchewing. But whichever of these is *your* main interest, back off every now and then and take a look at what some of the other hams do. Perhaps you will find that you are missing out on something.

Ham radio is a grand hobby, one that can be shared by every member of the family. As your knowledge of the principles, construction and operation of ham radio increase, you will find that you can have more and more fun. We hope that this book has brought you more understanding of amateur radio.

The R-S-T System
READABILITY
1—Unreadable.
2—Barely readable, occasional words distinguishable.
3—Readable with considerable difficulty.
4—Readable with practically no difficulty.
5—Perfectly readable.

SIGNAL STRENGTH
1—Faint signals, barely perceptible.
2—Very weak signals.
3—Weak signals.
4—Fair signals.
5—Fairly good signals.
6—Good signals.
7—Moderately strong signals.
8—Strong signals.
9—Extremely strong signals.

TONE
1—Extremely rough hissing note.
2—Very rough a.c. note, no trace of musicality.
3—Rough low-pitched a.c. note, slightly musical.
4—Rather rough a.c. note, moderately musical.
5—Musically-modulated note.
6—Modulated note, slight trace of whistle.
7—Near d.c. note, smooth ripple.
8—Good d.c. note, just a trace of ripple.
9—Purest d.c. note.

If the signal has the characteristic steadiness of crystal control, add the letter X to the RST report. If there is a chirp, the letter C may be added to so indicate. Similarly for a click, add K. The above reporting system is used on both c.w. and voice, leaving out the "tone" report on voice.

The American Radio Relay League

In these pages of *Understanding Amateur Radio* we have from time to time made references to "ARRL" or "the League." The American Radio Relay League is a nonprofit association of radio amateurs, founded in 1914. It is the spokesman for amateur radio in the United States and Canada, and is dedicated to the service of its members. It represents the amateur in legislative matters. It publishes the monthly magazine *QST* and numerous technical manuals, of which this volume is one. It promotes and coordinates amateur traffic handling and emergency communications networks. It is a noncommercial organization strictly of, by and for radio amateurs.

The operating territory of the League is divided into one Canadian and fifteen U. S. divisions. Members in these divisions elect directors who serve without pay on the League's Board of Directors and formulate the policies under which the League operates. These divisions are further broken down into sections, and other amateurs are elected by the Members in the sections to serve as Sections Communications Managers and coordinate the operating activities of amateurs in their sections.

The League has a headquarters office in Connecticut staffed with a number of licensed amateurs, whose function is to administer the various services rendered by the League. Code practice transmissions are sent over the headquarters station, W1AW. Training aids are provided for affiliated amateur radio clubs. A technical information service is available for members. Technical equipment is designed and constructed in the League's laboratory, and described in *QST*. Various contests and awards are administered, with certificates being presented to leading participants. Close liaison is maintained with national and international groups that regulate radio communications.

The League has grown and prospered through the years because of the enthusiasm and number of its members. Elsewhere in this book you will find an invitation to *you* to join and add your voice to the spokesman for amateur radio, the American Radio Relay League.

Amateur Radio

OPERATING ABBREVIATIONS

ABBREVIATIONS FOR C.W. WORK

Abbreviations help to reduce the length of transmissions. However, make it a rule not to abbreviate unnecessarily when working an operator of unknown experience.

AA	All after	OB	Old boy
AB	All before	OM	Old man
ABT	About	OP-OPR	Operator
ADR	Address	OSC	Oscillator
AGN	Again	OT	Old timer; old top
ANT	Antenna	PBL	Preamble
BCI	Broadcast interference	PSE	Please
BCL	Broadcast listener	PWR	Power
BK	Break; break me; break in	PX	Press
BN	All between; been	R	Received as transmitted; are
B4	Before	RAC	Rectified alternating current
C	Yes	RCD	Received
CFM	Confirm; I confirm	REF	Refer to; referring to; reference
CK	Check	RIG	Station equipment
CL	I am closing my station; call	RPT	Repeat; I repeat
CLD-CLG	Called; calling	RX, RCVR	Receiver
CUD	Could	SED	Said
CUL	See you later	SEZ	Says
CUM	Come	SIG	Signature; signal
CW	Continuous wave	SINE	Operator's personal initials or nickname
DLD-DLVD	Delivered	SKED	Schedule
DX	Distance, foreign countries	SRI	Sorry
ECO	Electron-coupled oscillator	SVC	Service; prefix to service message
ES	And, &	TFC	Traffic
FB	Fine business; excellent	TMW	Tomorrow
GA	Go ahead (or resume sending)	TNX-TKS	Thanks
GB	Good-by	TT	That
GBA	Give better address	TU	Thank you
GE	Good evening	TVI	Television interference
GG	Going	TVL	Television listener
GM	Good morning	TX	Transmitter
GN	Good night	TXT	Text
GND	Ground	UR-URS	Your; you're yours
GP	Ground plane	VFO	Variable-frequency oscillator
GUD	Good	VY	Very
HI	The telegraphic laugh; high	WA	Word after
HR	Here; hear	WB	Word before
HV	Have	WD-WDS	Word; words
HW	How	WKD-WKG	Worked; working
LID	A poor operator	WL	Well; will
MA, MILS	Milliamperes	WUD	Would
MSG	Message; prefix to radiogram	WX	Weather
N	No	XMTR	Transmitter
ND	Nothing doing	XTAL	Crystal
NIL	Nothing; I have nothing for you	YF (XYL)	Wife
NM	No more	YL	Young lady
NR	Number	73	Best regards
NW	Now; I resume transmission	88	Love and kisses

Abbreviations Commonly Used for Technical Terms

a.c.—alternating current
a.f.—audio frequency
a.f.c.—automatic frequency control
a.f.s.k.—audio frequency-shift keying
a.g.c.—automatic gain control
a.l.c.—automatic load control
a.m.—amplitude modulation
a.n.l.—automatic noise limiter
a.v.c.—automatic volume control
b.f.o.—beat-frequency oscillator
c.f.m.—cubic feet per minute
c.o.—crystal oscillator
c.p.s.—cycles per second
c.r.—cathode ray
c.r.t.—cathode-ray tube
c.t.—center tap
c.w.—continuous wave (telegraphy)
d.c.—direct current
d.f.—direction finder
d.p.s.t.—double-pole single-throw
d.p.d.t.—double-pole double-throw
d.s.b.—double sideband
e.c.o.—electron-coupled oscillator
e.m.f.—electromotive force (voltage)
f.d.—frequency doubler
f.m.—frequency modulation
f.s.—field strength
f.s.k.—frequency-shift keying
g.d.o.—grid-dip oscillator
h.f.—high frequency
h.t.—high tension
h.v.—high voltage
i.d.—inside diameter
i.f.—intermediate frequency
l.f.—low frequency
l.o.—local oscillator
l.s.b.—lower sideband
m.f.—medium frequency
m.g.—motor-generator
m.o.—master oscillator
n.b.f.m.—narrow-band frequency modulation
n.c.—normally closed; no connection
n.o.—normally open
o.d.—outside diameter
p.a.—power amplifier
p.e.p.—peak envelope power
p.i.v.—peak inverse voltage
p.o.—power output
p.p.—push-pull
p.s.—power supply
r.f.—radio frequency
r.m.s.—root-mean-square (effective) value
RTTY—radioteletype
s.b.—sideband
s.e.o.—self-excited oscillator
s.f.—standard frequency
s.h.f.—superhigh frequency
s.l.c.—straight-line capacitance
s.l.f.—straight-line frequency
s.n.r.—signal-to-noise ratio
s.p.s.t.—single-pole single-throw
s.p.d.t.—single-pole double-throw
s.s.b.—single sideband
s.w.—short-wave
s.w.r.—standing-wave ratio
t.r.—transmit-receive
TV—television
u.h.f.—ultrahigh frequency
u.s.b.—upper sideband
v.c.—voice coil
v.f.o.—variable-frequency oscillator
v.g.—voltage gain
v.h.f.—very high frequency
v.l.f.—very low frequency
v.o.m.—volt-ohm-milliammeter
v.p.—velocity of propagation
VR—voltage-regulator
v.s.w.r.—voltage standing-wave ratio
v.t.v.m.—vacuum-tube voltmeter
v.x.o.—variable crystal oscillator
VOX—voice-operated control
w.v.—working voltage
w.w.—wire-wound

Metric Prefixes Used in Radio Work

Name	Abbreviation	Multiplier
kilo	k	10^3 (1000)
Mega	M	10^6 (1,000,000)
Giga	G	10^9 (1,000,000,000)
milli	m	10^{-3} (1/1000)
micro	μ	10^{-6} (1/1,000,000)
micromicro / pico	$\mu\mu$	10^{-12} (1/1,000,000 × 1/1,000,000)

Schematic Symbols

Most-Often-Used Circuit Symbols

Standard circuit designation in parenthesis after name.

ANTENNA (E)	DIODE SEMICONDUCTOR (CR)	LOUDSPEAKER	TERMINAL (E) SINGLE
BATTERY (BT)	CHASSIS	METER (M) *	TRANSISTOR (Q) N.P.N.
CAPACITOR (C) FIXED	HANDSET (HS)		P.N.P.
VARIABLE	HEADPHONES (HT)	MICROPHONE (MK)	TRANSFORMER (T)
SPLIT-STATOR	INDUCTORS (L) FIXED	PLUG, PHONE (P) 2-CONDUCTOR 3-CONDUCTOR	FIXED ADJUSTABLE COUPLING
FEEDTHROUGH	VARIABLE	RELAY (K)	IRON CORE ‡
CONNECTORS (J) CONTACT, MALE CONTACT, FEMALE COAXIAL	IRON CORE ‡	RESISTOR (R) FIXED	TUBE (V) ** DIODE
GENERAL	JACK (J) OPEN-CIRCUIT	ADJUSTABLE	TRIODE
MOVABLE (P) GENERAL	CLOSED-CIRCUIT	SHIELD (E) GENERAL	
COAXIAL	PHONO	AROUND WIRE	TETRODE
A.C. TYPE MALE	KEY (S)	SWITCH (S) TOGGLE, S.P.S.T.	PENTODE
FEMALE	LAMP (I) DIAL	TOGGLE, S.P.D.T.	
CRYSTAL, QUARTZ (Y)	NEON	ROTARY	VR

*Type of meter to be designated by appropriate letter or letters inside circle; e.g., V, voltmeter, MA, milliammeter.
**Indirectly-heated cathodes shown; directly-heated type (inverted V-shaped symbol) can be substituted.

‡Optional; i.e., the double-line core symbol may be omitted if desired. Practice in this book is to show the core symbol only in the case of coils and transformers for power and audio frequencies.

Note: Symbols may be separated into component parts when desirable. For example, heaters may be shown separately from remainder of tubes; relay coils may be separated from contacts; sections of ganged switches (or other mechanically-linked components) may be placed where convenient in the diagram. Individual sections of a multiple-unit component are identified by adding letter subscripts to the circuit designation; e.g., the sections of a ganged switch would be numbered S_{1A}, S_{1B}, S_{1C}, etc.; relay coil would be numbered K_2, with separate sets of contacts designated K_{2A}, K_{2B}, etc.

Electrical and Radio Units			
Quantity	Fundamental Unit	Auxiliary Unit	To Convert to Fundamental Unit Multiply by
Voltage	volt	kilovolt (kv.) millivolt (mv.) microvolt (μv.)	10^3 10^{-3} 10^{-6}
Current	ampere	milliampere (ma.) microampere (μa.)	10^{-3} 10^{-6}
Power	watt	kilowatt (kw.) milliwatt (mw.) microwatt (μw.)	10^3 10^{-3} 10^{-6}
Frequency	cycle per second	kilocycle (kc.) megacycle (Mc.) Gigacycle (Gc.)	10^3 10^6 10^9
Resistance	ohm	kilohm* megohm**	10^3 10^6
Reactance Impedance	ohm ohm	Same as for resistance	
Inductance	henry	millihenry (mh.) microhenry (μh.)	10^{-3} 10^{-6}
Capacitance	farad	microfarad (μf.) micromicrofarad (μμf.) or picofarad (pf.)	10^{-6} 10^{-12}

*In circuit diagrams, the letter K written after the number is used to represent 1000; e.g., 15K indicates 15 kilohms or 15,000 ohms.
**Generally written "meg." in values given on circuit diagrams; e.g., 2.2 meg. = 2.2 megohms or 2,200,000 ohms.

On Purchasing Parts

The components specified for the circuits shown in this book should be readily available through radio parts distributors. If you can visit a distributor's store you should be able to buy them over the counter.

However, such distributors usually are to be found only in the larger cities, and beginners not thus located often are at a loss as to how to get the components they need. If this is your situation, you can order by mail. Houses specializing in mail-order parts business on a national scale advertise regularly in QST. Nearly all will furnish catalogs on request. It is important to realize that these catalogs do not always list the *complete* components line of a manufacturer, so the fact that a particular part you need is not listed does not mean that you can't get it. A mail-order house that carries a given brand can get *any* component sold under that brand name if you order it. Look at the index of manufacturers in the catalog to see whether the distributor carries that line.

If a particular brand name does not appear in the catalog, an inquiry addressed to the manufacturer should bring you information on where the parts can be purchased. Most manufacturers of components made particularly for amateur applications advertise regularly in QST.

Military surplus, which forms the basis for several of the items of equipment in this book, is not a stock item with mail-order houses. It is obtainable on a "when available" basis from dealers who specialize in it. Consult the advertising pages in radio magazines for up-to-date information on equipment of this nature.

Changing Crystal Frequencies

The frequency stability of a crystal-controlled oscillator is a great asset, especially for the beginner, but it is also a handicap—the frequency can't be changed appreciably by any transmitter adjustments. To work on a variety of frequencies you need a variety of crystals. Military surplus crystals in or close to an amateur band often can be picked up quite cheaply, and can be moved to a desired spot in the band with relatively little trouble. Also, it is sometimes useful to be able to shift a crystal you now have; after buying it, you may have discovered that some other spot would have been more desirable.

The frequency of crystals used by amateurs is determined by the thickness of the quartz plate, as described earlier in this book. You can't add to the quartz; all you can do is grind it thinner. This means that the frequency can only be *raised*, not lowered. So if you buy surplus crystals in anticipation of moving them to a frequency you want, be sure to buy frequencies that are *lower* than the frequency you expect to reach.

The most suitable crystals are those in the "FT-243" type holder shown in the drawing. They are usually square, about ½ inch on a side. For grinding, you need a piece of plate glass about 8 inches square (or larger), and some powdered carborundum. The 220 grit is about right for "coarse" grinding, when you have to move the frequency a considerable distance. For "fine" grinding, where you want to go slowly enough to avoid overshooting the mark, the 400 grit or emery powder should be used. Put a little water on the glass plate and mix in enough grinding compound to form a sludge—not too thick, but thick enough to do some cutting. Take the crystal from the holder and mark one side with pencil. Do all your grinding on the *other* side, and renew the mark occasionally, if necessary, to be sure that only the one side is ground. This will maintain the flatness of one surface—an important consideration, since the grinding must be quite accurate if the crystal is to continue to work in an oscillator circuit.

To grind, place your first and second fingers on opposite corners of the crystal and move it around in the sludge in figure-8 fashion over an area of four or five inches. Use light, even pressure. After making a few figure 8's, change the fingers to the other two corners and continue. Keep repeating this way, moving around on the glass plate so you don't follow a single track all the time. Never exert pressure on the center of the crystal, because if the center is ground thinner than the corners and edges the crystal probably won't oscillate. After making a dozen or so figure 8's, wash the crystal thoroughly in water, dry it with absorbent tissue, and put it back in the holder. Handle it only by the edges. Check the frequency to see how much it has moved; this will give you an idea of how much grinding you have to do to get to the desired frequency. When you get close, lighten the pressure and use finer compound in a thinner sludge—and go carefully!

If the crystal stops oscillating, the probable reason is that it has been ground too much in one spot. The two surfaces have to be essentially parallel and flat. The best way to check is with a micrometer, exploring the thickness all over the crystal. Over most of the surface the thickness should be uniform within a ten-thousandth of an inch, but the corners and edges should be a few ten-thousandths thinner than the center

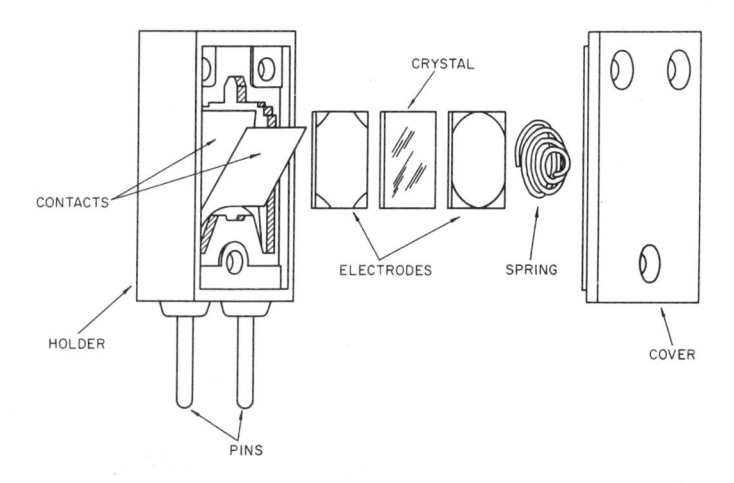

Exploded view of the FT-243 type crystal and holder. The metal electrodes have small raised areas, called "lands," at the corners. These clamp the crystal under pressure of the spring. The main body of the electrode does not touch the crystal.

UNDERSTANDING

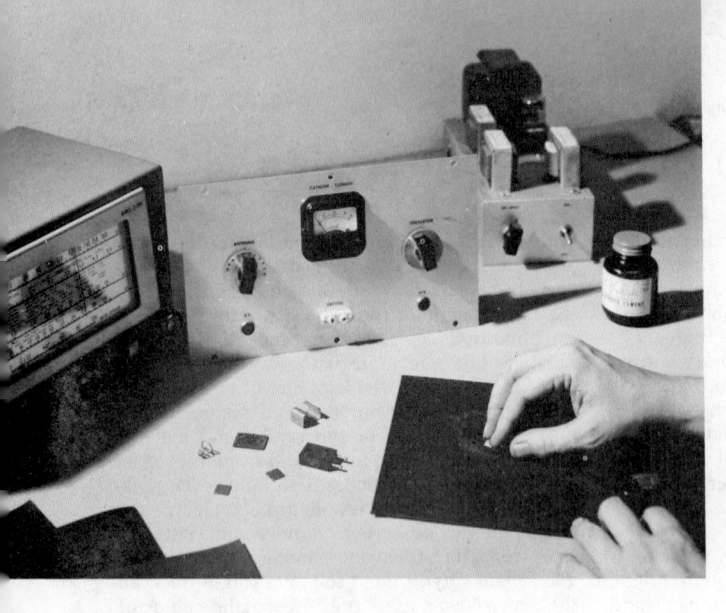

A setup for grinding crystals using the waterproof abrasive paper method. Note the quick-change adapter clip for putting the holder together. This is made from a piece of metal—aluminum, tin or copper—strong enough to hold its shape. The size of the adapter will depend on the size and shape of the holder.

area. If you have no micrometer, sometimes additional grinding with very light pressure will start the crystal oscillating again, and it is also helpful to grind *one* edge a little (one or two figure 8's with the crystal vertical), testing frequently to see whether the crystal has again become active.

The activity of the crystal can be checked by putting it in your oscillator—make sure it and the holder electrodes are thoroughly cleaned and free from lint—and measuring the grid current of the following stage in the transmitter. The larger the grid current the more active the crystal. It is a good idea to make such a measurement before starting to grind, so that a check can be kept on the activity as you move the frequency.

Estimating Transformer and Choke Ratings

Now and again you may acquire through trades, or otherwise "inherit," an unmarked transformer or filter choke that might be usable if its ratings were known. If the item is of the type that is often salvaged from old radio or TV receivers, it isn't hard to make a reasonably shrewd guess as to its capabilities.

The standard color code for power transformer leads is as follows:
    Primary: Black
    H. V. Secondary: Outside ends, red; center tap, red and yellow.
    Rectifier filament, 5 volts: Yellow.
    Tube heaters: Green

If there is a second heater winding, the leads are brown. Occasionally, the filament or heater windings will be center-tapped, in which case the center-tap lead will be striped, one color the same as the main leads and the other yellow except in the case of the rectifier filament, which will be yellow and blue.

Transformer voltages can easily be found by measurement, using a multimeter having a.c. scales. But before measuring voltages, measure the resistances of the windings; the transformer may be an ancient one made before the days of color coding. Filament windings will show almost zero resistance on an ohmmeter. The 115-volt primary usually will measure 2 or 3 ohms. High-voltage windings may run from 50 to a few hundred ohms each side of the center tap. In general, the larger the transformer the lower the resistances of all windings.

Having established which winding is which, apply 115 volts to the primary and measure the output voltage of each winding. *Use clip leads to connect the voltmeter and don't touch the terminals while the power is on!* These no-load voltages can be expected to be 10 to 20 per cent higher than the full-load voltages at which transformers usually are rated.

Next, measure the transformer with a ruler. Then get a parts catalog and compare the size and voltages with listed transformers of the same construction style. You may not find one that corresponds exactly, but you should be able to find something that is reasonably close. The probable current ratings for the various windings will be fairly close to those given by the nearest equivalent, in size and voltage, listed in the catalog.

Heater windings should be tested by paralleling enough tube heaters to get the current load you estimate the transformer should handle, connecting them to the winding, and measuring the voltage under this load. The voltage will not drop below the proper value for the tubes (e.g.,

Amateur Radio

6.3 volts) when the load is within the transformer rating.

With filter chokes, measure the resistance of the winding, measure the linear dimensions, and look up the nearest thing in a catalog. The ratings of this equivalent will give you an estimate of the inductance and d.c. current-carrying capacity.

In using such components at your estimated ratings, bear in mind that a winding isn't going to burn out immediately when you go just a little over what it was intended to handle. The ultimate test is how hot the transformer or choke gets after some hours of continuous use. Try your hand on the power transformer in your TV set after a similar period of use and you'll get an idea of how hot these components can run.

Winding Small Transformers

Burned-out transformers from broadcast and TV receivers are not always a total loss, if one is willing to make the effort to put on new windings that may be required. Usually only one winding, the high-voltage secondary, will have been ruined; in most cases the primary and filament secondaries will still be usable. At worst, the core can always be salvaged and an entirely new set of windings put on. Such salvaged transformers are of particular value when some nonstandard, but relatively low, output voltage is wanted. Replacing high-voltage windings is a very considerable chore, because of the large number of turns required, and not too practical for the home builder.

There are cases, too, where a perfectly good but obsolete transformer is at hand—for example, one having 2.5-volt filament windings instead of giving the 6.3 volts required by practically all modern tubes. Such transformers can be used by taking off the old filament winding and putting on a new one. The same scheme can be used for winding special heavy-duty filament transformers. A receiving power transformer having, say, a power rating of 75 watts (the sum of the rated power outputs of each of the secondary windings) can be rewound to give 75 watts of filament power at any desired voltage, by removing all windings except the primary and replacing them with one or more new ones as required.

Power transformer cores all use the "E-I" construction shown in the accompanying drawing. There are several standard lamination sizes, with increasing "window" space as the width of the center leg of the E part is made greater.

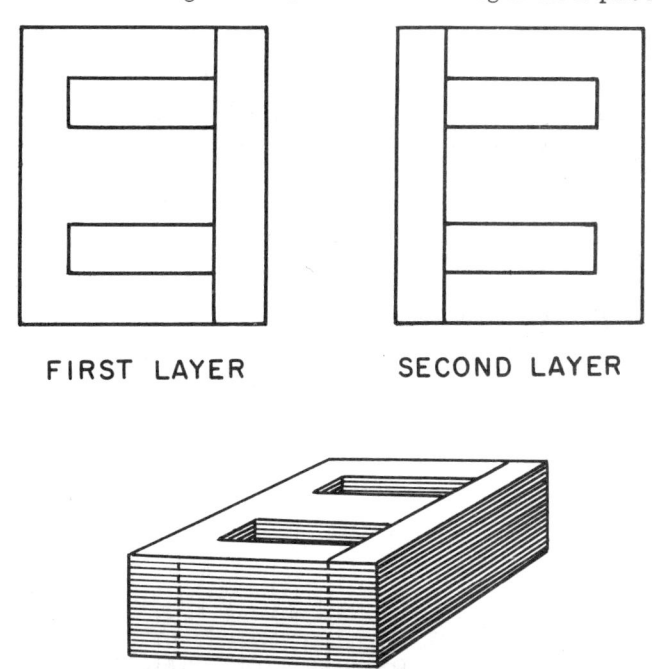

FIRST LAYER   SECOND LAYER

Transformer core construction. Note that alternate E laminations face in opposite directions in the stack. The laminations are interleaved so the magnetic path will offer the least opposition to the magnetic lines of force. In practical construction, two or three laminations may be stacked in the same direction, as a group, the groups being interleaved in the complete core.

Transformer Design

Input (Watts)	Full-Load Efficiency	Size of Primary Wire	No. of Primary Turns	Turns Per Volt	Cross-Section Through Core
50	75%	23	528	4.80	1¼″ x 1¼″
75	85%	21	437	3.95	1⅜″ x 1⅜″
100	90%	20	367	3.33	1½″ x 1½″
150	90%	18	313	2.84	1⅝″ x 1⅝″
200	90%	17	270	2.45	1¾″ x 1¾″
250	90%	16	248	2.25	1⅞″ x 1⅞″
300	90%	15	248	2.25	1⅞″ x 1⅞″
400	90%	14	206	1.87	2″ x 2″
500	95%	13	183	1.66	2⅛″ x 2⅛″
750	95%	11	146	1.33	2⅜″ x 2⅜″
1000	95%	10	132	1.20	2½″ x 2½″
1500	95%	9	109	0.99	2¾″ x 2¾″

It is general practice to build up the core by using enough laminations so that the cross section of the center leg of the completed core is approximately square. Transformer design data for various such cross sections are given in the accompanying table. These figures are conservative. The number of turns per volt is based on a more-or-less rule-of-thumb figure for 60-cycle transformers of 7.5 turns per volt per square inch of core cross section. Thus if the core has an area of 2 square inches, the required number of turns per volt is 3.75, and a 115-volt primary winding would require 3.75 × 115 = 430 turns, approximately.

When the existing primary winding is to be used, the number of turns per volt used in that transformer design can be found by unwinding a low-voltage secondary, counting turns as they come off. The turns per volt then can be found by dividing the counted turns by the rated output

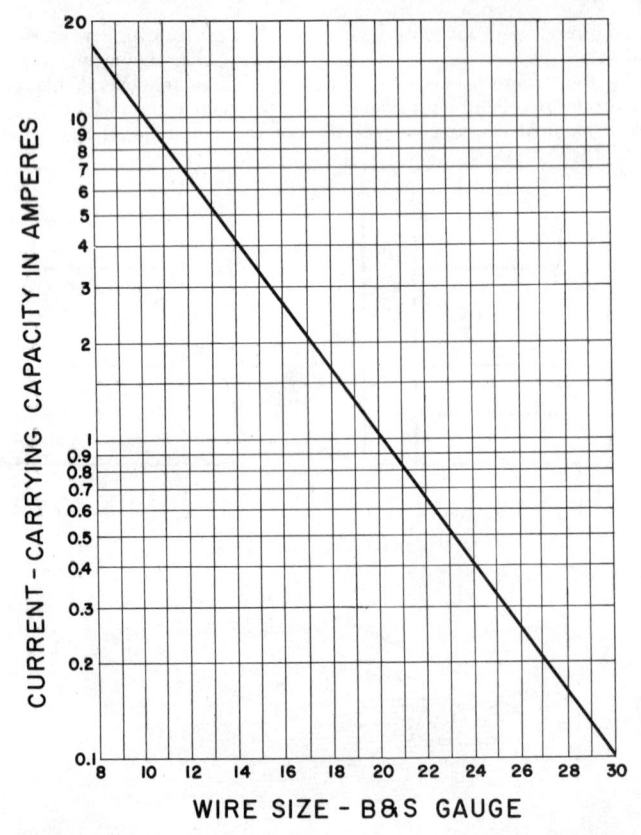

Current-carrying capacity of copper wire in transformer coils. The values given by this curve are conservative, and may be increased up to 50 per cent if the wire size indicated will not fit in the available window space.

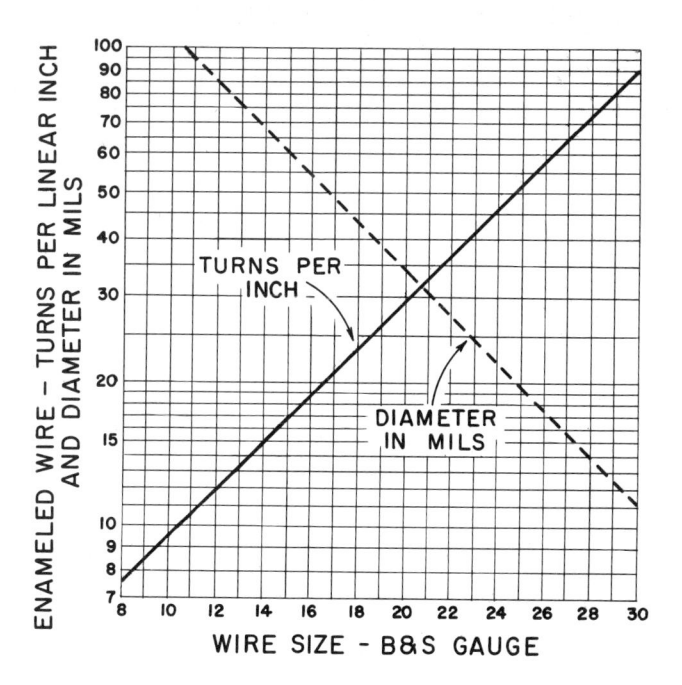

Turns per linear inch and diameter in mils (thousandths of an inch) for commonly used sizes of enameled wire. These are average figures; there will be some variation in practice because of enamel thickness and manufacturing tolerances.

voltage of the winding. This same number should be used in determining the number of turns required for the new winding.

The two charts give useful wire data. The safe current for various wire sizes is conservative; some small-transformer designs use as little as half the wire cross-section allowed by the chart shown. In general, either of the next two smaller sizes than indicated can be used if the transformer window size will not accommodate enough turns for the voltage desired. The winding will run a bit hotter and the voltage regulation will not be as good, but the transformer will stand up in intermittent amateur service.

To arrive at a satisfactory wire size, first determine (by measurement) the area available for the winding. Select the wire size for the current to be drawn, using the chart. From the other chart, find the number of turns per inch for that wire size, and from that and the length of the window space determine the number of turns per layer. Make allowance for the fact that the ends of the layer must be insulated from the core. The chart also gives the diameter of the wire in mils (a mil is 0.001 inch). Add to the wire diameter the thickness of any insulation to be used between layers. Usually, each layer is covered with a turn of paper both for insulation and mechanical support. Ordinary waxed paper, which is quite suitable for the purpose, has a thickness of a little over 1 mil; regular bond paper runs 3 to 4 mils, as a rule; and medium-weight kraft wrapping paper has a thickness of about 5 mils. Divide the sum of the wire diameter and paper thickness into the window width to determine the number of layers. This times the number of turns per layer will give the total number of turns possible with that wire size. It is best to decrease the number so found by at least 20 per cent, to allow for inaccuracies in winding. If the required number of turns is not obtained, using this method, with the wire size originally selected, try the next smaller size.

In putting on entirely new windings, divide the available window space approximately equally between the primary and all secondaries. Allowance should be made for insulation over the center core leg (heavy cardboard is ordinarily used for this) and over the entire winding assembly. As each layer of a coil is wound, coat it with lacquer or shellac, allowing the turns to "set" before going on to the next layer.

The laminations should be handled carefully, to avoid bending them (which prevents stacking them as closely as possible) and nicking the edges. In assembling the core in the windings, force as many laminations into place as possible. The last ones can be hammered into place if a block of wood is used between the core piece and hammer.

INDEX

A

Abbreviations, metric	306
operating	305
technical	306
Absorption wavemeter	265
Accessories, station	13
Accuracy, of measurements	256, 262
Acoustic feedback	9
Alternating current	18
Amateur's Code	Cov. II
Ammeter, r.f.	261
Amplification	34, 47
factor	32
in receiver	49
of modulated signal	36
power	34
voltage	34
Amplifier, buffer	75, 201
cascade	35
classifications	35
Class A	36
Class B	36
Class C	36, 72
gain	35
grounded-grid	43
grounded-plate	43, 201
i.f.	56
linear	89
modulated	84, 87
neutralization	75
power	40
push-pull	40
r.f.	60
resistance-coupled	39
speech	236
stabilization	74
stage	34
transformer-coupled	39
voltage	38
Ampere	16
Amplitude	48
limiting	159
modulation	82
Antenna, beam array	116, 292
change-over	9
construction	277, 293
current and voltage distribution	97
dummy	67, 221
gain	117
grounded	98, 114
impedance	98
long-wire	116
matched	110
multiband	112, 277
random-length	286
receiving	48, 115
resistance	98
supports	277, 293
v.h.f.	290
vertical	287
Antinode	97
Audio frequency	80
limiter	159
selectivity	161
Automatic gain control	61, 165
Awards, operating	300

B

Backlash	137
Balanced modulator	90
Bandpass	30
Bandset	50, 136
Bandspread	50
Bandwidth	26, 53
coupled circuits	30
Base	46
Beat	53
oscillator	63, 254
Bias, back	46
cathode	37
fixed	38
for modulator	243
grid	33
grid-leak	38
protective	174
Birdies	54
Bleeder	20, 231
Blocking	9
capacitor	38
Break-in, c.w.	10
Buffer amplifier	75, 201
Buying parts	308
Bypassing	38, 157, 235
cathode	38
plate	39
screen	42

C

Capacitance	19
distributed	157
grid-plate	42
input	42, 157
interelectrode	41
output	42
stray	159
Capacitor	19
blocking	38, 45
bypass	38, 39
filter	229
trimmer	157
types	129

Carrier	82
controlled	88, 252
reduced	92
reinserted	91
shift	251, 254
suppressed	90
Cathode	31, 46
follower	43, 201
Catwhisker	46
Characteristic, tube	32
Charge	15
Chassis layout	121
Choke, radio-frequency	45
filter	230, 310
Coax, for receiver	10
Collector	46
Component substitution	128
Contests, operating	301
Control grid	36
Controlled carrier	88, 252
Converter, frequency	60
Coupling, between tuned circuits	29
capacitive	31
coefficient	30
common	40
critical	30
inductive	27
losses	100
overcoupling	30
to line	113
top and bottom	31
variable	28
Critical coupling	30
Cross-modulation	64
Crystal, quartz	77
frequencies	169
grinding	309
overtone	170
tolerance	169
Current, alternating	18
charging	19
direct	16
line	25
peak	228
rectifier output	228
Cutoff	34
remote	37
sharp	37
Cycle, a.c.	18

D

D'Arsonval movement	257
Decibel	35
Deflection, cathode-ray	276
Decoupling	39
Demodulation (see detection)	
Detection, detector	31, 48, 61
f.m. (slope)	66
product	66
s.s.b.	65, 91
Dielectric	24
Difference of potential	15
Diode	31
detector	48, 55

semiconductor	45
symbol	46
Dipole	95
center-fed	277
end-fed	281
folded	111, 283, 290
matched	110, 281
multiple	283
Direct current	16
Directivity	100
patterns	100
Director	117
Dissipation, power	31
plate	32, 91
Distortion	34
Double conversion	54
Double sideband	92
Doublet, terminals on receiver	10
Drift, frequency	59, 155, 200
Driving power	36, 67, 72
DX work	298

E

Effective value	18
Efficiency, of frequency multiplier	77
plate	32, 69
transformer	28
Electron	15
secondary	36
Emergency work	301
Emitter	46
Energy	16
in inductance and capacitance	23
magnetic	19
stored	19

F

Feedback, capacitive	44
inductive	44
magnetic	44
negative	38
positive	44
r.f.	239
Feeder (see line)	
Field	19
electric	19
magnetic	19
Filter, capacitor-input	226
choke-input	226
harmonic	213, 215
half-wave	214
high-pass	216
low-pass	215
power-supply	226
resistance-capacitance	227
section	227
Fire protection	12
Flattening	85
Folded dipole	111, 283, 290
Frequency	19
audio	80
calibration	270
converter	60
drift	59, 155, 200

local oscillator	53
image	53
intermediate	53
markers	268
measurement	263, 271
multiplication	76
shift	254
stability	136
standard	269
tolerance	169
Front end	54
Fusing	13, 234
Frequency modulation	66, 92
incidental	203

G

Gain, amplifier	35
antenna	117
automatic control	61
manual control	62
power	35
voltage	35
Ganging	60
Grid bias	33
current	34, 72
leak	38
return	37
Grid-dip meter (g.d.o.)	272
Grinding crystals	309
Ground	12
chassis	16
potential	16
Guys	295

H

Half-power point	26
Harmonic suppression	73, 213
Heat	17
Heterodyne	53
High-C	200
Hole	45
Hop	103
Hum	238, 253

I

Image	53
Impedance	23
characteristic	104
complex	24
matching	29
parallel-tuned circuit	26
resistive	24
transformation	28
Indicator, r.f.	220, 280
Induced voltage	19
Inductive coupling	27
Infinite line	103
Injection, oscillator-mixer	59
Intermediate frequency	53
amplifier	56
transformer	56
tunable	57, 139
v.h.f.	151

Ionosphere	102

J

Junction, semiconductor	45

K

Key, telegraph	7
Keying	201
chirps	202
clicks	202
differential	179, 202
monitor	11

L

Leakage, capacitor	19
Level, power	35
Lightning protection	12
Limiter, automatic	165
c.w.	159
Line, transmission	103
coaxial	109
fittings	128, 278
infinite	103
installation	280
length	278
losses	108
open-circuited	105
parallel-conductor	109
short-circuited	106
terminated	105
velocity factor	109
Linear amplifier	89
Linearity	57, 85
Link	27
Load	27
amplifier	68
resistance	27
Loading coil	100, 115
in lines	105
Local oscillator	53
Log keeping	302
time standards for	303
Loop	97
Low-C	157

M

Mast	294
Matching, impedance	29
lines	105, 107
Message form	300
Microphone, carbon	80
ceramic	81
crystal	80
Milliammeter	256
Mixing	53, 58
circuits	58
Modulating impedance	88
Modulation	58
amplitude	82
characteristic	85
checking	253
controlled-carrier	88, 252

current.............................	251
frequency......................66,	92
percentage........................	82
phase.............................	94
plate..............................	87
power...................83, 87, 240,	243
screen............................	84
sidebands.........................	81
splatter...........................	253
talk power........................	85
transformer...................88,	242
Modulator, balanced................	90
current.......................247,	252
grid bias..........................	243
plate..............................	239
reactance.........................	93
screen............................	237
screen supply for..............244,	245
Monimatch....................14,	209
use of............................	210
Monitoring, c.w.................11,	217
Multiplier, frequency................	76

N

N-type semiconductor................	45
Negative............................	15
Neutralization.......................	75
Node................................	97
Noise limiter........................	165
Nonlinearity........................	58

O

Ohm.................................	16
Ohm's Law..........................	16
Operating...........................	296
abbreviations.....................	297
awards...........................	300
c.w...............................	297
contests..........................	301
emergencies......................	301
phone............................	296
Oscilloscope....................14,	275
Oscillator.......................42,	44
capacitive feedback................	44
Colpitts...........................	44
crystal............................	77
drift..................59, 155,	200
electron-coupled...................	78
harmonic output...................	79
Hartley...........................	44
high-C...........................	200
inductive feedback.................	44
local..............................	53
magnetic feedback.................	44
Pierce............................	78
pulling............................	59
tickler............................	44
tubes.............................	77
tuned-plate, tuned-grid.............	42
variable-frequency.................	198
Overcoupling........................	30
Overloading.........................	64
Overmodulation.....................	83

P

P-type semiconductor................	45
Parallel feed.........................	45
Parallel impedance...................	26
resistances.......................	17
Parasitic element....................	116
Parasitic oscillation..................	75
Passband...........................	26
Peak-envelope power.............84,	89
Peak-inverse voltage.................	46
Phase...............................	20
antenna..........................	97
Phase modulation....................	94
reversal..........................	97
Phonetic alphabet....................	296
Pi network..........................	71
Pitch control........................	65
Plate, beam-forming.................	37
Plate resistance......................	32
Polarity.............................	15
antenna..........................	97
electrolytic-capacitor...............	229
relative...........................	15
Positive.............................	15
Potential............................	15
difference of......................	15
Power...............................	16
apparent.........................	20
dissipation........................	31
driving.....................36, 67,	72
grid..............................	34
level.............................	35
modulation................83, 240,	243
peak-envelope............84, 89,	242
voice........................84,	240
wattless..........................	20
Primary, of transformer..............	28
Pulling, oscillator....................	59
Purchasing parts....................	308

Q

Q..................................	24
capacitor.........................	24
circuit bandwidth..............25,	26
coil...............................	24
effect on coupling..................	30
tank circuit.......................	73
tuned circuit......................	25
Q signals.....................Cov. III	
QSL cards..........................	302

R

R.M.S..............................	19
Radiation resistance..................	99
Ragchewing.........................	298
Ratings, choke......................	310
diode.............................	46
peak-inverse......................	46
transformer......................	310
tube..............................	42
Reactance......................19,	20
capacitive........................	21
inductive.........................	21
modulator........................	93
series and parallel.................	21

Rectification	31	Series resistances	17
a.g.c.	61	reactances	21
bridge	225	Shield, shielding	29
circuits	225	Shunt, for milliammeter	257
detection	48	Sidebands	81
Rectifier, high-vacuum	228	Sidetone	10
mercury-vapor	228	Signal reporting	304
point-contact	46	Signal-to-noise ratio	49
semiconductor	229	Skip distance	103
Reflection	96	Slug, iron	24
earth	101	Single sideband	92
ionosphere	102	reception	65
Reflector	116	Soldering	124
Regulation, voltage	230	Splatter	253
Reradiation	117	Spreaders	278
Resistance	16	Spurious responses	54
antenna	98	Stage, amplifier	34
apparent	17	Stand-by	9
characteristic	104	Standard-frequency transmissions	271
combining	18	Standing-wave ratio	96
equivalent	17	bridge	14
internal	23	on lines	105, 108
load	27	Static	49
measurement	258	Substitution of components	128
plate	32	Superheterodyne	51
radiation	99	double-conversion	54
Resistor	16	Suppressor grid	37
bleeder	20, 231	Symbols, circuit	307
cathode	37		
decoupling	40		
equalizing	229	**T**	
screen-dropping	42		
Resonance	21	T.R. switch	10
harmonic	98	TVI	215
in lines	106	Committees	216
parallel	23, 25	filter	13, 215
series	22	Table, operating	7
tank-circuit	68	Tank circuit	45, 68
with tube	42	Temperature compensation	155
Ripple	226	Termination, line	105
		Tetrode	36
		beam	37
S		Thermocouple	261
		Time constant	20
S meter	167	Tools	119
S.W.R. (see Standing-Wave Radio)		care of	126
Safety	11, 235	Traffic handling	299
Saturation, plate-current	36	nets	299
Schematic symbols	307	Transconductance	32
Screen grid	36	Transformer	28
Secondary, of transformer	28	b.f.o.	64
Secondary emission	36	design and construction	311
Selectivity	51, 137	i.f.	56
adjacent-channel	51	low-frequency	28
adjustable	57	modulation	88, 242
audio	162	power	28
coupled-circuit	30	r.f.	60
single-circuit	25	tuned	30
skirt	53	turns ratio	28
Semiconductor	45	Transistor	46
Send-receive relay	9	Transmatch	13, 113, 281
switch	8	adjustment	210
Sensitivity	137	v.h.f.	292
control	10	Trouble-shooting	132
signal-to-noise	49	chart	134
Series feed	45	Tube characteristic	32

voltage drop	32
Tuning rate	136

U

Units, electrical and radio	308

V

V.H.F.	150
V.O.M.	14, 258
V.T.V.M.	14, 263
Variable-mu	37
Velocity factor	109
Volt	15
Volt-ampere	20
Voltage divider	233
doubler	227
drop	17
induced	19
injection	59
peak-inverse	46, 228
regulation	230
regulator	156, 231
striking	232
tube drop	32
Voltmeter, a.c.	260
d.c.	256
r.f.	252, 261

W

WWV, use of	271
Watt	16
Wave paths	102
Wavelength	95
electrical	96
Wavemeter	14, 264
Wire, B & S gauge	313
current-carrying capacity	312
Wiring techniques	123

Y

Yagi antenna	116, 292

Equipment Descriptions

Antennas: See Chapter 15

Measuring Equipment:

Band-finder wavemeter	265
Frequency standard, 50-100 kc.	268
Grid-dip meter	272
Volt-ohm-milliammeter	260

Receivers and accessories:

ARC-5 receiver modifications	139
automatic noise limiter and a.g.c.	165
S meter	167
Audiofil (selective audio filter)	161
Audio limiter	159
Crystal-controlled converter, for WWV	163
for 7–50 Mc. bands	143
for 14, 21 and 28 Mc.	247
for 50 and 144 Mc.	150

Transmitters and accessories:

Amplifier, 150 watts, 3.5–28 Mc.	183
ARC-5 transmitter modification for Novices	187
Band checker (wavemeter for coax line)	267
Crystal switch assembly	223
Dummy antenna	221
Filter, half-wave for 3.5 and 7 Mc.	213
same with switching	285
low-pass	215
Keying monitor and code-practice oscillator	217
Matchtone (keying monitor)	219
Modulator, for 120-watt transmitter	247
Monimatch	209
Transmatch, simple	208
universal	211
for random-length antennas	288
Transmitter, 60 watts, 3.5–28 Mc.	169
100 watts, 3.5–28 Mc.	177
15 watts, 50 and 144 Mc., with modulator	191
Transmit-receive switch, electronic	221
TVI trap, 50 and 144 Mc.	216
Variable-frequency oscillator, 3.5 Mc.	203

Here's a tip for the newcomer as well as the old timer:

the more you know about ham radio the more fun you can have hamming. And, it follows that the best way to learn about ham doings is through *QST*, the official magazine of the American Radio Relay League. Written and edited by amateurs for amateurs, *QST* is delivered twelve times a year to League members throughout this country and in many foreign countries.

In the pages of *QST* you meet the fellows you hear and work on the air. You get operating data, available nowhere else, to help you become a better operator and to get more fun from your activities.

No matter what your principal interest may be—rag chewing, traffic, DX, experimenting, contests, v.h.f.—you will find EXTRA space in *QST* devoted to your speciality.

You get the newest and the latest technical developments. Throughout *QST*'s forty-five year history it has consistently presented the important developments in the radio art—FIRST. Many come from the League's own lab, and many come from the workshops of members.

QST is the amateur's own magazine. Unlike commercial magazines, all of *QST*'s income, after normal expenses, goes into services for members. Fact is, through your ARRL membership you will be one of the owners. You will have a warm, friendly, personal sense of pride in *QST*— YOUR OWN MAGAZINE.

QST and ARRL Membership $5, $5.25 in Canada, $6 elsewhere

Please enroll me as a member of the ARRL and begin *QST* with the.................issue. My membership dues are enclosed. Check..............Money Order..............Cash..............
NAME..CALL....................
STREET..
CITY...ZONE........STATE..................

UAR-2

The American Radio Relay League, Inc.
225 MAIN STREET   NEWINGTON, CONN.